SPECLAB

SPECLAB

Digital Aesthetics and Projects in Speculative Computing

JOHANNA DRUCKER

University of Chicago Press | Chicago and London

Johanna Drucker is the Martin and Bernard Breslauer Professor in the Graduate School of Education and Information Studies at the University of California, Los Angeles.

The University of Chicago Press, Chicago 60637
The University of Chicago Press, Ltd., London

Printed in the United States of America

18 17 16 15 14 13 12 11 10 09   1 2 3 4 5

ISBN-13: 978-0-226-16507-3 (cloth)
ISBN-13: 978-0-226-16508-0 (paper)
ISBN-10: 0-226-16507-8 (cloth)
ISBN-10: 0-226-16508-6 (paper)

Library of Congress Cataloging-in-Publication Data

Drucker, Johanna, 1952–
SpecLab : digital aesthetics and projects in speculative computing / Johanna Drucker.
p. cm.
Includes bibliographical references and index.
ISBN-13: 978-0-226-16507-3 (cloth : alk. paper)
ISBN-13: 978-0-226-16508-0 (pbk. : alk. paper)
ISBN-10: 0-226-16507-8 (cloth : alk. paper)
ISBN-10: 0-226-16508-6 (pbk. : alk. paper)
1. Humanities—Research—Data processing. 2. Humanities—Digital libraries. 3. Image processing—Digital techniques. 4. Information storage and retrieval systems—Humanities. 5. Metadata. I. Title.
AZ105.D785 2009
025.06′0013—dc22

2008041462

♾ The paper used in this publication meets the minimum requirements of the American National Standard for Information Sciences—Permanence of Paper for Printed Library Materials, ANSI Z39.48-1992.

For Bethany and Jerry,
with affection and appreciation

CONTENTS

ILLUSTRATIONS

TABLES

FIGURES

INTRODUCTION

The Background to SpecLab

At the core of this book are convictions derived from both theoretical investigations into problems of knowledge production and experimental projects conceived under the general rubric of speculative computing. Speculative computing arose from a productive tension with work in what has come to be known as digital humanities. That field, constituted by work at the intersection of traditional humanities and computational technology, uses digital tools to extend humanistic inquiry. Computational methods rooted in formal logic tend to be granted more authority in this dialogue than methods grounded in subjective judgment. But speculative computing inverts this power relation, stressing the need for humanities tools in digital environments. The goal of SpecLab, then, was to challenge the conceptual foundations of digital humanities through aesthetic provocation. The relevance of the arguments I make here—for the importance of aesthetics, subjectivity, and speculative work—is not restricted to projects undertaken in electronic environments. But these insights arose in the process of working out problems in knowledge representation and interpretation that were central to digital humanities in the late 1990s and early 2000s.[1]

Digital humanities has taken on an important task: addressing the methods and implications of the migration of our cultural legacy into digital form and the creation of new, born-digital materials and tools. Nowhere was digital humanities more highly developed in the 1990s than at the University of Virginia.[2] I was a latecomer to that dynamic arena but one with an established interest in aesthetics and the use of new media in fine arts, poetry, graphic arts, and information design.[3] With Jerome McGann and Bethany Nowviskie, I created several projects, beginning in 2000, that became the core of SpecLab.

Under that rubric (short for Speculative Computing Laboratory) we undertook collaborations with uncertain outcomes, intent on contesting the emerging conventions of the digital humanities community. The happy circumstance of having, within a community where practices and methods were not yet fully consolidated, a small cohort intent on introducing experimental projects created a fertile institutional context within which the undertakings on which this book reports were conceived and executed.

I now believe that the lessons of SpecLab are as vital to the future of the humanities as digital humanities is to the continuation of scholarship and research across the humanistic disciplines. We learned many lessons about knowledge and subjectivity, about information design and representation, about creating conditions of use, and about designing instruments to show the complex activity of interpretation. These have recast my understanding of traditional print materials as radically as my encounter with critical theory, deconstruction, and poststructuralism did twenty-five years ago. In many ways, SpecLab allowed those theoretical constructs to be applied to practice, even as they were combined with insights from quantum physics, 'pataphysics, systems theory, and cognitive studies.

This book is thus addressed to several communities. Among digital humanists, reflection on our specific experience may yield lessons for future research. To the wider field of humanist scholars, it offers an introduction to the theoretical implications of the practical changes being wrought by digital activity in the way we do our daily business. The field of digital humanities is not simply concerned with creating new electronic environments for access to traditional or born-digital materials. It is the study of ways of thinking differently about how we know what we know and how the interpretative task of the humanist is redefined in these changed conditions. SpecLab's projects were attempts at designing ways to model and demonstrate new conceptions of that work and its fundamental assumptions. For me, the questions guiding this activity have always been the same: What is the relation between aesthetic expression and knowledge? And how do we take such relations into account in modeling our interpretative approaches so that they expose the ideological as well as epistemological workings of complex cultural activities? From my very first encounters with digital media, I have been convinced that the powerful cultural authority exerted by computational media, grounded in claims to objectivity premised on formal logic, can be counterbalanced through aesthetic means in which subjectivity is central to the concept of knowledge as interpretation.

Aesthesis, as will become clear in the pages ahead, is the term by which I refer to a theory of partial, situated, and subjective knowledge—a theory whose aims are ideological as well as epistemological. Digital media, through a curious combination of capabilities, can be the site of demonstrating simultaneously the exclusion of subjects (persons constituted in their relation to social, cultural, historical circumstances) *and* their presence, inscription, and participation in the production of knowledge. I think this is what brought home so strongly the realizations that form my argument.

These are large claims. The concept of aesthesis engages basic questions about knowledge and its representation, and interpretative acts and the values assigned to them within a cultural frame. Insofar as form allows sense to appear to sentience, to paraphrase Aristotle, the role of aesthetics is to illuminate the ways in which the forms of knowledge provoke interpretation.[4] Insofar as the formal logic of computational environments validates instrumental applications regarding the management and creation of digital artifacts, imaginative play is crucial to keeping that logic from asserting a totalizing authority on knowledge and its forms. Aesthesis, I suggest, allows us to insist on the value of subjectivity that is central to aesthetic artifacts—works of art in the traditional sense—and to place that subjectivity at the core of knowledge production.[5]

Formal logic, with its grounding in *mathesis* and claims to objectivity, can be challenged only by an equally authoritative tradition of aesthetic works and their basis in subjective forms of knowledge production. Conceived in such a framework, neither "works" nor "forms" are self-evident entities. They are emergent phenomena constituted by shifting forces and fields through productive acts of interpretation. Thus "forms" (texts, images, complex expressions of any kind) are coded artifacts, constrained and specific, that provoke a reading or interpretative event.

This "interpretative event" is an intervention in a probabilistic or discursive field linked to a sphere of cultural production that has historical and temporal, as well as bibliographical, cultural, and other, dimensions. An individual reader, however, often accesses the field only through an encounter with the artifact. (Scholarly work, research, and criticism, by contrast, deliberately expand beyond the "artifact" into the broader discourse of production, reception, and so on.)

Conceived in this way, knowledge forms are never stable or self-identical but always situated within conditions of use. Knowledge, then, is necessarily partial, subjective, and situated. Objectivity, we have long recognized, is the wish dream of an early rational age, one that was mechanistic in its approaches. The persistence and success of that rational tradition is realized in the extent to which our contemporary administered culture builds its own authority upon the formal procedures computational logic enables and makes instrumental.

Aesthesis challenges the authority of this systematic rationality by questioning its founding assumptions, particularly its totalizing concepts of knowledge. In a curious historical coincidence, the very era that witnessed the dismantling of truth claims by poststructuralist practice and deconstructive theory witnessed the rise of the cultural authority of computational media. Digital technology has insinuated itself into the infrastructure and rituals that form the basis of daily life to such an extent that, despite the availability of a philosophical base for undoing its authority, there is a pervasive tendency to bracket any critiques in the interest of getting on with business. Nowhere was this contradiction more evident than in the struggles to keep humanistic theory central to the digital humanities. Time after time, we saw theoretical understandings subordinated to the practical "requirements of computational protocols." As one of my digital humanities colleagues used to remark, we would go into the technical discussions as deconstructed relativists and come out as empirically oriented pragmatists.

Thus the single most important challenge we gave ourselves in SpecLab was to design representations that modeled subjectivity within knowledge production. Making visible these subjective acts of interpretation, and the role of imaginative play, served to challenge the authority claims of formal logical systems. The event of interpretation in a digital environment includes many steps: creating a model of knowledge, encoding it for representation, embodying it in a material expression, and finally encountering it in a scene of interpretation. Each is part of a performative system governed by basic principles of second-generation systems theory, in particular, codependence and emergence.[6] These can be used to describe an aesthetic experience grounded in subjective judgment just as surely as they can be used to describe formal systems.

When I arrived at the University of Virginia, I found that a culture of design and visual knowledge was conspicuously missing.[7] What interest there was in visual or graphic design was grounded in information design and its ideals of transparency in the representation of data.[8] Critical editing, corpus linguistics, translation, and archive building had been central to early digital projects, but attention to subjectivity and the rhetorical properties of graphical aesthetics was not part of the design process at UVa (or anywhere in the humanities at that time). Graphic design and interface design were often regarded as window dressing, a skin to be grafted, at the last minute, onto an already formed information structure.[9] The understanding that *design is information* was not part of the approach.

My background combined historical and practical approaches to graphical forms of knowledge production, as well as a theoretical disposition to produce a critical metalanguage for describing these approaches. I had everything to learn about digital humanities when I arrived at UVa. I didn't know the basics of HyperText Markup Language (HTML, the code used to specify the graphic characteristics of information in an on-screen/browser environment). I was far from initiated into the mysteries of the Text Encoding Iniative (TEI, a set of conventions for standardizing tags across projects) or Extensible Markup Language (XML, a generic form of tagging and structuring data).[10] But my training as an artist-practitioner and art historian had given me a deep conviction about the ways graphical forms of knowledge embody subjective inflection. The specificity and variety of graphical expressions, and their relatively unstable, informal codes, combined with the rhetorical force of presentation—constitute an argument in any information display.[11]

I joined the digital humanities community at UVa at a fertile moment. Under the visionary leadership of John Unsworth, the Institute for Advanced Technology in the Humanities (IATH), had established itself at the forefront of the field. Its international reputation, justly deserved, had fostered an atmosphere of heady engagement with questions of metadata, display features and functionalities, and such now-quaint but still persistent topics as overlapping hierarchies. In the early 1990s, when a gift from IBM was used to establish the research center, Unsworth had the insight that the future of digital humanities was on the Web. IATH created pioneering projects with a core group that combined the talents of computational humanists, philosophers of information science, and some senior scholars who were early and eager adopters. The Rossetti Archive, established by Jerome McGann as a demonstration of the capacity of computational technology to provide an environment for scholarly editing and archival research, and The Valley of the Shadow, Ed Ayers's showpiece of historical interpretation, in which all primary materials would be made available as part of the scholarly work, were two of the initial undertakings.[12] Along with other major projects at UVa and elsewhere (the Blake Archive, Project Muse, the Crystal Palace, Voice of the Shuttle, and the Perseus Project, among others), these became the testing ground on which first-generation digital humanities scholarship came of age.

In 1999 the Web was only a few years old, though the Internet backbone on which it was built had been in existence for decades. Palm Pilots, iPods, CD burners, and DVDs were still future technology or just on the horizon. Critical studies in digital media were beginning to appear, with a handful of serious works on the cultural impact of new technology, particularly in the arts.[13] Real-time interactions in virtual space using text-only display had, already in the 1980s, demonstrated the addictive quality of social networking and online environments in multiplayer games. MUDs (Multi-User Dungeons) and MOOS (MUDs Object-Oriented) had proved so seductive that undergraduates would go without sleep, food, sex, and face-to-face social interaction in order to keep playing.[14] Then in the 1990s, the graphical interface that had made desktop computing so user-friendly began to be translated into vivid new displays. Text-only screens, blinking green or amber against dull black, were replaced by full-color monitors. Search engines sprang up and competed: AltaVista, Jeeves, Google, and others now vanished from the scene. Amazon and eBay were already well-established brands, but online news and day trading were still primitive. So much of what is now established habit was then barely in view.

As an early adopter, the University of Virginia had invested in creating an infrastructure to encourage the delivery of services in electronic form and in laying the foundation for a digital library. Most importantly, it had fostered the development of models of electronic scholarship. These projects were using new technology to ask research questions that were not viable using traditional print-based materials. Many of these questions had a meta aspect to them, encouraging reflection on models of knowledge, rather than simply focusing on objects, artifacts, or scholarly inquiry. Staying up nights discussing classification systems and thinking of ways to structure data may sound drab—and tame alongside the debates then raging about the right to self-determination of intelligent machines and the possibilities of silicon-based life replacing carbon forms—but for those of us engaged in the dizzying tasks of "disambiguation" and "content modeling" required by digital methods, these activities were engrossing and stimulating.[15]

Not since my days in the poetry world of the Bay Area in the 1970s, or the film theory circles at Berkeley in the 1980s, had I experienced such intellectual camaraderie and exuberance. The generosity of colleagues, their willingness to engage in serious conversation about information structures, computational language, and the cultural and ideological implications of technological transformation, and the common commitment to figuring out what digital humanities had to teach us about our traditional approaches and unexamined assumptions as scholars was striking. We shared readings and projects without reserve, in an atmosphere of generative collegial contention. Within SpecLab in particular, we had the rare opportunity to develop a specialized insight and understanding that has yet to be fully documented and described. This book aims to communicate the spirit and substance of that activity as I experienced it.

This book is neither a history of digital humanities nor an introduction to its tenets or practices.[16] Nor is it a history of IATH and related ventures at the University of Virginia. That story is not mine to tell. Anyone interested in the intellectual frameworks of that community and its development would do well to read *Radiant Textuality* by Jerome McGann, who participated from the outset. A fascinating journalistic account could be written documenting the curious history of digital humanities at UVa. But at this moment, it seems more pressing to communicate the intellectual substance of what we learned and use it to envision the next phase of work.

From the very beginning of my engagement with digital humanities I have benefited from a community of colleagues of exceptional generosity and vision. The aforementioned John Unsworth, Matt Kirschenbaum, Kim Tryka, Mike Furlough, Bethany Nowviskie, Daniel Pitti, Thorny Staples, Worthy Martin, Geoff Rockwell, and Andrea Laue made substantive contributions to our work and thought. More recently, Bess Sadler, Bradley Daigle, Nick Laicona, and Eric Rettberg provided their unique skills. Others who passed through our orbit—Nathan Piazza, Steve Ramsay, Annie Schutte, John Maeda, and John David Miller, among others—had their own impact.

In the last decade, I have had the good fortune to enjoy an ongoing dialogue with Jerome McGann. He is the other of the "we" that appears frequently throughout this text. Rare indeed to have so kindred a spirit for so unusual an undertaking. Our talents are complementary rather than overlapping, though our shared interests and sensibilities, and common frames of reference, make our exchanges highly fruitful. The spirit of play and invention in the service of imagination is crucial to our vision, and I cannot imagine that this work would exist as it does were it not for his engagement in the conception and execution of many of its ideas and arguments.

The conviction that led me to write the essay "Can Graphesis Challenge Mathesis?" in the late 1990s, however, had been forming for a decade before I met Jerry. I came to subjectivity from a studio practice and the critical study of graphical objects, an approach that brought an emphasis on visuality and design into our UVa conversations and projects as a crucial component of aesthetic insight. The issues that were formulated in a rudimentary way in that essay, and later expanded from graphical issues in knowledge representation to the larger question of subjectivity and the design of conditions of use and interpretation, are of quite a different kind and sensibility than those that come from textual scholarship. They overlap in the fundamental and crucial interest in designing electronic instruments to engage and demonstrate the subjective character of knowledge as interpretation. My contribution was to take what I had learned about subjectivity through visuality and aesthetics into our collective labors in speculative computing.

This book has evolved considerably. Originally conceived as a collection of essays, on which it still draws heavily, it was simply going to replay the development of our thinking and projects at SpecLab. Under the influence of my judicious readers and with the support of Susan Bielstein at University of Chicago Press, it has become a more synthetic book. My central argument is that subjectivity and aesthetics are essential features in the design of digital knowledge representation as that terrifying but very real prospect comes to fruition—the migration of our cultural legacy into electronic environments and the instrumental processing of nearly all aspects of daily life through digital media. The lessons of this book are not confined to insights into how to make things in a digital environment. They spring from that source, but I hope they provide insights into how to think within the broader culture. Where, how, and through what means can we model our understanding of knowledge as a humanistic endeavor within the structures and strictures of our increasingly administered and digitally instrumentalized world?

The first section of the book, "Speculative Computing," provides an introduction to digital humanities in order to contrast with it the distinctive character of speculative computing. The second section, "Projects at SpecLab," describes the development of our work and traces the way hands-on design and production are integrated with theory as a working process in order to imagine environments for subjective knowledge production. In the third section, "From Aesthetics to Aesthesis," various aspects of the materiality, specificity, and implications of the study of digital media are discussed. The section begins with a discussion of my initial impulse to examine graphical codes and the challenge posed by analog images to the logical premises and assumptions that underlie much digital work. This discussion extends into the examination of texts and codes, insights that shifted from mechanistic to probabilistic approaches to materiality and led to investigations of higher-order intellectual structures in metadata and modeling. A discussion of the aesthetic properties of digital media from historical and contemporary perspectives is followed by a discussion of ideology and virtuality. A concluding note sketches a few thoughts on lessons of SpecLab for digital media studies, current and future design practices, and humanistic inquiry.

The spirit of play with which we imagined these projects is an essential aspect of generative insight. Around conference tables or in public presentations, our projects often provoked the query "Are they serious?" The discussion and design of Temporal Modeling, Ivanhoe, and our sketches for the 'Patacritical Demon or my Subjective Meteorology all generated this response. The discomfort caused by our challenges to the cultural authority of computational methods registered the significance of subjective approaches and the threatening aspect of playfulness as a generative engine of imagination. That was crucially important. That moment of questioning disbelief showed that we were creating a gap between familiar ways of imagining what we know and unfamiliar possibilities for reimagining them. In that gap we created the projects of SpecLab.

1.0

Speculative Computing

1.1 From Digital Humanities to Speculative Computing

Our activities in speculative computing were built on the foundation of digital humanities. The community at the University of Virginia in which these activities flourished was largely, though not exclusively, concerned with what could be done with texts in electronic form. Early on, it became clear that aggregation of information, access to surrogates of primary materials, and the manipulation of texts and images in virtual space all provided breakthrough research tools. Projects in visualization were sometimes part of first-generation digital humanities, but the textual inclination of digital humanities was nurtured in part by links to computational linguistics whose analyses were well served by statistical methods. (Sheer practicality played a part as well. Keyboarded entry of texts may raise all kinds of not so obvious issues, but no equivalent for "entering" images exists—a point, as it turns out, that bears on my arguments about materiality.) Some literary or historical scholars involved in critical editing and bibliographical studies found the flexibility of digital instruments advantageous.[1] But these environments also gave rise to theoretical and critical questions that prompted innovative reflections on traditional scholarship.

The early character of digital humanities was formed by concessions to the exigencies of computational disciplines.[2] Humanists played by the rules of computer science and its formal logic, at least at the outset. Part of the excitement was learning new languages through which to rethink our habits of work. The impulse to challenge the cultural authority of computational methods in their received form came later, after a period of infatuation with the power of digital technology and the mythic ideal of *mathesis* it seemed to embody. That period of infatuation (a replay of a long tradition) promoted the idea that formal logic might be able to represent human thought as a set of primitives and principles, and that digital representation might be the key to unlocking its mysteries. Naïve as this may appear in some circles, the utopian ideal of a world fully governed by logical procedures is an ongoing dream for many who believe rationality provides an absolute basis for knowledge, judgment, and action.[3] The linguistic turn in philosophy in the early decades of the twentieth century was fostered in part by the development of formalist approaches that aspired to the reconciliation of natural and mathematical languages. The intellectual premises of British analytic philosophy and those of the Vienna Circle, for instance, were not anomalies but mainstream contributions to a tradition of mathesis that continued to find champions in structural linguistics and its legacy throughout the twentieth century.[4] The popular-culture image of the brain as a type of computer turns these analogies between thought and processing into familiar clichés.[5] Casual reference to nerve synapses as logic gates or behaviors as programs promotes an unexamined but readily consumed idea whose ideal is a total analysis of human thought processes, as if they could be ordered according to formal logic.[6] Science fiction writers have exploited these ideas endlessly, as have futurologists and pundits given to hyperbole, but widespread receptiveness to their ideas shows how deeply rooted the mythology of mathesis is in the culture at large.[7]

Digital humanists, however, were interested, not in analogies between organic bodies and logical systems, but in the intellectual power of information structures and processes. The task of designing content models or conceptual frameworks within which to order and organize information, as well as the requirements of data types and formats at the level of code or file management, forged a pragmatic connection between humanities research and information processing. The power of metalanguages expressed as classification systems and nomenclature was attractive, especially when combined with the intellectual discipline imposed by the parameters and stringencies of working in a digital environment. A magical allure attached to the idea that imaginative artifacts might yield their mysteries to the traction of formal analyses, or that the character of artistic expressions might be revealed by their place within logical systems. The distinction between managing or ordering texts and images with metadata or classification schemes and the interpretation of their essence as creative works was always clear. But still, certain assumptions linked the formal logic of computational processes to the representation of human expressions (in visual as well as textual form), and the playful idea that one might have a "reveal codes" function that would expose the compositional protocols of an aesthetic work had a compelling appeal. At first glance, the ability of formal processing to manage complex expressions either by modeling or manipulation appeared to be mere expediency. But computational methods are not simply a means to an end. They are a powerful change agent setting the terms of a cultural shift.

By contrast, speculative computing is not just a game played to create projects with uncertain outcomes, but a set of principles through which to push back on the cultural authority by which computational methods instrumentalize their effects across many disciplines. The villain, if such a simplistic character must be brought on stage, is not formal logic or computational protocols, but the way the terms of such operations are used to justify decisions about administration and management of cultural and imaginative life based on the presumption of objectivity.[8] The terms on which digital humanities had been established, while essential for the realization of projects and goals, needed to be scrutinized with an eye to the discipline's alignment with such managerial methods. As in any ideological formation, unexamined assumptions are able to pass as natural. We defined speculative computing to push subjective and probabilistic concepts of knowledge as experience (partial, situated, and subjective) against objective and mechanistic claims for knowledge as information (total, managed, and externalized).

: : :

If digital humanities activity were reduced to a single precept, it would be the requirement to disambiguate knowledge representation so that it operates within the codes of computational processing. This requirement has the benefit of causing humanist scholars to become acutely self-conscious about the assumptions under which we work, but also to concede many aspects of ambiguity for the sake of workable solutions. Basic decisions about the information or substantive value of any document rendered in a digital surrogate—whether a text will be keyboarded into ASCII, stripping away the formatting of the original, or how a file will be categorized—are fraught with theoretical implications. Is *The Confessions* of Jean Jacques Rousseau a novel? The document of an era? A biographical portrait? A memoir and first-person narrative? Or a historical fiction? Should the small, glyphic figures in William Blake's handwriting that appear within his lines of poetry be considered part of the text, or simply disregarded because they cannot be rendered as ASCII symbols?[9] At every stage of development, digital instruments require such decisions. And through these decisions, and the interpretive acts they entail, our digital cultural legacy is shaped.

Because of this intense engagement with interpretation and epistemological questions, the field of digital humanities extends the theoretical questions that came into focus in deconstruction, postmodern theory, critical and cultural studies, and other theoretical inquiries of recent decades. Basic concerns about the ways processes of interpretation constitute their objects within cultural and historical fields of inquiry are raised again, and with another level of historical baggage and cultural charge attached. What does it mean to create ordering systems, models of knowledge and use, or environments for aggregation or consensus? Who will determine how knowledge is classified in digital representations? The next phase of cultural power struggles will be embodied in digital instruments that model what we think we know and what we can imagine.

Digital humanities is an applied field as well as a theoretical one, and the task of applying these metaconsiderations puts humanists' assumptions to a different set of tests. It also raises the stakes with regard to outcomes.[10] Theoretical insight is constituted in this field in large part through encounters with application. The statistical analysis of texts, creation of structured data, and design of information architecture are the basic elements of digital humanities. Representation and display are integral aspects of these activities, but they are often premised on an approach influenced by engineering, grounded in a conviction that transparency or accuracy in the presentation of data is the best solution. Blindness to the rhetorical effects of design *as a form of mediation* (not of transmission or delivery) is an aspect of the cultural authority of mathesis that plagues the digital humanities community. Expediency is the name under which this authority exercises its control, and in its shadow grow the convictions that resolution and disambiguation are virtues, and that "well-formed" data behaves in ways that eliminate the contradictions tolerated by (traditionally self-indulgent) humanists. The attitude that objectivity—defined in many cases as anything that can be accommodated to formal logical processes—is a virtue, and the supposedly fuzzy quality of subjectivity implicitly a vice, pervades the computation community. As a result, I frequently saw the triumph of computer culture over humanistic values.[11]

Humanists are skilled at complexity and ambiguity. Computers, as is well known, are not. The distinction amounts to a clash of value systems, in which fundamental epistemological and ideological differences arise. Digital projects are usually defined in highly pragmatic terms: creating a searchable corpus, making primary materials for historical work available, or linking such materials to an interactive map and timeline capable of displaying data selectively. Theoretical issues that arise are, therefore, intimately bound to practical tasks, and all the lessons of deconstruction and poststructuralism—the extensive critiques of reason and grand narratives, the recognition that presumptions of objectivity are merely cultural assertions of a particular, historical formation—threaten to disappear under the normalizing pressures of digital protocols. This realization drove SpecLab's thought experiments and design projects, pushing us to envision and realize alternative possibilities.

Digital Humanities and Electronic Texts

Digital humanities is not defined entirely by textual projects, though insofar as the community in which I was involved focused largely on text-based issues, its practices mirrored the logocentric habits endemic to the academic establishment.[12] Even so, many of my own convictions regarding visual knowledge production were formulated in dialogue with that community. Understanding the premises on which work in the arena of digital humanities was conceived is important as background for our design work at SpecLab—and to heading off the facile binarisms that arise so easily, pitting visual works against texts or analog modes against digital ones, thus posing obstacles to more complex thought.

Textual studies met computational methods on several different fields of engagement. Some of these were methods of manipulation, such as word processing, hypertext, or codework (a term usually reserved for creative productions made by setting algorithmic procedures in play).[13] Others were tools for bibliographical studies, critical editing and collation, stylometrics, or linguistic analysis. Another, mentioned briefly above, was the confluence of philosophical and mathematical approaches to the study of language that shared an enthusiasm for formal methods. The history of programming languages and their relation to modes of thought, as well as their contrast with natural languages, is yet another. All of these have a bearing on digital humanities, either directly (as tools taken up by the field) or indirectly (as elements of the larger cultural condition within which digital instruments operate effectively and gain their authority).

Twenty years ago a giddy excitement about what Michael Heim termed "electric language" turned the heads of humanists and writers. Literary scholars influenced by deconstruction saw in digital texts a condition of mutability that seemed to put the idea of differential "play" into practice.[14] The linking, browsing, combinatoric possibilities of hypertext provided a rush to authorial imagination. Suddenly it seemed that conventions of "linearity" were being exploded. New media offered new manipulative possibilities. Rhizomatic networks undercut the apparent stasis of the printed page. Text seemed fluid, mobile, dynamically charged. Since then, habits of use have reduced the once dizzying concept of links and the magic of being able to rework texts on the screen to the business of everyday life. But as the "wow" factor of those early encounters has evaporated, a deeper potential for interrogating what a text is and how it works has come into view within the specialized practices of electronic scholarship and criticism. In particular, a new order of metatexts has come into being that encodes (and thus exposes) attitudes toward textuality.

Early digital humanities is generally traced to the work of Father Roberto Busa, whose *Index Thomisticus* was begun in 1949. Busa's scholarship involved statistical processing (the creation of concordances, word lists, and studies of frequency), repetitive tasks that were dramatically speeded by the automation enabled by computers. Other developments followed in stylometrics (quantitative analysis of characteristics of style for attribution and other purposes), string searches (matching specific sequences of alphanumeric characters), and processing of the semantic content of texts (context sensitive analysis, the semantic web, etc.).[15] More recently, scholars involved in the creation of electronic archives and collections have established conventions for metadata (the Dublin Core Metadata Initiative), markup (the Text Encoding Initiative), and other elements of digital text processing and presentation.[16] This process continues to evolve as the scope of online projects expands from creation of digital repositories to peer-reviewed publishing, the design of interpretative tools, and other humanities-specific activities. The encounter of texts and digital media has reinforced theoretical realizations that printed materials are not static, self-identical artifacts and that the act of reading and interpretation is a performative intervention in a textual field that is charged with potentiality.[17] One of the challenges we set ourselves was to envision ways to *show* this dramatically rather than simply to assert it as a critical insight.

The processes involved in these activities are not simply mechanical manipulations of texts. So-called technical operations always involve interpretation, often structured into the shape of the metadata, markup, search design, or presentation and expressed in graphic display. The gridlike structures and frames in Web browsers express an interpretive organization of elements and their relations, though not in anything like an isomorphic mirroring of data structures. Features such as sidebars, hot links, menus, and tabs have become so rapidly conventionalized that their character as representations has become invisible. Under scrutiny, the structural hierarchy of information coded into buttons, bars, windows, and other elements of the interface reveals the rhetoric of display. Viewing the source code—the electronic equivalent of looking under the hood—shows an additional level of information structure. But this still doesn't provide access to or reading knowledge of the metadata, database structures, programming protocols, markup tags, or style sheets that underlie the display. Because these various metatexts actively structure a domain of knowledge production in digital projects, they are crucial instruments in the creation of the next generation of our cultural legacy. Arguably, few other textual forms will have greater impact on the way we read, receive, search, access, use, and engage with the primary materials of humanities studies than the metadata structures that organize and present that knowledge in digital form.[18]

Digital humanities can be described in terms of its basic elements: statistical processing, structured data, metadata, and information structures. Migrating traditional texts into electronic form allows certain things to be done with them that are difficult, if not impossible, with print texts. Automating the act of string searching allows the creation of concordances and other statistical information about a text, which in turn supports stylometrics. The capacity to search large quantities of text also facilitates discourse analysis, particularly the sort based on reading a word, term, or name in all its many contexts across a corpus of texts.

Many of the questions that can be asked using these methods are well served by automation. Finding every instance of a word or name in a large body of work is tedious and repetitive; without computers, the basic grunt work—like that performed by Father Busa—takes so long that analysis may be deferred for years. Automating narrowly defined tasks creates enormous amounts of statistical data quickly. The data then suggest other approaches to the study at hand. The use of pronouns versus proper names, the use of first person plural versus singular, the reduction or expansion of vocabulary, the use of Latinate versus Germanic forms—these are basic elements of linguistic analysis in textual studies that give rise to interesting speculation and scholarly projects.[19] Seeing patterns across data is a powerful effect of aggregation. Such basic automated searching and analysis can be performed on any text that has been put into electronic form.

In the last decade the processes for statistical analysis have grown dramatically more sophisticated. String searches on ASCII (keyboarded) text have been superceded by folksonomies and tag clouds generated automatically by tracking patterns of use. Search engines and analytic tools no longer rely exclusively on the tedious work of human agents as part of the computational procedure. And data mining allows context-dependent and context-independent variables to be put into play in ways that would have required elaborate coding in an earlier era of digital work.[20] The value of any computational analysis hangs, however, on deciding what can be expressed in terms of quantitative or otherwise standard parameters. The terms of the metric according to which any search or analytic process is carried out are framed with particular assumptions about the nature of data. What is considered data—that is, what is available for analysis—is as substantive a consideration as what is revealed by its analysis. I am not making a simple distinction between what is discrete and can be measured easily (such as counting the number of *e*'s in a document) and what cannot (quantifying the white space that surrounds them). Far more important is the difference between what we think can be measured and what is *outside that conception* entirely (e.g., the history of the design of any particular *e* as expressed or repressed in its form). The critique that poststructuralism posed to structuralist formalisms exposed assumptions based in cultural value systems but expressed as epistemological categories. The very notion of a standard metric is ideological. (The history of any *e* is a complicated story indeed.) The distinction between what can be parameterized and what cannot is not the same as the difference between analog and digital systems, but that between complex, culturally situated approaches to knowledge and totalized, systematic ones.[21]

Metalanguages and Metatexts

The automated processing of textual information is fundamental to digital humanities, but so is the creation and use of metatexts, which describe and enhance information but also serve as performative instruments. As readers and writers we are habituated to language and, to some extent, to the "idea of the text" or "textuality." Insights into the new functions of digital metatexts build on arguments that have been in play for twenty years or more in bibliographic, textual, and critical studies.[22] Metalanguages have a fascinating power, carrying a suggestion of higher-order capabilities.[23] As texts that describes a language, naming and articulating its structures, forms, and functions, they seem to trump languages that are used merely for composition or expression. A metatext is a subset of metalanguage, one that is applied to a specific task, domain, or situation. Digital metatexts are not merely commentaries on a set of texts. In many cases they contain protocols that enable dynamic procedures of analysis, search, and selection, as well as display. Even more importantly, metatexts express models of the field of knowledge in which they operate. The structure and grouping of elements (what elements are included in the metadata for a title, publication information, or physical description of an artifact?) and the terms a metadata scheme contains (are graphical forms reproduced, and if so in what media and format, or just described?) have a powerful effect. Indeed, metadata schemes must be read as models of knowledge, as discursive instruments that bring the object of their inquiry into being, shaping the fields in which they operate by defining quite explicitly what can and cannot be said about the objects in a particular collection or online environment. Analysis of metadata and content models, then, is an essential part of the critical apparatus of digital humanities.

One tenet of faith in the field of digital humanities is that engaging with the constraints of electronic texts provides insight into traditional text formats.[24] Making explicit much that might elsewhere be left implicit is a necessity in a digital environment: computers, as we are often reminded, cannot tolerate the ambiguity typical of humanities texts and interpretative methods. Because digital metatexts are designed to *do* something to texts (divide elements by content, type, or behavior) or to do something as metatexts (in databases, markup languages, metadata) they are performative.

The term *structured data* applies to any information on which a formalized language of analysis has been imposed. In textual work in digital humanities, the most common mechanism for structuring data is known as markup language. "Markup" simply refers to the act of putting tags into a stream of alphanumeric characters. The most familiar markup language is HTML (HyperText Markup Language), which consists of a set of tags for instructing browsers how to display information. HTML tags identify format features—a <header> is labeled and displayed differently than a
 (break) between paragraphs or text sections. While this may seem simplistic, and obvious, the implications for interpretation are complex. HMTL tags are content-neutral. They describe formal features, not types of information:

<italic>This, says the HTML tag, should be rendered in italics.</italic>

But the italics used for a title and those used for emphasis are not the same. Likewise a "header" is not the same as a "title"—they belong to different classification schemes, one graphical and the other bibliographical. While graphic features—bold type, italics, fonts varying in scale and size—have semantic value, their semiotic code is vague and insubstantial.

XML (Extensible Markup Language), in contrast, uses tags that describe and model *content*. Instead of identifying "headers," "paragraphs," and other physical or graphical elements, XML tags identify titles, subtitles, author's names or pseudonyms, places of publication, dates of editions, and so on:

<conversation>
<directquote>"Really, is that what XML does?"</directquote> she asked. <directquote>"Yes,"</directquote> he replied, graciously, trying to catch her gaze.
</conversation>

All this seems straightforward enough until we pause to consider that perhaps this exchange should take a <flirtation> tag, given the phrases the follow:

[or perhaps <flirtation> starts here?]
<conversation>
<directquote>"Really, is that what XML does?"</directquote> she asked. <directquote>"Yes,"</directquote> he replied, graciously, [or should <flirtation> start here?] trying to catch her gaze.

</conversation>
<flirtation> [or start here?]
His glance showed how much he appreciated the intellectual interest—and the way it was expressed by her large blue eyes, which she suddenly dropped, blushing. <directquote>"Can you show me?"</directquote>
</flirtation>

Can we say with certainty where <flirtation> begins and ends? Before or after the first exchange? Or in the middle of it? The importance of defining tag sets and of placing individual tags becomes obvious very quickly. XML tags may describe formal features of works such as stanzas, footnotes, cross-outs, or other changes in a text. XML tags are based on domain- and discipline-specific conventions. The tags used in marking up legal or medical documents are very different from those appropriate to the study of literature, history, or biography. Even when tags are standardized in a field or within a research group, making decisions about which tags to use in a given situation involves a judgment call and relies on considerable extratextual knowledge. In the example above, the concept of <flirtation> is far more elastic than that of <conversation>.

XML documents are always structured as nested hierarchies, or tree structures, with parent and child nodes and all that such rigid organization implies. The implications of this rigidity brought the tensions between mathesis and aesthesis to the fore in what serves as an exemplary case. The hierarchical structure of XML was reflected in a discussion of what was called the OHCO thesis. OHCO stands for "ordered hierarchy of content objects." The requirements of XML were such that only a single hierarchy could be imposed on (actually inserted into) a document. This meant that scholars migrating materials into electronic form frequently faced the problem of choosing between categories or types of information to be tagged. One recurring conflict was between marking the graphic features and the bibliographic features of an original document. Did one chunk a text into chapters or into pages? One could not do both, since one chapter might end and another begin on the same page, in which case the two systems would conflict with each other. Such decisions might seem trivial, hairsplitting, but not if attention to material features of a text is considered important.

Returning to the example above, imagine trying to sort out, not only where <flirtation> begins and ends, but how it overlaps with other systems of content (<technical advice>, <XML queries>, <social behavior>). The formal constraints of XML simply do not match the linguistic complexity of aesthetic artifacts.[25] Even saying that texts *could be considered* ordered hierarchies for the sake of markup, rather than saying that they *are* structured in modular chunks, registers a distinction that qualifies the claims of the formal system. But despite these philosophical quarrels and challenges, the process of tagging goes on and is widely accepted as necessary for pragmatic work.[26]

Because XML schemes can be extremely elaborate, a need for standardization within professional communities quickly became apparent. Even the relatively simply task of standardizing nomenclature—such that a "short title" is always <short title>, not <ShortTitle> or <ShtTtl>—requires that tags be agreed upon. Creating significant digital collections would require consensus and regulation, so an organization called the Text Encoding Initiative was established. TEI, as Matt Kirschenbaum once wittily remarked, is a shadow world government. He was right of course. An organization setting standards for knowledge representation, especially standards that are essentially invisible to the average reader, is indeed a powerful entity. Protocols and practices that require conformity are the subtle, often insidious, means by which computational culture infiltrates humanist communities and assumes an authority over its operations. One can shrug off a monopoly hold on nomenclature as a smoothing of the way, akin to standard-gauge rails, or suggest that perhaps transportation and interpretation involve similar power struggles. But standards are powerful ideological instruments.

Discussion of tags is a bit of a red herring, as they may disappear into historical obsolescence, replaced by sophisticated search engines and other analytic tools. But the problem raised by XML tags, or any other system of classifying and categorizing information, will remain: they exercise rhetorical and ideological force. If <flirtation> is not a tag or recognized category then it cannot be searched. Think of the implications for concepts like <terror> or <democracy>. A set of tags for structuring data is a powerful interpretative grid imposed on innately complex and ambiguous human expression. Extend the above example to texts analyzed for policy analysis in a political crisis and the costs of conformity rise. Orwell's dark imaginings are readily realized in such a system of explicit exclusions and controls.

Paranoia aside, the advantages of structured data are enormous. Their character and content make digital archives and repositories different in scale and character from static websites and will enable next-generation design features to aggregate and absorb patterns of use into flexible systems. Websites built in HTML hold and display information in one fixed form, like objects in a display case or shop window. You can read it in whatever order you like, but you can't repurpose the data, aggregate it, or process it. A website might contain, for instance, a collection of book covers, hundreds of images that you can access through an index. If it has an underlying database, you might be able to search for information across various fields. But an archive, like Holly Shulman's Dolley Madison letters, contains fully searchable text.[27] That archive contains thousands of letters, and theASCII text transcription of each is tagged, marked up, structured. Information about Madison's social and political life can be gleaned in a way that would be impossible in a simple website.

Through the combined force of its descriptive and performative powers, a digital metatext embodies and reinforces assumptions about the nature of knowledge in a particular field. But the metatext is only as good as the model of knowledge it encodes. It is built on a critical analysis of a field and expresses that understanding in its organization and the functions it can perform. The intellectual challenge comes from thinking through the ways the critical understanding of a field should be shaped or what should comprise the basic elements of a graphical system to represent temporality in humanities documents. The technical task of translating this analysis into a digital metatext is trivial by contrast to the compelling exercise of creating the intellectual model.

Models and Design

Structured data and metatexts are expressions of a higher-order model in any digital project. That model is the intellectual concept according to which all the elements of a project are shaped, whether consciously or not. One may have a model of what a book is or how the solar system is shaped without having to think reflectively about it, but in creating models for information structures, the opportunity for thinking self-consciously abut the importance of design is brought to the fore.

A model creates a generalized schematic structure, while a representation is a stand-in or surrogate for some particular thing. A portrait, a nameplate, a handprint, and a driver's license are all representations of a person. None are models. All are based on models of what we assume a portrait, a name, an indexical trace, or an official document to be. The generalized category of "official document" can itself be modeled so that it contains various parameters and defining elements. A model is independent of its instances. A representation may be independent of its referent, to use the semiotic vocabulary, but it is specific and not generalizable. A model is often conceived as a static form, but it is also dynamic, functioning as a program to call forth a set of actions or activities. The design of the e-book, to which we will return in a later chapter, provides a case study in the ways a model of what a common object *is* can be guided by unexamined principles and thus produce nonfunctional results.

A textual expression may encode all kinds of assumptions yet not explicitly schematize a model. Text modeling, however, creates a general scheme for describing the elements of a text (form, format, content, and other categories each of these, as will become clear below, ask us to think about a text differently), but it is also a means of actively engaging, producing an interpretation. Modeling and interpretation can be perilously iterative—and the creation of metadata can involve innumerable cycles of rework. Even when metadata remains unchanged, its application is neither consistent nor stable. Just as every reading produces a new textual artifact, so any application of metadata or text models enacts a new encounter.

Many information structures have graphical analogies and can be understood as diagrams that organize the relations of elements within the whole. But the models of these structures are often invisible. An alphabetical ordering is a model. So is a tree structure, with its vocabulary of parent-child relationships and distinct assumptions about hierarchy. Matrices, lattices, one-to-many and one-to-one relationships, the ability to "cross walk" information from one structure to another, to disseminate it with various functionalities for use by broadly varied communities, or to restrict its forms so that it forces a community to think differently—these are all potent features of information architecture.

All of this work, whether it is the design of a string search, the graphical presentation of a statistical pattern, the creation of a set of metadata fields or tags, or the hierarchical or flat architecture of a data structure, is modeling. It is all an expression of form that embodies a generalized idea of the knowledge it is presenting. The model is abstract, schematic, ideological, and historical through and through, as well as discipline-bound and highly specific in its form and constraints. Different types of models have their origins in specific fields and cultural locations. All carry those origins with them as an encoded set of relations that structure the knowledge in the model. Model and knowledge representation are not the same, but the morphology of the model is semantic, not just syntactic. On the surface, a model seems static. In reality it is, like any "form," a provocation for a reading, an intervention, an interpretive act. These statements are the core tenets of SpecLab's work.

The ideological implications of diagrammatic forms have been neutralized by their origins in empirical sciences and statistics where the convenience of grids and tables supercedes any critique of the rhetoric of their organization. The arrangement of arrival and departure times in a railway schedule, as closely set columns of numbers, emphasizes their similarity over their difference. But is the predawn departure from a cold, deserted station really commensurate with the bustle that attends a train leaving from the same platform on a holiday afternoon? What, in such an example, constitutes the data? Formal organizations ignore these differences within the neutrality of their rational order.

The cultural authority of computing is in part enabled by that neutrality. Likewise, the graphical forms through which information is displayed online are shot through with ideological implications. The grids and frames, menu bars and metaphors of desktops and windows, not to mention the speed of clicking and rules that govern display, are all ripe for a new rhetoric of screen analysis. The graphic form of information, especially in digital environments, often *is* the information: design is functionality. Information architecture and information design are not isomorphic. Therein lies another whole domain of inquiry into the rhetorical force of digital media.

When design structures eliminate any trace or possibility of individual inflection or subjective judgment, they conform to a model of mathesis that assumes objective, totalizing, mechanistic, instrumental capability readily absorbed into administering culture. Why is this a problem? Because of the way generalizations erase difference and specificity and operate on assumptions that instrumentalize norms without regard for the situated conditions of use. Collective errors of judgment constitute the history of human cultures, and when the scale at which these can be institutionalized is expanded by electronic communications and computational processing, what is at stake seems highly significant. The chilling integration of IBM technology into the bureaucratic administration of the extermination machines of the Third Reich provides an extreme example of the horrific ends to which managed regimes of information processing can be put. That single instance should be sufficient caution against systematic totalization. On a smaller scale, the darkly comic narrative of the film *Brazil* turns on a bureaucratic typo: the confusion of the names Tuttle and Buttle has dire consequences for the bearers of those names—a scenario too frequently replayed in the real-world (mis)management of medical, insurance, and credit records. One needn't have exaggerated fears of a police state to grasp the problematic nature of totalizing (but error-ridden) systems of bureaucracy—or of subscribing to the models on which they base their authority. The ethics and teleology of subjective inflection, and its premise of partial, fragmentary, nontotalizing approaches to knowledge, cannot, by contrast, be absorbed into totalizing systems. On that point of difference hangs what is at stake in our undertaking. Finding ways to express this in information structures and then authoring environments is the challenge that led us to speculative computing.

1.2 Speculative Computing: Basic Principles and Essential Distinctions

With speculative computing, we moved beyond the instrumental, well-formed, and increasingly standardized business of digital humanities. We used the computer to create aesthetic provocations—visual, verbal, textual results that were surprising and unpredictable. Most importantly, we inscribed subjectivity, the basis of any and every interpretative and expressive representation, into digital environments by designing projects that showed inflection, the marked specificity of individual voice and expression, and point of view as a place within a system. We wanted to show interpretation, to expose its workings. We wanted to force questions of textuality and graphicality to the fore. To do this, we (a small core of SpecLab participants) created a series of experimental projects that ranged in their degree of development from proof-of-concept to working platforms for use. But we also created a theoretical and methodological framework.

Our readings and conversations led us to develop a specialized vocabulary that borrowed from disciplines concerned with bringing issues of interpretation into a new intellectual framework. Most important among these for my development were radical constructivism, as articulated in the work of Ernst von Glasersfeld, and the work of second-generation systems theorist Heinz von Foerster. A reformulation of knowledge as experience, based on these two sources and the work of biologists/cognitive scientists Francesco Varela and Umberto Maturana, allowed us to shed the old binarism in which subjectivity is conceived in opposition to objectivity.[1] In this reformulation, knowledge is always interpretation, and thus located in a perceiving entity whose position, attitudes, and awareness are all constituted in a codependent relation with its environment. The system is always in flux, and thus has the complex heterogeneous character of a cultural field shot through with forces that are always ideological and historical. Because we are all always simultaneously subjects of history and in history, our cognitive processes are shaped by the continuous encounter with the phenomenal and virtual world such that we constitute that world across a series of shifting models and experiences. These are familiar concepts within cognitive studies and constructivist theories of knowledge, as well as within critical theory (though implementing these ideas within knowledge production environments poses new challenges). Alan MacEachren's synthesis of an information-processing model of visual perception (which sounds far more mechanistic than it is) incorporates the constructivist approach touched on above, provided another methodological touchstones.[2] Integrating these precepts into the design, or at the very least, the conception of the design of digital environments meant to expose models of interpretation, was a challenge that may still have eluded our technical grasp, but it motivated the projects at SpecLab from the beginning.

In addition to the basic concept of codependent emergence and constructivist approaches to knowledge we incorporated the ideas of probability from quantum theory and various tenets of the turn-of-the-twentieth-century poet Alfred Jarry's 'pataphysics. Our sources for creating a probablistic rather than a mechanistic concept of text came from the work of Heisenberg and Schrödinger, largely through McGann's influence.[3] Adopting their theories of probability and potentiality, we shifted from mechanistic models of text to quantum ones. A text became defined as a field of potentialities, within which a reading intervened. We conceptualized a text, thus, not as a discrete and static entity, but a coded provocation for reading; constrained by those codes, a text is formed anew with each act of interpretative intervention. Here again the echoes of deconstruction are perceptible, but shifted into problems of modeling and representing such activities within an electronic space. The *n*-dimensionality of texts, to use McGann's term, engages their social production and the associational matrix through which meaning is produced in a heteroglossic network, combining the dialogic method of Mikhail Bakhtin with influence from the thick readings of a generation of French textual theorists.[4] But speculative computing is neither a rehash of poststructuralist theory nor an advanced version of either dialogic or dialectical approaches. Speculative computing is grounded in a serious critique of the mechanistic, entity-driven approach to knowledge that is based on a distinction between subject and object. By contrast, speculative computing proposes a generative, not merely critical, attitude.

My approach to graphical knowledge production came from the semiotic studies of Jacques Bertin, the history of visual languages of form, work by MacEachren, and an extensive and systematic reading of materials in information visualization, psychology and physiology of vision, and cultural history of visual epistemology.[5] This last is an enormously underdeveloped field, one that calls for serious study, now in particular, when visualization is becoming so ubiquitous within digital environments. Finally, our reading of Charles Peirce provided a method of interpretation based in abduction as well as a tripartite theory of signification (a sign stands for something to someone and does not operate merely in the formal signifier/signified structure outlined by Ferdinand de Saussure).[6] This theoretical foundation provided a platform on which to elaborate a theory of enunciation and subjectivity.

Within our definition of the concept of subjectivity, we considered both structural and inflected modes. McGann's reference for the structural approach is Dante Gabriel Rossetti's idea of "the inner standing point." This idea posits subjectivity in a structural way, as the inscription of point of view within the field of interpretation. This maps readily onto linguistic theories of enunciation and the contrast of speaking and spoken subjects within a discursive field.[7] The second aspect of subjectivity is that of inflection, the marked presence of affect and specificity, registered as the trace of difference, that inheres in material expressions. I once referred to this as the "aesthetic massage coefficient of form." Though this was a deliberately wry and overwrought phrase, its compact density contains a real description of the relation between differential traces and material expressions—that is, "forms." Bringing in aesthetics links the concept to perception, or the idea of knowledge as experience. All of these concepts became working keywords for us, shorthand used in our elaborate conversations: subjectivity, the inner standing point, inflection, graphical knowledge production, aesthetics, and experiential rather than totalized approaches. I've referred to this attitude as "post-

Cartesian" to indicate the leap from subject/object distinctions and the mind/body split to a conceptualization that escapes such binarisms. In a post-Cartesian frame, subjectivity is not opposed to objectivity but instead describes the codependent condition of situated knowledge production informed by poststructuralist and deconstructive criticism, second-generation systems theory, probabilistic approaches, and radical constructivism.

So while speculative computing builds on certain competencies developed in digital humanities, its theoretical polemic overturns the latter's premises in many respects. The humanistic impulse has been strong in its dialogue with "informatics" and "computing" but has largely conformed to the agenda-setting requirements set by computational environments. Our goal at SpecLab, by contrast, has been to push against the logical constraints imposed by digital media. This is not to deny that such constraints have provided many advantages. Scratch a digital humanist and they'll tell you everything they've learned by being subjected to the intellectual discipline of a field grounded in formal logic (even if, as in the markup debates described above, they disavow the implications). SpecLab projects, though, sought deliberately to challenge the authority of such formality. For anyone familiar with digital humanities, this reads as a radical move. To understand how radical, the contrast has to be sketched explicitly.

As I stated earlier, at the crux of work in digital humanities was a willingness to engage with the task of disambiguation required to process information in digital form.[8] The job of explaining what we do to a "machine" (the quaint colloquial term by which computational and digital technologies are identified in common parlance), in step-by-step procedures, leaves no room for judgment calls or ambiguity of any kind. An intense self-reflexivity results. Even the most apparently simple task—naming or classifying—is immediately revealed as a complex interpretive act. Basic categories of textual activity—the title of a work, name of an author, place or time of publication—suddenly reveal their uncategorizable nuances. Some of these complications are technical (terms in translation versus transliteration, problems of orthography or nomenclature). But some are conceptual: what constitutes the "work" in a piece that exists in many versions, in forms ranging from notes, scrawls, and mentions in correspondence to manuscripts, corrected proofs, and publications?

The accomplishments of digital humanities have been notable: establishing technical protocols that allow texts to be processed in a meaningful way and using various tools of quantitative analysis, pattern recognition, or stylometrics for analysis. The benefits of aggregation and large-scale data processing are immediately apparent to those involved. Scholars can now present original materials in facsimile form online, create linked and hyperlinked text bases and reference materials, and aggregate much that was peripheral or geographically distributed within a single working environment. Translation, searching, collation, and other methods of text manipulation have been automated with different degrees of success or useful failure.

The extent to which the method of digital humanities embodies assumptions about, and thus constrains, the objects of its inquiry is apparent, however, when we look at the terms that delimit its principles: calculation, computation, processing, classification, and electronic communication. Each deserves a momentary gloss to establish the distinctions between digital humanities and speculative computing.

Calculation, based on numerical information and the ability to perform certain functions that can be readily automated through mechanical and digital means, is limited by its inability to represent any content other than quantitative values. Charles Babbage's nineteenth-century devices and more recent office machines are not computers, only automated calculators.[9]

Computation links automated processing to the symbolic realm. Computation makes use of signs whose values can represent *any* information, although their ability to be manipulated through a fixed set of protocols is still determined by a succinct formal logic. The addition of an extra level of articulation in coding symbolic values onto binary entities that could be processed electronically made the leap from automated calculation to computation possible. This leap disconnects semantic values (what symbols mean) from their ability to be processed (as elements in a formal system). Information, as Claude Shannon famously demonstrated, is content-neutral.[10]

Digital processing enacts that logic through step-by-step algorithmic procedures, many specified by programming languages and their different functionalities and syntax. Alan Turing's design for a universal computer transformed this basic capability into an inexhaustible computational engine.[11]

Classification systems build a higher-order linguistic signifying structure on that formal base. The use of digital surrogates (themselves digital artifacts inscribed in code) reinforces the disposition to imagine that all code-based objects are self-evident, explicit, and unambiguous. Schemes for nomenclature and organization come out of library sciences and information management, as well as long traditions of typologies developed in the encyclopedic and lexicographic traditions across the sciences and humanities.[12]

Electronic communication assumes that information functions in a transmission mode—encoded, stored, and then output. The fungibility of information and the function of noise are certainly taken into account in theoretical discussions, even when these are transmission-based and highly technical approaches to electronic communication, such as those of Shannon and his collaborator, Warren Weaver.[13] Bit by byte, the digital approach reinforces a mechanistic understanding of communication and representation. Knowledge in this context becomes synonymous with information, and information takes on the character of that which can be parameterized through an unambiguous rule set.

This basic terminology is premised on the cultural authority of code and an engineering sensibility grounded in problem solving. But the code base of computational activity is full of ideological agendas, which go unquestioned because of the functional benefits that flow from its use. The formal logic required becomes naturalized—not only as a part of the technical infrastructure but as a crucial feature of the intellectual superstructures built to function on it. I've reiterated this several times because this is the crux of our motivation to differentiate SpecLab intellectually from digital humanities.

Now, of course, many digital humanists have raised such questions. Considerable self-reflexive thought about objects of study has pulled theoretical philosophers of all stripes back into discussions of digital projects.[14] Calling assumptions into question is the name of the digital epistemological game as much as it is the standard of conference papers and publications elsewhere in the humanities. Likewise, the study of artifacts in a digital environment is just as apt as conventional scholarly research to prompt discussions of race/class/gender in hegemonic practices, critiques of imperialism and cultural authority, power relations and disciplinary measures. In this regard, SpecLab is part of a larger critical phenomenon, even if its approaches are deliberately pitted against certain aspects of the base on which it builds.[15] Our focus, however, is not on the artifacts themselves, but on the design of the environments and the way ideological assumptions built into their structure and infrastructure perpetuate unexamined concepts.

Speculative computing distinguishes itself from digital humanities on the basis of its sources of inspiration and intellectual traditions. If, as I have said before, digital humanities is grounded in an epistemological self-consciousness through its encounter with disambiguation, speculative computing is driven by a commitment to interpretation-as-deformance in a tradition that has its roots in parody, play, and critical methods such as those of the Situationist International, Oulipo, and the longer tradition of 'pataphysics with its emphasis on "the particular" over "the general."[16] Speculative computing torques the logical assumptions governing digital technology. It pushes back in the dialogue between the modes of interpretation native to the humanities and code-based formalism. Obviously, any activity functioning in a digital environment continues to conform to the formal logic of computational instruments on the processing level. But the questions asked are fundamentally different within the theoretical construct of speculative computing, as summarized in table 1.2.1. The digital humanities community has been concerned with the creation of *digital tools* in humanities contexts. The emphasis in speculative computing is instead the production of *humanities tools* in digital contexts. We are far less concerned with making devices to do things—sort, organize, list, order, number, compare—than with creating ways to expose any form of expression

Table 1.2.1. Attributes of digital humanities versus speculative computing.

Digital humanities	Speculative computing
Information technology/ formal logic	'Pataphysics/the science of exceptions
Quantitative methods (Problem-solving approaches) (practical solutions)	Quantum interventions (Imagining what you do not know) (imaginary/imaginative solutions)
Self-identical objects/entities (Subject/object dichotomy)	Autopoiesis/constitutive or configured identity (Codependent emergence)
Induction/deduction	Abduction
Discrete representations (Static artifacts)	Heteroglossic processes (Intersubjective exchange/ discourse fields)
Analysis/observation (Mechanistic)	Subjective deformance/intervention (Probabilistic)

(book, work, text, image, scholarly debate, bibliographical research, description, or paraphrase) as an act of interpretation (and any interpretive act as a subjective deformance).

Here, then, is a brief elaboration of the binary pairs in table 1.2.1:

Information versus 'Pataphysics. The use of information technology in digital humanities has supported development of tools for corpus linguistics: counting, measuring, and thinking differently about the instances and contexts of word use. Other tools allow for searching, storing, retrieving, classifying, and then visualizing data. All of these procedures are based on assumptions of the self-identity of the object, rather than its codependence on those processes. Computational processes constitute their objects of inquiry just as surely as historical, scientific, or other critical methods. Informatics is based on standard, repeatable mathematical and logical procedures. Quantitative methods normalize their data in advance, assuming a system that conforms to standardizable rules. Information conforms to statistical methods.

By contrast, 'pataphysics derives from the study of exceptions and anomalies in all their specificity—the outliers often excluded by statistical procedures. Only a punning method suffices, thus the invention of our term *'patacritical*. If norms, means, and averages govern statistics, then sleights, swerves, and deviation have their way in the 'pataphysical game. Adopting a 'patacritical method is not an excuse for the abandonment of intellectual discipline. Rather, it calls for attention to individual cases without assumptions about the generalizations to be drawn. In short, it takes exceptions as rules that constitute a de facto system, even if repeatability and reliability cannot be expected. Deviation from all norms and constant change dictate that the exception will always require more rules.[17] Such an approach privileges bugs and glitches over functionality. Not necessarily useful in all circumstances, exceptions are valuable to speculation in a substantive, not trivial, sense.

Quantitative method versus quantum intervention. The idea of interpretation as a quantum intervention is based on the insistence that any act of reading, looking, or viewing is by definition a production of a text/image/work. For this to be the case, the work under investigation can't be conceived as static, self-identical, or reliably available to quantitative methods. Instead, speculative methodology is grounded in a quantum concept of a work as a field of potentiality (*poetentiality* might be a better term). The act of reading/viewing is an intervention in the field, a determining act that precipitates a work. The spirit of indeterminacy goes against the engineering sensibility. Problem-solving methods do not apply in a quantum field. Practical solutions have no bearing on the exposure of interpretation as intervention. Humanistic research takes the approach that a thesis is an instrument for exposing what one doesn't know. The 'patacritical concept of imaginary solutions isn't an act of make-believe but an epistemological move, much closer to the making-strange of the early-twentieth-century avant-garde. It forces a reconceptualization of premises and parameters, not a reassessment of means and outcomes.

Self-identicality versus codependent emergence. Another tenet of the speculative approach, common sense once it is familiar but oddly disorienting at first glance, is that no object (text, image, datum, quantity, or entity) is considered self-identical: $A = A$ if and only if $A \neq A$. McGann cites philosopher George Spencer-Brown's *Laws of Form* as the source for this formulation within the field of logic.[18] The poststructuralist critical tradition is premised on the analysis of contingency. Every text is made as an act of reading and interpretation.[19] When this is combined with recent theories of cognitive studies and radical constructivist psychology, it returns the interpretative act to an embodied, situated condition and the object of inquiry becomes a constituted object, not an a priori thing.

Maturana and Varela's theory of autopoiesis, or codependent emergence between entity and system, also changes mechanistic concepts of subject and object relations into dynamic, systems-based reconceptualization.[20] In an autopoietic description, subject and object are not discrete but interrelated and codependent, and an entity's identity (whether it is an organism or an object of intellectual inquiry) emerges in a codependent relation with its conditions, not independent of them. The conventional distinctions of subject and object are not blurred; rather, the ground on which they can be sustained disappears because there is no figure/ground, subject/object dichotomy, only a constituting system of codependent relations. The subject/object dichotomy that structures text/reader relations was as mechanistic as Newtonian physics. To reiterate the quantum method cited above, and integrate it here, the intervention determines the text. The act of reading calls a text (specific, situated, unique) into being.

Induction versus Abduction. Under such circumstances, scientific techniques of induction and deduction don't work. They are structured and structural, assuming self-identicality in themselves and for their objects, and inscribe relations of causality and hierarchy. By contrast, methods of comparative abduction, taken from Charles Peirce, do not presume that a logical system of causal relations exists outside the phenomenon and supplies an explanation for their relations. Configured relations simply (though this is far from simple) produce semantic and syntactically structured effect independent of grammatical systems. The contingently configured condition of form—a hefty phrase indeed—points to the need to think about forms as relations, rather than entities. This requires rewiring for the Anglo-analytic brain, accustomed as it is to getting hold of the essence and substance of intellectual matter. Even the mechanisms of the dialectic tend to elude the empiricist sensibility, despite its device-driven process and hierarchy of thesis, antithesis, and higher-order synthesis. Abduction adheres to no single set of analytic procedures. Every circumstance produces its own logic—as description rather than explanation.

Discrete versus heteroglossic. Traditional humanistic work assumes its object. A book, poem, text, image, or artifact, no matter how embedded in social production or psychoanalytic tangles, is usually assumed to have a discrete, bounded identity. Our emphasis was instead on the codependent nature of that identity. In this conception, a book is a sort of snapshot, an instantiation, a slice through the production and reception histories of the text-as-work. The fiction of the "discrete" object is exposed as a function of its relation to what we call the discourse field. This comprises the object's composition and distribution, including its many iterations and versions, as well as its multidimensional history of readings and responses, emendations and corrections, changes and incidental damages. A discourse field is indeterminate, neither random, chaotic, nor fixed, but probabilistic. It is also social, historical, rooted in real and traceable material artifacts.

The concept of the discourse field draws directly on Mikhail Bakhtin's heteroglossia and the dialogic notion of a text. Heteroglossia tracks language into a semantic field and an infinite matrix of associations. It brings about shifts and links across historical and cultural domains. Behind every word is another word, from every text springs another, and each text/word is reinvigorated and altered in every instance of production (including every reading, citation, and use). These associations are constantly remade, not quantifiable and static in their appearance, meaning, or value.

The dialogic approach underpins a theory of media (including texts and artifacts) as nodes and sites of intersubjective exchange (rather than as things-in-themselves). A film, poem, or photograph or other human expression is a work spoken by someone to and for others, from a situated and specific position of subjectivity. Its very existence as a work depends on its provoking a response from a viewer or a reading from their own conditions and circumstances. The work is not an inert or fixed text or image, no matter how stable it appears in print or on a screen. It is not an information delivering system but a medium of exchange in social space, the instrument for creation of value through interpretative activity.[21]

Analysis versus deformance. Finally, speculative computing draws on critical and theoretical traditions to describe every act of interpretation as *deformative.* It is not merely performative, bringing to life, or replaying, an inert text, but also generative and productive. Like Situationist *détournement,* intepretation is charged with social and political mobility, but also aesthetic mutation and transformation.

The theoretical agenda of speculative computing may seem to be thick with unsupported claims, but it supplies a brief for project development. Our thinking developed with the projects, not in advance, and the relation between theoretical work and practical design was and remains generative and fluid, necessarily so. Speculative computing takes seriously the destablization of all categories of entity, identity, object, subject, interactivity, process, or instrument. In short, it rejects mechanistic, instrumental, and formally logical approaches, replacing them with concepts of autopoiesis (contingent interdependency), quantum poetics and emergent systems, heteroglossia, indeterminacy and potentiality, intersubjectivity, and deformance. Digital humanities is focused on texts, images, meanings, and means. Speculative computing engages with interpretation and aesthetic provocation. Like all computational activity, it is generative (involved with calls, instructions, encoding), iterative (emergent, complex, nonlinear, and noncausal), intra- and intersubjective (dealing with reference frames and issues of granularity and chunking), and recursive (repeating but never identical deformances).

Speculative computing struggles for a critical approach that does not presume its object in advance. It lets go of the positivist underpinnings of the Anglo-analytic mode of epistemological inquiry. It posits subjectivity and the inner standing point as the site of intepretation. It replaces the mechanistic modes of Saussurean semiotics, with their systems-based structures for value production, with Peircean semiotics, which includes a third term (a sign represents something to someone for some purpose). It attends to the moment of intervention as deterministic of a phenomenon within a field of potentiality. It attempts to open the field of discourse to its infinite and peculiar richness as deformative interpretation. How different is it from digital humanities? As different as night from day, text from work, and the force of controlling reason from the pleasures of *delightenment*.

2.0

Projects at SpecLab

Projects form the core of SpecLab. Putting theory into practice by building things has forced our ideas to become concrete. The route from idea to product is neither passive nor direct; the idea of a game becomes very different when one gets into the nuts and bolts, the step-by-step distillation of its moves and rules. This attitude connects SpecLab to other digital humanities projects, but also to the design or art studio and even the traditional print shop. Making things, as a thinking practice, is not only formative but transformative. Our SpecLab projects brought design to the forefront of our intellectual activity and intensified my understanding of what it means to think about the design of intellectual projects and expressions in a humanities context.

In the design of our SpecLab projects, iterative conceptualization, visualization, production, and rework are the means by which intellectual work takes shape (literally and metaphorically). Thus my focus here will be on the way argument is presented through design, by creating an infrastructure through which the content of individual examples comes to have functional value. As we learned to design the conceptual primitives, the screen and interactive spaces, and the relations of content models to functionalities, I came to understand their integral connection and interpretative force as elements of a text or a data set. Design in this larger sense forms the substantive core of these projects and thus of the lessons learned at SpecLab.

The full theoretical shape of SpecLab was not clear at the outset. As in any substantive research investigation, its dimensions could not have been known in advance. The issues that came under discussion with each project were informed by our reading and theoretical inclinations. But the work also pushed on these theoretical ideas until our ideas consolidated in a specialized vocabulary of concepts and principles. Only as the projects followed on each other did their common features come into real focus: the design of environments for knowledge production that supported the principles of subjectivity, codependence, and emergence and put interpretation in the foreground.

Another aspect of our learning experience came from the necessities of figuring out how to work in groups and through relationships other than the traditional student-teacher or supervisor-employee structures. Institutional and administrative restraints are also factors in any complicated, long-term humanities research. Consultants, collaborative work spaces—such logistical matters are outside the habits of humanistic scholarship. Rather than detail such challenges, however, I will focus on the intellectual and aesthetic aspects of the projects. We actually built Temporal Modeling and Ivanhoe, so they have the most substantial documentation and histories. For Subjective Meteorology, I made drawings, animations, and a complete study. This is an imaginative art project carried out to demonstrate an idea, not functioning software, though it could and may be built. Artists' Books Online (AbsOnline) is ongoing. Its innovations have to do with using structured metadata to try to shape critical discourse in a field that has almost none. In that sense, for all its apparent modesty, it is a radical attempt to transform scholarship and critical practice within a community by using networked capabilities. I present it as a study of the ways metadata models critical thinking. As for the 'Patacritical Demon, speculating on its design remains important, if only because its elusive nature shows that our horizon of conception continues to expand.

: : :

Temporal Modeling was the first SpecLab project. In fact, it came into being in advance of SpecLab's founding and helped establish the viability of an approach to the design of digital projects that emphasized subjectivity and interpretation.

In 1999 or early 2000, John David Miller of Intel (or JDM, as he prefers to be known) came to the University of Virginia to demonstrate a project named Grand Canyon, which he had developed at MIT with designer John Maeda. This was a beta version of timeline software intended to make it easy to organize and display (mainly graphic) information on screen. JDM's primary role at Intel was to scout interesting activities in universities and to help fund experimental projects at early stages. I responded to JDM's demo and also to his official role and sent him a proposal to rework Grand Canyon's design using humanistic premises. I was lucky enough to get two years of funding from Intel, beginning in spring 2001. By the time the funding cycle came to an end (Intel had gone through some reorganization and was no longer providing support for these experimental projects), the team I assembled had created a proof of concept of our "playspace"—an environment for creating graphic timelines.

Temporal Modeling was built with a team of players. Bethany Nowviskie guided the design process in technical, conceptual, and graphical ways. She educated me in the realities of digital humanities and helped keep our goals in focus and the project on track. Jim Allman, a freelance Flash designer who had worked with other projects at UVa, particularly at IATH, created the programming structure for the project. To compensate for the lack of a graphic design culture at UVa, we coordinated with designer Louise Sandhaus at Cal Arts, along with a group of her students. We gave the students small assignments dealing with time and temporality, and they created designs. The young designer whose approach offered us what we were looking for, Petra Michel, came and worked with us briefly and helped give the project an elegant look and form. A conference-workshop in early summer 2001 brought scholars and designers together for conceptual and technical discussions of the project. Much of the really imaginative exploration of subjectivity as inflection, as individuated and highly specific notation, was never developed, however, since our efforts focused on a workable proof of concept. Also, the display space, the piece that would have taken XML files and created a display based on their parameters using our visual system, remained unbuilt. Still, the project taught us a great deal and demonstrated a crucial principle to a community that had previously been almost entirely text-based: that a visual theater for knowledge production could create primary information and analysis, not merely serve as its display.

Ivanhoe, the second SpecLab project, arose from an e-mail exchange between Jerry McGann and myself. The Walter Scott novel of the same name was the point of departure for the game, but any work could have filled that role, for the guiding principle was that any act of interpretation is an intervention within the discourse field that constitutes a work's existence. In May 2000, McGann had given me a copy of *Ivanhoe,* proclaiming its many virtues. A fan of nineteenth-century fiction, I began reading with enthusiasm, only to be discouraged by the grim dullness of a text flattened by pageantry that seemed hopelessly clichéd. The antihero postured, the heroine advanced toward her unfortunate fate—it was all weirdly nonclimactic.

Registering my lack of interest, McGann protested. Look at this amazing scene, he said, and that engaging character. Look, particularly, at the scene between Bois Guilbert and Rebecca on the balcony, when he could, should, might have swept her away to Arabia Deserta, there to indulge their mutual passion. Huh? Yes, yes, he replied. And we set about a series of exchanges involving such a rewriting of the tale.[1] My relation to the novel changed dramatically. From a dull, unengaging text it turned into a territory I was eager to know intimately. Charged to identify points at which I would intervene and turn the story to a new advantage, I became focused on its structure and design. We began to play, making what we would later call "moves." Every text we generated was an alternative to the existing one, deforming or transforming it. We each wrote from a point of view, unacknowledged at first, that later became formalized as a role. Our exchange became the basis of Ivanhoe the project, a game of interpretation. We structured the design to reveal what we felt was at stake in exposing assumptions about texts and textuality, reading and production, the "discourse field" as a rich, ongoing assembly of artifacts of which the text in question was but an instance. And then we set about building Ivanhoe as a real space for play.

Subjective Meteorology extended the idea of Temporal Modeling into an art project that allows subjectivity to be mapped and marked. The project was sponsored by the Digital Cultures Instititute at the University of California, Santa Barbara. Bill Warner and Alan Liu were kind enough to let me spend a month with them and their group drawing, writing, and creating animations and proof of concept sketches. The project makes use of the vocabulary and graphical system of traditional meteorology as metaphors and templates for graphing individual subjective experience. One might, for instance, represent a morning storm of anger generated by a front of frustration colliding with a cloud of anxiety. An idiosyncratic project, unapologetically imaginative, Subjective Meteorology resonates with many individuals. Giving form to such experience as a way to apprehend it, and in some versions of the project to develop a predictive or therapeutic dimension, is all part of the goal. Realized as drawings, a manual, and animations, this project is another demonstration of the possibilities of visual knowledge production.

The 'Patacritical Demon has been envisioned as many things, but above all it is the essential interpretation-modeling device, the means of exposing the process of interpretive activity in its many dimensions. It serves to demonstrate ideas about signification and subjectivity by expressing the transformed and deformed versions of texts produced anew in every reading. The Demon is thus a way to express the way we can understand a "text" or other artifact as an indeterminate field of potential within which a reader intervenes. It is the text/work that is produced as a projection (imagine a hologram) in the spaces between a reader and the planes of discourse and reference. I made sketches of the Demon. We made notes from conversations about its design and form. These may, ultimately, constitute a project. I'm certainly willing to imagine that one SpecLab project might remain purely speculative, emblematic of the always receding horizon of the possible. Our projects sought to bring into being things that seemed just out of reach, but as soon as they became realized, the energy of imagination, like some errant and unruly spirit, would dash off to another corner of the room and again hover beyond our grasp. I'm content to let the Demon be that energy and its sign.

SpecLab's days as a forum for experiment are probably done, even as our projects continue to develop in various ways. Its working unit, Applied Research in 'Patacriticism (ARP), is now given over to various tools projects, especially Collex (an online collections development environment) and Juxta (a textual collation tool). These are part of McGann's ambitious large-scale project Networked Infrastructure for Nineteenth Century Electronic Scholarship (NINES), which aims to demonstrate the viability and necessity of building an online scholarly community. The fissionable energy that charged our conversations has been put to other activities, at least for now. The game of "Designing Ivanhoe," or conceiving of the parameters for Temporal Modeling, or designing the document type definition (DTD) for ABsOnline, were all highly compelling. The act of making, designing, bringing these adventurous projects into being, is where the learning occurs. Other projects may come along, but I hope that the imaginative and once seemingly strange energies of SpecLab will serve as an example of work that began without any clear outcome, highly risky and much laughed at—only to be realized and recognized as useful in fact as well as concept.

The lessons of SpecLab are substantive. Most importantly, they show the possibility of a genuine synthesis of high-level theoretical concepts and digital humanities projects. More specifically, they demonstrate that the design of environments for knowledge production has to be based on a foundation of subjective and partial approaches if humanistic values are to operate within the otherwise instrumental and administered terms of digital ideology and its cultural practices. But perhaps the ultimate lesson of SpecLab is that all forms of interpretation and scholarship are *design* problems premised on models of knowledge that make assumptions about what their object of study is. Discourses, as is well known, constitute their objects; they do not simply apprehend the world—or a text—as it *is*. In our current working lives, we are all digital humanists, and the task of modeling knowledge is part of our daily business. We work within the models embodied by digital environments and instruments, and we ignore the implications of this at our peril. The legacy of SpecLab seems vital to the next phase of our collective endeavors—creative, critical, and scholarly.

2.1 Temporal Modeling

Temporal Modeling provided the first test of our conviction that humanistic principles could be used in the design and implementation of digital projects, and that graphical means could serve as a primary mode of knowledge production. The project, as mentioned above, began as a response to a demonstration of an interface for the display of images and texts designed by John David Miller and John Maeda. Though their software was clever in its use of screen space and creation of conventions for ordering materials, it was based on what I considered nonhumanistic, objective conventions. Such timelines are derived from the empirical sciences and bear all the conspicuous hallmarks of its basis in objectivity. They are unidirectional, continuous, and organized by a standard, nonvarying metric. They are therefore almost useless for describing the experience of time in humanistic documents where retrospective, simultaneous, and crosscut temporalities are discontinuous and move at very different rates (from flash forward to the stilled moment). Temporal Modeling was designed to create a visualization scheme appropriate to the analysis and study of such experiences.

We began with research into the ways the experience

of time had previously been understood and represented. We expected that if we read across a wide range of fields and disciplines, we would discover striking cultural and historical differences. But the basic concepts and conventions we unearthed comprised a small and unified array, most of them central to the ways time and temporal relations (the difference between these being quite significant, as will be seen in a moment) are used in computational models.

The basic approaches to measuring and marking time traced back to ancient mappings of the sun's movements, seasonal cycles, and planetary activity, with little alteration of basic methods or units since the Babylonians. The structure of the year, the counting of days, and other metrical devices might vary in particulars, but conventions for their representation were standardized: lines and grids, or circles divided radially. By the time a fully rationalized system of timekeeping appeared, with water clocks and hourglasses, the conceptual foundation of time was well established. The authority on which Western empirical sciences draw to establish the linear, regular, continuous parameters of time for statistical analysis and data gathering was supported by philosophical and mathematical assumptions unchanged from the classical period.

The distinction between *time* and *temporality* is among the basic principles bequeathed from classical philosophy. The Greeks understood time as an absolute, a priori condition or field but conceived of temporality as a description of relations among elements that constitute that field and its values. Most of the issues still attached to these two concepts (such as the crucial problem of "the dividing instant") were described in classical literature. When relativity and quantum theory challenged mechanical models in physics, they introduced new concepts into the scientific and mathematical understanding of temporality for the first time in several millennia. Writers of fiction and fantasy have long developed their fluency with elastic models of time, unexplained simultaneous occurrences at a distance, and the conventions of flashbacks, crosscuts, jump cuts, and foreshadowing. But the disciplinary lines between aesthetic work and empirical analysis were well-defined and defended. One of our primary goals in this project was to bring these worlds together.

As I stated at the outset, in the mechanistic, empirical worldview of traditional mathematics and natural sciences, timelines and graphs are premised on common assumptions of time as unidirectional, neutral, and homogenous. We based Temporal Modeling on counterassumptions. First, we set out to model temporal relations—not time. Rather than approach time and space as already existing boxes into which events or things are put, we chose to embrace the premise that temporality and spatiality are constructs grounded in relations among phenomena. A phenomenological approach is better suited to modeling the temporal relations contained in the aesthetic artifacts and documents that comprise the basic materials of humanities scholarship.[1] These materials are often fraught with complexities and contradictions regarding the ordering of elements in a temporal scheme. The experience of events and their interpretation is grounded in subjective perspectives. The simple fact that any human-authored document represents an individual and inherently fragmentary point of view from within events, rather than an objective record from a presumed external stance, suggested that our counterassumptions were essential if our designs were to serve a humanities community.

We did not, however, want to proliferate idiosyncratic or novel concepts without justification. The challenge was to develop our novel, graphical system for representing the subjective experience of temporality while at the same time situating our project within an existing literature. In the first phase of our work, this entailed a literature review on which we drew for the outline of our conceptual primitives—the basic elements of the Temporal Modeling system. In the second phase, we distilled a content model and designed a space in which it could be used.

Defining the Project in Conceptual and Technical Terms

We knew that we wanted to design a notation scheme that would allow us to represent such notions as anticipation or regret, since retrospective and prospective ways of conceiving of future and past inherently involve transformation of the record and representation of events. To do this, we needed a system that supported the representation of multiple narratives simultaneously, even narratives based on contradictory accounts, since this is often characteristic of the ways individual memory works against the backdrop of official history. We tried to develop a set of metaphors and templates that would accommodate mutable and inflected timescales and be useful as a research tool not only for interpretation and analysis of temporal data, but for their display. The challenge was to create a graphical communication scheme capable of representing a subjective, inner standing point within temporality in a legible manner.

The project was framed, therefore, within these assumptions:

- Time may be experienced as a unidirectional flow within human perception, but the interpretive ordering of temporal events has forward-

branching (prospective) and backward-branching (retrospective) options.
- Temporal relations are inflected by emotions, mood, atmosphere. Not all moments can be measured on the same scale, or on a homogeneous scale; the shape of time intervals (granularity, scale, and metric) varies according to subjective perception.
- Temporal relations are not necessarily continuous. Breaks, ruptures, repeats, and overlapping events occur within different points of view from a single event or within relations among events.

The technical problem was to create an interactive tool set for representing and modeling temporal relations from humanities data, *in advance of* creating a database, document type definition (DTD), or XML markup scheme. That is to say, instead of first making a model of temporal relations in a text or group of documents and then displaying it, we wanted to make a space in which visual tools could be used for primary representation and analysis that would then give rise to interpretation.

In a typical digital humanities scenario, a set of letters or family papers, or a set of incidents in a narrative text, might be analyzed. This analysis would give rise to a hierarchical scheme in which different levels of a time-based system would structure the organization. The representation of the events would follow, conforming to the already established conceptualization (modifications to the content model would require going back, changing the scheme, and repeating the subsequent steps). Events (information, texts, other data) would then be marked with a set of XML tags. Finally, these items could be displayed on a timeline according to parameters already fixed within the hierarchy.[2]

This practice of developing the content model initially in XML is a methodology imported from data management. Information structures are essential for organizing materials in archives, collections, or any digital repository. They are powerful interpretive instruments, but they don't always behave according to the principles of humanistic interpretation and its many theoretical approaches. The habit of creating an elaborate content model in advance of display had come to be a conceptual limitation. Because information display was always a second phase in that approach, visual representations were always secondary. They might be convenient and efficient ways of showing information, but digital humanists rarely thought of graphical displays as ways of *generating* information.

We wanted instead to design a system capable of representing the complex and fragmentary information typical of human records *before* designing the data structure. If we read a group of family letters and documents, our goal would be to place them in some temporal relation to each other in a graphical scheme, and let that activity determined the model of chronology among them. The idea was that if we designed the composition space within sufficient technical constraints (so that every mark, line, and point could be parameterized), it could be used to give rise to a formal knowledge representation scheme.[3] But we were also keen to return interpretation to the field of digital humanities, which had in many ways subjected itself to the mindset of analytic and empirical approaches as it borrowed the technical methods of data capture and information organization.

Our readings included works from a considerable range of disciplines: humanities fields (philosophy, narratology, structuralist discourse analysis, history, knowledge representation), social sciences (particularly anthropology and religious studies), informatics (formal logic, linguistic analysis, temporal database development), the natural sciences (biology, geology, physics and relativity theory), and visual design (art history as well as graphic methods for information design). We imagined we would eventually graduate to topological mathematics and the spatial modeling of events (the "rubber sheet" metaphor always seems appropriate to the distortions of subjective experience), the analysis of temporal elements in narrative and linguistics (including deixis and tense modalities), and the field of diagrammatic reasoning and semantics.

From our literature review, we distilled a set of conceptual primitives for the representation and modeling of elements in temporal relations. The review itself is worth summarizing briefly, since the act of culling basic concepts into the smallest possible usable set of conventions was one guiding principle of our work at the formative stage.

Literature Review of Time and Temporality

Though the literature on time and temporality cuts across humanities, social sciences, natural sciences, and informatics, our survey yielded a surprisingly concise set of terms and basic concepts. Philosophical concerns focused on issues of ontology and metaphysics. Logicians devised formal systems that were useful for informatics and met their requirements for instrumental and practical applications. Discourse analysis and narratology provided a basis for thematic description and material encoding of concepts of time and temporality in natural language. With the exception of twentieth-century developments in relativity, the ideas about time and temporality shared across disciplines had been understood and established by the early centuries of the Common Era.[4] More specialized terminology and more elaborate scholarly schemes of analysis have emerged in recent decades, but the fundamental conceptual underpinnings have remained remarkably consistent across historical periods and fields of intellectual inquiry.

We found that in almost every discipline an important distinction was made between absolute and relational time. *Absolute time* is a given, conceived as a structural container of events, while *relational time* emphasizes temporality as a product of the relative sequence and duration of events within a frame of reference. These distinctions, however, are not always clearly observed. In many cases, the assumptions on which they operate are inherent in a disciplinary perspective. For instance, the idea that time preexists events has a strong foothold in the natural sciences, where the ontological existence of time goes largely unquestioned. Even the most intuitive interpretations of the subjective experience of temporality are often framed in relation to this a priori concept and the empirical premises it reinforces. We wanted to be aware of these assumptions but focus on the ways temporality is understood thematically and encoded in representations such as language and other symbolic forms. Our goal was to create an interface for *interpretation* of temporal relations in humanities data. To do so, we had to jettison the idea that time or temporality in themselves were going to be modeled in our system.

Beginning with philosophy and metaphysics, we situated our inquiry within what computer scientist Fabio Schreiber terms the study of temporal ontologies or "the major issues in the nature and structure of time."[5] An empiricist bias was evident even in the simple assumption that "the nature and structure of time" could be described as a singular, homogenous entity. Working within the field of informatics, Schreiber had pragmatic reasons to establish such parameters for temporal consideration—such as the need for synchronization of distributed computational systems. But his survey of the literature at the intersection of philosophy, history, and informatics provided us with a useful list of descriptive approaches to understanding what time *is*. These begin with a distinction between linear and circular conceptions of time. The linear conception reinforces the idea of the unidirectional flow of time's arrow, while the circular suggested the repetition of life cycles, circadian rhythms, and other apparently identical replications within temporal sequences. But the idea of ontological understandings (what time is) can be confused with the ways it can be represented. For instance, no intellectual or mathematical support exists for the idea that time could take a circular or cyclic form even though recurrent activities in human or mechanical realms are often casually referred to as cycles.[6] The "cycle" is actually a flattened view of a spiral in which repeating activities flow around the same set of milestones or markers but the whole process unfolds along a unidirectional axis. Schreiber's inventory also included other, more self-evident concepts: the contrast between a belief in infinity and the human experience of the finiteness of time; the experience of discrete moments or units of time as against its perceived continuity and flow; an absolute sense of time described as past, present, and future; and a relative sense of time described in terms like before, after, or during.

For Schreiber the flow of time is an objective feature of the physical world, and this provides Western science with philosophical support for its assumptions. This flow can be understood in the language of formal logic and linguistics. The very idea that temporal measures are arbitrary (hours, minutes, seconds) reinforces the conviction that time "itself" exists as a container for events. Conventions for measuring time, marking its divisions and subdivisions according to named intervals, follow calendrical, horological, and other extrinsic systems—sidereal, physical, biological, or time-stamped and dated—each of which is bound to historical and cultural realms. (Religious and sacred times overlay and interpenetrate secular calendars even when the same system of dates was used as a scaffolding for both.) Anthropological research offers evidence of temporal schemes that mark complex, parallel multiphase systems, but no matter how many different patterns they entail, these systems' premises do not challenge the a priori existence of time or its unidirectional flow.

Schrieber has a pragmatic agenda, distilling basic information about temporal ontologies for applied use. The scholar J. T. Fraser, on the other hand, systematically examined the ways time was understood from various disciplinary perspectives. His list of descriptive rubrics differs dramatically from the information-based categories into which Schreiber organized his survey. Fraser's list includes:

- eotemporality: the rational progression of temporal events in an apparently sequential form;
- nootemporality: time experienced by the human mind;
- psychotemporality: perceived time, psychologically inflected;
- sociotemporality: time proper to a specific social system or condition;

- biotemporality: temporal distinctions operating within a continuous, organic present (with apparently cyclic and other purely linear patterns);
- atemporality: the temporality of physics, in which the universe is simultaneous, unordered, chaotic; and
- prototemporality: undirected, discontinuous, primary.

In Fraser's discussion, these concepts also assume that time is an a priori condition, available to description either as a sequence of events in human experience or as events that can be ordered within a descriptive schema. Even such a subjectively oriented concept as psychotemporality would be measured against a normative extrinsic temporality, defined, that is, as a contrast with "time" as an absolute.

Fraser takes the concepts and systems as descriptions of time itself, not as intellectual constructs to be analyzed, and thus neglects the materials most important for humanities work and interpretation—the linguistic, visual, or symbolic systems in which concepts of time are encoded. So we added a single category to Fraser's list:

- discursive temporality: the representation of time in discourse.

We also modified Fraser's discussion by making a clear distinction between the assumption of an objective perspective (in either metrics for charting time or the assumption of time as an a priori given) and the recognition of subjective experience with temporal dimensions. By distinguishing the intellectual representation of concepts of time and temporality from a conviction regarding the a priori ontological existence of time as a thing in itself, we based our work on a self-conscious attention to representation and its interpretative contingencies, rather than a presumption of external realities and their absolute, unconditional existence.

Fraser provided a panorama of temporal schemes designed to suit individual disciplines. This extended the terminology we had derived from Schreiber and others, including logicians who describe the relations of intervals and events in a linear system that emphasizes the relative ordering of temporal events. Their work is different from what we called "discursive" temporality in one significant respect; it is not grounded in analysis of the specific qualities of linguistic expression. Formal logicians, such as James Allen, though focused on language, are characterized as "de-tensers" because of this feature of their approach. They have a vocabulary for describing relations among time intervals rather than focusing on the language in which temporal experience is marked and represented.

In "Time and Time Again," an essay much cited in the liteature, Allen's relational diagrams offer a logical framework for all possible orders and sequences of events.[7] These events are assumed to exist outside of their representation, and the formal scheme is a way of elaborating a typology of these relations. Allen's approach is useful for analyzing such relations (temporal logics), particularly when they can not be correlated to an absolute or extrinsic dating system (calendar or clock time), or when they can be linked only with what are referred to as pseudodates, an intrinsic dating system. The formality of Allen's logical system allows for a fully disambiguated description of such temporal relations. It also accommodates forward-branching options, a desirable feature for computational situations in which a single, determinate past might connect to a multiplicity of future options.

Allen's logical relations are defined by a succinct set of terms (and their complements): before, meets, overlaps, during, starts, finishes, equals. These lend themselves to representation as sets of arrows whose formal, schematic relation precisely matches the temporal relation and corresponds to its verbal description. For Allen, the concept of tense is cast entirely within formal language, which has made his work useful for the requirements of informatics. The concepts of temporality needed for time-stamped database operations make use of similarly formal logic in distinguishing the moments at which a fact is stored in a database, the moment of a query, or the moment at which a fact might be true within a modeled reality. These systems depend upon internal clock mechanisms, intrinsic systems of highly formal, unambiguous temporal relations. They therefore lend themselves to formal description rather than either correlation with extrinsic systems or subjectively inflected and ambiguous tense modalities.[8]

In contrast to the formal approach provided by logicians, linguists and scholars of language in literature and narrative offered us terms appropriate to the analysis of fictional, historical, or other documentary narratives. Their work focuses on the encoding of assumptions about temporality in symbolic representation in natural language, whether in an utterance, document, or narrative. In such an approach, the first problem is to identify the linguistic markers of tense or other temporal features. The next is to understand the cultural, psychological, or other symbolic value by which the temporal system is inflected.

Mark Steedman's study "The Productions of Time" provided an extensive catalog of tense modalities or tense logic in language, incorporating classic work in discussion of speech, reference, and event points within linguistic representation, as well as a summary of contemporary work in this area.[9] Rather than attempt a description of events grounded in formal relations of intervals, Steedman and his colleagues sought to elucidate the semantic implications of distinctions embedded in linguistic terms. *Achievements,* measured at or in a particular moment were contrasted, for instance, with *accomplishments,* which were extended in time, and *activities,* which endured for a set period. These sorts of descriptive categories clarify the means by which natural language encodes cognitive concepts about time and temporal relations.

Extending our research into narrative theory, we encountered the realm of constraint logic programming. From this we derived analytic and interpretive tools for defining narrative elements within a system of internal references that describe temporal relations; each element is analyzed and its temporal identity constrained within a formal system in order to extract an ordered sequence of referenced events out of the language of experience, action, or description in the narrative.[10] These approaches are dependent upon the careful analysis of tense indicators in syntax and discourse structure.

In one such study, Pamela Jordan, a linguist studying narrative, made use of tense markers to demonstrate distinctions among narrative reference frames.[11] Tense markers such as "here" and "now" not only describe relative time frames, but also link the representation of time to individual subjectivity. The concept of deixis, derived from structural linguistics and applied to narrative theory, refers to the way subjectivity (individual speaker identity and position) was structured in language. Though classical narrative, as defined by Aristotle's unities of character, action, and location, assumes that time and space are universal, continuous, and coherent, such assumptions are certainly not part of all narrative frameworks. Self-conscious manipulation of these unities has been part of modern literature and its theoretical and interpretive approaches, and can be brought to bear on the analysis of documents in historical studies.

Historians and anthropologists note that ideological and cultural values often inflect time systems. Herbert Bronstein, in "Time Schemes, Order, and Chaos: Periodization and Ideology," points to the repetitive cyclic conceptions inherent in a notion of an eternal being and the radically contrasting ways this concept has operated within Jewish and

Christian approaches to historical chronology.[12] The difference between believing that the Messiah is still to come or has already appeared serves as an organizing feature of all historical events, and casts a markedly nonneutral interpretive frame on the description of human experience. Bronstein's example shows dramatically that any historical scheme embodies a worldview laden with a sense of movement toward or away from a culturally sanctioned goal such as progress, salvation, enlightenment, or rebirth. The very division of history into discrete epochs or periods—ancient, medieval, modern—reflects assumptions about shifts in cultural paradigms along an irreversible temporal axis.

Cross-cultural perspectives demonstrate the bias inherent in concepts of temporality that are taken to be intuitive or to organize social relations into a network of cultural activities. These distinctive formulations are most conspicuous in the use of various timekeeping schemes but also extend to notions of dream time, ideas of the present as a point floating within a nonlinear past and future, and other alternatives to the rational system of logical, unidirectional order in Western time concepts.[13] These approaches are in many ways more appropriate to our subjectively organized approach than those that derived from formal language and empirical sciences.

At the end of our study of tense and alternatives to linear, unidirectional time-arrow frameworks, we looked briefly at the literature on relativity and its influence across a wide spectrum of cultural activities.[14] Fiction and narrative, as well as scientific discussions of event modeling, historical patterns, and events within the realms of physics, all lend themselves to description according to models derived from what is termed space-time.[15] Scientific debates about the absolute existence of a time arrow focus on the second law of thermodynamics (the tendency of chaos to increase in the physical universe along an apparently asymmetrical temporal axis). But in narrative imagination, the theory of relativity provides suggestive starting points for the reordering of perception.[16] The multiple temporalities available in such systems fragment the unity of time as well as its illusory order in human experience.

In summary, we could see that the apparent order of time as a given, a priori container for experience, had a counterpoint in the conception of temporality created and shaped by the ordering of events, objects, elements, and effects in increasing layers of complexity. Sharp differences existed between objective and subjective conceptions of temporality, and among variously inflected interpretations of the value of events within temporal orderings. This literature provided us with a stable

OBJECTS/ELEMENTS (basic temporal elements to be represented) line or axis calendar grids clock faces points intervals events granularity tics metrics (intrinsic and extrinsic) notations and inflection markers start and stop points now and the now-slider
RELATIONS/STRUCTURES (attributes or connections among elements) order (or temporal direction?) rupture multiple and/or inflected granularities the dividing instant visual positioning of elements certainty of temporal position determinacy of boundedness alternative iterations (now-slider–generated lines) degrees of inflection and relation among inflected elements
ACTIONS/OPERATIONS (activities a user should be able to perform) generating and viewing time slices positioning and labeling elements ordering and reordering attaching and detaching a metric choosing/inflecting/zooming a metric defining intrinsic granularities now-sliding (generating alternative iterations) inflecting temporal relations

Table 2.1.1. Initial conceptual scheme of objects, relations, and actions

nomenclature of concepts. These in turn informed the elaboration of our "temporal primitives"—the basic elements that comprised our conceptual scheme (tables 2.1.1, 2.1.2).

Graphical Conventions

In parallel to this literature review, we conducted a survey of visual conventions for the representation of time and temporally marked information and quickly became aware of how limited the graphic conventions were for picturing data in time.[17] These conventions also shared the assumptions that time is unidirectional, neutral, and homogenous.

The two fundamental elements of any temporal diagram are the *reference frame* through which it is structured and the *notational vocabulary* with which temporal relations are expressed. Reference frames make the expression of temporal relations possible by defining the rules under which the visual system operates. These frames are either extrinsic to the data (assuming an objective time framework against which the absolute temporal position of an element can be measured) or intrinsic to it (based solely on the relations among the elements themselves but traditionally assuming a linear chronology). In some instances, reference frames present a combination of intrinsic and extrinsic measures (e.g., the perceived time of an experience and the actual time of the event measured against a standard timekeeping device). Conventional vocabularies for temporal notation contain three types of markers: *points,* or discrete instants in time; *intervals,* or segments of time; and *events,* which are occurrences in time.

Diagrammatic representations of temporal relations fall into three basic categories: *linear, planar,* and *spatial.* Linear diagrams, or timelines, are by far the simplest and most prevalent forms. Almost every diagram we found was, in essence, linear, by virtue of the way it used its axes and metric scales. The archetypal timeline consists of a single axis on which a stable metric and a sequence of markers or labels representing the progression of events in time are organized. The timeline is a linear spectrum with homogenous granularity. On a linear diagram data can exhibit only three relative temporal conditions: *earlier than, later than,* or (sometimes awkwardly) *simultaneous with* (or overlapping).

Planar diagrams chart temporal relations on two axes. Sometimes, as in calendar grids, which mark days against the larger structure of weeks, both of these axes are temporal. Often, however, time is marked according to a uniform metric along a single axis and data representing some

Time	absolute time: container of events relative time: relations among events
Temporality	system constructed as a way to visualize temporal relations
Axis or **line**	time arrow
Point	(no extensible duration) **start point** **end point**
Interval	demarcated segment of time
Event	occurrence in time Linguistic vocabulary for modal expressions of events (Mark Steedman, "The Productions of Time") achievements (at or in a particular period of time) activities (for a set period of time) accomplishments (extended in time) Formal logic (James Allen, "Time and Time Again") extrinsic (absolute dating system) intrinsic (pseudo-dating system) forward branching (multiple future options) logical relations of temporal intervals: before during meets overlaps starts finishes equals
Metrics	**extrinsic metric**: conventional measure (e.g., hours, days, weeks, years) physical measure (e.g., quartz clock cycles) **intrinsic metric**: in relation to lived experience (e.g., birthday) **chronon**: smallest unit in any time system
Ordering	sequencing without regard to a metric
Iterations	versions of temporal sequence reordered through subjective perception

Now-slider	fundamental reference point within the field of interpretation
Granularity	change of scale of a fixed order or chosen metric
Slice	state of elements at a specific temporal moment
Date-stamped	element with certain and determined form
Dividing instant	point at the intersection of two segments
Vocabulary inflections (apply to points, intervals, events to give character attributes)	
	determinate/indeterminate (with respect to start and end) **certain/uncertain** (with respect to date-stamped accuracy) **rupture** **user-definable terms** mood, atmosphere importance
Grammatical inflections (structural relation of elements)	
	prospective effects foreshadowing causality anticipation user-defined relations **retrospective effects** causality regret user-defined relations

Table 2.1.2. Nomenclature scheme for conceptual primitives
Note: Terms in bold type were adopted for use in our conceptualization.

other quantitative value is charted against the other axis. Diagrams in this category—which includes the familiar bar and line graphs—may present information about a single data type (bivariate graphs) or about multiple information streams (multivariate graphs). In most cases, this form does not emphasize temporal relations but rather the evolution of a specified attribute over time. Like linear diagrams, the planar form presents the flow of time as unidirectional and asymmetric.

A spatial diagram, the least common representational scheme, attempts to map data on multiple axes—sometimes literally tracking events as they move in time and through geographical space, and sometimes modeling data in a three-dimensional format in which none of the coordinates measure literal space. Digital spaces offer new opportunities for three-dimensional diagrams and fourth-dimensional progressions. But the conception of *n*-dimensional space-time would need to be invoked if the rich conceptual potential of modeling chaos, complex fluid dynamics, and even topological forms and processes were brought into play. The difficulty with such complex visualizations is graphic legibility. The ideas and the notations are unfamiliar, and legibility depends on familiarity and habits of reading. New conventions no doubt lie ahead but have yet to be designed and tested for the kind of project we are describing. We did play with sketches and schemes for such representations and sought graphical models of relativity, topology, and quantum effects. But realistically, such models were beyond our computational and conceptual capabilities, not to mention being difficult to produce except as schematic indicators, provocative and suggestive rather than useful. Such horizons point toward future development for this project.

Modeling Temporal Modeling

Our modeling scheme challenged the three basic assumptions of conventional models: unidirectionality, continuity, and homogeneity. We felt we could create a visual scheme in which alternatives could be represented for purposes of basic research and visual display. We proceeded with the conviction that humanities scholars, dealing with many variables in the temporal relations expressed in documents and accounts, need a less rigidly empirical and more flexible system of representing these relations.

Instead of a time arrow, we designed our system to include branching narratives that could go back as well as forward in time. A map of

Figure 2.1.1. Preliminary design sketches for Temporal Modeling

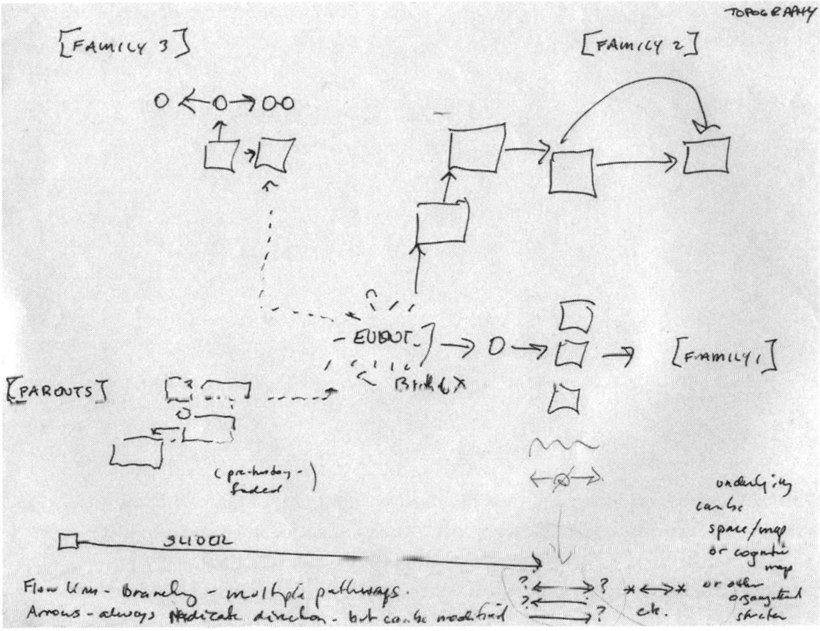

2.1.1a. The original conception of the design included multiple narratives and branching paths. Here the central event ("Birth of X") is shown in the center of a field of documents, each of which is positioned within the temporal space. The central event creates a common reference for different sequences of reactions, responses, or other information registered in documents. As this is meant to serve humanities research purposes, all information is linked to some kind of artifact or reference (hence the little boxes). Note that the now slider is present, even in this early sketch, allowing forward and backward movement.

past events may change dramatically in response to new information or occurrences that do not merely recast our interpretation of events but alter our conviction about what actually occurred. (For example, the development of theories of geological history in the nineteenth century subjected biblical accounts of past events, until then taken seriously as metrics by historians as well as theologians, to radical reconfiguration in order to conform to empirical evidence.) Similarly, anticipation of future events and the degree to which this anticipation shapes the present, a major aspect of narrative practice in prose and drama, is difficult to chart on a standard time line. These shifts, in our model, gave rise to branches, which were linked through what we called a "now-slider" (figure 2.1.1).

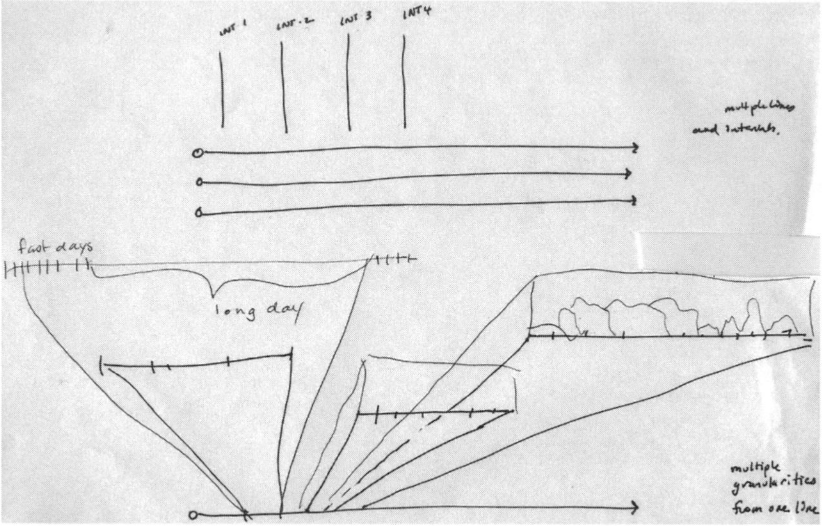

2.1.1b. Our interest in varying the granularity of the temporal metric prompted a scheme for expanding selected segments; the aligned vectors in the upper register allow these different scales to be read against each other using a standard metric. The stretchy timeline concept is shown at center left, where the unevenly spaced tick marks (labeled "fast days" and "long day") register the experience of time passing more quickly or more slowly along a single continuum.

Instead of treating time as a neutral, containerlike setting for events we wanted to be able to show (and manipulate) the tensions and pressures exerted by events that inflect temporality with subjective qualities. The idea of "the distant future" or "someday" or "after my lover comes back"—all quite logically compatible with subjective experience of temporality—resist being absorbed into a neutral concept of time with a stable, extrinsic metric. The relation among events separated by time, rather than an experience of time itself, is the focus of such experience.

Finally, we wanted an alternative to the standard metric. The idea that temporal relations can be mapped on a single scale is based on the supposition that time is homogenous and consistent. In much humanities-based research, as in lived human experience, subjective notions of time differ depending upon circumstances and emotional or other investments. The perception of granularity changes with context. (Clearly the appropriate granularity for a historian documenting the burning of Atlanta during the Civil War, for instance, is quite different from that used in the narrative of *Gone with the Wind* in either its film or book versions.) And the relation of parts to each other, parts to a whole, or metric scales to each other cannot always be unified within a single homogenous frame. Breaks, inequities, and discrepancies in pacing are all elements of the lived experience of time and its record in humanistic documents. These ruptures or lacunae are often the periods of greatest interest to the humanities scholar and lay user of time-based digital media alike. One of our solutions was an elastic timeline that introduced malleable and variable metrics into a single line. Another was to change granularity from one segment of a line to another, or to introduce different metrics within a single representation.

With this conceptual framework in place, our work became focused on the elaboration of an effective visual design for the interactive tool

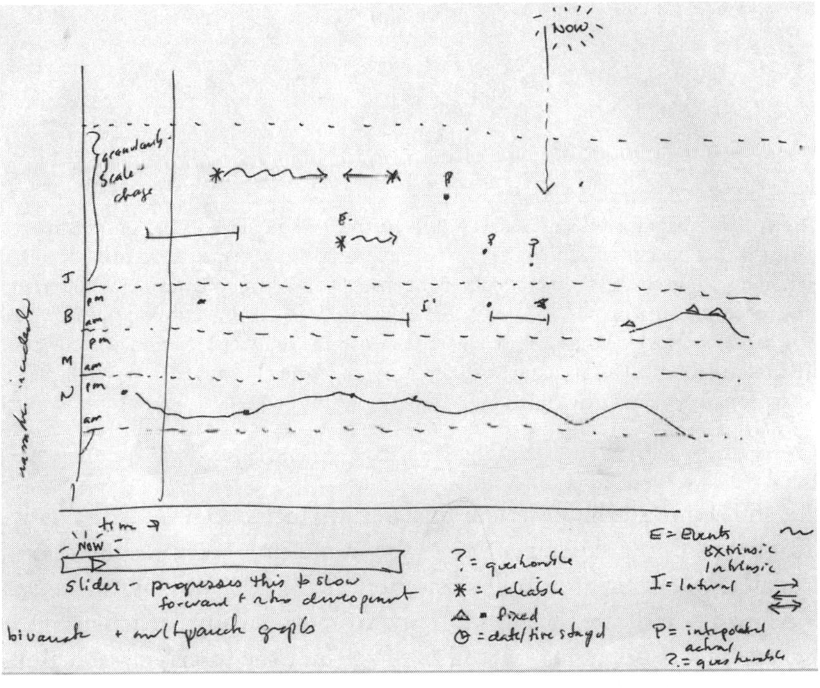

2.1.1c. Thinking about various ways the now slider could work within varied narratives of a single event led us at first to a single moving slider that would advance or replay a sequence of events. "Now" was conceived as a moment in the overall temporal scheme, not linked to a particular, individual point of view. We also explored graphic effects that would indicate emphasis and inflection. A well-defined legend was always a requirement, since conventions of legibility for affect are not established in any existing form. Customized and customizable graphic modes were deliberately designed to undercut the notion of an external or transcendent authority exercising objective judgments.

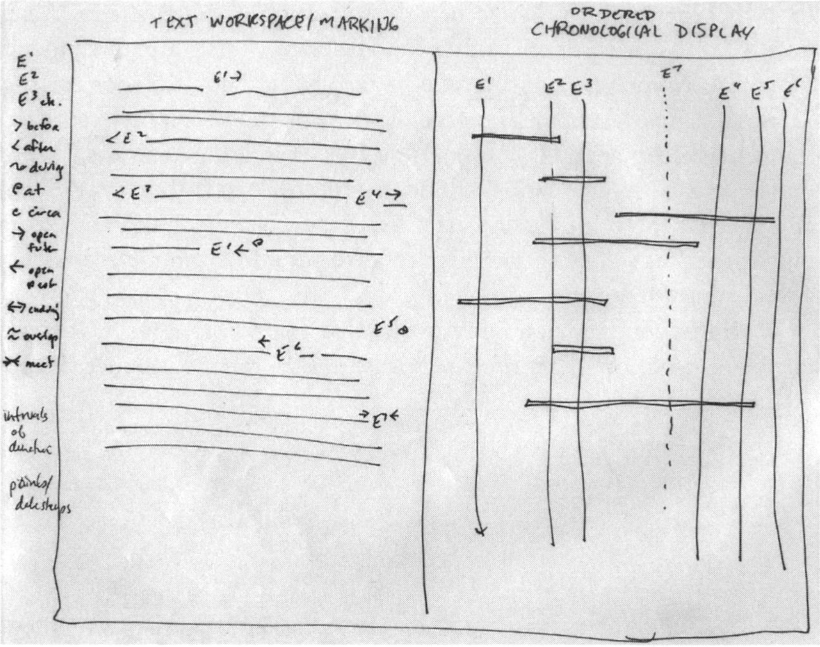

2.1.1d. This diagram links a textual analysis and a chronological display of information that has been marked up in the text. The idea was to generate structured data that could be exported and then repurposed for analysis and display. By marking different references within a text as "before," "after," "during," and so on, according to the basic logical categories from our research, we could map the temporal relation of the events (E in the markup). This coded set of references could then be moved into the display space for analysis and manipulation.

kit and composition space. Our final designs included several distinctive features: the now-slider mentioned above, multiple and multidirectional timelines, and semantic and syntactic inflections. The now-slider was designed to indicate the point of view from which any particular representation was occurring, as well as to advance the interpretation along its own temporal axis. This innovation became more developed in our designs of Ivanhoe, where we allowed each player/role to have a point of view within the game space. But its appearance in the Temporal Modeling space marks a definitive commitment to marking subjectivity as the place within which the system registers all of the information it displays. The use of timelines that branched, broke, or, in various experiments, bent or sagged, was a way to include prospective and retrospective views. Semantic inflections allowed for a customized legend of themes, characteristics, or other values to be attached to individual points, lines, and events in the system. Syntactic inflections were designed to indicate relations of influence or interdependence among elements.

The design process itself involved paper-based sketches and, as I mentioned before, subsequent coordination with a group of students at Cal Arts, followed by work with a Flash programmer. The results were attractive, and the labeling system for the composition space allowed for much information to be entered and used. The crucial shift we enacted was to move from picture making (rich depictions of relations, concepts, situations, circumstances, and effects) to designing software. This was an

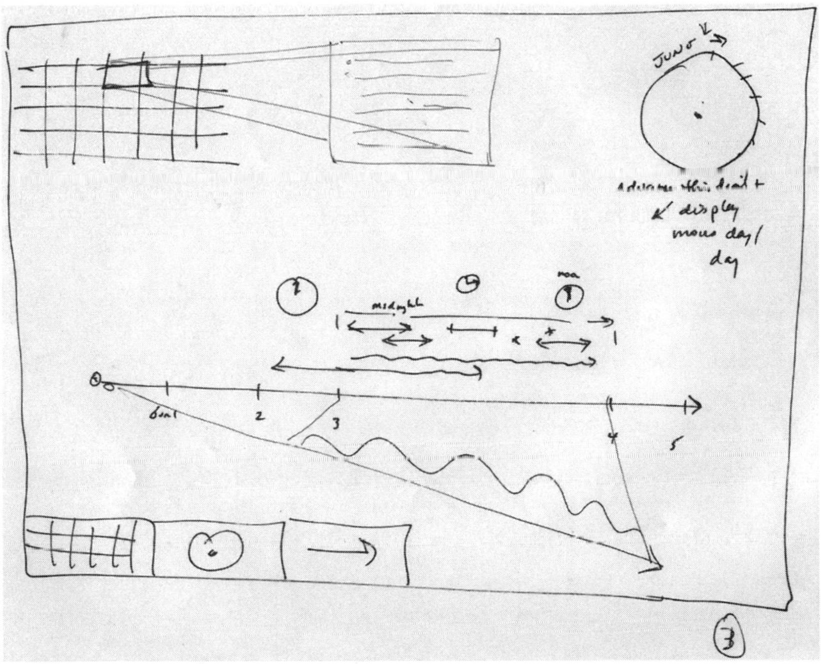

2.1.1e. In the early stages of design, we used projections from one line or plane to another as a way of displaying influence or effect. Our aim was to create a language of affect for semantic and syntactic relations (foreshadowing and anticipation, for instance, were conceived as forces that had their own impact on the unfolding of events). In this sketch event segments of different duration are shown floating above a time line under which a vector shoots downward. The projections from events above can be expanded upon in the graphics display, their relative importance and impact shown by the area, density, and frequency of graphic inflections. Correlation between that vector and the timeline is arbitrary. The menu bar (bottom left) was meant to indicate that one could choose various methods of display: grid, dial, or line.

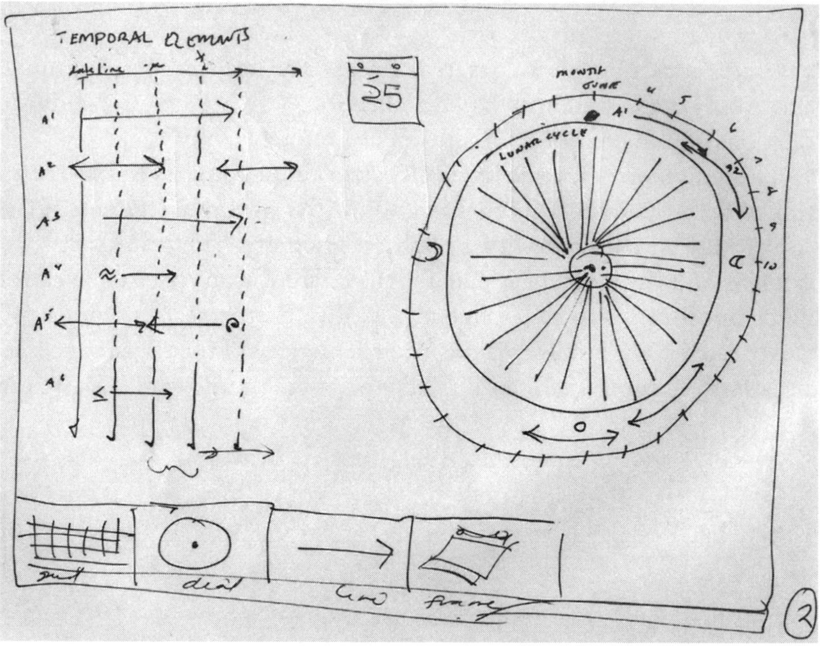

2.1.1f. This sketch shows various systems correlated with each other, including a calendar, dial, grid, and line.

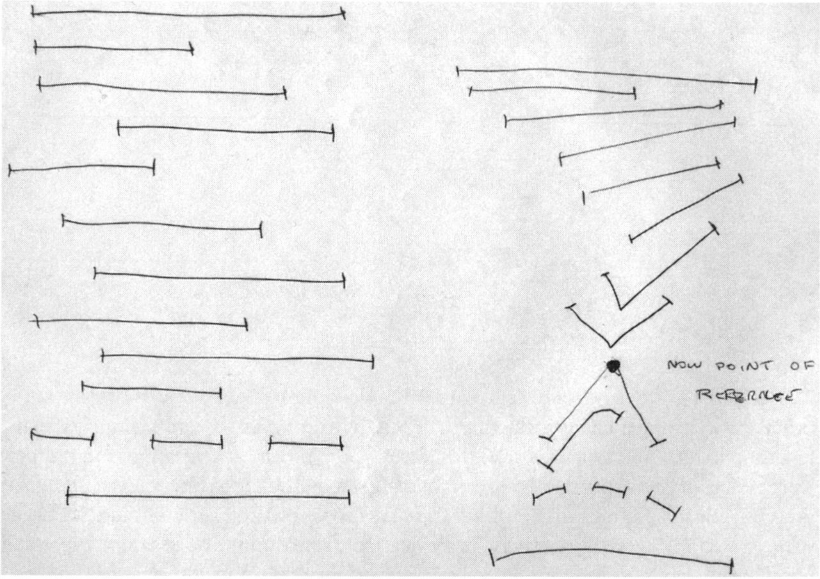

2.1.1g. We experimented with multiple now sliders and "rubber sheet" timelines to suggest the effect of subjective perception on a set of elements and their relations. The neutral, exteriorized, "objective" view of events (left) is reconfigured when point of view is taken into account (right).

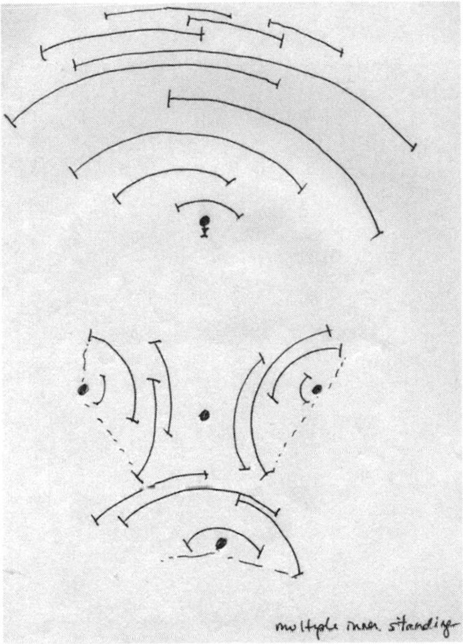

2.1.1h. An image could also register multiple viewpoints and their effects within a field of events.

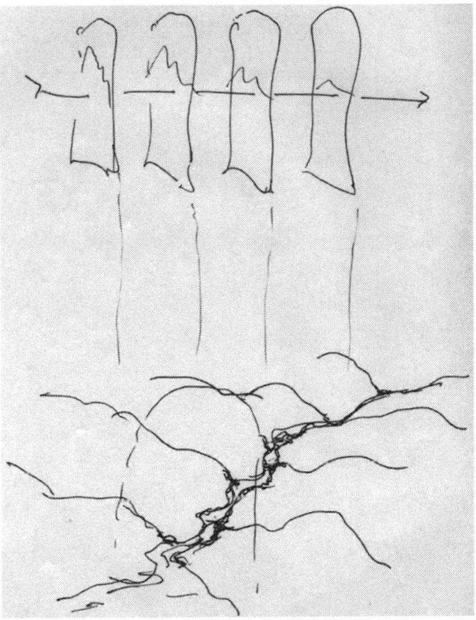

2.1.1i. We experimented with spatial, topographic images of temporal events—a time landscape—with the idea of being able to map experience. The time-slice offered a way to cut through such complex fields for analysis. The difficulty in these conceptions is assigning a value to the z-axis.

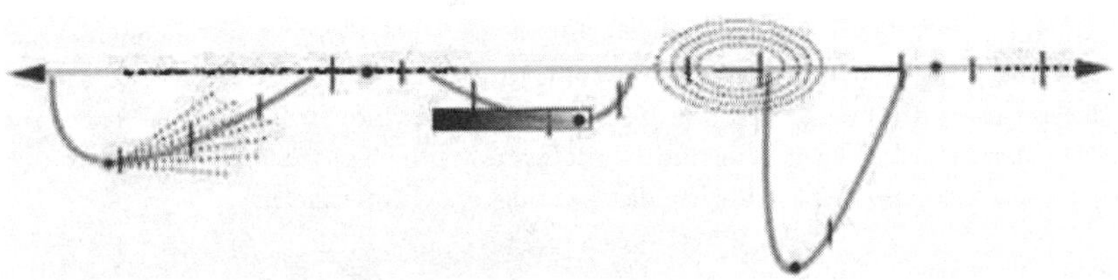

2.1.1j. The warped timeline, created by Bethany Nowviskie as one of our early experiments, had much aesthetic and imaginative appeal. Although "stretchy" timelines were part of our conceptual vocabulary from the outset, we did not develop this feature. Anomalous experiments such as this often demonstrated possibilities that could be usefully incorporated into future designs. In Subjective Meteorology, such anomalies and affective expressions are the basis from which the system is generated (see chap. 2.3). Thanks to Bethany Nowviskie for permission to use this image.

Fig. 2.1.2. Screen images of composition space

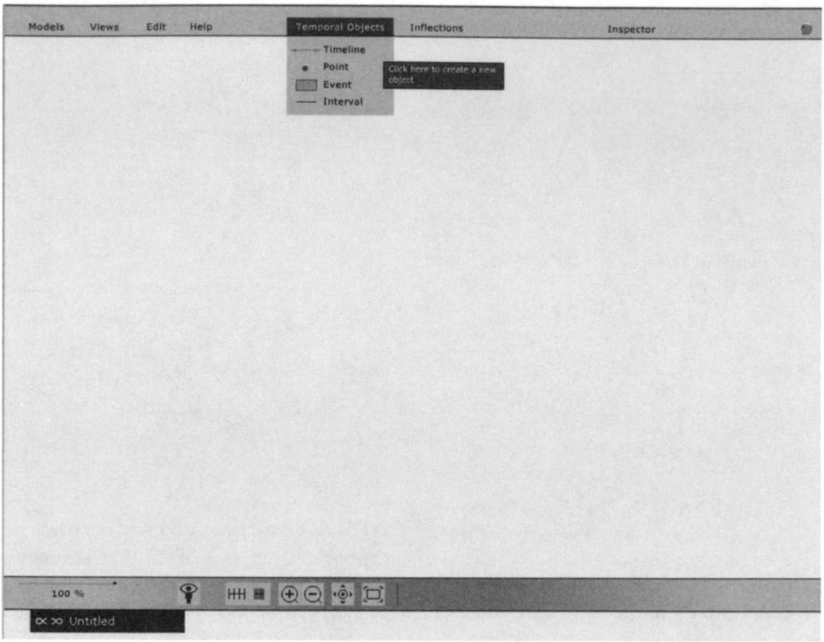

2.1.2a. The Temporal Modelling composition space was very clean and elegant. Basic temporal elements—line, point, event, and interval—could be repeated indefinitely, renamed, annnotated, assigned labels, colors, and intensity, and manipulated on many distinct layers. The menu bar at the top provided access to existing models, views, editing tools, help tips, inflections, and the inspector (see figure 2.1.2e). The bottom bar held a tool that allowed different layers to be foregrounded or made to recede, a now slider (the eyelike figure), a compression/expansion feature for horizontal display, and adjustments for focus and scale. Individual layers could be named and manipulated independently and displayed with different degrees of transparency.

enormous leap and involved a major tradeoff between aesthetic richness and functionality (figure 2.1.2).

The constraint on the design imposed by the technical requirements of making the composition space generate XML output ultimately proved very limiting. A relentless linearity remained in the structure of the lines and the parameterization of the space. We were not able to render *n*-dimensional space, or to create the warps and breaks and ruptures essential to the subjective nature of phenomena. We did, however, produce a usable system, one that was capable of modeling data directly from visual input. That demonstration of visual activity as a primary mode for

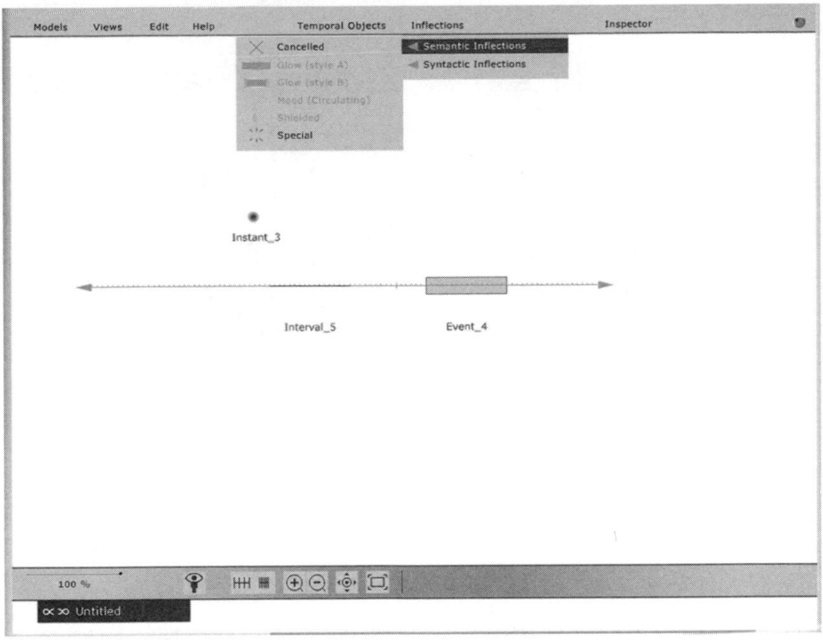

2.1.2b. Simple single-layer model with some of semantic inflections displayed. This untitled layer, identified in the bar at lower left, is unlinked from other layers at this point.

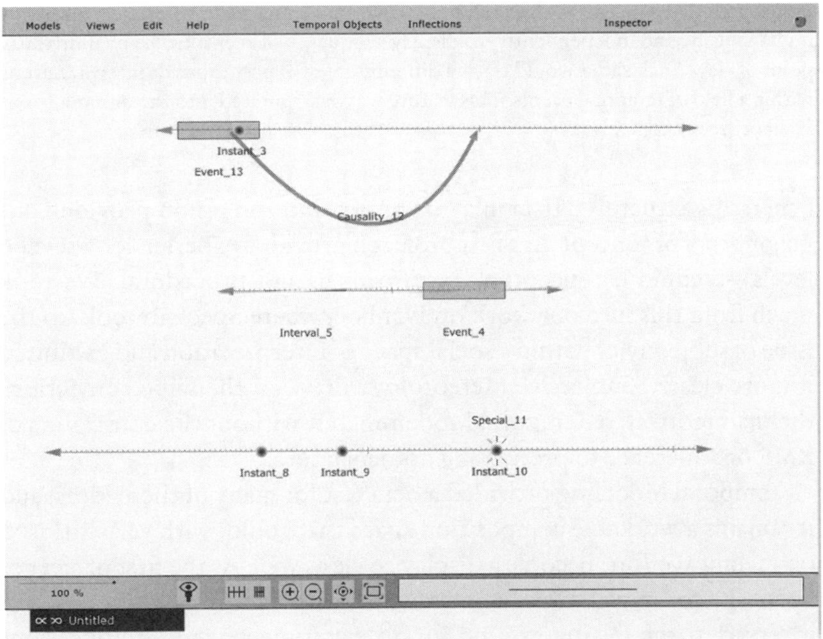

2.1.2c. More complicated single-layer model showing causality arrow. Causality was one of the basic syntactic inflections, while the glow around "instant 10" is a semantic inflection. Syntactic inflections always involve a relation between one or more elements, while semantic inflections can be attributed to any single element.

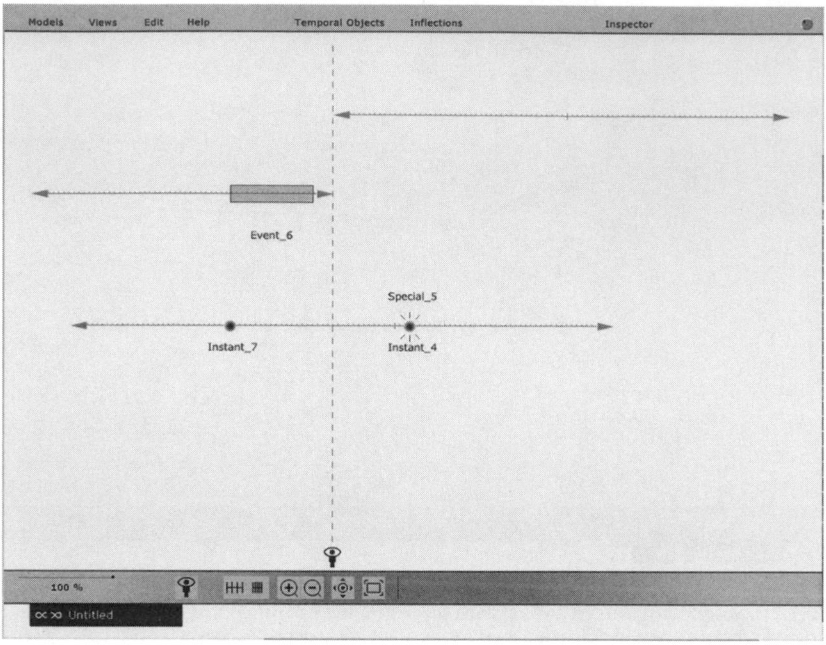

2.1.2d. Model showing a now slider positioned within a sequence of events. The design included plans for multiple sliders so that branching paths could be used simultaneously (and independently) to display a sequence of events from any individual point of view. Each slider would track a different set of expectations or interpretations within a field of temporal events. This feature was incorporated into the design of Ivanhoe (see chap. 2.2).

generating structured data in a humanities interpretation provided our major proof of concept. As a first project it provided experience on several levels—technical, conceptual, institutional, and procedural. We took much from this into our work on Ivanhoe, where SpecLab took up the issue of subjectivity within a social space of interpretation and exhibited it more clearly. Subjective Meteorology drew on the same convictions that had motivated Temporal Modeling, but without the constraints of XML or adherence to preexisting nomenclature.

Temporal Modeling provided a beta test for many of these ideas, and it remains a workable composition space that could, with very little extra technical effort, become a display space as well. As the first project of SpecLab, conceived even before that entity formally came into being, it served as the testing ground for an experiment whose implications were not fully clear at the time but, refracted through the experience of designing Ivanhoe and Subjective Meteorology, have since become

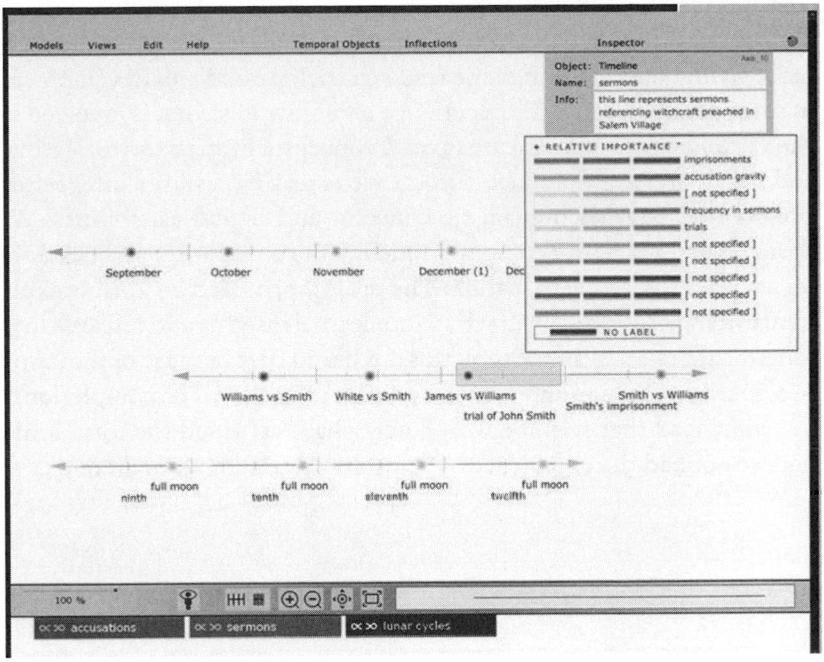

2.1.2e. A view of the inspector, a labeling system for using color values, entering data and labels, and creating a legend that can be stored and used within a model or exported for use with other models.

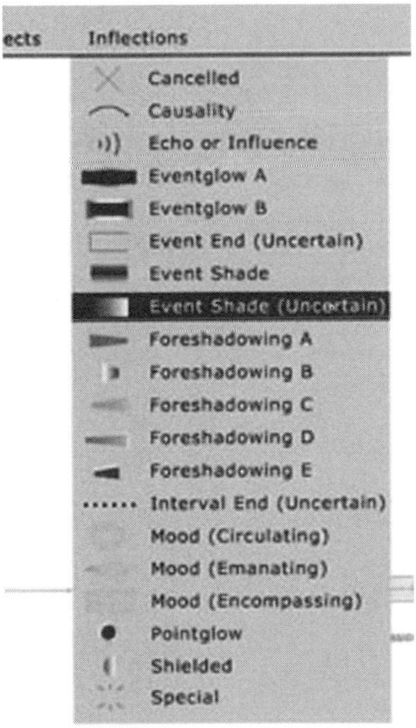

2.1.2f. Detail of the palette showing some of the semantic and syntactic inflections included in the prototype.

more vivid. I have described the research background and development in some detail, because the experience demonstrates what is involved in the various stages of literature review, conceptualization, investigation, and iterative design practice necessary to creating a viable intellectual project that engages humanistic concerns and digital capabilities. We spent several years on Temporal Modeling. Its conception and development cycles took time to mature. The project provided invaluable experience in the process of abstracting intellectual insights and research into functional design. Ivanhoe took this further, in part because of the influx of resources—human and monetary—that propelled it to completion.[18] I'm convinced that Ivanhoe would never have achieved the form it did had we not had the experience of Temporal Modeling to build on.

2.2 Ivanhoe

Temporal Modeling succeeded in demonstrating that visualization could serve as a method of creating interpretative analysis, and not merely of displaying it. To the extent that the semantic and syntactic inflections were designed, they embodied a vocabulary for indicating values and relationships among elements that carried interpretative judgment. The now-slider feature, though still primitive at the time we stopped working on the project, embedded a subjective point of view within the system. We could not make the stretchy timelines, with their variable metrics, work within the technical constraints imposed by the design, and the overall linearity and underlying rational grid structure undercut some of the premises the sketches and original conception had envisioned.[1] Ivanhoe continued some of these ideas. Other aspects of our initial project were extended in Subjective Meteorology.

Ivanhoe came into being through a number of impulses. Chief among these was the desire to design a project that could embody and demonstrate critical principles while providing a model digital environment for next-generation pedagogy and scholarship.[2] We were keenly aware of the disconnect between the experiences our undergraduate students brought to the classroom and the academic environment. Broader social issues with regard to literacy, democratization of education and access to knowledge, and changing patterns of reading, writing, and use were present in our minds, and our projects at SpecLab engaged with the altruistic goal of fostering social change.

We found that educators seemed immune to the siren call of theory, while theoretically inclined colleagues often seemed baffled by the role of digital technology. Ivanhoe may prove, ultimately, too esoteric in its aims and design for broad adaptation— the idea of a game of interpretation to save the humanities gained few converts—but its virtue was that it was a toy and a tool, not packaged content, and its design was premised on a theoretically sophisticated set of premises. When finally built, it successfully demonstrated its theoretical principles.

We had been asking whether digital media could be used to provoke critical modes of reading within literary studies. Ivanhoe, a game of analysis and interpretation, was our answer. It was designed to integrate traditional bibliographical materials and other artifacts of literary study within an online environment while pushing to the fore critical issues in literary studies: the non–self-identical condition of texts, the relation of any text to its field of production, the intersubjective activity of readership and scholarly work, and the development of reflective self-consciousness as a fundamental goal of humanities activity.[3] Our goal was to create an electronic environment in which these issues can be integrated into practice.

The bulk of digital humanities projects had focused on library and information management systems for the administration and delivery of materials. While these uses of technology were important, they were not sufficient to engage with either the cultural or theoretical agenda we had in mind. Ivanhoe was meant as an imaginative, provocative space that would move beyond instrumental management and statistical processing of text-based materials. It was designed in response to the question of how the future of literary and humanities scholarship might be provoked by electronic instruments, and on what foundations adequate tools could be established. We were also keen to be sure that surrogates and virtual facsimiles for the study of print-based materials be included in their design.

Deconstruction left a legacy of metaquestions that informed our research. For instance, in a paper titled "From Work to Text," one of the touchstones of the literary theoretical turn in the 1960s, Roland Barthes set out an approach to reading and interpretation that established terms of play and difference as productive in their capacity to generate a text-as-reading.[4] Within the sphere of deconstruction and poststructuralist approaches, the term "text" came to stand for a tectonic shift in approaches to the task of interpretation. Readings, not meanings, were the focus of this method, with the text as a field of signifiers in play. As we conceived Ivanhoe, we made an obverse, but not reactionary, move to reconceive the "work" as a constituted field.

Just as earlier twentieth-century literary critics addressed questions of style, authorship, attribution, meaning, and interpretive relations to ideology, politics, or culture, we drew on the developments of the last quarter of a century, creating Ivanhoe so that it would provoke questions about the ontology of texts, the intersubjective condition of their production and reception, and the ways their material existence is contingent upon a discourse field—an aspect of their capacity to function as elements within a signifying practice. The metaphors of networked culture find a corollary in the dispersed condition of discursive practice, and in the contingent condition of texts within a diffuse field of artifacts. Nonetheless, students of literature and even many scholars of renown seem to forget these lessons when they sit down to the daily activity of interpretation. Students regularly come to the classroom intent on finding the "meaning" of a poem within an apparently stable text, as if it were a self-evident and self-identical work. Such attitudes prevail in the visual arts and media studies as well, with a strong vernacular strain of criticism in the popular press. Visual artworks are regularly subjected to description-based analysis (hardly worthy of this final term) and film to narrative recounting based on story, plot, and character description. The ideological and epistemological interlinkings of deconstruction get little play in such circumstances. While lip service to the theoretically informed agendas of critical inquiry persist in the research university, literary studies or pedagogical work that genuinely engage theoretical principles remained exceptional.

In conceiving Ivanhoe, then, we combined our cultural concerns about the place and perception of the humanities with our commitment to critical issues in literary studies in the early twenty-first century. We explored technological possibilities for facilitating humanistic research and the potential of aesthetic provocations in the design of digital reading environments. But we also kept our theoretical goals in the foreground.

Our cultural concerns were fairly straightforward, even if in 2000–

2001 they still seemed slightly heretical to our colleagues (many of whom still think of themselves as nondigital in their orientation as teachers and scholars, in spite of their daily use of electronic tools and resources). We were and are aware that a considerable gap exists between the activities and artifacts of mass media culture and academic life. The very acts of reading required by traditional humanities seem alien to many of our students, whose daily experience of interconnectivity and interactive media involves e-mail, online games, networked information systems, and other small-scale, short–attention span environments. Literary texts, particularly historical and experimental works, appear to be peculiar artifacts—remote, antiquated, or esoteric. More crucially, dialogue and self-conscious reflection are barely present in the structure of media discourse, with its emphasis on the commodification of information and experience.

The role of the humanities, with its focus on the creation and preservation of cultural artifacts, particularly works of imagination and subjective experience, felt imperiled, at risk of being swept away by a rising tide of seemingly philistine cultural influences. In some sense, the humanities are equally threatened by several forces: popular disinterest in cultural traditions (beyond the banal "produced" culture of the entertainment industry), the esoteric self-involvement and obliviousness of academic institutions (with those at the highest level of research seeming most immune to changes in the cultural context of their activity), and an inadequate engagement with the critical issues that broke theoretical ground in the last decades.[5]

All of this high-minded rhetoric would never have brought Ivanhoe into being, I think, if not for a playful yet highly charged collegial exchange between Jerome McGann and myself. That exchange, as I mentioned in the introduction to part 2, turned on the novel by Sir Walter Scott from which the project took its name. McGann had suggested that we see the book in terms of an opportunity for rewriting—for all of the possibilities within the book that it held out as potential tales. Tasked with the challenge of thinking about where and how we might change it, our relation to the text shifted radically. My own motivation as a reader suddenly spiked, fueled by an investment in finding the right place for an intervention. This realization of the power of shifting from passive to active reader, from spectator to participant in a project-based exercise became the basis of my commitment to Ivanhoe.

Ivanhoe was not intended as a recast version of Clue, or a dinner mystery for the amateur scholar. It was a game, but it was also meant to promote collaborative and intersubjective fields of exchange. McGann and I played a round of Ivanhoe, largely through an e-mail exchange of rewritten endings to Scott's tale. From that epiphany came the outlines and basic principles of the game:

Role Playing. To begin the game, each player had to assume a role, or "embodiment metaphor." This provoked a self-conscious identification of subjectivity, making explicit the usually implicit framework of critical writing. Every move, every comment, remark, or research gesture was clearly identified with the point of view or position from which the player was writing—including historical and social circumstances, education, gender, motivation, professional credentials or other interests. No neutral articulation of critical positions could be assumed.

Roles could be specific. For instance, in one game played with Emily Brontë's *Wuthering Heights,* I chose to "be" Isabel Arundel, the woman who married the nineteenth-century author and traveler Richard Burton. I chose to engage her persona at the time of her betrothal, when Burton was traveling and Arundel was imagining her own escape from the confines of convention. In her character, I rewrote the attitude of young Catherine Earnshaw as Arundel's exploration of feminist fantasies of independent adventure while she waited for the wandering Burton. The levels of embeddedness were often multiple, since we were players taking on personae in order to enter the characters in the various tales. The game of masks was part of the pleasure of the social exchange in play.

Other roles were more vaguely defined, with characters coming into focus over the sequence of plays. I played the *Turn of the Screw* in the persona of an Oulipo-inspired graduate student assistant working with the compilers of a concordance of the text. In that character, I chose to rework the text at every occurrence of the name of the character Flora. I reconfigured the work as a feminist protest of Henry James's conception of the girl's sexual imagination. In this case, I had a sense of the strategy I would use to generate moves but not of the specific outline of the persona through which the texts would be enacted. More and less erudite engagements were possible, but every move had to be accompanied by a journal entry justifying the intellectual basis of the contribution from the point of view of the assumed role. The point of such self-conscious masks was to debunk the myth of authorial neutrality. We worked to enact our authorial conceits and constructed subjectivities, in accord with lessons from critical theory, and to enact the social production of scholarly or critical work.

Discourse field. Every text, document, or artifact that came into play had to be introduced deliberately (or "called") into the discourse field; each document was conceived, not as "original" or "primary," but as part of a social history that included its production and reception. Exposing this field though connections, links, associations, and readings was part of the task of Ivanhoe. In order to show that no text was self-identical or self-evident, the game registered the alterations rendered by each interpretation. The text literally changed, according to the interventions of the players, who could alter the document or refract the text through commentary, links, or other glosses. By making an environment where a text was constantly altered or deformed, we created a discourse field in which reception and production were integrated and registered in the material structure of the text and game space. Every act of interpretation remained part of the structure and display of the document.

Moves. Ivanhoe was conceived as a writing game. Though "moves" were not limited to text-based activity, and could include visual, and in principle, time-based audio or video material, the primary mode of intervention in the game was through text. Each act of interpretation, whether the creation of a note, node, or link or the introduction of a new document or commentary into the discourse field, was accompanied by a justification or explanation in the player's log. (In one version of our design, a player could be challenged to justify a move, and if no log existed, or it didn't make a strong enough argument, then the player would "lose" the challenge.) Ivanhoe's role-playing structure and open-ended discourse field allowed for creative and imaginative writing as well as critical or scholarly production.

Point of view. All features of the game—creation of roles, introduction of elements into the discourse field, moves, comments, player logs, and so on—were linked to particular players and roles. This structured subjectivity into the game space, since the game was always seen from the point of view of one of the players/roles. No "outside" view existed. A now-slider feature, marked with ticks that showed each player's contribution to the game in sequence, could be advanced to show the progress of the game from someone's point of view. An Ivanhoe game is always read from within the space of game-play.

Social space. We quickly became aware of the importance of the social space of Ivanhoe as an impetus for motivating game-play. The motivation to make a clever move to delight or pique one's fellow players upped the ante for reading considerably. This was an important aspect of Ivanhoe's effectiveness. We were intent to demonstrate that scholarship and criticism, as well as authorship of creative and imaginative works, take place in social space, but also to make that part of the game in a substantive way. Skills in bibliographical work, wit, aesthetics, or composition were rewarded in versions of the game in which points and scoring systems are put into play.[6] Though we later abandoned the idea of a scored game model, we noted that Ivanhoe fostered some of the competitive dimensions that motivate performance in social space, simply by virtue of the public nature of players' moves.

: : :

To reiterate, the basic principles on which Ivanhoe was conceived were the non–self-identicality of texts; theoretical engagement with a discourse field; attention to bibliographical artifacts and their materiality; attention to documentary evidence and the trail of works through their production histories; the transformation of a text through its reception (marked in responses and versions); the situatedness of every reader within a role whose historical and social conditions had to be made explicit; and the social space of play. The task of designing Ivanhoe arose in parallel discussions and experiments.

Ivanhoe went through several technological and design iterations. Each had an impact on the reading experience and on the ways computational capabilities could be engaged in the game-play. The fundamental game can be played with pen and paper as well as in electronic space (as was done in a middle-school version), and the lower technological requirements refocus social and personal goals toward a classroom experience.[7] The use of various Web-based tools for creating an environment for geographically distributed and asynchronous play increased the capacity for participation among a wider group of players in the *Wuthering Heights* and *Turn of the Screw* games. But the tools we were adapting had significant drawbacks, relying on long scrolling screens of accumulated exchange to log the progress of the game. These environments were difficult to navigate and harder yet to conceptualize in any cognitive gestalt. We realized we had to design a customized environment.

Designing Ivanhoe

Creating designs for Ivanhoe's interface advanced our critical thinking about the project, perhaps even more so than in the case of Temporal Modeling, in part because we talked about the design in so many contexts and with such a varied cast of characters.[8] Key ideas emerged from visual sketches, and implementations of critical and technical issues derived from the way activities were visualized. But grappling with this design also reinforced my understanding of the ways an interface exists at the intersection of two distinct practices—engineering and information design. Each discipline has its own priorities and values. Engineering emphasizes functional implementation, while information design draws on the capacity of graphic expressions to communicate clearly to a user. Yet both approach visuality with certain shared assumptions about communication and visual forms. And both operate far from the influence of critical thinking about visual representation. Though it may seem a reach to connect engineering-based approaches to human-computer interface and poststructuralist criticism, that is precisely what designing Ivanhoe required. Promoting serious dialogue between the traditions of critical thought and applied knowledge was crucial to Ivanhoe, and reflection on the place of interface design within the larger concerns of visual studies seems useful as a critical frame of reference.

Engineering approaches to interface, such as those perfected by Ben Shneiderman or Stuart Card, are grounded in certain assumptions that serve the task at hand but go unquestioned at the level of ontology. Their approach is pragmatic, drawing on cognitive psychology, with its attention to the problem of designing environments that work with the operative limitations of human intelligence rather than against them, and with increasing computational speed and capabilities. Such design is guided by principles of perception and cognition that can be tested and codified, such as the number of items that can be held in short-term memory, expectations about real-time interactivity, hand-eye coordination, and so forth.[9] No one would argue with the soundness of this approach as a basis for the design of everything from air traffic safety systems to operating room feedback mechanisms and ATM machines. After all, less considered approaches are the source of innumerable frustrations. Think, for instance, of being asked to enter information into a space too small to display it.[10] (Try typing "Charlottesville" into the "city" space in most standard forms, and hope that the arbitrarily truncated name doesn't result in your tax documents being sent to North Carolina instead of central Virginia.) Or recall searching for the "enter" button to respond to a question on the gas pump display—only to realize that it happens to be labeled "OK" on this particular keypad. For most practical purposes, the engineering approach to design of human-computer interface is essential.

Engineering and cognition-based approaches place a lower premium on aesthetics than on what they consider functionality. Engineering solutions often stop with a design that works adequately, rather than seeking solutions that emphasize the rhetorical benefits of seductively engaging or rewarding a viewer. Sometimes such literal notions of functionality can prove so restrictive that they undermine the results—as is famously demonstrated in cross-cultural instances in which a machine interface is developed with no regard for the social rituals that would allow it to work effectively in context.[11] (ATM machines in Japan were almost ignored before the introduction of animated figures that greet the customer.) Information designers are well aware that there is no such thing as "mere information"—organization, sequence of access, and relations among parts of an information system all contribute to the success or failure of communication in an interface.

Overall, information designers rate clarity above beauty, as if the two were mutually exclusive, or even separable. The work of Edward Tufte is a notable exception, hence the high regard for his elegant designs. But no matter where they fall on the aesthetic spectrum, information designers—whether we are referring to Tufte, a consummate professional like Richard Saul Wurman, or the producer of garden-variety presentation graphics—share a core belief system with their engineering colleagues.[12] They believe that the formal properties of graphic presentation can create a stable image of data. The quality of *transparency*—the ability to reveal information—is premised on a belief in *apparency*, the conviction that formal structures communicate directly through visual means.

Intent on creating effective means of communicating information in visual form, information designers almost entirely ignore the substantive theoretical problems, posed by iconographical studies, semiotics, and poststructuralist theory, that touch on the identity of images themselves or the cognitive function of aesthetics.[13] An empiricist assumption that what you see is what is there underpins their practice. The self-evident character of graphic entities—lines, marks, colors, shapes—is never itself brought into question, however much the parameters on which they are generated or labeled might be criticized. That images themselves might be dialectical, produced as artifacts of exchange and emergence, is an idea foreign to the fields of engineering and information design. (While information displays can be interactive, and results produced through variable input, they are not imagined to have been brought into being through dialectical relations.)

Even the idea that diagrams or graphics have a cultural history and resonance carries little weight unless the issues have the kind of impact of the Japanese problems with ATMs.[14] Press an engineer or information designer on these issues, and you will likely be told they are irrelevant. Presentation graphics, though produced with a keen awareness of formal, material properties, are still premised on the notion of appearance as a means of revealing information rather than on a cognitive, performance-oriented model. As if "information" existed a priori and independent of human subjectivity, visual forms are arrived at through a series of design decisions that present the "best"—that is, most transparent—image of that information. Such approaches are consistent with a structuralist semiotics, in which a sign system comprises two related elements in a simple binary relation, both at the micro level (signifier/signified) and at the next higher order of organization (plane of discourse/plane of reference). Such binarisms, and the stable-seeming sign systems they employ, are the legacy of a structuralist tradition that is formal and descriptive (transcendent) rather than dialectical and dynamic (emergent). Ivanhoe's design was premised on the idea that an image is a structure created through an act of intervention in a potential field and that this image calls forth a performance. An image is not a stable form revealing a fixed meaning (or predetermined possible meanings, in the case of interactive display). The implications of this distinction are profound.

We need not resort to the deconstruction of visuality to critique approaches to information design that are based on faith in the a priori existence of data. It is intellectual child's play to conceive of misguided statistical methods that produce inaccurate quantitative results that nonetheless pass for empirical data. Nor is it difficult to demonstrate that the visual form in which information is presented has a great impact on how that information reads and what it is assumed to communicate.[15] But the assumption remains that the rhetorical distortions introduced by an ill-conceived or overly expressive visual presentation can be "corrected" to make the image a clearer, more transparent instrument for revelation of the "truth" of the data. Despite decades of work subjecting truth claims to critical scrutiny, mathesis has had a strong resurgence of cultural primacy thanks to digital technology. The statistical character of data has asserted the validity of quantitative approaches all over again.[16] These are issues I have already mentioned and to which I will return in the essays on aesthetics.

But even if we consider information design on its own terms, many critical issues could be raised about the relation between information and its presentation. These arguments would demonstrate, ultimately, that the presentation does not embody information that exists elsewhere in another form. Presentation in graphical form creates a structure to engage the cognitive production of meaning. Some of the visualizations we imagined for Ivanhoe operated in familiar ways, serving to create a compact, highly legible display of quantitative information. In other cases, however, we deliberately selected an arbitrary-seeming display format in order to be suggestive or provocative. These theoretical issues arose from discussions in fields other than information graphics and engineering, drawing on traditions that critique the idea of "presence" and the apparently self-evident character of visual images.

The presumption of visual presence, or of graphical form as self-evident, is similar to the attitude toward textuality that construes a literary work to be equivalent to its words—or, worse, its "meaning." (Readings of materiality that emphasize formal characteristics and the discernment of meaning, as if the literal surface were transparent, are equally plagued by the shortcomings of the information-delivery model of graphical presentation.)[17] Ivanhoe's interface design attempted to use visual and graphical means to make critical awareness central to the game, while also, incidentally, raising issues about visuality that complement those underlying its conceptualization from a textual studies perspective. The graphic vocabulary of Ivanhoe thus calls attention to emergent, generative, iterative, procedural, and transformative activities. These dynamic characteristics are conspicuous properties of digital media and, once they are really understood, of *any* artifact, no matter what the medium. Electronic interface design, in our approach, is premised on the idea that a visual form *does* something, rather than that it *is* something. In Ivanhoe, this principle was foremost. As will be clear in my discussion of e-books in chapter 3.3, this insight is relevant to understanding paper-based and print artifacts as well as electronic ones. We borrowed from systems theory and cognitive science, rather than engineering and formal graphics. But we also drew on the theoretical context of visual studies in formulating the aesthetics on which the design of Ivanhoe is based.

Our theoretical conversations were informed by information-processing theories of vision that have displaced older, mechanistic models of perception. Here, again, cognitive science and psychology combine to create an iterative conception. Instead of imagining vision as a one-way communication channel—an eye receiving stimulation and sending a signal to the brain—cognitive approaches describe an optical system. A feedback loop connects a learning eye and a continually revised cognitive model. In this system, neither image nor idea exists a priori, and sensation is an effect of cognitive capability. This shift in models of vision has implications for the way images are understood. The work of the biologists Humberto Maturana and Francisco Varela, pioneers in the cognitive approach to human knowledge, demonstrates that we constitute the objects we perceive through our capabilities, and they, in turn, act on and transform our capabilities as well as our understanding.[18] Vision is an emergent activity. An image is an entity constituted through a perceptual act. In other words, as Alan MacEachren says, the information processing model of vision has undermined previous ideas about the autonomy of images, sensations—and of individuals as discrete entities simply reacting to or perceiving preexisting elements as a set of stimuli-response mechanisms.[19] Instead, we have to understand all of these as components of a dynamic system in which interaction among elements produces effects. Such an approach doesn't disregard the intrinsic properties of, for instance, texts, graphics, and images. But it emphasizes that these formal and material properties define a set of contingencies, conditions from which an intervening perception can be produced. An image is constituted by this act as well as giving rise to it as a performance of its structured codes and possibilities. The idea that an eye "learns" through exposure to various kinds of stimuli lends support to arguments for aesthetic agency and the formative power of expressive means.

The idea of autonomy, undermined by this cognitive turn, came directly out of modernist art and aesthetic theory.[20] Indeed, one distinctive characteristic of modernism was its insistence on autonomy, defined, first and foremost, as the insistence that images are self-sufficient presences, rather than representations. ("You present a baby," Picasso famously stated, making an analogy between paintings and other progeny, "you don't represent it.")[21] On this belief are built ideas of autonomous art as a form of cultural expression, but the founding premise is that images are self-evident and apparent. This concept of self-evident autonomy is qualified by the recognition that many signs are legible only with specialized knowledge of their codes; nonetheless, visual *forms,* it is asserted, can be grasped directly by the eye in their full and replete self-sufficiency. This conviction is integral to the still-persistent tenets of structuralist semiotics, in which the apparency of the signifier is never up for question. Formalist approaches to visual images are based on these assumptions, and Ferdinand de Saussure's lectures of 1911–1912, famously transcribed as the founding texts of structural linguistics, incorporate the same formalist precepts.[22] We can see evidence of the idea of visual autonomy in every critical articulation of modern art. Emile Zola, writing of Edouard Manet's painting in the mid-nineteenth century, stressed the "thereness" of visual art as "nothing but simple facts"—with an emphatic insistence on the formal presence of images.[23] This idea of aesthetic autonomy can be traced to mid-nineteenth-century critical writings, but it reaches a crucial turning point in the early twentieth century. In the 1910s and 1920s, theories of representation—visual, linguistic, and semiotic—align under the banner of full-fledged modern formalism. Since this is also the historical epoch in which graphic design as we know it came into being, it is hardly surprising that a field like information design continues that sensibility into the present. The terminology of visual communication, the so-called "language of design," is itself a direct legacy of the work of artists like Wassily Kandinsky and Paul Klee, whose attempts to fix the rules of abstract composition, color, and form had such a influence on early-twentieth-century art and design.

Other intellectual traditions lend credibility to notions of formalist autonomy. The formalist turn is a part of the larger "rationalization of sight" described by print historian William Ivins. The virtue of printed images, their capacity for "exactly repeatable" replication, contributed to the stable representation of knowledge and its dissemination in standardized form. Ivins argued that standardized, conventional, stable representation in graphical or pictorial formats gives visuality a unique role in modern epistemology. Like the work of imaginative artists of the early twentieth century, Ivins's work is based on Cartesian principles of rationality. Principles of post-Cartesian graphics—non-Euclidian geometry, nonlinear analysis of event-formations—have yet to serve the daily business of information graphics or to become a staple of display design. For current purposes, the simple critique of formal autonomy should be understood in relation to the way assumptions about presence—a legacy of modernism—persist in current electronic information design. When put to specific use in the display of empirically gathered data, an image is considered a stable entity whose materiality is conflated with its presence. The tendency is to collapse the materiality of images with their formal value. An image *is* by virtue of its formal properties. But just as a text is a field of possibilities that engages a reader, so an image—and graphical forms of text are included in this term—should be seen as a work to be performed through interaction and response. This approach to vision cycles us back to the information-processing model described above and contains the suggestion that the very field of visual presentation should shift in response to the active engagement of a viewer. Ivanhoe's interface isn't designed merely to "represent" individual subjectivity, but to provide the space in which it can be performed.

The applied aesthetic challenges of Ivanhoe were just as daunting as the theoretical ones. First, we faced the challenge of making visible critical concerns that were almost intractably abstract. How could we present a "discourse field," conceived as encompassing a social and production history and field of associations, within which any particular text is simply one snap-shot instantiation? Was it possible to make evident through graphical means the "non–self-identicality" of a text? Could the dynamics of play be given a configured form as a visualization that becomes a primary site of activity rather than simply a display? The technical challenges were nontrivial, as were the design tasks. What were the conceptual primitives of a schema for such a design? What set of objects, relations, and behaviors defined the structural foundation of this system, free of specific content but able to provide a framework for the activity of critical studies? Our ideas became increasingly concrete as we proceeded. We moved from named and identified areas of a screen subdivided into windows to a fluid, activity-based, space of activity zones—in other words, from a rigid, formal structure with a priori labels to a dynamic field configured to show emerging relations and contingencies.

The earliest versions of the interface design, hand-drawn in the summer of 2000, used conventions of software design based on windows, icons, and pull-down menus. This allowed us to schematically represent all the functionality we wanted to include in the design. We dealt with the limited screen real estate by collapsing many of the activities of Ivanhoe into spaces that could be clicked open. The sense of "thereness" in this design was overwhelming. To begin with, the interface was organized around a "source text" (a term we subsequently discarded in favor of "called text" and "declared edition"), which dominated the screen. This text, and the workspace below it, were strongly reified by framing devices that fixed their relationship into a hierarchy while making it almost impossible to display any other documents from the discourse field. While this scheme worked well as a sketch, as a design it was flawed. The software interface wasn't Web-friendly, for one thing. And the windows structure was at odds with the basic premise that entities will define each other through contingent relations. By creating fixed spaces, labeled in advance for each activity and type of text or move, this interface embodied many ideas we had set out to counter. It provided a useful point of departure, however, for all those reasons (figure 2.2.1)

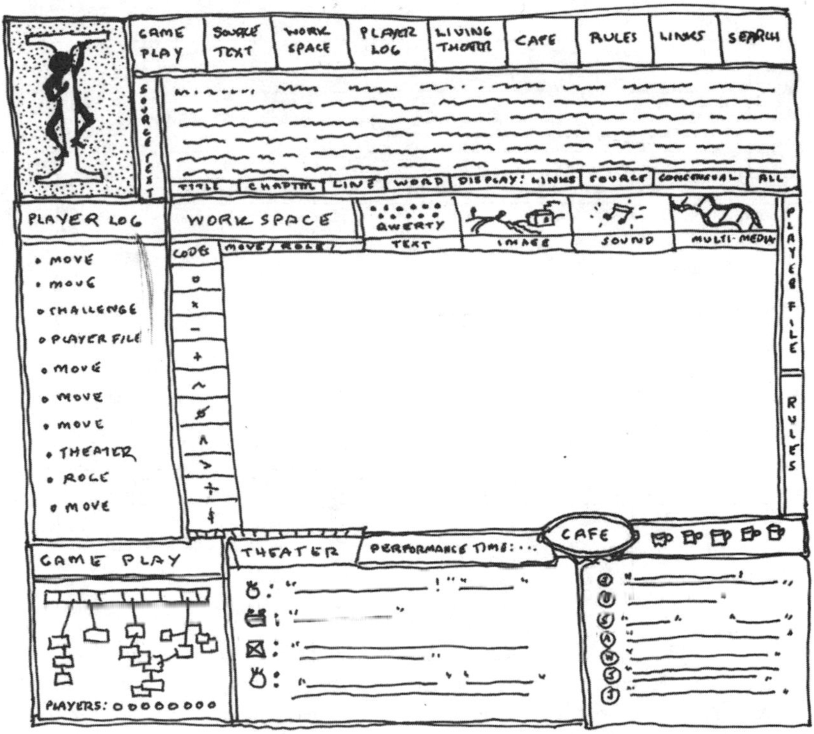

Figure 2.2.1. This hand-drawn image of Ivanhoe, the first vision of the design of the game space, uses the idiom of existing software, with menus and bars and spaces for work all fixed into a grid. A remarkable number of these elements ended up in the final design: access to a source text and the capacity to modify it; spaces for players/roles to engage the text, log their moves, and record exchanges with other players; tools for visualizing game play; a list of rules , search capabilities; and outside links. The figure of Ivan was redesigned, made more streamlined and robotic, but his figure remained as well (he is shown here embracing the *I* on the upper left). Our struggles with moving from codex-style texts to online display are marked in the small menu bar below the source text where title, chapter, line etc. are aligned. We also envisioned the possibility of different media—visuals, sound, video, and other time-based artifacts—being brought into the discourse field.

We formalized that initial design as a storyboarded exercise so that we could see how it would work in step-by-step user scenarios. This interface included visualization spaces for various aspects of the game that lent themselves to iterative and procedural presentation, such as the game-play diagram and a space for showing linked elements of the discourse field (figures 2.2.2, 2.2.3). From the beginning we knew that visualizations had to play a key role—not only in providing the graphical

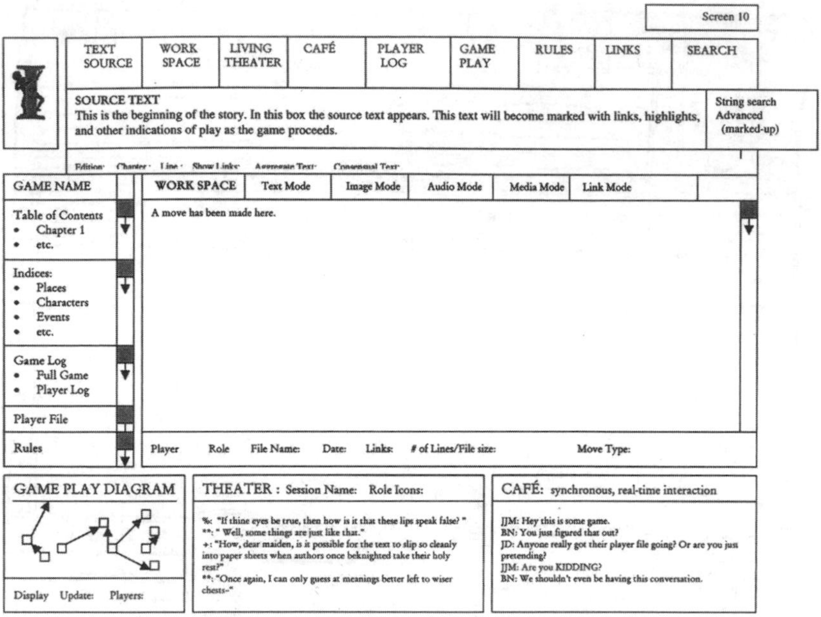

Figure 2.2.2. This translation of the hand-drawn image into a screen design has greater clarity but introduces no substantive changes. The image shown is Screen 10 from a series of storyboards tracking the game from an opening menu to this state. Storyboarding was part of our design process in the early stages, before we did any coding or programming.

form of texts, but as a means of analyzing the dynamics of an emergent work and social field of activity.

As we struggled with alternatives to the windows-based interface, we considered the possibility of the archive, library, or codex as a foundational metaphor. In one version, we took this idea to a literal extreme, conjuring an interface design that "looked like" a library space (figures 2.2.4, 2.2.5). Abstracting function and activity from literal representation of spaces took some time. The tendency is to imagine that a simulated space, because it can be visualized on a screen, will function in the same way as a real space. The specific properties of a digital environment, however, its irremediably flat surface and limited screen real estate, are fundamental to its display capabilities. Thus very different design conventions were needed. We imagined a desktop environment capable of flexible arrangements and interlinked documents (figure 2.2.6). So we overturned this rather literal metaphor and turned our attention to imagining ways to create a deep-space, nonrepresentational topography

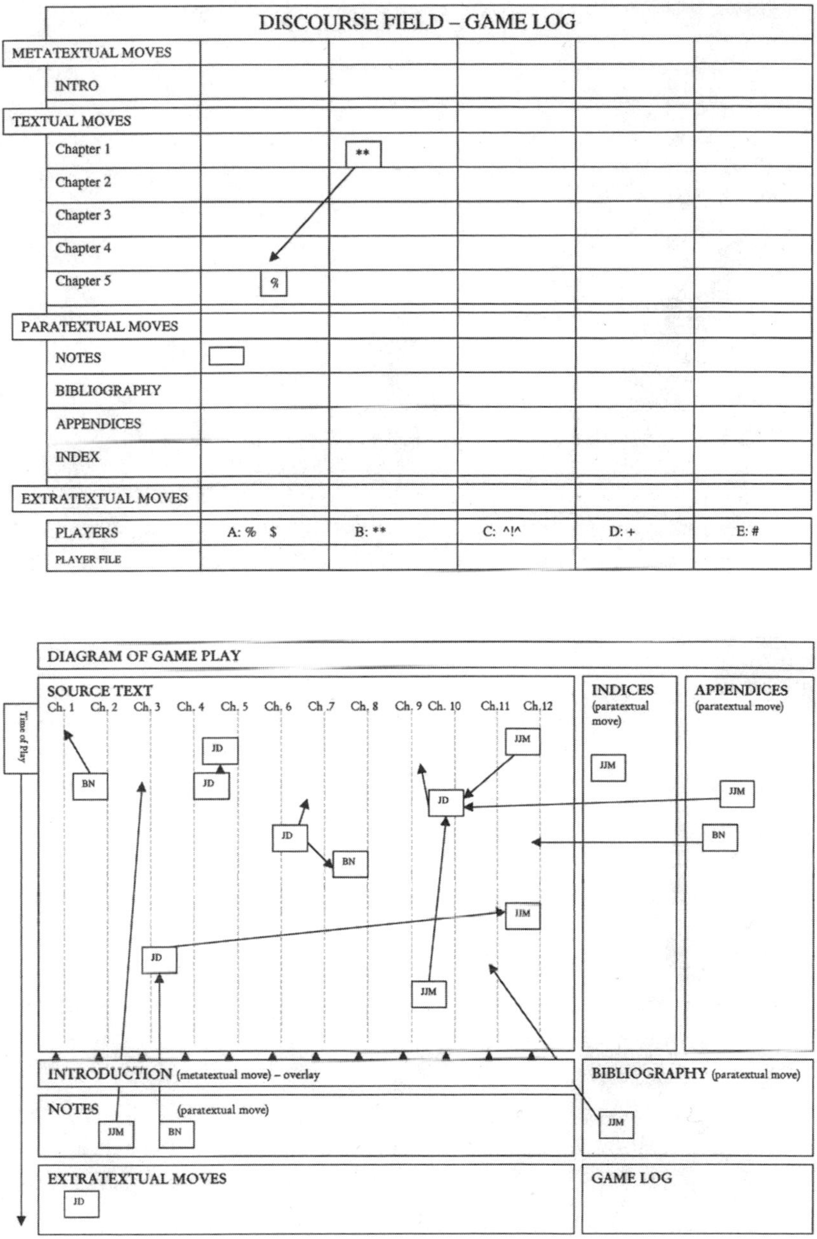

Figure 2.2.3. Details of the game play are shown in these diagrams. These visualizations, not part of the functioning software, were modified only slightly in the final designs, changed more in look than in functionality. The mapping of the social space of interpretation was central to our designs from the outset.

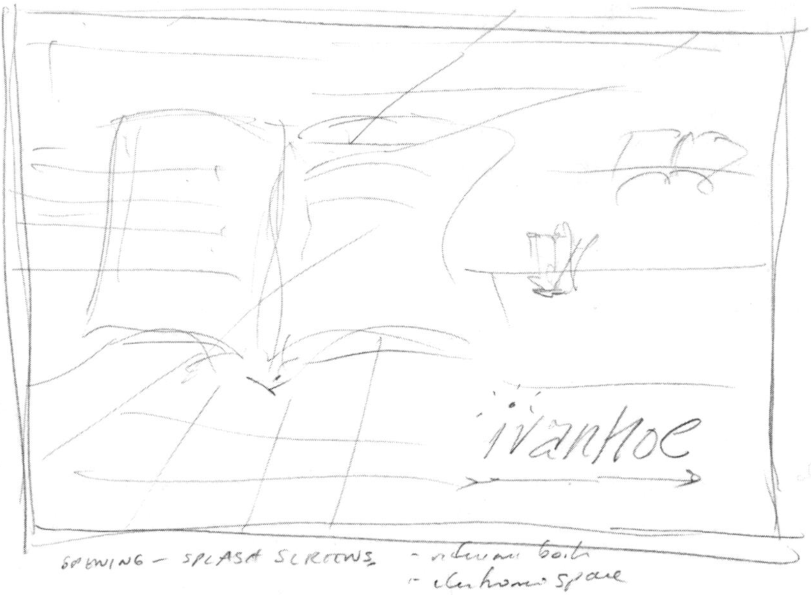

Figure 2.2.4. This sketch of books and papers in play pushed the issue of literal representation in the interface design into focus. Even in its preliminary form, it was dismissed by the design group, but it occasioned a useful discussion about the distinction between imitation of form and replication of activity or functionality.

organized by coordinate axes. The question of parameterization, and of what semantic values to attach to these axes, came up immediately. We may well concede that a book is a three-dimensional object that we encounter along a fourth, temporal axis. But we don't navigate a text spatially, at least not in the kind of three-dimensionality that is used for fly-through views. Though we jettisoned the idea of creating artificial conventions for spatialized display, we preserved the use of dimensional illusion in some of the designs.

A persistent feature of Ivanhoe designs was the presence of visualizations of data generated by player activity. We were interested in the process of abstraction that allowed displays to be created from such seeming intangibles as the choices of an individual player, or the character of their engagement with a particular text or selection of elements. The idea was to produce images of an "emergent work" as it might be generated through the intersubjective exchanges among participants and artifacts. The arbitrariness of assigning values for display was evident. But we were not trying to model any a priori evidence; rather, we sought to support a model as an emergent manifestation of activity. The idea of the visual presentation as an aesthetic provocation, as a primary interpretive act, was at work, along with a willingness to suspend allegiance to empirical models of statistical information gathering.

Our experiments with screen display included exploring the ways resizable elements, tabs, stacking and layering, and careful variation of

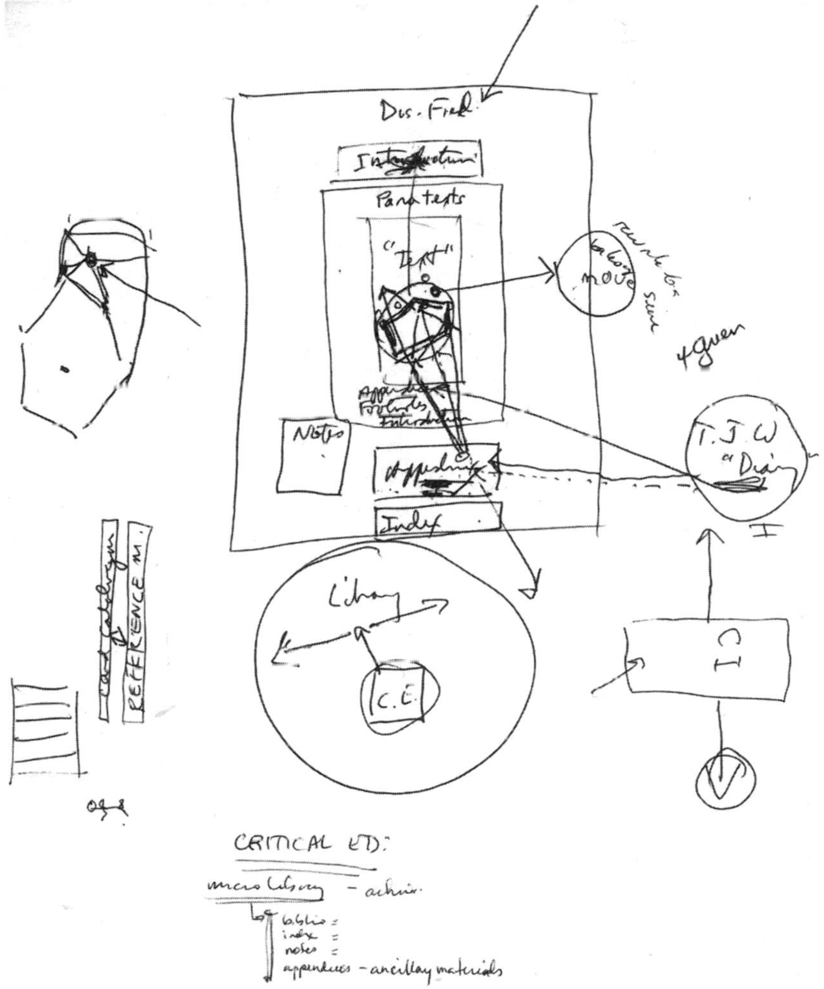

Figure 2.2.5. This is another literal diagram, but instead of simply copying the image of the books (as in figure 2.2.4) it combines images of a discourse field and a library workspace. The exercise was to make clear to ourselves how we understood the research process as a series of specialized zones and activities. This kind of modeling created an abstract scheme on the basis of which we could design an environment that arose from within the specific constraints of digital media.

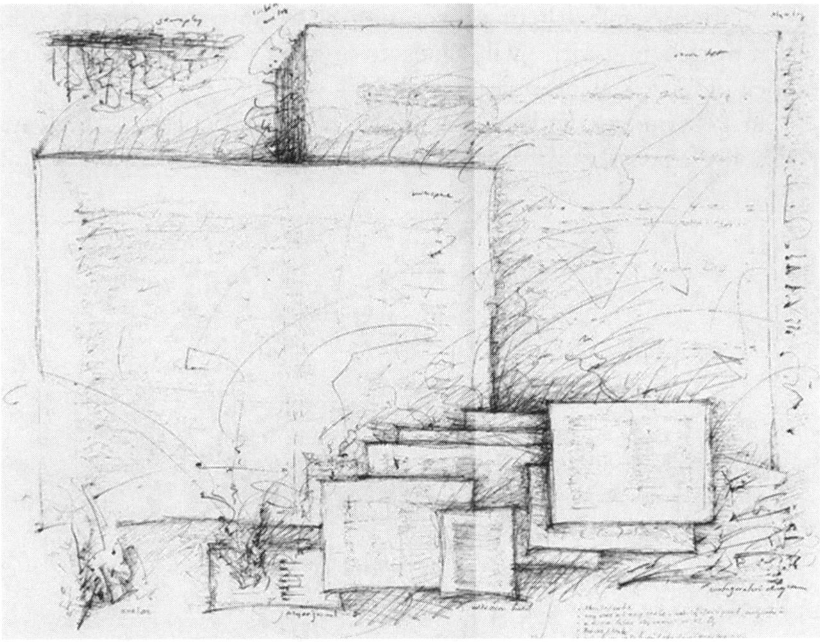

Figure 2.2.6. This image of documents in a discourse field shows the production of a reading through relations and association trails and traces (very much in the spirit of the 'Patacritical Demon; see chap. 2.5). The crucial idea was to design a space in which connections among documents and the readings they provoked would be visible. A small game-play diagram (upper left) tracks interactions among players in the game space. Though not directly applicable to our designs, drawings such as this helped keep our visualizations from shutting down into already established interface conventions.

transparency and opacity might take full advantage of electronic environments. Bethany Nowviskie's visualizations of emergent avatars gave form to on-the-fly characterizations of play, creating abstract figures, in a prototype demonstration of the aesthetic provocations originally sketched by hand. Many of these elements found their way into the "frames-based" hand-drawn sketches from which our final designs were derived.

Other electronic renderings included a modified version of a blog, created by Nowviskie for playing *Turn of the Screw* in spring 2002, and a Web-based windows version designed by Nathan Piazza. Nowviskie's design, though quite simple (no dynamic, on-the-fly diagrammatic features or elaborate navigation), provided a legible way to separate the several areas of game-play. The source text, moves, player journals, evaluations,

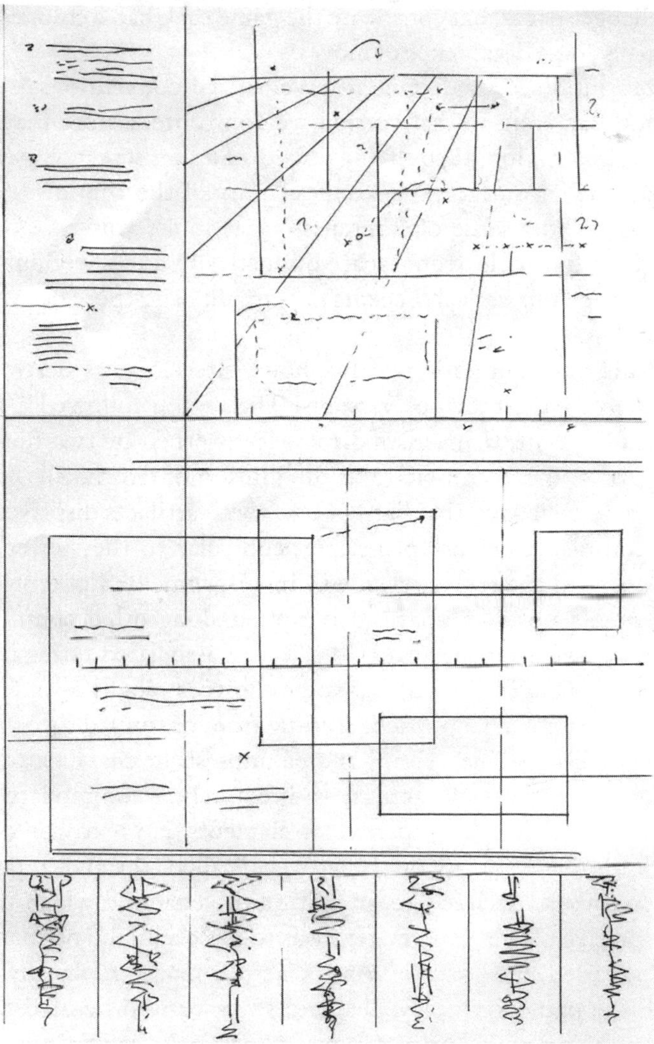

Figure 2.2.7. This schematic visualization of areas of a workspace grew directly out of the previous image. The perspectival lines (top) are meant to chart a discourse field. Document-based features appear in the overlapping planes of a work/play space below. And images of emerging avatar/players are arrayed along the bottom edge. We imagined that these figures would be spontaneously generated by the computer and that players would adjust their games to change the shapes of their avatars, thus responding to their own "look" and "style" as manifestations of their approach to the game.

and challenge spaces that constitute the game each had a color code and individually logged sequence of moves.

Piazza's interface used standard Web-space conventions (sidebars, navigation bar, etc.). In that version we confronted, more than in any previous visualization, the reifying effects of an on-screen presentation of a text. The flat surface, the seamless unity of the windows environment, reinforced a sense of "thereness" that spoke volumes about the need to modify our electronic space dramatically, and to rethink the relation between theoretical precepts in textuality and those derived from visual studies.

The subsequent iteration of Ivanhoe's interface was derived from reflection on these previous versions. The design followed a few basic principles. First, that screen display is governed by two fundamental properties: the flat surface and the illusion of depth. All display in Ivanhoe acknowledged that flat surface—with artifacts displayed in the convention of the picture plane, perpendicular to the viewer's point of view. Within the screen, even within a document, a potentially unlimited number of deep-spaces could open along other coordinate or perspectival axes. For instance, if a series of associated terms (e.g., the heteroglossic field of a word) was to be shown, it could open from any place in the text as a deep space receding from or toward the viewer for purposes of display. The layering and palimpsestlike character of textual interpretation and bibliographical study could be accommodated by adjusting the transparency of particular elements. No text would simply appear; rather, every text would have to be "called" through a "discourse portal" and then "declared" as an edition or version in which to work. As interpretive play began, the text was to be "claimed" and marked, its "codes" revealed, and its structuring principles made graphically evident through the patterns of play. Size and scale were to be used to facilitate the stacking of documents, support materials, palimpsest versions, moves, workspaces, journals, logs, and other materials. And no fixed windows or frames would unify the space. All the elements in play at any time, both within an individual player's space and in the game as a whole, were to be represented in some visualization, as were the configured relations among these (figures 2.2.8, 2.2.9).

These premises counteracted the idea of display as the extrinsic visual manifestation of already present or known "information." We intended to take advantage of the efficiencies of visual modes of gestalt for complex, large-scale sets of information. Visualization was instead conceived as an aid to intellectual and imaginative thought, not simply a means to

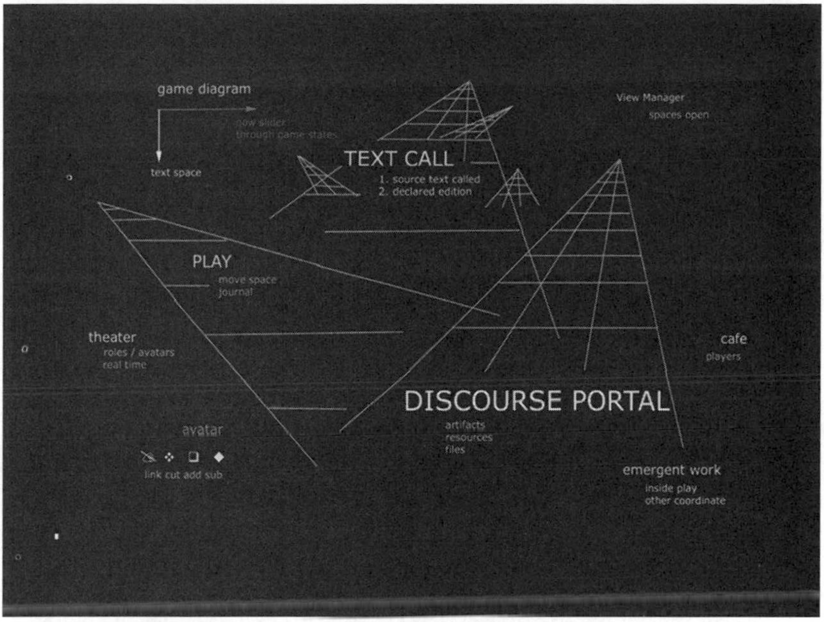

Figure 2.2.8. This diagram, a translation of the schematic features of our conceptualization into a free-floating field, was an enormous leap forward. We had determined that no text would be specified as a "source text" since that would give it a priori identity and authority. Instead texts would have to be "called" and then "declared" the working copy or edition for the game. This element became part of the game space of Ivanhoe in the sense that a player's bringing a text into the game constituted a "move." The discourse portal was the place where all the texts called into the game were versioned and accessed. The "emergent work" was the text being constituted within the game. Café space and theater spaces allowed for in-game role playing and discussion. And the two axes of the game diagram allowed the now slider to progress through game states even as the text space continued to evolve. Link, cut, add, and substitute functions for altering a text hover near an avatar, since it is likely that the avatar image will be formed by the sum of the activities of the player. This schematic, though it couldn't be realized directly, did create a model of the game as a set of intersecting zones of activity. It functioned as a visual model of the conceptual primitives (content types and functions) in Ivanhoe.

provide access to a fixed set of structured relations but a primary mode of theoretical query. The design called for a fluid, dynamic, and highly iterative and emergent interface, one that allowed for transformation of the information within the field of play at the level of material. Visualization was a means for intervention as well as display. The interface included basic zones of activity, rather than rigidly defined areas, a priori subdivisions, of its limited screen real estate. No activity had a predefined

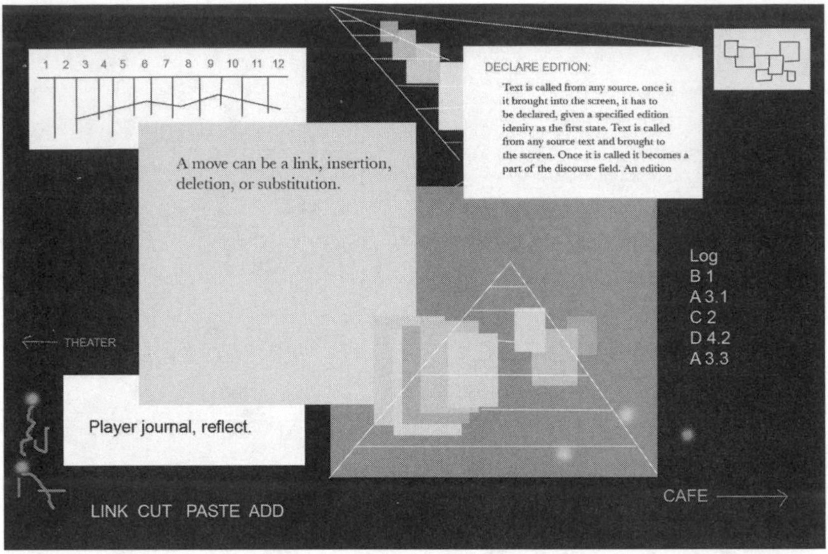

Figure 2.2.9. This visualization took the schematic diagram from 2.2.8 and made it into a screen-based work environment. The ghosts of the previous diagram show in the background, and degrees of transparency allow the visualization to register some of the crucial aspects of versioning, game-play, available functions, avatar development, and so forth. This was the last of the sketches and visualizations made before we actually created Ivanhoe.

area or space, though the basic elements of "text call," "discourse portal," and areas of "play" and "commentary" were assigned zones within the overall screen. As the process unfolded, a work emerged from the elements brought into play and the relations configured among them and the participants. The dynamic web of relations could be viewed from any number of subjective positions—no view existed outside the game-play, just as no work preexisted its performance in the electronic space.

At that stage of design and theorization, when actual software development took over from the conceptual design process, Ivanhoe's interface drew on a host of concepts from the history and theory of information in emergent spaces. These are touched upon in the following list of attributes. (See figure 2.2.10 for images of Ivanhoe as implemented in working prototype.)

Dialogic and networked. Ivanhoe was created at the intersection of individual subjectivities in dialogue with each other through a work and its interpretive field. The interface was meant to permit the mapping of these interrelations, and emphasized the social nature of the production of imaginative work and the collaborative character of interpretive acts. The critical foundation for this approach drew on conceptions of the Web as a social space. Envisioned in H. G. Well's prescient vision of the world mind, this notion of virtual communication and exchange was the impetus for J. C. Licklider's work in the 1960s on human-computer symbiosis as the basis of virtual communities.[24] Reinforcing the idea of interface as a portal to social interaction, Ivanhoe was designed to discourage solipsistic play and to encourage the recognition that creative and scholarly work takes place in a social space.

Figure 2.2.10. Elements of the game in action

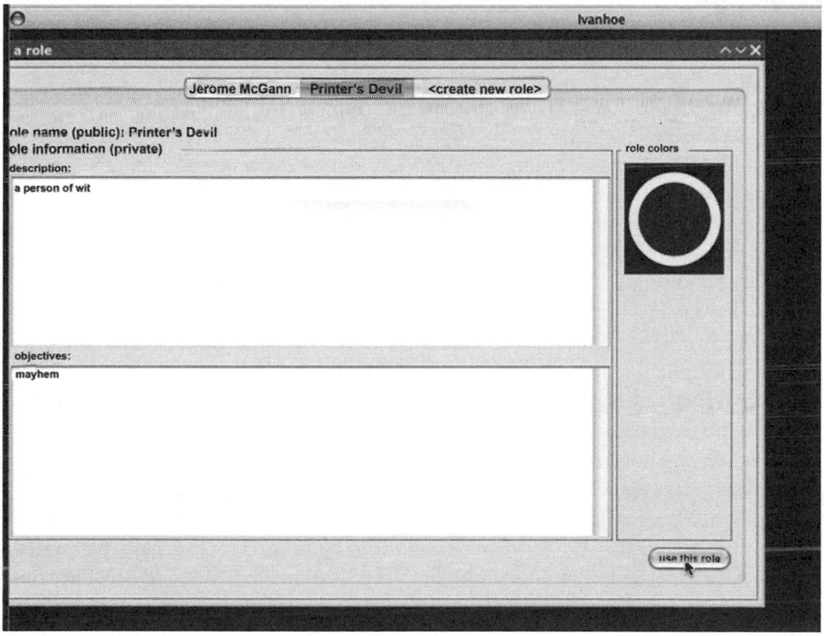

2.2.10a. Role definition. The first task for a player wishing to enter the game is to create a role and then to define the game by giving it a description and objectives. Here player Jerome McGann created the role "Printer's Devil." Though the name was public, the description and objectives were not. These were elements that were scripted into the game either for pedagogical purposes or for use in group tasks where assessment or self-assessment might be useful. But even within the realm of criticism or research, they serve to increase self-consciousness about the tasks and approaches the player is taking on as a role. A player could have more than one role. (Thanks to Jerome McGann for permission to use these images.)

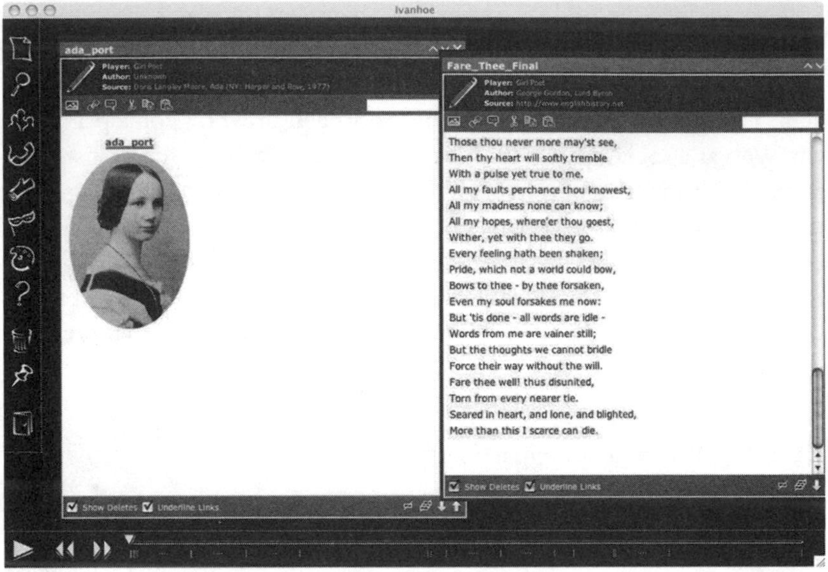

2.2.10b. Initial contributions to the discourse field. This screen shot shows some early moves in a game we played with Lord Byron's poem "Fare Thee Well." The examples that follow will be from this game. Every action in the game space is called a "move," whether it involves introducing a document, adding a comment or link, or making some other change to the game state. Girl Poet (my role) has introduced two elements, one the photograph of Ada Lovelace, whose author is "unknown," and the text of the poem, whose author is identified. Sources are given for both objects. Note that each window has a pencil in its upper left, indicating that Girl Poet can edit or work in this space. The menu bar on the left side of the screen shows the various activities the game allows: adding a document, searching, communicating with the other players, sending a personal message, entering information in the player log, creating a role, changing colors in the play space, getting help, discarding moves, posting moves, and leaving the game. On the bottom of the screen a now slider with tick-marks shows every move made by every player, with ellipses indicating that time has elapsed between these that is not marked in the metric of the line. Every object entered into the discourse field has its own identity but is open to versioning by any player. The menu bar in the window of each object shows that it can have links, comments, cuts, and other actions performed.

Generative and procedural. Max Bense's discussion of generative aesthetics in the early 1960s established the idea that visual forms, even those he defined as artistic, have an algorithmic foundation.[25] Extended through the study of complex systems, the idea of rule-based, procedural production of imaginative works or interpretations is grounded in computational methods. Bense's vision was limited by his mechanistic conception of form, but certain features of his premise remain useful. Generating

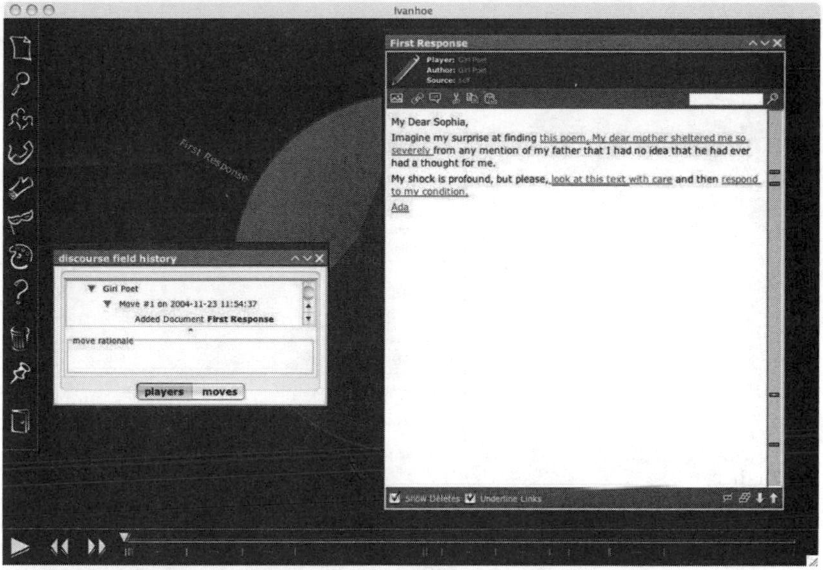

2.2.10c. This game, like most of our test games, involved playful invention. Here the Ada character, as envisioned by Girl Poet, is writing to a girlfriend, Sophia, in order to justify the discoveries she is making about herself, her father's feelings about leaving her mother, and her own changes to the poem he wrote. Every underlined area is a link, or double link, and tick marks in the side bar indicate changes or additions to the text in the window.

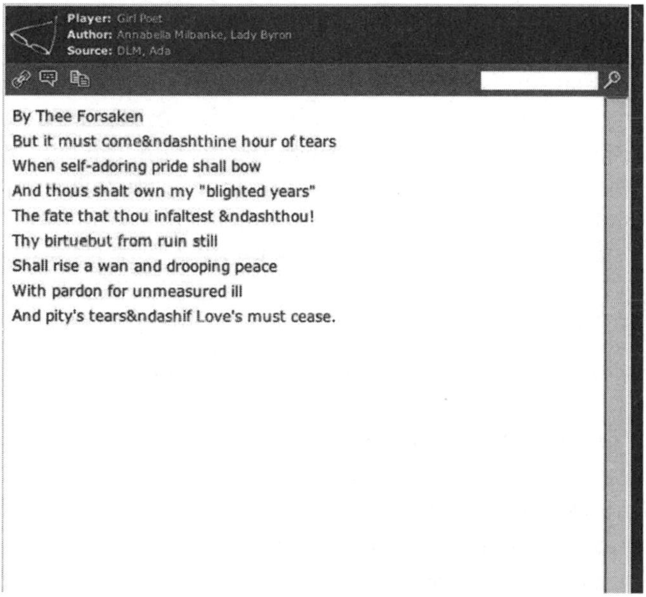

2.2.10d. Annabella Milbanke's poem was written by Girl Poet, though another source is speciously identified. All kinds of display bugs show in this artifact, especially where special characters were to be set. This was an early design glitch.

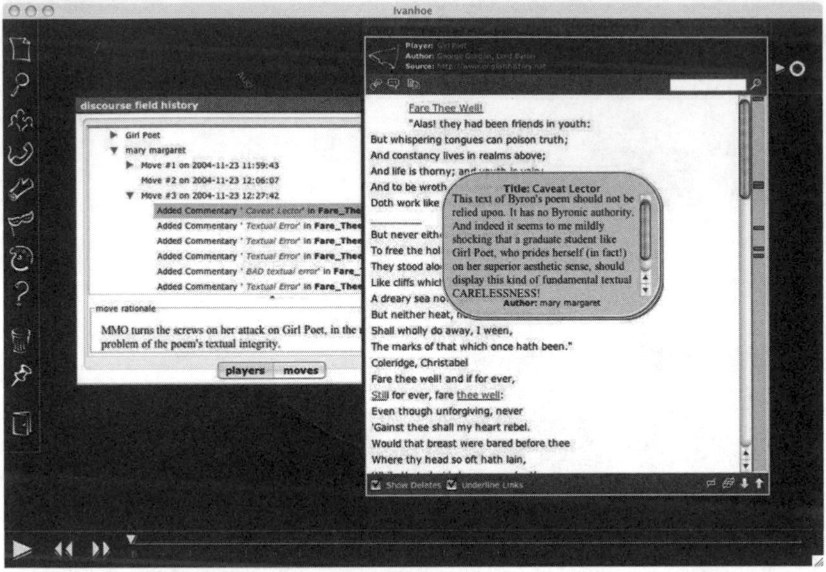

2.2.10e. As the game progressed, the character Mary Margaret, pretending to be a student, made comments and interventions. Her icon in the marble tray (upper right) is highlighted, indicating that the screen represents her point of view. Her move is being entered in the discourse field and recorded in the discourse field history.

visualizations, or moves, from prescribed procedures is one aspect of this approach, but reversing that process and generating rules and procedures from visualizations is the other. The latter implements the potential of an image as a primary, first-order expression of knowledge, whose algorithmic foundation can be revealed. In addition, the notion of generative aesthetics applies to every graphical visualization of text or other artifact within the space of play, since these involve "calling forth" a document and then rendering it through an algorithmic transformation of the data.

Emergent. Visualization through on-the-fly processing of information that is itself constantly changing manifests degrees of complexity not contained within or accounted for in the first generation of instructions. Emergent behaviors, such as those generated by swarm systems, or even by simpler artificial intelligence engines that use probabilistic methods to produce statistically varied results, were used to create player avatars and other game-play diagrams and representations. These were meant to return to the game as aesthetic provocations. The shape of game-play would produce the "emergent work" that was the ongoing outcome of interpretive actions among players. This would necessarily be a continually evolving form.

Relational. Ted Nelson's earliest notions of the Web as a space of associative meaning extended Vannevar Bush's concepts of Memex from the mid-1940s.[26] Ways of thinking about knowledge as an interlinked field have been a part of the mythology of networked knowledge systems since their invention, and earlier, paper-based diagrammatic organizations of knowledge and argument can be traced to Ramus and his method in the late Middle Ages. Reconfigured conceptions of this approach are part of Renaissance formalizations of knowledge (classification systems, textual and paratextual apparatuses, and well as graphical modes of information representation), but the electronic environment has awakened aware-

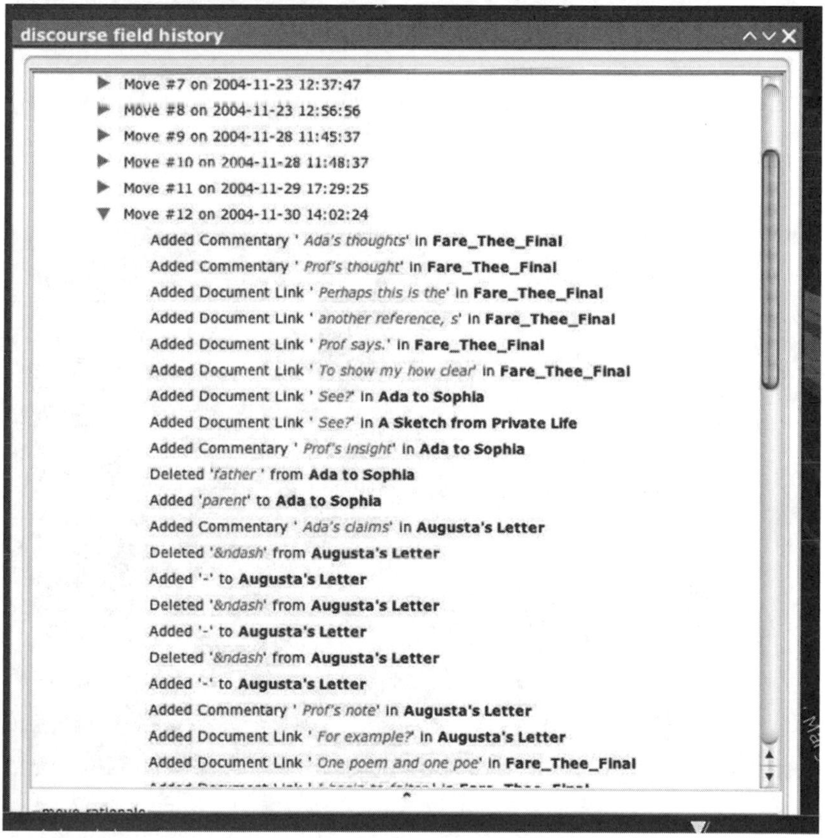

2.2.10f. The discourse field history is date stamped and shows every alteration, comment, link, or move made by the player/role in the game space.

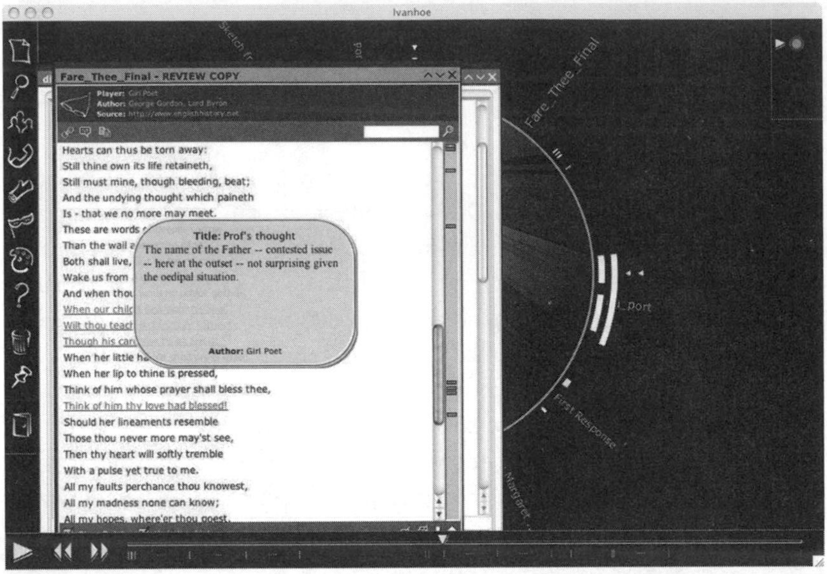

2.2.10g. The version of "Fare Thee Well" shown here is the final copy of the poem from Girl Poet's point of view. Underlined text indicates changes and comments, such as the one that is in the bubble in the window. Girl Poet's marble is highlighted, and the discourse field (background disk) is coded with her colors.

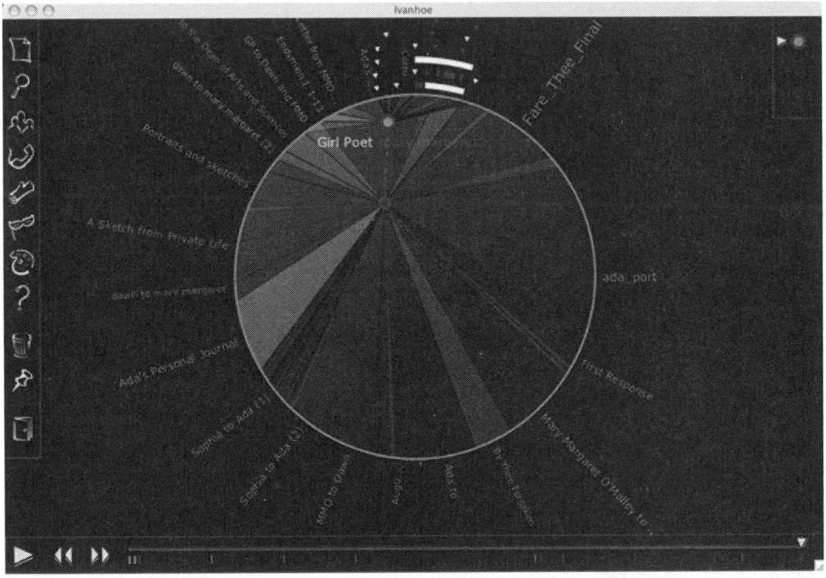

2.2.10h. View of the game in its final stages, from the point of view of my character, Girl Poet. The size of the poem's title shows its importance, measured in terms of the attention given it by players in the game. The productive, generative dialogue in the game is shown by the relative positions of the players, their dotted lines of influence, and the many rays of connection and relation to each other.

ness of the living character of associative thought processes.[27] The notion of a text as a single, incidental instance of a larger discourse field—as a snapshot in the sequence of production events—was one central theme of Ivanhoe. The bibliographical or genealogical relation of artifacts to each other was presented according to production and reception histories within recursive, nonhierarchical, and contradictory models.

Iterative and manipulative. The interface permitted versioning, and the palimpsest of meaning production could be perceived as a thick field of interpretive activity and meaning production. Game-play states, their continual transformation and relation to each other, to the initial state of the game, and the basic work and rework of a series of recursive interactions were all available for presentation and analysis. The lessons of responsive interface and direct manipulation (or its illusion) in work by Ben Shneiderman and others established conventions for a reversible, navigable, legible design.[28] Going beyond menu-driven options, or combinations of preset data, Ivanhoe's interface was designed to allow manipulation of the elements of the game at the information level (coded data), not only that of display. Display and visualizations provided a point of direct access to the structured data of the emergent work or the game-play.

Dialectical. The interface demonstrated the non–self-identical character of visual artifacts. No work existed a priori in Ivanhoe. Images or text had to be called and then declared, that is, given a presentation form. This reinforced the realization that any visual form is a constitutive intervention in a field of potential, rather than the display of an inert or fixed artifact. This principle, of calling and declaring, echoed the process of intervention crucial to quantum mechanics, where the act of intervening determines an outcome from a probability distribution. Ivanhoe's interface called artifacts, game-play, player profiles, activities, and behaviors into being so they could be configured as visual entities. On another level of display, the continually shifting configuration of elements in play constituted the "work." The design emphasis was on contingency and relational production of configured form rather than on any formal or a priori structures. Through algorithms that respond to a participant's activity, the computer became an active player, engaged in a dialogic exchange of activity and display.

Transformative. The transformation of information at the level of material instantiation in code and visual presentation was perhaps the key over-

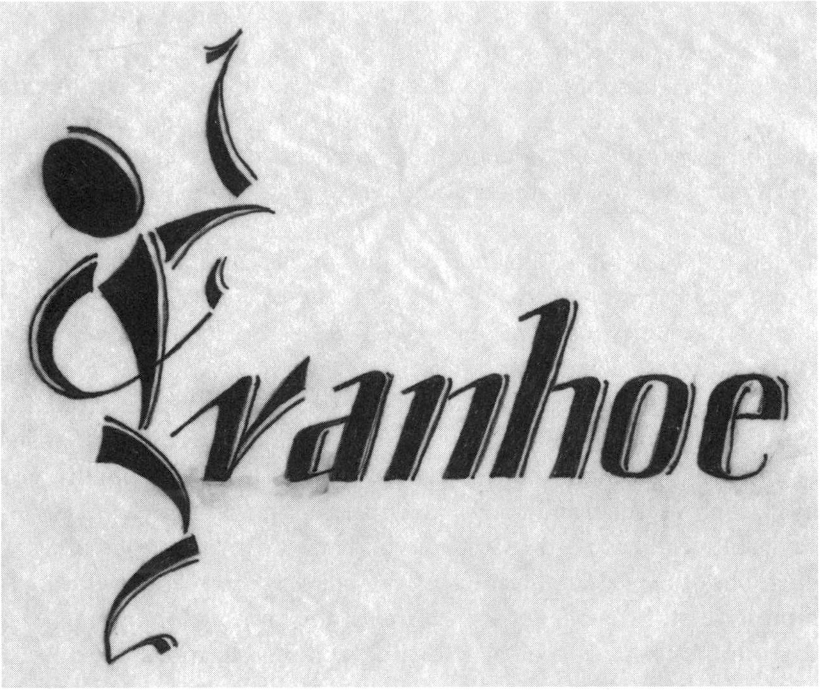

Figure 2.2.11. The Ivanhoe logo, used for the game splash page, is shown here in its hand-drawn original.

arching concept of the interface design as a whole. This emphasized our conviction that interpretation is a performative transformation of the material condition of an artifact. This is always the case, and the electronic environment extends, rather than innovates, in this tradition. The dramatic possibilities for making evident the effects of interpretation as acts of deformance were drawn from a legacy of such approaches. Some of these have their roots in esoteric practices, such as gematria and kabala, some in the ludic sensibility that governs combinatoric visual and verbal artifacts (volvelles, movable books, mobile and kinetic works of art), and some in the critical traditions of potential literature (Oulipo). Ivanhoe's interface intended to make this graphically evident.[29]

In summary, then, Ivanhoe's design was conceived to counter the idea of self-evident materiality—the idea that an object or artifact simply is what it is—by offering instead a set of conditions for the creation of what can be. This set of contingencies was structured into the interface so that every game state was clearly generated as a possibility within a potential field. The interface was responsive and emergent, perceiving the participant as part of the information of the system, while insisting that an emergent work is always constituted at the intersection of the participants' aggregate activity within a social sphere of production.

Ivanhoe's generative aesthetics opened the screen as field of play, of ludic invention. Iterative visualization provoked an emergent, rather than self-evident, representation. Ivanhoe's initial design recovered an alternative tradition of inventive and generative approaches to visual epistemology and representation that is at odds with rational modernism and traditions of visual epistemology that derive from fine art and scientific visualization. The modern sensibility made critical approaches to literature into what McGann terms a "spectator sport." Both textual and visual fields were governed by the assumption that materiality was a stable fact, unproblematic, a priori, and self-evident. By contrast, Ivanhoe assumes a complex system in which a work is produced by the dynamic interplay of an individual interpretation and a set of possibilities structured and encoded in an emergent field. Ivanhoe's interface was designed to make these principles part of the experience of play, as well as to open the horizon of research onto wider application of what I term post-Cartesian approaches to graphesis, or subjective, situated approaches to visual knowledge production.[30]

2.3 Subjective Meteorology: A System of Mapping Personal Weather

Subjective Meteorology was created entirely as an act of aesthetic provocation and a work of imagination. It mades no concession to the standards of digital humanities or to the exigencies of disambiguation or any other technical constraints. My intention was to create a work that demonstrated the capacity of graphical forms to be a primary mode of capturing subjective experience for later analysis and understanding. In Temporal Modeling we conceived of subjectivity as a combination of *position* and *inflection*—the point of view from which any experience was represented and the affective or qualitative values attached to the elements of these experiences—which are structured into the design as a "now-slider" and other elements. In Ivanhoe, subjectivity is again structured into the design as a position within the game, a point of view or inner standing point. But in Subjective Meteorology, subjectivity is marked at the level of inscription, in the hand-drawn traces of pencil lines and forms that show the mark of individuation in their material production. These marks contribute to a higher-level system, in which emotive and personal experience comprises the entire representational code. What makes Subjective Meteorology distinct, and in some sense more extreme and unfamiliar, is that the *content* of its expressions is subjective experience. Thus, it provides a system for mapping individual psychic experience—"personal weather."

Could Subjective Meteorology be built in a computational environment? Absolutely. The software for modeling atmospheric systems and conditions would provide a foundation. The elements would need to be renamed and parameters for their behaviors specified. But the principles and methods would be the same as in traditional meteorology, with its complex and subtle ability to describe (even predict) shifts in mood and atmosphere.

This project has deep roots in my graphical work, going back into the 1970s, when I first did a series of "event" drawings. This theme has manifested itself in numerous projects I've undertaken in the intervening decades, usually as series of drawings concerned with the capture of experience as energy fields, flows, and forces and the deconstruction of apparent entities into these dynamic components. With Subjective Meteorology, I set out to be extremely systematic. I had the benefit of being a fellow at the University of California, Santa Barbara, in the Digital Cultures Institute in the spring of 2004. During the month I spent there I did a series of experimental/experiential drawings, studied the nomenclature and representational modes of traditional meteorology, and created a graphical system of notation and nomenclature that would make use of traditional methods for the innovative system of mapping personal weather. The project was conceived within the larger context of work on visual epistemology and the authority of graphical forms of knowledge production, on which I was working at the time.[1] The project extended critical questions about inscribed and inflected subjectivity while insisting on the ability of graphical systems to produce partial, situated, and aesthetic knowledge. I see this work as a part of SpecLab, especially as some simple digital animations were produced to demonstrate the feasibility of creating dynamic representations of activity over time within the metaphoric language (verbal and visual) of the system. Although it was undertaken independently of SpecLab, where the focus had shifted to applied tool-building and development, Subjective Meteorology is intimately related to the other projects in which I played a part, and the design of the project builds on lessons learned, while raising the level of imaginative expectation within our digital activities.

: : :

Mathematician and philosopher René Thom asserted that there were only two stable, reliable modes of knowledge representation: natural language and mathematics. Graphical methods seem equally compelling and useful for creating and analyzing knowledge, and for calling into question issues of epistemology and representation. Do we know something and then draw it? Or can we make a graphical record and through it come to know something?

Subjective Meteorology makes a case for graphical knowledge as a primary method of producing as well as analyzing phenomena. But the particular, even peculiar, theme of this project pushes the capturing of knowledge through graphical means toward an investigation of subjective experience. Several levels of abstraction and translation are involved in this activity. As stated above, Subjective Meteorology is a system to describe "personal weather" or the psychic moods and atmospheres of subjective experience. But what does this mean? The system takes the metaphors of traditional meteorology and uses them to represent the dynamics of lived experience. Not such a strange notion, and one that many viewers resonate with intuitively. The conception of Subjective Meteorology is radical, given longstanding habits of subjugating the possibly unruly character of visuality to the rules of information design or logocentric disciplines (Thom's mathematical and natural languages). Yet the approach—pushing perception through imaginative, rather than rational, processes—is well within the conventional bounds of image making in the fine arts. It is certainly proper to characterize Subjective Meteorology as an aesthetic project that makes an argument for affective rhetoric.

To create the basis for a workable digital project, I used the drawings and notes as a preliminary stage of content modeling. This allowed me to create the basic elements of the generative system from within the recording of subjective experience across a series of days. This approach was a deliberate experiment, controlled and delimited by time and the number of drawings I set out to do (ten). The project intentionally countered the repression of this micro level of inscribed subjectivity (at the level of the mark) in the design of works of digital humanities (and traditional humanities, as well) in that it was not based on the constraints built into rationally grounded systems of representation. Whereas mathematical and natural languages are relatively stable, as Thom makes clear, with regard to the relation of notation to value, in graphical systems—systems that rely on handwriting, analog notation, or other variable but materially rich qualities—this relation is unstable.[2] The highly charged term "disambiguation" defines the border along which traditional humanities approaches encounter the digital environment but can have no bearing on an asystematic condition dependent on marks and traces that are not part of a limited, finite notation system. Subjective Meteorology stretched digital humanities, enamored as it had become with the idea of clarifying one's thought in formal and systematically formalizable terms. As the field strove to shift humanities onto firmer ground, closer to that of other, more formal knowledge systems, it had moved very far from the fundamental premises of the humanities, in particular the humanistic commitment to subjective and partial knowledge. Subjective Meteorology used the digital environment differently, to construct a system of analysis grounded entirely in subjective experience—and focused exclusively on representing it.

Graphesis, aesthesis, and speculative computing find common ground in the conviction that visual knowledge production can be driven by affective rhetoric. Using image production as a primary means of grasping experience as form creates a base of material expressions from which a systematic scheme of content modeling can be elicited as the basis of a generative system in a digital environment. In other words, the drawings come first. They are intuitive, not formalized, and serve as source material from which to generate analytic principles. They do not serve as display of preexisting rules or formal structures. Subjective Meteorology is thus an experiment in knowledge creation, artistic insight, and imaginative work.

The anxieties that this work seems to engender are indicative. The specters of Thom's natural language and mathematical notation arise regularly, potent and demanding: Will there be a stable notation system—a *legible* one that stabilizes the ambiguities of visual form in natural language? And how will I parameterize experience, that is, make use of a mathematical metric? Systems of knowledge constitute their objects, bringing into view only what the system can conceive. These drawings, with all their particularity, complexity, ambiguity, and subjectivity, show graphical methods as a primary mode of knowledge creation. Objectivity is the watchword of empirical science, and the distinction between observed phenomena and observing subject is assumed. Subjectivity invoked as the counterpoint term does not distinguish the humanities from the sciences but, rather, calls the basis of objective knowledge into question. Subjective Meteorology uses the metaphors and templates of meteorology (fluid dynamics of the atmosphere), to create a system for representing the always situated and partial dynamics of subjectivity (personal weather, in vernacular parlance).

In this regard, Subjective Meteorology, like other SpecLab projects (to varying degrees), is conceived in the spirit of Alfred Jarry's 'pataphysics, the science of exceptions and imaginary solutions. Keenly aware of the way scientific knowledge was constituted by the premises on which it conceived of its objects, Jarry offered 'pataphysics as an alternative based on particulars. Jarry's 'pataphysics had a few basic tenets: syzygy, clinamen, ethernity.[3] Syzygy, a term used to describe planetary conjunctions, came to signify the reification of relations into figures or forms. The concept of the clinamen is derived from the work of the natural philosopher Lucretius, who suggested that deviation in the activity of atomic particles was essential to the creation of the universe. Lucretius's approach was opposed to that of Democritus, for whom, as for many observers of the natural world to the present day, the atomic universe was to be understood as regular in its behavior. When twentieth-century theories of quantum mechanics began to grapple with phenomena that did not fit the classical models of physics, Lucretius's clinamen found some scientific credibility. Jarry, aware of these scientific writings at the turn of the twentieth century, took clinamen as a fundamental force for creativity in the universe. And ethernity conjures a transcendence of traditional notions of space-time into a continuum whose elasticity is limited only by the imagination. (The punning play in itself makes the shift in language into a performative gesture of invention, and thus provides a demonstration as well as a tenet of Jarry's beliefs.) As a serious approach to the pursuit of knowledge, 'pataphysics remains largely the province of poets, rather than of physicists or mathematicians.[4]

An aesthetic undertaking, Subjective Meteorology is serious in its conception. As an attempt to create a systematic understanding of the ways experience can be grasped and analyzed, it begins with the conviction that intuitive investigation can give rise to systematic study. This is neither a top-down, rule-based approach to knowledge nor a bottom-up, sensation-based one but a subjective, immersive, embodied approach. What we know, and how we know, depends upon the cognitive schema we build. What we experience through embodied perceptual means can be transformed through processes of attentive, concentrated study and imaginative leaps, remodeling the basis of knowledge.

The working premise for my approach to this project, as I stated above, is that graphical activity can be used to capture and analyze subjective experience in a primary set of drawings from which a notation system can be derived. This served in turn as the basis of a generative system of making and showing dynamic principles of "personal weather."

The drawings are the primary research, and they stand on their own,

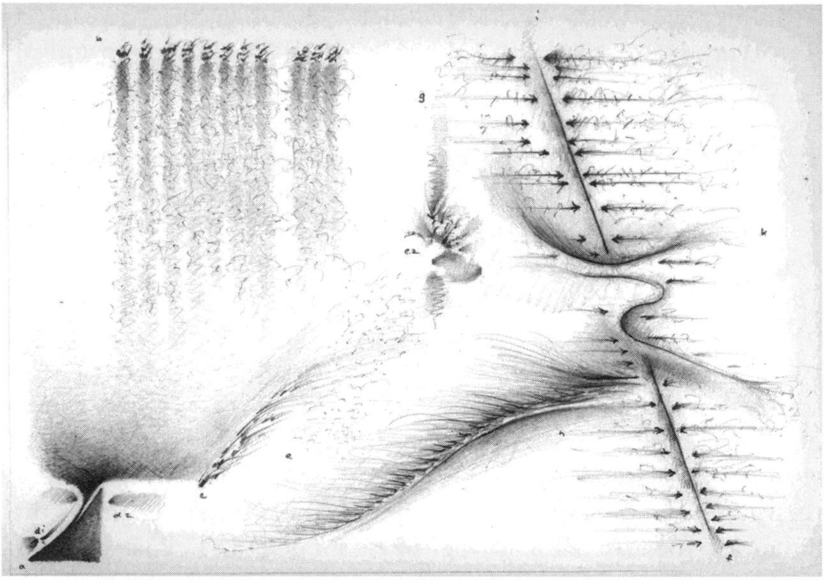

Figure 2.3.1. This was one of my first drawings plotting the events of a day. The task was to capture experience and give it form without any preconceived code or system of signs: I drew first and analyzed after. The final system of graphical elements—and to some extent, the conceptual elements, was derived from these studies put into dialogue with research and readings in traditional meteorology. In this image, the temporal sequence unfolds from right to left. The vertical axis was used to map degrees of proximity, with somatic events at the bottom, remote or distant forces (including telecommunications) at the top, and social activity and moods in between. Legend: (a) compression of exhaustion on rising; (b) communications at a distance; (c) rising energy; (d) residual dream associations; (e) a field of work or task potentialities; (e-2) communications in proximate space; (f) anticipated interruption; (g) vectors of interest; (h) rising front of dynamic activity; (i) more vectors of interest; (j) ridge of resistance to interest and opinions.

with their annotations (figure 2.3.1). From them I distilled the working elements of a graphical system, and a table of equivalents between the language of fluid dynamics of the atmosphere, or standard meteorology, and the system of Subjective Meteorology (figure 2.3.2). This is basic and in its details can be elaborated to a very fine degree. The poetics of such a system are inexhaustible, since they arise from individual experience. The conceptual schema for a digital model of elements, forces, conditions, behaviors, and perceptual positions is finite, even as it allows for infinite variation in execution of any given condition (tables 2.3.1, 2.3.2).

: : :

Atmospheric (fluid) dynamics is an extremely complex systematic analysis of natural phenomena. The forces of wind, heat, temperature gradients, and relations to terrain and physical features vary at every level of granularity within the atmosphere. The subjective experience of daily life can only be described in such a system with all its inexhaustible repleteness. The shift from tradition to Subjective Meteorological systems is marked by a vocabulary change, but it also involves a conceptual shift. The fluid dynamics used in the analysis of atmospheric systems can be grounded on classical physics and Euclidian geometry. But topological mathematics, quantum mechanics, and 'pataphysics offer alternatives. Subjective Meteorology is premised on a conception of space-time that it shares with Temporal Modeling. Its coordinates and metrics can be nonlinear (multidirectional and multidimensional), heterogeneous

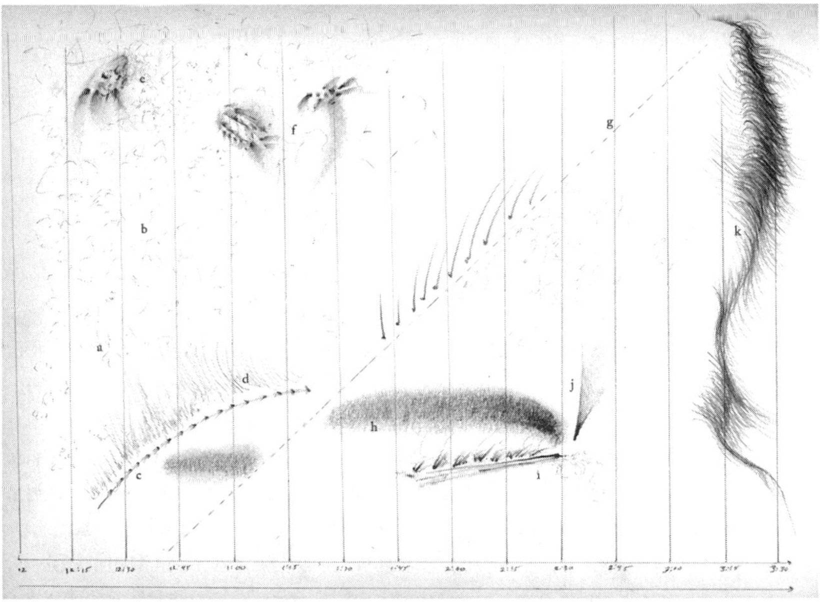

Figure 2.3.2. Graphic forms for a working system. These elements were distilled from a series of day drawings that charted changes in mood, activity, events, formations, conditions, etc. Legend: (a) charged field of potentialities showing (b) areas of perturbation with (c) activity line giving rise to (d) a dynamic wave; (e) singular entity at a distance causing disturbance; (f) complex entities in dynamic exchange; (g)line of anticipation; (h) clouds of anxiety; (i) concentrated task energy breaking at (j) interruption behind (k) front of concentration; (l) temporal grid.

Table 2.3.1. Table of equivalents

Traditional meteorology	Subjective meteorology
atmosphere	atmosphere (charged field of potentialities)
radiation	energy
heat	activity (anxiety)
boundary layers	zones, ruptures, limits, event breaks
moisture	emotional intensity
stability	contingency index
cloud formation	mood formation
precipitation	productivity/consumption cycle
dynamics	interactions
local winds	entities and presences
global circulation	social activity
air masses and fronts	lines of break (attention, intention, etc.)
cyclones	interactions and other events
thunderstorms	intense events and dramatic state changes
hurricanes	catastrophic transformations
air pollution	illness or malaise
climate change	change of personal, social or cultural circumstance

(with variable density, granularity, or intensity and variable metrics), and noncontinuous.

The principle of *nonlinearity* permits simultaneous, ruptured, broken, replicative, redundant, and other synchronous and asynchronous, continuous and noncontinuous event spaces. That of *heterogeneity* permits varying (rubber-sheet) degrees of intensity, as well as variable scales of temporality and metrics with elastic coordinates. And the *noncontinuous* nature of the system permits discrete zones of differentiated activity. These are the premises on which the system of Subjective Meteorology is based. Though they echo the tenets of Temporal Modeling, the design principles of the project are different. In Temporal Modeling, we designed the general principles first and from those created a system of representations and graphical notations. Our content model became the basis of a working system of elements that could be used to represent humanistic temporality. In the case of Subjective Meteorology, by contrast, the system arose from specific examples and their study, from which the

Table 2.3.2. Conceptual categorization of equivalents

Forces (dynamic rather than substantive, providing energy for change)	energy: perceptible and potential emissions: propagative distribution effect: activity = manifest evidence of energy conviction: free/forced inevitable: gravity/levity mutable: momentum/perturbation systemic/local: circulation, mean flow, forcing remote: actions at a distance pervasive: drives and desires dynamic: expansion/contraction of moods pressures: gradient force (tension/relaxation)
Conditions (aspects of the field within which elements and forces engage)	rigidity/flexibility, buoyancy/stability in terrain or circumstances turbulence, drag, stress, friction, roughness, free and forced conviction somatic factors (food, sleep, temperature) climatic factors (weather) psychic factors (recollected dreams)
Behaviors (ways in which elements, forces, and conditions interact)	approach, avoidance, autonomy, stability, communicative exchange, reflection, refraction, tangents and unexpected turbulence or perturbation, occlusion, vorticity/spin, feedback/response/reaction, contraction, contradiction
Perceptual positions (point of view from which situation is produced/perceived)	Self: anticipation (anxiety, trepidation), presentness, mirages, retrospection (relief, nostalgia, longing, mourning), critical angle, scattering, distraction Other(s): parallax, contradiction, complication, alignment

general principles, always subject to revision, were then elicited. The descriptive vocabulary draws on specific instances. Animated, digital versions of the system, also created as proof of concept, brought this project to a certain point of design completion, and in that sense the images that accompany this chapter embody the principal arguments made by the text, whether or not Subjective Meteorology is ever realized as a working digital environment.

2.4 Modeling a Critical Approach: Metadata in ABsOnline

Subjective Meteorology, like Temporal Modeling, had been an exercise in creating a system of graphical notation. In the process of their creation, a content model had been designed for each that was a nomenclature scheme and a template for design. In Ivanhoe, we had modeled and designed spaces for critical intervention and theoretical engagement with texts. In creating the metadata scheme for Artists' Books Online, I took what I had learned from these activities and created a template to model criticism for a field that is sorely lacking in such discourse. ABsOnline is a digital collection of facsimiles and metadata meant to provide a resource for access to and study of artists' books. Defined as original works of art made in the book format, artists' books are often created in very limited editions and are usually held in special collections. Criticism and research in this field has been slow, and a larger picture of collections development, publication patterns, and other large-scale historical patterns is limited. Because of my own long involvement in this field as a practitioner, scholar, and critic, I felt the need to design metadata to provoke scholarship and criticism. Thus ABsOnline is an exercise in using metadata as a modeling device within an intellectual project where the design of the intellectual field draws on digital techniques to shape and structure a critical approach.

The metadata for ABsOnline is structured on bibliographical principles, and contains fields at three levels: work (idea or conception), edition (material expression), and object (individual instance). In each level, the information called for is organized to provoke a particular set of readings of specific aspects and features of an artist's book. The metadata scheme was put together with the input of several working groups comprised of librarians, catalogers, curators, bibliographers, artists, critics, and scholars. We drew on existing controlled vocabulary and created local terminology culled from shared sources to create a list of descriptive terms for such fields as binding, production methods, materials, and so on. But the overarching project is designed to demonstrate that metadata can bring critical discourse into being by the way the fields model a reader's relation to a book work. In that sense, ABsOnline is a case study in the way metadata function as criticism and interpretation. It was conceived very differently from the graphic projects in which subjective experience and point of view were paramount.

Electronic metatexts are more dramatically performative than print texts with respect to way they model content and configure conditions for use. Because these metatexts actively structure a domain of knowledge production in digital projects, they are crucial instruments in the creation of the next generation of our cultural legacy. No other textual form will have more impact on the way we read, receive, search, access, use, and engage with the primary materials of humanities. So rather than focus on the display and representation of texts and artifacts, I'm going to examine the ways in which the metatexts *model* digital texts and artifacts, with specific reference to ABsOnline.

Metalanguages and Metatexts

A metatext is a subset of metalanguage, one that is applied to a specific task, domain, or situation. In ABsOnline, a collection of virtual representations of artists' books, the metadata deliberately attempts to shape a field of scholarly inquiry.

In creating the metadata structure for ABsOnline, I came to understand more fully the ways in which digital metatexts are dynamic and performative.[1] All textual production is rule-bound and code-based, in oral and print culture, but these constraints become dramatically apparent when texts are created in or migrated into electronic environments.

Metatexts not only express such rules, they describe and encode the rules that govern the composition of texts and their use in textual systems. They also contain instructions that call forth behaviors, prescribe and delimit domains, and set out the parameters on which knowledge is shaped and bounded. The epistemological power of a metatext in a digital environment comes from the way software parsers determine when a text is well-formed—when it fits the requirements and outline of information set by the metatext—or from the constraints it places on what can be entered as data and in what order and form. Because conformance is rigorously enforced (and nonconformant documents rejected) the relation of metatext to document (rules of expression to instances) is quite explicit. Information, interpretation, knowledge—everything has to fit the model encoded in the metatext. Print formats are far more forgiving. A "stanza" written in nonstandard form won't be rejected by the page, for example, but malformed metadata does not "parse." The relation of rules and expressions is thus very explicit.

In conventional print formats, paratexts and metatexts assist in navigation as well as providing interpretative materials and explanations about the shape and content of textual works. Paratexts are often structural (tables of contents, headers and footers, and indexes, for instance, provide signposts even though they encode sharply defined arguments and assumptions, while footnotes, marginal notes, and other elements are charged with analytic functions). Metatexts have *descriptive* power that, wittingly or not, becomes a model for the texts they describe. The model is comprised of types of textual elements (semantic and syntactic but also bibliographic, graphic, semiotic, social, pragmatic, etc.) and the order of elements in the structure. The model fully articulates the shape and system of these typologies and their orderings in a generalized schema. The description is an interpretation, selectively emphasizing certain features and characteristics of a text. Digital metatexts thus act upon other texts to perform an analysis, generate a display, act out a search, or make other interventions within the textual field.

A digital metatext also embodies and reinforces assumptions about the nature of knowledge in a particular field. And it is only as good as the model of knowledge it encodes. The metatext is built on a critical analysis of a field and expresses that analysis in its organization and the functions it can perform. The intellectual challenge comes from trying to think through the ways the critical understanding of artists' books should be shaped or what should comprise the basic elements of a graphical system to represent temporality in humanities documents. The technical task of translating this analysis into a digital metatext is trivial by contrast to the compelling exercise of creating the intellectual model.

Models

As I noted in chapter 1.1, a model creates a generalized schematic structure while a representation is a stand-in or surrogate for a particular thing. A table of contents is in effect a content model of a work, as is an index. Each provides an abstracted version of a text's contents that includes some things, excludes others, and has a structure of its own that is in certain respects isomorphic to the text (the order of chapter titles in a table of contents) and in others utterly independent of it (the order of the items in the index). But insofar as a table of contents or index exists as part of a document, it exists on the same level, as another element of the text, and this muddies the model/text distinction. A generalized scheme that outlines what a table of contents is (and what a header is, and a text page, a text block, a footnote, an index, etc.) is a content-type model for a print text.

In electronic environments the process of creating content models creates several new kinds of documents and textual artifacts that are not quite of the same order as those familiar in print culture. Content modeling for digital artifacts depends on creation of a metastructure (an XML file known as a DTD or document type definition) to which every other file of a given sort must conform. The classification schemes created for structuring data in a digital collection thus encode a model of that field of knowledge. In my own experience, creating the DTD for the Artists' Books Online project did much to clarify a critical approach to these objects. At another level of granularity, this gesture was meant to control the critical discourse in the field and its assumptions about how to constitute artists' books as an object of study.

A DTD is a specialized kind of metatext that expresses the rules for a potentially infinite number of XML documents that conform to its outlines. It is built of XML tags, a set of labels identifying various elements (for instance, the tags <altTitle> and <altTitle> would surround a word or phrase that was an alternate title for a work). But a DTD structures the relationships among these elements in specific ways, and this constrains the documents to which it is applied.

For an XML document to "parse" it must conform to the rules of the DTD. The DTD requires that the correct elements must be present in the correct order within its hierarchical scheme, just the way a game's rules constrain moves or a recipe stipulates the order in which ingredients are combined. You cannot hit a home run in football or make a mousse by first scrambling the eggs. The rules are a model; the game is an instance, a specific, concrete, and particular expression of those rules. Thus a DTD governs the files that are created in its image. The DTD is the site of content modeling for textual artifacts.

To model content for a field one must first consider the elements that constitute its specialized domain of knowledge. In ABsOnline, we created a set of fields that call for examination of an artist's book in terms of various aesthetic design features. What are the typographic features of the work? The pictorial elements? Are the turnings (movement from one spread to another as a page is turned) particularly well used? If so, how? Is the graphic organization of the work a conspicuous aspect of its effects? How does the development of the book register within the work as a whole? Identifying and naming these areas as fields into which text will be entered makes it possible to aggregate information across the corpus of files in a collection as a whole. A minor act, perhaps, to ask scholars, critics, artists, curators, and catalogers to look at and attempt to describe these features, but it is a deliberately provocative act. These are the ways to think about an artist's book, the metadata asserts. To fill in the specified fields, a person has to attend to various features of these artifacts. Thus the metatextual documents in ABsOnline propose to model a critical discourse by creating protocols for enhanced cataloging and description.

In addition to descriptive metadata—extended cataloging records, such as those just described—there are metadata files about who made the electronic files, scanned the books, and contributed to the data. Yet another kind of metadata document consists of a style sheet to selectively transform data in the files into displays. Metadata schemes create the file structures that hold all text and image files within several levels of hierarchy—the files for each book work (structured as work/edition/volumes/objects/images), each collection and each exhibit, essay, and resource. Metadata schemes may be designed at a later date to aggregate information and records from outside ABsOnline's home source files with those authored explicitly for ABsOnline. Others may be brought into being as style sheets capable of customizing different functionalities or views of the information with ABsOnline.

We began with the basic problem of making a DTD for the metadata that describes an artist's book. This seemingly simple task took a year from initial consideration to completion. The elaboration of a descriptive scheme for a book, at first just laid out in text as a set of fields to be filled in, turned out to be very different when it came time to implement. Calling for description of various elements and features, a production narrative, and detailed analysis of texts, images, their relation, and so forth, as per the first iteration of the metadata scheme, turned out to be prohibitively difficult, even though I was both the author of the scheme and the user. Elaborating the conceptual framework for "what a book is as a work" was one thing, keeping in mind one's place in that conceptual framework while describing a book in hand quite another. The urge to tell the story of the book—its background, production details, human and artistic history—clashed with the structured metadata, which only asked for a Note on the Title of the Work in the first fields. (The very idea of the Work as a larger category of conceptual project within which the book came into being was confusing and unfamiliar. The tendency was to want to describe the Edition, the instantiation that one held in one's hands.) Turning the "natural" sequence of observational events—picking up, handling, looking at, reading about, and reflecting on the book—into a form that could be organized as structured data was an enormous challenge. Especially since the desire to structure the data logically, categorically, from the largest to the smallest detail (the speck of dirt on the inside back cover, the personal inscription, an isolated instance of overprinting) so contradicted the associative patterns of reading and encountering an artwork.

This conflict was never resolved. But by reorganizing the sequence in which the metadata was structured, we were able to put a descriptive field near the top, thus allowing for someone entering data who was familiar with the book to pour out everything they knew at the outset, instead of pulling out requested bits of information piecemeal. Specific details—binding, paper, or place of publication and printing history—are subsequently called out in order to provide search capabilities across the collection (all books with spiral bindings, all works made at Nexus Press, etc.).

The metadata for ABsOnline is performative in a social sense as well as a technical one. By insisting on the need for critical discussion of the specific properties of artists' books, it makes a strong statement to a field that has been without gatekeepers and critical discourse of the sort that makes for professional standards. Ideally, the project would, in the long term, provide information about the production and collection of artists' books that would push scholarship by assembling information on which new kinds of questions could be based. The great advantage of

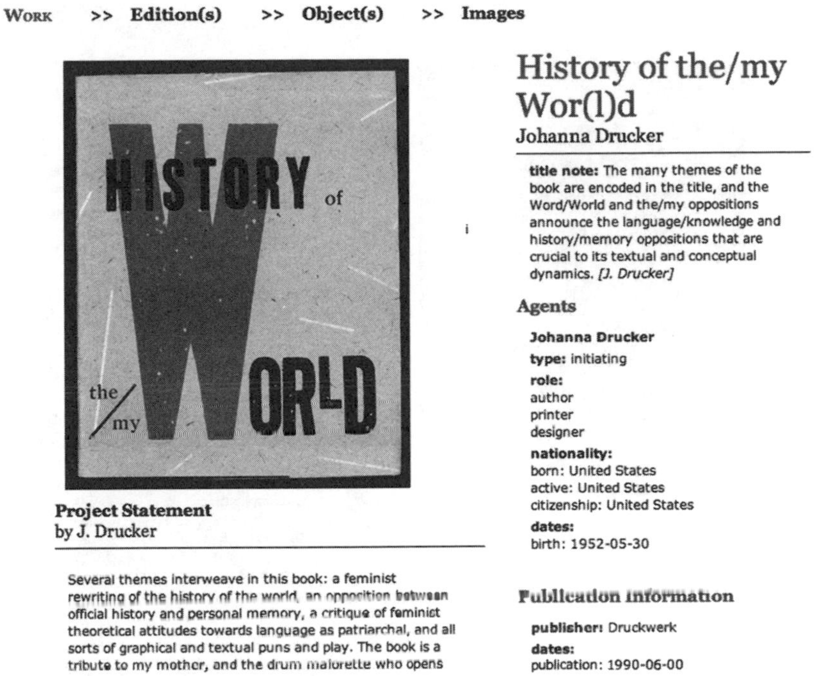
Work >> Edition(s) >> Object(s) >> Images
HISTORY of
the/my
WORLD
Project Statement
by J. Drucker
Several themes interweave in this book: a feminist rewriting of the history of the world, an opposition between official history and personal memory, a critique of feminist theoretical attitudes towards language as patriarchal, and all sorts of graphical and textual puns and play. The book is a tribute to my mother, and the drum majorette who opens
History of the/my Wor(l)d
Johanna Drucker
title note: The many themes of the book are encoded in the title, and the Word/World and the/my oppositions announce the language/knowledge and history/memory oppositions that are crucial to its textual and conceptual dynamics. [J. Drucker]
Agents
Johanna Drucker
type: initiating
role:
author
printer
designer
nationality:
born: United States
active: United States
citizenship: United States
dates:
birth: 1952-05-30
Publication information
publisher: Druckwerk
dates:
publication: 1990-06-00

Figure 2.4.1. Screen shot showing metadata scheme at the Work level. We used standard bibliographical description in AbsOnline, breaking our organization of information into three levels: Work, Edition, and Object. The Work level is meant to include the concept of the project in the largest sense, from the germination of an idea to its many and various executions in sketches, editions, mock-ups, research, study, and so on. The Edition is any instantiation of the Work in published or issued form. The Object is the single entity that a cataloger, student, researcher, or scholar has in hand and from which the description is being made. While these are useful intellectual constructs, they proved counterintuitive and difficult to reconcile with the experience of looking at and describing a book. Also, since we wanted a representative image at the Work level, we used the cover of the first edition, and this created an additional level of confusion and ambiguity.

electronic processing is the aggregation of data, distributed participation, and the capacity to collect in virtual space artifacts that are separated in physical space. All of this functionality is enabled by the metadata and the metatextual apparatus (including the scripting languages that bring dynamic capabilities to the Web) (figures 2.4.1, 2.4.2).

: : :

ABsOnline uses metadata in several ways. It organizes a critical approach to the field. It organizes the file system and structure of textual and

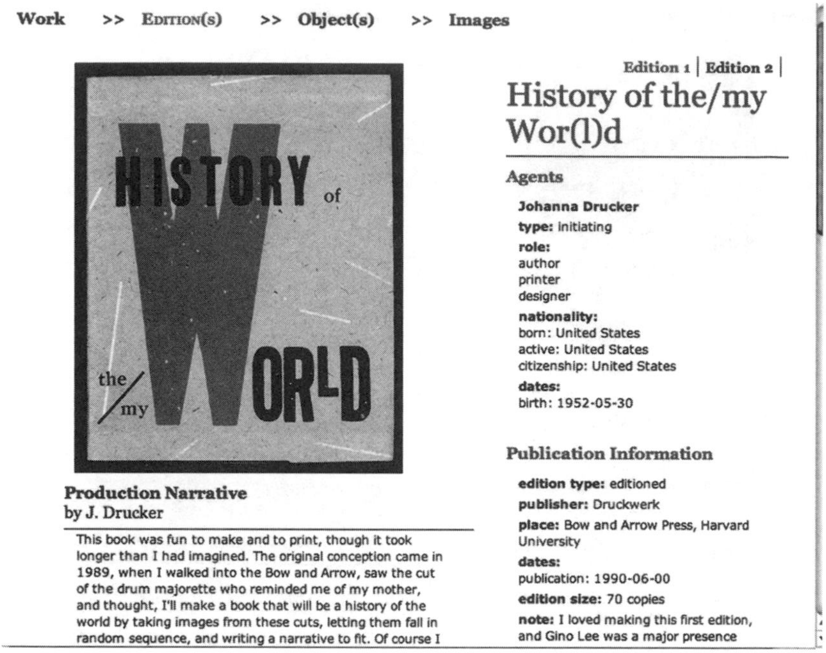

Figure 2.4.2. Screen shot showing metadata scheme at the Edition level. The information about the Edition is more specific than that at the Work level, with details of the physical form, the production narrative, and other particulars relevant to the actual issue of an edition in print.

graphic artifacts in the collection. It records information about the production of files and contributions to the collection. And it could eventually facilitate integration of the collection with other records in electronic file formats that live on the Web.[2] Metadata is both the repository of and the instrument for enabling these activities.

Metadata thus models texts and artifacts, collections, and behaviors at several levels. It models texts through markup and DTDs. It models content for display, search, use, analysis, and interpretation. It models an organizational scheme for the storage, access, and presentation of artifacts that may be text-based or may be images, sound files, video, animation, or statistical files. Through its performative capabilities, metadata enacts the non–self-identical character of texts, creating them anew in each iteration. Metadata encodes the very condition of potentiality, of a text as a field of possibilities called into being in each instance. The social conditions of use, the situatedness of access, and particularly of purpose to which these means are put immediately returns us to the cultural condition of our own practice—personal, institutional, and historical. Early-twenty-first-century textual studies and cultural studies are most emphatically, and necessarily, digital studies as well. The seductive force of intellectual engagement with shaping knowledge, creating forms for the preservation and study of our cultural legacy, depend upon metatexts and the performative capacities built into them by our expectations. The task of using metadata to give intellectual shape to a field is a charge to our interpretative energies more than it is a labor to produce the protocols with which to execute them. Modeling criticism or knowledge in any field is an iterative process of dynamic exchange between the metatexts that encode our epistemological assumptions and our ability to reflect (individually and collectively) upon these sufficiently to be aware of how they shape our understanding of knowledge and its ideologies. The process of designing a metadata scheme for ABsOnline was meant to engage a community of users, a task that proved more difficult. I wanted artists' books, in all their rich particularity, to have their aesthetic properties codified in order to move the study of printed artifacts beyond literal, descriptive materiality into dynamic, constitutive, and contingent materiality.

2.5 The 'Patacritical Demon

Though the 'Patacritical Demon has never been built, it has been the object of considerable discussion and collective imagining over the course of our SpecLab activities. Its very unrealized condition may be what's essential—in keeping open a space for future engagement, the Demon is the image of the speculative spirit. Indeed, this project goes to the core of SpecLab's concerns: the drive to represent the activity of interpretation, to give form to the very process of intervention that produces a work. Even now, we continue working to visualize the dimensionality of interpretation that can be sustained in electronic space.

The term *'pataphysics* has woven its way through these projects and texts. As in the work of Alfred Jarry, who coined the term for "the science of exceptions and of imaginary solutions," its use in our domain, though distinctly ludic, isn't exclusive of serious intellectual purpose.[1] Just as Jarry was sincere in declaring the principles of his new science, so we have been seriously engaged in devising a figure that, like James Clerk Maxwell's demon, sorting molecules into zones of entropy and order, (dis)orders the world of knowledge. Our Demon has been given graphical expression as well as discursive description. Insofar as the theory of interpretation on which it draws is an extension of poststructuralist approaches, it has a distinct alliance with theories of enunciation and the intellectual framework they provide for describing reading as a productive and generative act. In such a theoretical formation, the Demon becomes the figure through which the operation of aesthetics as a practice of situated, subjective, and partial knowledge is enacted and given expression. Schematic though its outline may be, it seems useful to offer at least a note or two about the Demon's conception, and a few drawings that conjure its activity. In the next section of this book the connection between these concepts of aesthetics and the tenets of speculative computing and digital media will be made explicit.

In November 2002 I made a sketch of the activity of interpretation as we then conceived it (figure 2.5.1). We had been talking about Charles Peirce and the idea of a "third term" that distinguishes his semiotics from that of Saussure. Where structural linguistics relies on a two-part sign (signifier/signified) within a finite system where value is determined by finite constraints, Peirce relies on an interpretant, a person for whom and in whom signification is produced. This keeps signs and signification from being cast as transcendent. Instead, Peirce's sign—representing something to someone for some purpose—is premised on a situated condition of value production through use within a subjective context.

Even when designing Ivanhoe, we had discussed the problem of picturing or representing the constitutive processes of text production. We had elaborated a scheme that follows enunciative theory but combines it with bibliographical notions of the social production of texts, a probabilistic approach to text/image as a field to be intervened, and the rest of the theoretical apparatus I have touched on repeatedly. I sketched this enunciative system in what I called "the double parallax"—a construct in which the discourse field and the intervening subject exist in a codependent relation. The "text" (whether an image, literary text, musical work, or performance) is produced in the space of projection that is the intersection of perceiving subject and probabilistic field. The text "looks back" at the reader-subject in an act of projection that mirrors the subject's act of reading, but the two cones of projection never meet or match. The double-parallax notion is meant to emphasize that the text so produced is never self-identical—within a discourse field, a text is only a possibility, and within an intervening subject, it cannot be separated from the experience of reading. In such a construct, the text has no

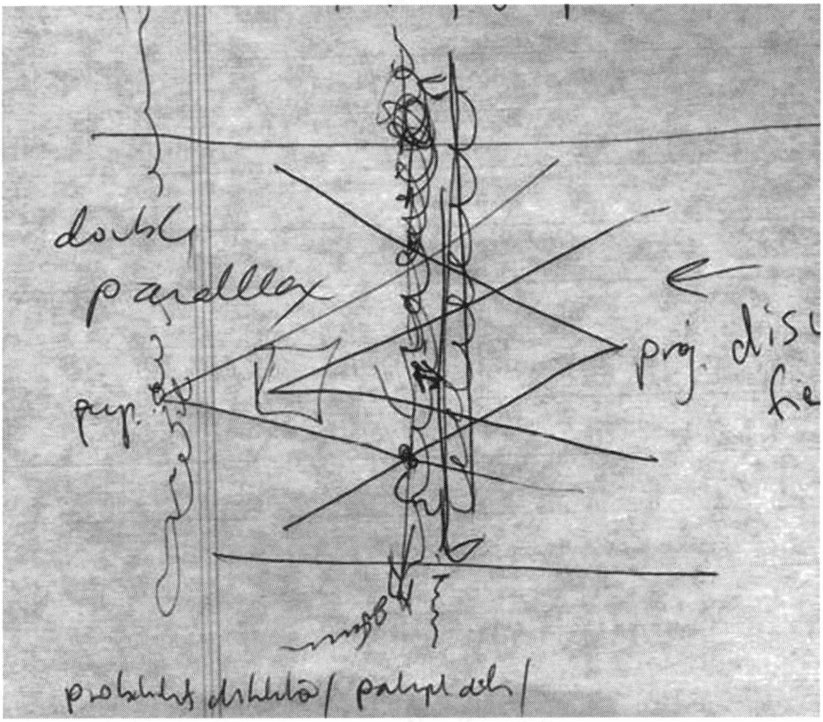

Figure 2.5.1. This sketch illustrates the conceptual and theoretical model of the 'Patacritical Demon in schematic form. The phrase "double parallax" describes the relation of a viewer ("percp," at left) and the discourse field (right). Projections of the discourse field and of the viewer's vision/perception are shown as cones. Where the two projections cross is an area of intervention, the constituted "text," which is neither self-identical nor equivalent to either the material or the virtual text. In the center of the diagram is a sketch of a classic structuralist model of signification. The sketch shows two lines, one a chain of signifiers/signifieds (the plane of discourse) and the other a plane of reference. The "text" is a projection in the intervening spaces, provoked by the two planes and a situated reader. That the text is itself part of the larger field of discourse, of whose social and historical production it is an instance, is indicated by the small square on the left, at the center of the inside cone of perception, which represents the *illusion* of a text as a fixed entity; the viewer's gaze is met by the projection of the discourse field and is always opening outward through the association of texts to each other, even as the reading opens outward.

inherent stability, even if the material object that provokes the reading can be held, looked at, described, and annotated (figure 2.5.2).

In 2003 I was invited to create a drawing for an exhibition of "books that had never existed," and at that moment our group's shared enthusiasm for Lucretius as a proto-'pataphysician was very much in mind. So I

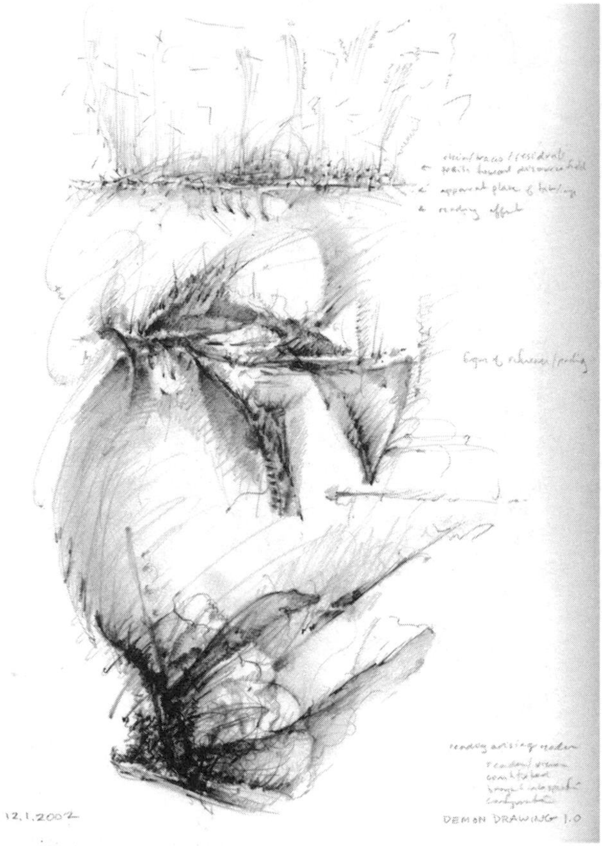

Figure 2.5.2. Demon drawing. The structuralist model of signification (see figure 2.5.1) recurs in the upper part of this drawing, oriented horizontally instead of vertically. The fissure or break along the center of activity is the space between signifying elements, and the links, loops, and energy that project out from this realm show the asynchronous connections made in the process referred to as "reading arising" (where "reading" indicates a "constituted text" or object). The "apparent plane" of the text image (central area) and the complexity of its design show that the signifying process is not a mechanical stringing together of signifying elements but a charged field of possible relations. In the area above that central charged line, trails and traces of association move toward the discourse field. The whole scheme shows the "reading effect" in diagrammatic form. In the lower part of this drawing the "demon" is the "figure" of reference, of the virtual text as an imaged configuration, dynamic, changing, morphing constantly but with rich, specific, particularity.

created a version of the Demon that combined the image of the double parallax with the image of a projection of a "text" and "reader" produced through their mutual intervention of a reading subject in a probabilistic field. I accompanied the image with this caption: "Trialectics—fragments of Lucretius—57 B.C. a discourse intervened within a dynamic field of potentialities—a treatise on the third term, work constituted as a relation of subject, object, interpretation-n-dimensional arising—shift from metalogic to meta-rhetoric in a discourse of non–self-identicality—entangled condition of the word—algorithmic unfolding of production within constraints, speculative methods and quantum poetics—autopoiesis and codependent arising—deformance as production, constitutive method—" (figure 2.5.3).

The text breaks off, leaving everything else unsaid and unspecified. The visual image supplies the real information. The largest section of the drawing is taken up with showing what *a reading looks like*. The sketch borrows from René Thom's topological models of catastrophe, events drawn forth and projected simultaneously in a dynamic hologram. A work figured as interpretation comes into being in a third space between reader and text. Subject, object, interpretation—this tripartite structure also depends on Peirce's formulation of the sign. In this image, a book, familiar and iconographic, is shot through with dynamic vectors. The action of reading is called forth by the text, as a provocation, but the text is produced within the encoded activity of reading. The idiosyncratic trajectory of a specific encounter (different in each case) is figured by the wandering lines. Networked into a series of interacting force fields they create the basin of activity that constitutes the "book" as a perceived and experienced interpretive event. A line of "reading" streaks across the bottom of the image space, like a monitor registering a heartbeat, brain activity, or some other vital function, rising and falling in pitch as the baseline text provokes response.

In the upper right corner two diagrams demonstrate the distance between a classic structuralist reading and a 'patacritical one. The structuralist linguistic model is binaristic, containing a plane of discourse and plane of reference. The plane of discourse is constituted by a series of bipartite signs, signifiers and signified in an interlocking chain, which is itself interwoven in the plane of reference to which it gives rise.

In the 'patacritical version the model is complicated by the reader's projective acts. Again, this critiques the structuralist model, not allowing any element to be taken as stable or self-identical, and the text arises as a result of a combined provocation and projection. The space between

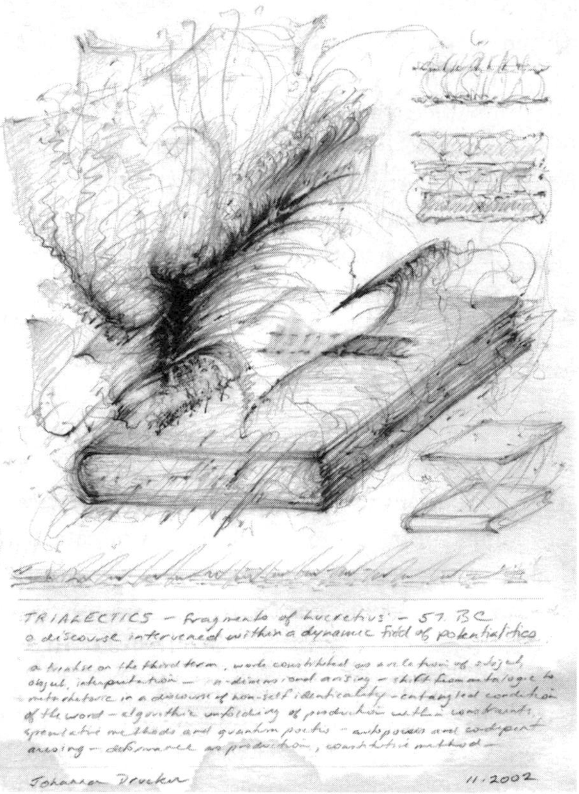

Figure 2.5.3. *Trialectics* is my title for an imagined book by Lucretius about a dynamic system based on three, rather than two, terms. Where a dialectic (thesis/antithesis/synthesis) acts in an abstract mode, the trialectic incorporates an intervening agent. The image was drawn in response to a call for works for an exhibition of "lost" books—books that had never existed. This provided an excuse to project the figure of the 'Patacritical Demon onto the image of a book whose purported content was this same theoretical construct. The book is shown as a material, literal artifact, but one that is crisscrossed with associative trails, making the plane of discourse into a plane of reference (on the lower right a virtual book floats above the literal book in a plane of projection). The figure of reading, the interpreted or constituted work, rises like the figure of the demon in figure 2.5.2 from the book form, taking over the space. Within its central dynamic field, one sees vague references to the sequence of pages, a flipping action, constant referencing of language, and lines of text across the literal space of the book. The result is the *espace* of configured interpretation, the unique and particular reading intervention into the book/artifact and text/field. The diagrams in the upper right show two stages of analysis, as described in figure 2.5.1. The signifier/signified chain and its asynchronous relations are shown at the top, then the relation between the plane of discourse (signs) and the plane of reference. "Deformance as production, constitutive method" reads the final phrase, suggestive of the dynamic activity of provocation that the demon figures.

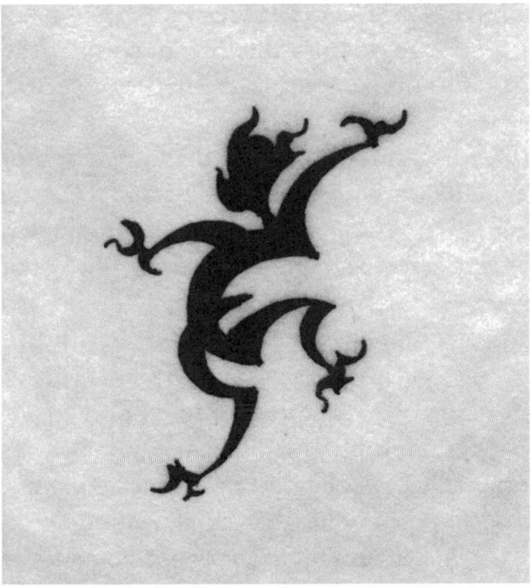

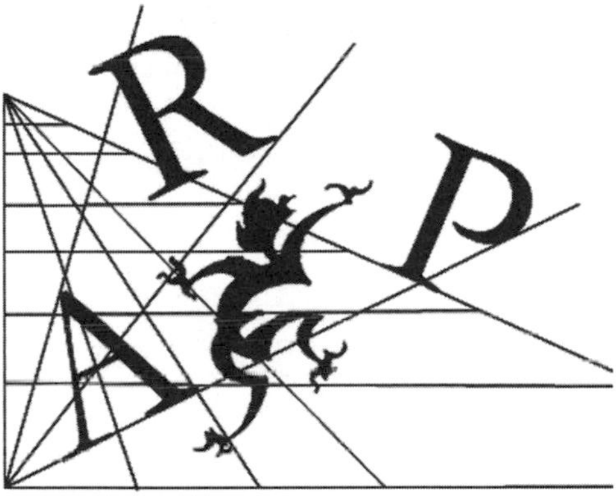

Figure 2.5.4. The 'Patacritical Demon. Original sketch of the logo for ARP, Applied Research in Patacriticism.

reader and text that is the "work" is constituted through a dynamic process. Quite simply, the book space depicted in the larger image is to be understood as a series of instructions for action and reading, rather than as a thing or object to be apprehended. The small drawing in the lower left simply diagrams the distinction between the literal book and the phenomenal book, showing the lines of force, vectors of interpretive activity, that keep the two in productive tension. The image marks the shift from a dialectical mode, in which thesis and antithesis generate a new synthesis, toward a trialectic. This trialectic, a space of enunciation rather than a system, embodies the spirit of the 'Patacritical Demon, destabilizing the order of logical things through the introduction of the third term—so that any thesis or its outcome is located in the subjective experience of an interpretant.

The Demon figured in this diagram is an initial demonstration of the event of reading as an experiential and subjective dynamic. On the one hand, it is a kind of game, a dare to the interpretative community. On the other hand, it is a sincere attempt to get hold of the process of interpretation.

Next-phase iterations involve a graphic scheme for showing the refraction of an object through the dimensions of interpretation. The resulting renderings are meant to dissolve the apparent unity of the text, revealing the aspects, sources, and multiplicities of frames and lineages latent in any artifact. These are, in effect, the other possible configurations that might be exposed within the potential field. The infinitely extensible, flexible space of electronic environments, traversed by lines of access or thought, would be charged for and with the tasks of display and authoring. In this way the Demon would demonstrate the modeling of textual interpretation as an event, an act of intervention in a discourse field of poetic potentialities.

3.0

From Aesthetics to Aesthesis

As the branch of philosophy concerned with perception, aesthetics provides a useful foundation for thinking about knowledge as partial and subjective. From an aesthetic perspective, knowledge is neither total nor systemic. Nor is it concerned with the apprehension of self-evident entities, structures, or ideas that constitute themselves as autonomous or transcendent. Rather, aesthetic knowledge is constituted in cognitive processes that are situated within interrelated and codependent systems. Our projects at SpecLab were an initial attempt to apply such models of knowledge—building on decades of theoretical work in the field of cognitive studies—to the design of instruments of knowledge representation and production. To mark the specific cast I am putting on the concept of aesthetics as a form of knowledge, I use the term *aesthesis* as a rubric under which to gather these thoughts.

Because I came to digital humanities and speculative computing from the visual arts and contemporary art history, aesthetics was central to my approach to electronic media. This led to the conviction that the purpose of work in new media is to continue the longer project of fine art:

to provide embodied expressions of experience and knowledge. The quality that distinguishes works of electronic *art* from other digital projects is frequently their nonconformity with formal principles in matters of content, expression, function, or behavior. Aesthesis focuses on the generative perception and cognitive production of information and its material expressions in any medium. Aesthesis is distinct from the analysis of representation, but is dependent on recognition of the cultural and historical characteristics of visual forms, their materiality, and the rhetorical assumptions built into formal expressions of knowledge. In the context of digital activity, the examination of aesthetic properties includes discussion of code and its specific materiality, modes of production that are integral to digital media (interactivity, intersubjectivity, iterative and algorithmic principles for production), models and modeling processes, and the specific ideology of virtual artifacts.

The discussion of the way graphical forms of knowledge work in digital environments takes on added urgency because the connection between "information" and "visualization" is so readily enabled by digital instruments. As discussed in chapter 2.3, the engineering sensibility that under girds most information visualization seems to take scant interest in the rhetorical and ideological force of its operation. Exposing the ideology of graphical forms is crucial to our contemporary condition, extending traditional critical discussions of these issues from the study of visual art and language. Critical work in textual and art historical studies has long connected cultural and social issues and the materiality of aesthetic works, but work on visual epistemology, or knowledge expressed in graphical arrangements (diagrams, charts, and other structuring forms), has been less developed.[1] Attention to materiality has too often assumed a literal, mechanistic reading of formal features rather than an analysis of material codes as provocations for cognitive processing. Moreover, information design has commonly been regarded as an engineering field, not an artistic one, and thus has not been subjected to the ideologically oriented analysis essential to its critical review.

The path to this insight began with an argument about visual forms of knowledge and what specific characteristics distinguish them from language and formal systems rooted in semiology or logic.[2] Therefore, the papers grouped here begin with an argument for graphesis in which I suggest that the visual creation of information provides a counterpoint to mathesis, the assumption that all human thought might be able to be properly represented in a formal language. That argument throughout the pages that follow is a continual attempt to open up space for subjectivity, individual expression, and specificity as challenges to the cultural authority of alignment, totalization, and systematic approaches to knowledge. Without romanticizing deviation or idiosyncracy, this argument suggests that acknowledging the difference between subjective knowledge (situated and partial) and objective knowledge (transcendent and totalizing) carries a political as well as aesthetic charge.

My discussion builds on long-established arguments about the value of cultivating the ability to discriminate through informed or attentive differentiation. Recognizing distinctions depends upon nonalignment, a critical resistance to totalized absorption. Those familiar with the history of the field will recognize an allegiance to Alexander Baumgarten's views on the refinement of taste as a basis of knowledge, to Immanuel Kant's notion of judgment as the link between reason and sensation, and to the history of materiality and the specific properties of media as articulated by Conrad Fiedler, the modern aestheticists Clive Bell and Roger Fry, and their followers within high modernist criticism, as well as to Max Horkheimer and Theodor Adorno's notions of critical theory (rather than instrumental reason).[3] In struggling to articulate the specificity of digital media, the field of aesthetics has the virtue of calling our attention to the very forms that, as Aristotle put it, allow sense to appear to sentience. But unlike our classical predecessors, we have no faith in essence outside of difference. Identity must be constituted, not assumed. The critical tenets of Jacques Derrida's deconstruction and Michel Foucault's archaeology, their borrowings from Friedrich Nietzsche and Martin Heidegger, and the work of philosophers Gilles Deleuze and Jean Baudrillard, provided lessons on which cognitive studies and systems theory could readily be grafted.[4]

The ideas worked out in these papers record the theoretical work I drew on for SpecLab projects and discussions. I began writing pieces about digital aesthetics in the 1980s. One of my very first critical essays focused on electronic media and the status of writing as a cultural instrument, asking what would happen when handwriting, forgery, and material codes (bases of identity, authenticity, and history) shifted their foundation from the stuff of paper, stone, and ink to fungible notation.[5] That paper was rejected by a serious academic journal of aesthetics with a phrase that now strikes me as very droll.[6] They said it was "too speculative." Lenore Malen asked me to be on a panel addressing "the work of art in the age of electronic technology" at the School of Visual Arts in 1994, along with Bob Stein of Voyager, Graham MacIntosh, Charles Bernstein and other people who were actively using electronic media in the arts. Also at Malen's urging, I edited a special issue of the *Art Journal* focused on digital media and the arts. Encounters fostered by a series of Digital Arts Conferences, connections to the Electronic Poetry Organization, and involvement with various exhibitions and works broadened my reference base in this area.

Though inspired by works of individual artists, my interest soon shifted to an investigation of the design conceptions that model digital art and humanities. I studied graphesis, code storage, the generative materiality of digital media, the ontology of digital artifacts, and the specific characteristics of new media. From these investigations emerged a theory of aesthetic knowledge that attends to the conceptual and formal role of design at the level of model and implementation for use. My emphasis on design deliberately shifts from attention to form (or the discussion of content, theme, and form that are the usual stuff of art and literary criticism) to the analysis of models of knowledge production and representation situations. By introducing the term *situation,* as opposed to *system,* I'm emphasizing the codependent relation of user and network in conditions of use. These shifts of emphasis may seem subtle or radical, depending on the degree of familiarity, but all are intended to reframe my approach within cognitive, generative, iterative, and probabilistic models of knowledge and away from rational, logical, mechanistic ones.

Critical hyperbole abounded in the 1990s. The "newness" of digital media seemed to dazzle and confound many writers and curators. But the challenge of skeptical response brought continuities as well as novelty to the fore. I became increasingly interested in the ontology of digital images, and in seeing their material and fungible condition of inscription in critical terms. I still find some of the earliest experiments in digital arts the most compelling—works by Melvin Prueitt, Roy Ascott, and even the Bell Labs engineers (e.g., Kenneth Knowlton)—because they have the figuring-it-out attention to process that is lost in later pieces built using state-of-the-art, off-the-shelf software.[7] The basic properties of digital art are clear: it is iterative, algorithmic, and networked and, in many cases, procedural and generative as well. But many of its surface effects turn out not to be so very different from their print precedents. One of the strongest impacts of digital media has been to provide ways to think about traditional work in new ways—to see print artifacts, for example, as interactive and intersubjective instruments rather than inert forms. The differences between traditional and new media are most strongly registered in the rate at which they change, their relative stability with respect to inscription, and thus, their ability to incorporate intersubjective exchanges and constitutive activities within their material instantiation. Medieval codices, palimpsests of such intersubjective exchange, are, after all, as dramatically hypertextual as wikis, but their access protocols, the rates at which change was registered, and the character of their material instantiation differentiate them from their digital counterparts.[8]

Discussions of the aesthetics of digital environments have emphasized their formal properties from the point of view of user experience. Navigational, combinatoric features, the collage and pastiche sensibility, and the capacity to present multiple worlds and seemingly inexhaustible sequential permutations have all come in for their share of praise and enthusiasm. So have embodied and affective notions of experience. Many, like those of Lev Manovich, have been conceived in terms of the mechanics of older media, particularly film. Likewise, the structure of code and claims for its value as the essence of digital languages has found its champions.[9] Simplistic descriptions of the materiality of digital media, well intentioned but literal and often undeveloped, have also contributed to the field. But the distinction between the operational effects and the models of digital instruments has rarely been sustained to the end of addressing the latter as a design problem that is at once aesthetic and epistemological.[10] Recalling some of our experience in designing Ivanhoe, I suggest that this has to do with the cultural clash of engineering, interface, and codework in environments that often conceal their mechanistic assumptions under the effects of their representational surfaces.

I now see that understanding the design of digital projects as rhetorical instruments is essential to critical analysis. If imaginative play is to enter into the production and representation of knowledge with any cultural authority, the way we model these projects has to include a higher order of understanding about the ways rational logic legitimates itself through instrumentalization. As an intellectual project, my argument for aesthesis is to promote and legitimate a basis for thinking differently about the basis of this cultural authority. The shift from logical-total systems to subjective-partial situations is the crux of this approach. My discussion takes place within a digital frame, but the implications are relevant to the legacy of Western logical thought wherever it aspires to totalized control. Digital media instrumentalize that logic in a perversely successful way, but they are neither the source nor the simple technological effect of formalized approaches.

In short, my questions about the nature of knowledge and digital artifacts began with an interest in certain properties of visuality, familiar to me from years of drawing and graphic work, that seemed unassimilable into either traditional linguistic and mathematical knowledge systems or digital systems based on discrete, unambiguous entities. Given the role of visual form in the dissemination of knowledge, as well as its capacity for the production of new knowledge, this exclusion seemed peculiar and troubling. One of the mantras of SpecLab has been that nothing is self-identical—no text, no image, no object. All aesthetic objects are fields of potential. It is only through interventions and aesthetic provocations that a work is constituted as an act. Inherent to visual mark making, expression, are the qualities of infinite variety and great specificity, properties that allow graphical marks to register subjective inflection yet resist the premises of finitude and closure that are central to mathematical and linguistic notation.[11] Visual codes are notoriously unstable, and attempts to describe them in logical terms, semioticians have long recognized, yield at best a semiology of the visual—not a formal system but a rhetorical one that by analogy has some of the properties of a language.[12] Some (even among my colleagues) would argue that text acts similarly, and indeed, in the study of type and typography the finitude of the letters unravels into a plentitude, for the very reason that letters are visual elements: drawn, printed, inked, brushed, scribbled, or inscribed. To what extent is the insistent nonidenticality of both sign and system in graphical inscription a feature of all forms of expression? And what insights into the circumstances of knowledge production and interpretation might be produced by extending this insight to other media and modes?

I remain convinced that aesthetic activity has a crucial role to play in resisting the cultural authority of mathesis, because of its capacity for registering subjectivity (as position and inflection, as partial and situated knowledge) within the highly specific and infinitely variable circumstances of material expression. The development of this argument from the studies of graphesis to the larger questions of design will become clear in the papers that follow.

3.1 Graphesis and Code

The original title of this paper, "Digital Ontologies: The Ideality of Form in/and Code Storage—or—Can Graphesis Challenge Mathesis?," compressed considerable theoretical bulk into its boxcar phraseology.[1] Coming to terms with the basic idea of mathesis was an important phase in the development of my critical thinking about how the cultural authority of digital media is premised and how it might be challenged. So revisiting these matters is useful, even if the late-1990s debates about truth in photographic imagery that arose from digital works have subsided.[2]

The attempt to understand the connections that link human thought to its representation (in language, image or signs) has been central to Western philosophy of knowledge. In every generation, some version of this question has been posed: If it were possible to understand the logic of human thought, would there be a perfect representation of it in some unambiguous, diagrammatic symbol set? This question, informed by classical metaphysics and philosophy, persists not only in contemporary struggles within the very different domains of visual art, information design, and computer graphics, but also in early formulations of cognitive science, with its proximity to symbolic logic, and in debates over artificial intelligence.[3]

Because of the emphasis on a distinction between idea and matter or form and expression that pervades Western metaphysics, the question arises whether an idea can exist outside of material form and yet appear to human perception.[4] Many forms and ideas are grasped by the human mind and communicable to a community of persons even though they exist without material instantiation—abstract concepts of law, love, justice, or spirit, for instance, or more concrete-seeming notions within the language of geometry, art, or social behavior ("good form"). But does this question take on a new cast when posed with respect to the digital environment? Should our conception of an image be changed by its capacity to be stored as digital code?[5] Or does code storage, as the defining condition of digital processing, finally satisfy the Western philosophical quest for mathesis, eliminating once and for all any ambiguity between knowledge and its representation? The various misperceptions of digital media as lacking materialiality gain some of their credibility through connection to a tradition that idealizes the immaterial, even placing it in a theological frame, above embodied knowledge.[6] The argument that code is *material,* however, seems incontrovertible.[7] Digital code may be relatively unstable with regard to the bond between inscription and configured form (by contrast to a letter carved in stone, for instance), but the pattern of stored values on a silicon chip is ineluctably physical.

The argument can be made that computational media are overwhelmingly material—requiring rather large amounts of hardware to perform what was formerly done in rather minimal means (paper and pencil). But the perverse magnetism that draws concepts of immateriality toward the lodestone of code is provided by the curious belief (even desire to imagine) that perhaps, just perhaps, the configured form of code and the formal logic of configured thought might be analogous. At the very least, they might be made to conform to similar *rules* of logic, to be governed by, if not precisely part of, the same order of things.

A framework for this discussion comes from two disparate positions within twentieth-century philosophy: Edmund Husserl's notion of the "ideality of form" and Theodor Adorno's problematizing of the notion of self-identity of form because of the social-political implications that derive from alignment within totalizing systems.[8]

These two positions are useful as a means to address the formalist assumptions underlying the authority of digital media as construed in the popular imagination. The premise on which this authority is sustained is a mythic one, as I hope to demonstrate. By moving between Husserl's embrace of ideality and Adorno's critique of self-identity the link between the idea of "data" and the materiality of its existence in digital form can be interrogated critically. This link is often overlooked in the rhetoric of cybermedia, and data is commonly presumed to be value-neutral, pure or raw, and immaterial. This allows data-as-code to be misconceived as exemplifying self-identicality—the relation of information to itself. If code and data configure each other in a perfect, isomorphic relation, and if that relation is abstracted into "ideality" instead of rooted in "materiality," the argument goes, then data and code are one and the same. Adorno would be quick to warn us that such yearnings for ideality preclude the critical reception of material expressions within cultural frameworks, where they operate in more pedestrian guise, rather like gods in mortal form in Greek mythology.[9]

My concept of "ideality" is derived from Husserl's discussion of the origin of geometry. The original geometer, he suggests, was able to apprehend *form* intellectually, outside of material expression.[10] Mathematical forms, he goes on to say, become apparent to human sentience—but are not dependent upon it (by contrast, the "form" of the story of Emma Bovary is dependent on human authorship even if it can live as a construct outside of the text). Husserl even suggests that the peculiar specificity of geometric forms is that, although they become conventionalized within representational systems, the original condition of their existence is independent of human constructs. Because mathematical forms have a claim to objective, universal status, even if their authority varies in cultural circumstances, Husserl's decision to focus on geometry makes his discussion appropriate to current mythologies in which the cultural authority of mathesis is supported.

If, following Husserl, geometric forms exist independent of human perception and are not changed by that perception from their ideal form, then does that ideality necessarily fall into the category of "self-identity" or "unity" of form? The idea of self-identity is anathema to Adorno, who argues that when empirical or positivist logic invades culture to such an extent that representation appears to present a unitary truth, there can be little or no room for the critical agency essential to any political action.

These two positions provide the poles of reference on which I examine the premises by which mathesis functions in current conceptions of digital data. I suggest that there is an underlying, at times overt, ideological bias in the way the myth of digital code is conceived in the public imagination. Because mathematical forms of knowledge are presumed to lie outside of ideology, this conception validates digital representation in a way that forecloses interrogation. My double agenda is to disclose the ideological assumptions in the way the ontological identity of the digital image is posed and to suggest that graphesis (information embodied in material, and thus ambiguous, formats) can challenge mathesis. In other words, the instantiation of form in material can be usefully opposed to the concept of image/form and code storage as a unitary truth or, to use Husserl's term, "ideality." My argument bears on digital media in its basic operation and use, not merely in what it represents. I suggest that the possibility of critical cultural agency is linked to the assertion that the real materiality of code should replace the imagined ideality of code.

Digital photography presents a useful starting point. Many questions about the truth, fiction, or simulacral identity of digital imagery were prompted in the 1980s and early 1990s by its presumed distinction from traditional darkroom photography. Images by photographer and early adopter Peter Campus, for instance, provoked critical discussion around matters of ethics and illusion. Such work and its reception offers a useful comparison with the fictions produced by those early-twentieth-century adolescents, Frances Griffiths and her cousin Elise Wright.[11] The pair created paper cutouts of fairies, expertly photographed by them in a garden setting, that appeared sufficiently real to elicit great debates. *Alice and the Fairies* (1917) shows one of the girls in a garden setting, a "fairy" close at hand. In this image, deceit seems inconceivable, as much due to cultural expectations about the innocence of adolescent girls as to the plausibility of fairies' existence in English gardens. That it was a hoax is now readily obvious. That anyone believed in the image based on its use of photographic codes seems less credible. By contrast, Peter Campus's digitally manipulated *Wild Leaves* (1995) was more simulacral than fictional (its impact comes from the way a surface can create a reality effect, rather than from narrative credibility), but a mere half step separates the photographic antics of Griffiths and Wright from those of Campus.[12] Any number of critics have pointed out that there is much more continuity than discontinuity in the shift from darkroom to digital.[13] The notion of photographic truth based on a pure, unmediated representation of a "real" referent was illusory even before Griffiths and Wright's confabulations; multiple exposures, multiple negatives, and blatant reworkings of both plate and print were all tools of the photographer's trade almost from its origin in the early nineteenth century.

Critic Hubertus Amelunxen contrasts two types of mimesis, both defined by Plato: eikon/likeness and semblance/simulacrum.[14] The difference between these terms supports distinctions between features of the photographic imitation of light and the presentation of an image of life as truth. Likeness privileges the indexical traces of actual light and the codes of verisimilitude that dominate our ideas of what truth "looks like." But in a world of digital special effects, the ability to produce virtual and hallucinatory reality is continually evolving. Market forces and competition, as well as habits of viewing, all favor novelty and invention. The skills through which the entertainment industry successfully deceives (some) of the senses raise philosophically charged questions.[15] But my argument is focused on the simpler, more fundamental question of assumptions about the truth value assigned to digital images as code.

Unlike traditional photographic "truth," the truth of the digital image is not, I would argue, posed as an index to the instant of exposure or as encoding the experience of "natural" visual perception. The digital image, photographic or not, is removed from those mechanics of production in which the metaphysics of light is linked via a moment of revelation to reality. Nonetheless, the digital image is (popularly and fundamentally) conceived as another kind of truth, premised on a deep conviction about a rational link between mathematics and form that is supposed to be irrefutably present in digital code. This premise is the foundation of a digital ontology. It promotes the idea that mathematical code is self-identical, irrespective of its material embodiment. This is a potent myth.

For the sake of argument, I want to approach the representation of thought as form along another trajectory, in which truth and form are put into a relation of identity. In the first decade of the twentieth century, the psychic Annie Besant produced a series of drawings of "thought forms."[16] Though her work, conceived within a late-nineteenth-century sensibility that embraced telepathy, magnetism, and the role of the medium, has a distinct naïveté, it also has a striking purity because of her conviction that thought *is* form and thus be manifested directly in visual images. Unlike Husserl's first geometer, however, Besant suggests that the representation of thought must be situated within a human context to be intelligible. She classed her images through a typology of universals: radiating affection, animal, grasping affection, watchful anger, jealous anger. These categories are typical of her time, a legacy of a theory of types and forms combined with a vocabulary of late-nineteenth-century psychology.[17]

By virtue of their schematic abstraction, Besant's visual forms have a formal resonance with a number of early computer-generated graphics, such as those produced by Jack P. Citron in the 1970s.[18] In their minimal, skeletal appearance, Citron's graphics have a pristine innocence. The mathematics and logic of thought that created both algorithms and their manifestations were conceived of as thought beyond the philosophical frame of human subjectivity. *Geometric Digital Graphic from a Curve,* for instance, might be said to stand in relation to the algorithm that preceded it as the Copy does to Idea (eidolon) in a Platonic scheme. The image might even be consigned to the more debased category of Phantasm, a copy of a copy. But such a hierarchy presumes that Idea (and, by extension, algorithm) has a stable, fixed existence. Is Besant's original "thought," which her "form" presumably expresses, also such an algorithm? Do these artists create forms whose graphic identity, because it presumes to manifest an ideal form, shares a common belief about ideality?

As a digitally produced and manipulated entity, Citron's algorithm is also stored in material—in silicon—through a sequence of instructions and address codes. But here is the crux of the matter: like the ideality of Husserl's geometric forms, these algorithms seem to be capable of appearing to sentience, of being apprehended, outside of a material form—as thought.

Curiously, Citron's work is thematically engaged with these questions as well. He made several works that use algorithms to express and then distort a form. The images trace a process of deformation from the mathematical ideal of a geometric form through its distortion—by manipulation of its stored formula or code. This was a common theme in works by "digital artists" in the early 1970s, almost as if the problems of form as mathematical ideal and form as instantiation were paradigmatic issues for computer graphics. George Nees's *Random Number Generator Causes Swaying* maps the distortion in a regular pattern caused by introducing a random element, and the Japanese CTG group's 1971 *Return to Square* is almost a poster image for the comfortable fit between the ideality of the square as order and the process of debasement by which it is transformed into a (material) image.[19] If we imagine that the algorithmic representation of the geometry is the pure code, the ideality, then the material graphic representation will always be cast as the degradation, affirming the Platonic hierarchy of Idea, Copy, and Phantasm.

This opposition of algorithm and graphic manifestation, or of geometric idea and encoded algorithmic equivalent, entails a fundamental flaw. And this flaw, bound to the myth of the "immateriality" of digital artifacts, informs all celebration of "codework" as autonomous and transcendent.[20] The manifestation into substance, the instantiation of form into matter, is what allows some thing, any thing, to be available to sentience. Ideas are apprehended through expressions (the illusory transparency of language as a means of expression often renders this invisible in common perception). This is true for the ideal form of a square, as well as for the analytic visualizations made by scientists using computer-generated images. An image of a complex molecule, for instance, purportedly showing detail at the atomic level, may in fact be a visualization expressing a mathematical model. The presumed ideality of the molecular structure, here made apparent as an image, serves as a convenient fiction through which we can gain access to the mathematical "truth" of the image, or even of the model it expresses.

But a digital image of something that is fully simulacral, such as the hyperreal renderings common in early music videos (as an example, the monster from Peter Gabriel's video *Mindblender*), refutes any easy link between an ideal algorithm and visualized reality as a fundamental unity. The existence of the image depends on the display, the coming into matter in the form of pixels on a screen. If, in one instance, the graphic display is manipulated by an algorithm, then, in other instances, the display becomes the site for manipulation of the algorithm. After all, the image on the screen is not even identical to itself. Not only are no two pixels alike, but the material expression of any algorithm varies from screen to screen, from moment to moment, from viewer to viewer. Embodied materiality is always distinct from the code it expresses. Conditions of use and perception enter into the production of an image in a very real sense, since forms are neither immaterial nor transcendent.

This brings me to the heart of my argument. What are we to imagine constitutes the "information" invoked or suggested in any of these various expressions? The algorithm? An ideal form (geometric or not)? An imagined molecule modeled mathematically? A simulacral monster whose algorithmic reality, its code-based model or identity, follows from the manipulation of data as visualized on the screen? In the visual practice of information design, in which graphic artists create schematic versions of the history of philosophy using as motifs an imagined solar system, or map thermal conductivity with fine, schematic precision, the assumption is that the information precedes the representation, that the information is other than the image and can be revealed by it. But we see from these examples that form is constitutive of information, not its transparent presentation. And no constituted expression exists independent of the circumstances of its production and reception.

Perhaps the most compelling, chilling image that I have come across in thinking about these issues is a computer-generated graphic by a very early experimenter in this field, artist-scientist Melvin Prueitt.[21] It is a nocturnal image of a field of snow, unbroken and undisturbed. To my mind this is a terrifying image of the ideal of digital purity, the pristine visual manifestation of code. Nothing human or circumstantial disturbs its form. But it certainly is not pure, any more than any other image output by a plotting pen, laser jet, or Giclée printer. Any act of production and inscription, the scribing of lines that create the specificity of an image, demonstrates that an expressed form is different from the underlying code. Whatever the "ideality" of code may be, even if it were available to sentience in some unmediated way, the encounter of expression and matter produces thought as form. Any interpretive act returns to this initial inscription through its own productive and generative process, reinscribing a work as product within a specific situation of viewing.

In a very real sense, code lurks behind Prueitt's image of snow. In saying "behind" I mean deliberately to invoke an ontological and chronological anteriority. But this code can't be conceived as "pure" in the sense of being independent from a material substrate or instantiation into material. Code is itself always embodied, instantiated in material.

The digital encoding of form as information, as data, as patterns of binary code might be used to assert that our understanding of what a "form" is should shift toward the realm of mathesis. That tradition of logic, envisioned by Leibniz, still drives a quest for cognitive, epistemological, and technical certainty that seeks to reduce all formal, even material, expressions to a "higher" logical order of existence. But the ideality that Husserl envisioned for mathematical forms is generalized and reductive, a mere category and placeholder within human expression (even if assumed to exist in some ontological sense outside cognition). His geometries are not replete and specific forms capable of showing that the world is understood through experience and perception. Thus, we can define graphesis as knowledge manifest in visual and graphic form, and insist that it is based on understanding of form as replete, instantiated, embodied, discrete, and particular.

In Karl Fredrich Schinkel's eighteenth-century, neoclassical rendering *The Invention of Drawing,* the act of form-giving is depicted within the tensions between the lived and the ideal. Schinkel's image inverts Pliny's tale of Dibutades, in which the daughter of the potter traces the outline of her departed lover, changing the genders, so that female beauty is objectified as an ideal within a male gaze. This painting suggests that aesthetic form-giving is always an inadequate copy, a lesser truth than the real. By contrast, in a late-1990s advertisement for Johnny Walker Red Scotch, a young man sits in khakis and topsiders on a deck, beachside, with his laptop computer open in front of him. On its screen is a wireframe graphic image of a dolphin, and beyond the man, leaping up and out of the Johnny Walker Red sea, we see the beast itself. The image of the dolphin on the screen does not match the image of the dolphin leaping from the ocean. Their direction, temporal moment, and other details are out of synch. But which is bringing the other into being? The visual image confuses the hierarchies of original and copy. The computer graphic seems to generate reality or, at the very least, to function on an equal, autonomous level as a form-producing environment. In *The Vision Machine,* Paul Virilio raises the specter of a sightless visuality, one in which images exist only as signals in the electronic currents of a closed system, readable by machines but neither visible nor legible to humans.[22] In such a situation, "form" is nothing other than code, still material but accessible only to some other sentience than the human. The case demonstrates even more fundamentally the link between the materiality of code storage and formal expression, since the networks cannot grasp ideality, only pulse and flow within their circuitry.

What is at stake in asserting the authority of graphesis—the material expression of form as the condition of its existence—is not the viability of code that has no graphic manifestation, but the fact that it is stored materially. Code is not an immaterial ideal. This in itself calls the mythic status of the digital as the realization of mathesis into question.

Such realizations have implications for the transformation of form from traditional media and representational systems into digital formats. They suggest that decisions about what aspects of material forms to encode, and how, have to engage with broader conceptions of information. When "form" is conceived in mathematical terms, it can be absorbed into an absolute unity of essence and representation. But when it is conceived in terms of graphesis, it resists this unity, in part through the specificity imparted by material embodiment.[23] Materiality cannot be fully absorbed into ideality, nor can it be understood as a mechanical, self-evident literal identity. Something is always lost when, for instance, a text is translated into ASCII format. Digital media have their own materiality (and material history to be sure), but in the distinction between mathesis and graphesis the resistance to the totalizing drive of the digital can be articulated. This is the beginning of the place from which an argument about the ideology of code can be created, but also the place from which a literal approach to materiality can be critiqued.

I return, for a final moment, to Melvin Prueitt's digital snowfield, in which, as Amelunxen says of such work, the gap between the algorithmic-numerical image and its origin is so slight that it seems to cast "no shadow." But the gap does exist. The distinction can be made just as surely as in any conceptual work. There is always a space between expressed idea and expression of an idea. The ideas that drive conceptual projects are not immaterial—they are usually expressed as language, as coded procedures capable of generating any number of material instantiations. But even as procedural statements, they are already both code and matter. Unless we revert to the mystical concept of ether, the base materiality of all human expressions will need to be accounted for in any analysis of objects and artifacts, forms and ideas, that are part of human experience.

Thus the crisis introduced into aesthetic discussions by digital media is not, as commonly reported, a crisis of the copy, of originality, or of authenticity or truth. What is at stake is more poignant, since it depends upon the possibility of reinscribing form into matter as part of a human, cultural, and social system. If code is ideal form, it resists inflection, cannot register subjectivity in its production or interpretation. But the specific, particular character of materiality always registers the circumstances of production, expression, interpretation.

This argument against the immateriality of code fosters critical consideration of the ways it actually participates in and helps replicate cultural mythologies. It dispels the idea of code as either self-identical or transcendent, or as constituting a truth. The easy interchange of image into code and back into image becomes loaded with a myth of technosuperiority, as if the independence of code from matter were so fundamental it could never questioned. In a system premised on mathesis, code is presumed self-identical, unavailable to critical interrogation, and everything else is reduced to data and equivalents. When this claim is extended to the cultural realm of representation, its hubris needs to be challenged. Graphesis is always premised on the distinction between the form of information and information as form-in-material. It insists on recognition of the specificity and particularity that resists self-identicality.

Most important, this argument cannot be reduced to a distinction between digital and analog. Whether an artifact exists as print, code, digital file, or physical image, its material expressions are always undergoing changes, aging, crumbling, acquiring or resisting wear. All forms of expression are ontologically incapable of self-identicality. Graphesis is premised on the irreducibility of material to code as a system of exchange and equivalents without acknowledgment of its specific instantiation. The materiality of graphesis constitutes a system in which there is loss and gain in any transformation that occurs as a part of the processing of information. In that process, space to register subjective inflection creates a place within which Adorno's critical reason can operate and in which humanity, such as it is, can be expressed. Digital media are no different than traditional media in this regard, but the claims and mythologies they sustain have allowed aesthetic work to be used to justify a cultural authority in which logic and its formalisms trump other, experiential, forms of knowledge. Or try to. Digital media are not Prueitt's dead zone of insubstantial rendering, in which neither experience nor perception, human subjectivity nor or social experience register. This realization presents a far more optimistic outlook than if the code world were a realm of intangible remoteness, absolute and transcendent.

3.2 Intimations of (Im)materiality: Text as Code in the Electronic Environment

Analog graphical artifacts challenge the formal and logical basis of digital information. The specificity and particularity of their material inscription have an inherent ambiguity that is not readily translated. But it would be a mistake to imagine that only analog artifacts confound the authority of digital media, or that the challenge to formal logic is solely a property of traditional media and their materials. Texts in an electronic format align themselves with code in a deceptively simple way, as ASCII text that translates into a string of binary digits, as if the text were in fact equivalent to that code. But this presumes that *what a text is* is a linear sequences of marks or signs. A text, of course, is more than this.

Debates about the nature of materiality with respect to writing in digital formats are often premised on a false binarism: print artifacts are considered material, electronic formats immaterial. Attention to the actual characteristics of digital texts, from the level of the letter to more complex aesthetic expressions and organizations, reveals the fallacy of such a binarism.

As discussed in chapter 1.1, the appearance of "electric language" initially generated a utopian buzz among theorists. Electronic environments seemed to promise the realization of hypertextual potentials latent in print formats. Traditional linearity seemed poised for an explosive expansion, and the term "rhizomatic," much overused, turned up everywhere, as if performing its own meaning.[1]

A few decades of word processors and Web browsers later, the sense of continuity between traditional and electronic formats has become as apparent as the sense of rupture introduced by new technologies. We have come to see that many features of hypertext, hypercard stacks, and the threads linking one Web node to another have antecedents in print formats. The structure and form of traditional print media, once spuriously characterized as linear, have been newly scrutinized for their generative and dynamic properties. Stasis, it is worth repeating, is a *relative* property of the material conditions of inscription, not a characteristic of texts. Current textual studies have brought attention to the ways various nonlinguistic aspects of that materiality (type, paper, book structure, layout) participate in the production of semantic meaning.

Still, the notion of the "immaterial" text has become fixed in popular and even critical imagination. Why? Though digital information is far more fungible than physical inscription, the codes on which electronic texts are based are themselves material. More to the point, however, the graphical and dynamic organization of texts continues to function as textual information in the electronic format.

At the most basic level of textual matter, we might ask, what is the link between a letter and the binary codes of electronic storage?[2] What constitutes the identity of a letter as an information form? Is the essence of an *A* its graphical shape? Or is a letter merely an element in a finite system that is sufficiently distinct from all the other elements (e.g., an *A* is not a *B*) to allow the system to function as a graphical code? What is the basic relation between form and information in letters? Does the letter have a body? Does it need a material form in order to register to perception or to function as signification (two different issues, to be sure, one rooted in human communication and the other in semiotic systems)?

If we assume that the letter innately possesses a body, then its identity is bound up in some essential way with that form. If the letter merely *needs* a body, then the implication is that the letter could function through difference. The first concept of identity suggests an inherent essence. The shape of an *A* is substantive information that would be irrevocably lost if the letterform were altered past all recognition (as in fact occurs when a letter is stored electronically). The second concept is more clearly semiotic, since it requires only that letters remain distinct from one another (as, indeed, the electronic code for an *A* differs from that for a *B*). In that case, the identity is functional or operational, not graphical or visual, and any notion of formal essence can be discounted. The question of whether graphic form is substantive information cannot, of course, be answered with a simple yes or no. The answer depends on the circumstances and what information is being gleaned. In questions about the history of printing, writing, textual production and transmission, and other such matters, visual form is clearly substantive. Losing such information through electronic encoding registers as a substantive loss. The question of what constitutes substantive textual information replays at every level of electronic production. One thing, however, is clear: the stripping away of material information when a document is stored in a binary form is not a move from material to immaterial form, but from one material condition to another. The format and graphical design that are part of the presentation of information in every area of communication (including poetic expression) contribute substantively to the text.

Electronic media push the examination of form-as-matter to what seems like its limit because of the inherent character of binary code. But when information is stored as code, is it really pared down to its essential identity? Is data an ontologically pure condition for information or merely a convenient format for management and administration? What is the ontology of a text in "code storage" if the graphic features of the preexisting text are eliminated by the process of encoding? Such questions open a rift between form and associative meaning, between a letter and its graphical identity, between a text and its configured format—relations that seemed inextricably intertwined in print media. In the electronic environment, by contrast, it possible to imagine and even to encounter a letter or text that seems to exist independent of any specific embodied form. Such encounters lead one to believe that a text need not be inscribed in a material substrate. A document can be stored electronically, then output through a variety of devices. A text file can be used to generate musical notes, patterns of light, graphic forms, or letters on a page. This uncoupling of the relation between material form of input and material form of output is what makes information in electronic media fungible and creates the illusion of immateriality. But in fact the mutable condition of "code storage" is endemic to all textual transmission. The time lag between when a text is read and when it was typed, or set by hand in a composing stick, is also a gap in which the fungible quality of textual information can be registered.

Thus the electronic condition introduces a new self-consciousness about writing's past functions, dependencies, and relations to materiality. Code scintillates between material conditions only long enough to ask us what the substantive content of each material inscription might be. Any text is materially instantiated; the degree of stability in the relation of inscription to material varies. In physical, graphic media, it is high, in electronic media, far lower. But if the material information of a text—at the level of the letter, document, or artifact—is an integral part of textual information, then how does storage as code and the mutability that entails transform or undermine the content of electronic texts?

The curious history of language in relation to electronic media enters the picture here. That history involves a split between the logical and formal language used to integrate human communication with machine function and the analysis and interpretation of "natural," data-rich language by the machine. In each case, the concept of what constitutes information is subject to particular constraints and limitations, and meets with different problems in machine processing. The development of programming languages in the course of the twentieth century spawned a veritable babel of dialects. But such languages are as much mathematical writing as they are linguistics. Highly constrained and specific, they work on the principle of eliminating or avoiding ambiguity, nuance, or variable interpretations.

Natural language processing reached certain impasses in the first decades of serious computing, and early-1960s optimism about the possibility of parsing natural grammar into machine-readable (or machine-producible) forms foundered on the complexities of context dependence and the need for a cognitive frame of lived experience outside the language system. In the history of debates from artificial intelligence to cognitive studies, belief systems split between top-down, rule-governed programming and bottom-up, experience-based learning. These debates struggled to decide whether a logical or a data-rich system of representation more accurately mirrors human learning processes, and thus which might more productively be modeled in computational environments where language processing is to take place. But neither position took into account the elements of *configured* language—that is, the format, graphical organization, and structural relations that contribute substantively to textuality in traditional and electronic formats.

The properties of configured language are not those of an algorithmically programmable statement based on formal logic, nor are they the same as those of the context-dependent utterances of natural language. Configured meaning is an aesthetic, rhetorical, and substantive part of linguistic expression. Configuration constitutes meaning. Taking into account configuration revives the inquiry into the relation of sense and form, idea and expression, within human communication systems and mediated exchanges. The letter is a good point of departure for thinking about the value of configured meaning and the way the apparently immaterial text of electronic environments reveals the fallacy of conceiving of code as an ideal form outside of material instantiation.

The process by which any text can be stripped of its apparent materiality as it enters the electronic environment is familiar. But imagine the dilemma of the archivist or librarian deciding on the appropriate means of migrating a handwritten or printed document into digital format. Many such documents contain at least as much visual information as textual. Saving the document as an ASCII file, a sequence of strokes on a keyboard, records a bare record of linguistic information. Or the document could be saved as page images, preserving the rich visual information, even if the verbal content is not available for electronic searching and processing. For scholars intent on attending to the material properties of textual production, such matters are crucial. The Renaissance typographer, designer, and metaphysical philosopher Geofrey Tory, for example, designed his letter *Y* as a study in Pythagorean morals, contrasting the fat stroke of easy indulgence, hung with hams and other pleasures from which one drops into a flaming hell, with the thorny, thin stroke, the path to virtue plagued with wolves and other difficulties. The absurdity of rendering Tory's *Y* with a keystroke is obvious, but this exceptional example makes obvious what is less evident, and sometimes less pertinent, in other cases: that the graphical characteristics of letters are information.

In the late 1970s and early 1980s, mathematician Donald Knuth attempted to make a program that would describe the letters of the alphabet so that he could overcome certain technical difficulties in typesetting his work.[3] This brought him to the heart of the question of whether any algorithm could describe any and every instance of a letter. In other words, does a letter have a single identity, an essential configuration in which it is always expressed, albeit with varying degrees of deviation from the norm. The idea that a letter could be described by a formula that always and only resulted in that letter turned out to be a chimera. As Douglas Hofstadter observed, the individual letters do not constitute a closed set.[4] Any and every instance of a letterform adds to the set without distorting or destroying its delimiting parameters, just as every chair—regardless of height, material, number (or absence) of legs—adds to the category of chairs. The functional life of letters is obviously different from that of chairs, if only because letters' significance depends on their being recognized. To commonsense perception, the essence of an *A* seems incontrovertible. But in actuality, the conventions by which we perceive, read, and process these complex forms are system- and context-dependent. As with chairs, they cannot be defined in a fixed, formal description.

Knuth's dilemma becomes all the more clear when the problems of generating letterforms are contrasted with those of recognizing letterforms. Programs for optical character recognition have become increasingly sophisticated. But the assessment of symbol codes according to primary characteristics (what to look for in the elements of crossbars, downstrokes, x-height, descenders or ascenders) is always calibrated in relation to the fixed set of letters that are to be distinguished as the alphanumeric code. If a letter were simply and fundamentally algorithmic, ideal, and, in current parlance, immaterial, then essential shape and distinctive features could be prescribed as variations on a single formula. In scalable, multisize fonts, letters are described either as vectors, as bit maps, or as complex objects whose internal proportions must to be altered as the scale changes. These letters may be stored in various ways: as sets of instructions about the coordinates that determine the shape of strokes and curves, as records of ductal movements or gestures, as patterns of start and stop points in a raster display, or as pixel patterns in a tapestry grid. The object can be treated in different ways depending on its code identity—sloped, thickened, stretched, resized, and reproportioned—without losing the shape that is essential to communicating its form. Nonetheless, the identity of letters, Knuth found out, cannot be described in an essential or prescriptive algorithm capable of generating any and every instance of that letterform. Mathematical code can accommodate elegant description—information about pathways, vectors, and shapes— but cannot encompass the identity of the letter as a form.

The specific materiality of letterforms, as it turns out, links these particular ways of describing them mathematically to larger traditions. Most have been created in the context of particular belief systems—cosmological, semiotic, or stylistic—and can be described as either constructed, gestural, pictorial, or decorative. *Constructed* letters are based on ideal forms and follow the most precise mathematical prescriptions for proportions—though almost always with some slight variation, to create more dynamism than a perfect mathematical form allows. The algorithm for their creation may be precise except for these slight, and oh, so critical, adjustments, which take into account use and perception. *Gestural* forms are ductal, their stroke patterns described as vectors rather than as a set of fixed geometrical elements, but here again the specific properties of material expression involve the swell and pressure of a brush or pen as the hand varies in its path. And again, the materiality is difficult to recover. *Pictorial* and historiated initials, with moralizing vignettes painted within the counters of majuscules or biblical scenes depicted in the strokes, can best be rendered in digital form as bit maps, but such renderings are perversely distant from the rich content of the image. *Decorative* letters, whose stylishness can be described in terms of component parts, might be coded as units and modules for combinatoric purposes. Granting a value to each piece of a letter by virtue, for instance, of its proportion of curved to straight form in order to calculate its place with a scheme of feminine and masculine principles, could be effectively coded into a mathematical formula. But within any of these conceptions, analysis of what a letter is requires apprehension of the expression within a material instantiation—since that is where its properties become apparent. The properties of letterforms are not inherent, nor are they transcendent, and the material structures within which they are expressed as elements of belief bear on them as they are migrated into digital form.[5]

In the early days of low-resolution monitors and crude output devices, the technical limitations of display pushed the question of the essence of letters as shapes. An alphabet like Wim Crouwell's machine-friendly "New Alphabet," designed in 1967, put as much emphasis on criteria of differentiation among letters as it did on essential form. One has only to isolate a few letters from that alphabet and try to read them on their own to realize how much legibility and recognition depend on context within a sequence of characters. As letter designers have taken advantage of the permissive potential of electronic environments to do things that no calligraphic, print, or photographic medium could do, three-dimensional type designs, rendered fonts that challenge legibility, and other challenges to convention have enjoyed their vogue and vanished. But a mystical belief in essences, with all its kabbalistic undertones, never fully vanishes from the scene. The notion of letters as cosmic elements has great allure, and the idea of code as the key to this universe of signs is too seductive to be put aside completely. These activities all revive the question of materiality with renewed vigor, since it is the inscription of letters in forms and shapes that accord with the whim and styles of a historical cultural moment that allows them to realize the affective potential of their formal expression. The extent to which graphical and visual properties inflect a text with a meaning that is inseparable from its linguistic content is always a feature of its materiality. What the code encodes is always an instantiation, and the transmission from one state of inscription to another merely iterates material conditions, it does not eliminate them.

If this is true with regard to letters, marks, and even the white spaces that constitute a textual field, it is also true of configured texts at the secondary and tertiary levels of organization—as text (composition) and document (artifact). Formats are information, rhetorical and semiotic structures that offer instructions for the production of a text through reading. Outline forms and diagrammatic textual structures are prime examples of texts in which configuration carries semantic value, and understanding the ideological force of schematic forms is an essential critical tool for reading the material codes of information design. Elaborately configured texts use a graphical scaffolding in their organization and layout that carries semantic value: hierarchy, relations of dependence and inheritance, metaphors of branching, various modes of grouping elements, proximity, and so on. These features of the graphical space are integral to the textual field, even if they are completely nonlinguistic.

Classification systems are clearly content rich and ideologically complex. Outline formats, for instance, encode striking numbers of reading clues in their organizational structure. The treatment of headings, subheads, and sub-subheads, their diminishing size, degree of boldness, capitalization, or degree of indentation, are all graphic indicators of the importance of the information within the structure as a whole. In the magnificently elaborate schemes of late medieval cosmologies, Aristotelian rhetorical structures that were pervasive among the Schoolmen, these systems embodied an argument about a worldview.[6] Such schemes blossomed again in the hands of ambitious polymath scholars of the Renaissance, who were intent on describing the order of all things in a manner that could, in the tradition of mathesis, be mapped into a linguistic order. As one of the preliminaries essential for creating the much-sought-after grail of a universal and philosophical language, the rather dauntingly determined Bishop John Wilkins made a full outline of the natural, cultural, spiritual world. His *Essay towards a Real Character and Philosophical Language,* completed in 1668, includes that scheme in all its detail.[7]

Wilkins's obsessive energies may distinguish him from his peers, but his project was neither idiosyncratic nor anomalous. Others pursued similar outlines with the same goal of creating a language that was isomorphic to the order of the world, and thus able to encode knowledge within its formal, logical order and its inscriptional notation system. The similarity between these undertakings and those of later figures such as George Boole and Gottlob Frege is rooted in their shared quest for a comprehensive algebra of thought. For inspiration, these figures drew on the notes of René Descartes and Leibniz's plan for a rational calculus.[8] The work of the young Ludwig Wittgenstein and the attempts made by Noam Chomsky to show the rational order of language are further extensions of this tradition.[9] But the graphical ordering of relations finds less explicit attention among such practitioners than these comprehensive schematic orderings of entities. The elements of syntax are absorbed into structures that mimic the formulas of mathematics or formal logic, where order and notations are graphically inscribed and *semantic,* but visual *rhetoric* is rarely analyzed.

A glance at the organization of Wilkins's scheme, familiar in its outline form, shows clearly how the configured visualizations not only order the information in their systems but *are* themselves information. The wonder of Wilkins's outline is that it adapted the outmoded late medieval diagrammatic tendency to a new purpose—a modern system of classification and typologies. His hierarchies and divisions order the world into clusters and zones that rehearse binarisms between heavenly and earthly, animate and inanimate, vegetable and mineral elements. These divisions are forged in the format as much as the nomenclature, and the branchings of the organizational armature contribute their specific semantic value within the larger whole. Much more needs to be said about the visual rhetoric of diagrammatic forms and the force of visual structures as interpretations, but that is another project altogether.[10] Having made the point that structures are semantic, my emphasis in this context is on features of materiality within electronic texts.

These observations are applicable to texts in any format, but in an electronic environment, where formal relations can be abstracted into a template (through the simple Save As function) or encoded as metastructures that model texts (the Document Type Definition and XML schemas are most familiar), the self-consciousness that attends to their organization as formal systems can be readily apprehended in their graphic expression. But taking these relational, structural features literally, finding a lexicon of values for each organizational form, would replicate a mechanistic and reductive approach to the analysis of materiality. While relations of hierarchy can be read for their value, and it is demonstrably true that basic graphical features perform with somewhat predictable effect (following, for instance, the principles outlined in a gestalt analysis of graphics), the findings of cognitive approaches go beyond literal, mechanistic materiality and replace it with a generative, probabilistic understanding. Mary Carruthers's important corrective to earlier studies of memory theaters provides a dramatically useful example of this shift in approach.[11] Memory theaters, devised in antiquity and perfected in conceptual terms in Renaissance revivals, provide a striking instance of meaning structured in spatialized relations. Frances Yates's well-known study showed that in these structures, space was used in metaphoric and schematic modes simultaneously.[12] Place had value, as did the "contents" exhibited in any location. But in analyzing medieval structures and their role in encoding systems of thought as representations, Carruthers found that the elaborate spatial schemes were experienced as part of a cognitive process in which one cue after another was encountered in the spatial organization. These cues prompted thinking, or action, in a performative mode, not simply a mechanical repetition of memorized information.

The distinction between provocation to interpretation and the older idea of a fixed structure is crucial to the shift between mechanical understandings of materiality and probabilistic ones. The analysis of graphical organizations as material forms gives rise to a basic set of critical insights. Outline forms impose hierarchy, tree diagrams suggest organic models based in genealogy, graphs order their information in relation to axes that usually make use of standard and uniform metrics, and grid structures, like graphs, bear within them the stamp of rationalized, bureaucratic approaches to information and representation. French semiologist Jacques Bertin created a list of seven graphic variables (size, scale, position/placement, spatial arrangement, orientation, shape, and color) that is extremely useful for description and design.[13] To these we can now add the rate of change and refresh cycles, perceived and programmed movement, and other dynamic features that are components of the design or analysis of time-based electronic media. Similarly, the principles of composition that have long governed design of visual communication are amplified by navigational factors in an electronic medium (flow, continuity or rupture, etc.).

But the *reading* of any text and its graphical structure is always an interpretive act, an intervention in a field that is coded to constrain the possibilities of reading but works through provocation, not mechanical transmission. In an electronic environment, the organized structures of graphical space are extended by the multidimensional possibilities of hyperlinked architecture or arrays and protocols for calling the elements of any data set into play. The configurations of Wilkins's cosmological scheme hold the elements in relation to each other so that the reading articulates their value through terms of proximity, derivation, or other features. The importance of analyzing the way such structures and protocols model information and the conditions for its use is paramount in understanding the ways electronic texts function within the cognitive processes that Carruthers describes for medieval architecture. The difference is that the iterative capability that distinguishes information structures from those of the built environment operates much more rapidly and thus perceptibly than that at work in the transformation of a stone building through its relations of codependence and use.

Backtracking for a moment, recollect the way debates about the value of literal materiality came into focus early in the implementation of HyperText Markup Language. In the mid-1990s, as Web browser became capable of generating graphical displays, the question of what should be encoded as information in a text came into focus. Was the typeface, style, or format of a text to be encoded or only the alphanumeric sequence? The initial HTML tag set suggested that graphical information was irrelevant. Headers were organized by importance and size but little else, and features of display were rendered generically. Graphic expression was primitive, and typography (much to the dismay of designers) was simply deemed not to be information. If a text were to be able to be displayed on any platform and in any browser/monitor situation, then, for practical reasons, such nuances as Garamond or Baskerville couldn't be stipulated as part of the display. In the late 1990s, the insistence that HTML should be capable of registering design features—that design was in fact part of the information—coincided with the development of style sheets. Increasing bandwidth, diminished anxiety about file size, and other technological changes were accompanied by the development of fonts and display forms made for the electronic environment. But the earlier omission had already drawn attention to the significance of material information in type and format decisions, whether these were developed in the electronic environment or merely stored there.

The idea that lurks in the study of electronic textuality is that binary code reduces information to an essential condition, and that this condition, with its insistent logic, matches Descartes's original idea of mathesis. The idea of understanding configured texts as logical forms, and code storage as their essence, misses the point that a multiplicity of materialities enter into the production of any text, even in advance of the reader's probabilistic intervention. The dynamic mutability and flexibility of display modes is a constant demonstration that the "essence" of code storage has no self-identical hold on the semantic value of a text. Files are constantly reconfigured in reading and display, and in each instance and iteration, material form and structure contributes substantively to the configured meaning. A text rendered in a skinny column of six-point type with carefully chosen line breaks is not the same as one that screams across the monitor in a stream of blinking, six-inch-tall, neon pink letters. Sharing an alphabetic sequence of letters as stored code does not make these two texts the same at all. The chimera of a code that would register the immaterial trace of pure difference (binarism at its most hubristic)—and thus fulfill Leibniz's dream—is pure fantasy. Code storage is neither immaterial nor self-identical, any more than any other inscriptional or notational format. The iterative display of electronic texts shows off the limits of reading within a frame of literal materiality (and thus the need for critical analysis of these features) rather than the probabilistic materiality in which we conceive of texts as products of interpretative acts.

Charles Bernstein's *Veil,* first published in print form and then in electronic format, offers a useful contrast in two modes of materiality.[14] The printed *Veil* is based on a typewriter poem in which Bernstein overprinted line after line of letters. This created a scrim or screen effect that rendered the language of the text almost illegible. But this illegibility is the point of the text, the porousness of which permits scraps of meaning to surface through the dense field of letters, the fine mesh of its own self-produced screen thus veiling the linguistic transparency of language. The materiality of print form is inherent in the visual and verbal value of the work. In a dialogic synthesis, the two aspects of writing, visual and verbal, play equal parts in the production of the whole.

In transposing the work into an electronic format, Bernstein modified the text and visual production. The letters in the printed *Veil* are always fully present, each layer sitting on the next in an irrefutable maximization of information. In the electronic version, however, the letters and blocks merge. For each point on the screen a single value is assigned to the pixel (one can say the same for the printed version, a photographic reproduction of the overprinting in the original typescript, though the photograph retains some of the material information of the original). This single value averages the overlapping rather than registering several values simultaneously. Unusual effects are produced that are not present in the print artifact. Some letters lighten the dark field of overlap, rather than invariably increasing its darkness.

In some ways the electronic *Veil* has more transparency than the printed version, but the texts in the electronic version no longer retain any degree of autonomy. Even if they couldn't be recovered and read from the print version, the individual text layers remained evident. In the electronic version, the history of placement, displacement, and layering simply can't be discerned.The production history might be saved in any number of file formats, but in the flattened display, the material trace of the early medium is lost, a new material expression in its place. The new *Veil* is thus a screen between production and display, erasing the history of production and erasing traces of its encoding. The poem has gone from being a text-as-image-of-its-production to being a graphical display showing the end result of now absent manipulations. In the digital condition it lacks—or appears to lack—a recoverable history of its own production. The electronic text has become a configured pattern, a palimpsest both real and illusory. Is the essence of its language the inherent but unreadable semantic value or the newly configured form of visual effect? Neither, of course, and both, as well as the many other visible and invisible features of its production and reinscription in any and every reading.

Obviously any notion that "pure code" is immaterial is false. Matt Kirschenbaum has described the apparent paradox between the "phenomenological materiality" of a text and the "ontological immateriality" of its existence.[15] We perceive the visual form of a letter on the screen or on a page in all its replete material existence (font, scale, color, etc.), even though the "letter" exists as a stored sequence of binary code with no tactile, material apparency. But the electronic current, hardware, support systems, and substrate for such code are materially complex. Even at its most basic level, as Kirschenbaum knows full well, code is not immaterial.[16] It functions as a temporarily fixed and infinitely mutable sequence that always refers to a place within the structure of the machine. As a binary sequence, code is always constituted as substantive difference, not simply metaphysical *différance,* and is part of the topographic structure of the computer's configured spaces and mapped territory. As computer historian René Moreau has said, "No item of information can have any existence in the machine unless there is some device in which its physical representation can be held."[17] Code is material, and its materiality has implications at every level of inscription and display, as well as for its role in accounting for configuration as information.

So long as we eschew metaphysics, code cannot be read as transcendent, as ideal, or as comprising a universal set of independent and autonomous symbols (any more than the alphabet should be read as comprising the fundamental elements of the cosmos). The configured meaning within code formations should be read as part of the material world, in variously layered interpretations: from the coming-into-being-as-form that can be grasped as sense (the originary inscription) to a level where form is interpreted within any of the many complexities of iconography, symbolic imagery, textual dimensions, aesthetic inflections, and their attendant historical and cultural engagements with discourses of power and the social conditions of production in all their individual and collective dimensions.

Treating meaning as transparent and materiality as insignificant renders these ideological values unavailable. The "immaterial" gap of transformation—that moment in the movement of a text from one condition of inscription to another—whether in the typesetter's head at the case, the typist's mind, eye, and fingers at the keyboard, the electronic generation of display from a stored file, or transmission to another file format—precipitates back into material expression unless the text is lost in the ether. Language is never an ideal form, always a phenomenal form. The configured features of language in electronic formats are as substantive and significant as in printed artifacts. Reading these literal features of materiality reveals the rhetorical force of specific properties even as every individual interpretation of texts produces them anew from the probabilistic field.

Coda: The Quantum Leap from Literal to Probabilistic Studies of Materiality

The distinction between mechanistic or literal approaches to the study of materiality and the probabilistic or quantum approach, though it is not only a feature of electronic texts, merits further comment. One crucial move, highly relevant in the context of digital code and reading practices, is the shift from a concept of *entity* (textual, graphical, or other representational element) to that of a *constitutive condition* (a field of codependent relations within which an apparent entity or element emerges).[18]

We can take typeface, page size, headers and footers, and column width in any electronic or printed textual artifact as points of departure. These apparently self-evident graphical features of any textual work, whatever the material format, tend to go largely unnoticed unless they interfere with reading or otherwise call attention to themselves. Works by book artists and designers cleverly may exploit these codes to defeat or trick expectations provoked by familiar conventions.[19] The irreverence of a mismatch between elaborate stone-carved majuscles and the "OMG" or "LOL" of a text message can be obvious and funny. The practice of reading such material codes, what I call literal or mechanistic analysis of materiality, is indisputably valuable.

Most style choices are made to please the eye, make a text legible and presentable, or produce an "aesthetic" design—not as studies in historical understanding. But with just a little background in type history, a reader can register striking paradoxes. The realization that typographic descendants of the rational seventeenth-century Romain du Roi are pressed into service for personal ads or crass commercialism could be disturbing to the eminent design committee commissioned by the French king, were they around to see it. A distinctly ahistorical medievalism runs rampant in video and online games. The typographic construction of a pseudomedieval setting (and its association with "gothic" themes of vampirism, dark magic, and undead forces) offers prime material for cultural study. But it is only a single, conspicuous version of more familiar blindnesses. Only the most rarified typophilic readers, for instance, inflect their morning reading of the *New York Times* with reflections on the fate of Stanley Morison's judicious design expertise. Nor do many casual observers note the debased uses of once elite fonts like Baskerville or Park Avenue. And very few viewers pause to read the IBM logo, designed by Paul Rand, as the essence of modern corporate systematicity and global imperialism. The typographic codes are at once too familiar to be read and their origins too obscure to be available without special knowledge. Fewer yet will puzzle through the cultural implications of the genealogical relation between such designers, their training, and the transformation of the communicative sphere. Every material artifact embodies such aesthetics in its formal properties and history, carrying the legacy of its use and reuse. Too much attention to these graphical properties quickly becomes reductive, as if a dictionary of equivalents existed in which Neuland + Ezra Pound + wide spacing = fascism.[20] If it did, it would embody literal, mechanistic materiality *in extremis.*

Still, the discussion of the "meaning value" or "expressivity" of visual means, though unfamiliar in particulars, finds more or less ready acceptance as a general idea. With just a little prompting most readers will admit a begrudging preference for one font or another or admit to the inflecting effect of graphic styles on semantic value. A sample display of posters, type samples, or graphic instances makes clear that such graphical codes affect our reading. Early-twentieth-century journals, for instance, display marked differences meant to signal appeal to different audiences: The unbroken, measured columns of serious news journals were aimed at a masculine sensibility. For the female reader, imagined to suffer from a deficient attention span, chunks of type were interspersed with graphics, encouraging her wandering eye to fall at random on ads for domestic labor-saving devices, corsets, and pickling equipment. Exchanging *Wall Street Journal* headlines with those in the *Weekly World News* provides a similarly dramatic demonstration. "Bond Markets See Rates Drop by Slight Margin" takes on a screaming impact as a banner headline set in white sans serif type atop a lurid photo, and "Woman Gives Birth to Angel" tones down considerably once modestly set in the greyest and least exclamatory of formats.

Meaning is produced, after all, not exhumed, and such exercises are dramatic demonstrations of this principle. Most literate people are fully ready to believe that the massage of meaning goes beyond surface effects even if many of these same readers, including textual-studies and lit-crit scholars, tend to shrug off these observations as trivial. Most literary types (and common readers) are closet transcendentalists, harboring a not-so-secret belief that, after all, it is "sense" that really matters. (The use of the word "matters" here is notably perverse. What could *matter* more than material? But more on this in a moment.) We could likely agree, however, without too much dispute, that any instance of graphic or typographic form can be read as an index of historical and cultural disposition. Attention to the "character of characters" is laudable, maybe even useful, and similar observations could be made with regard to other elements of layout and design.

Typographic expertise used to be an esoteric art, the province of trained professionals. Desktop publishing changed this and broadened sensitivity to design as a set of familiar variables. The features of graphical expression, when enumerated and described, comprise a set of entities that are now listed in the menu bars of Word, Quark, InDesign, and other text- and page-description programs. The user is offered a set of choices for transforming font, format, point size, leading, alignment, tab settings, and so forth. If these graphic elements of a text's appearance are not specified, the default ("normal") settings kick in: twelve-point Times, single-spaced, unjustified, set in a single, 5.5-inch column with standard margins, word spacing, and letter fit. But this approach, like the discussion of the historical analysis of style, is premised on assumptions that limit the scope of a larger inquiry into graphical aesthetics. Why?

Manipulation of each of these graphical components assumes that it is an entity with ontological autonomy and self-evident completeness. This is misleading. Interested as I am in material histories, the crux of my argument is that the very conception of these elements as discrete entities is problematic. The menu of options extends an attitude I call "literal materiality"—a sense that a graphical entity is simply *there* and thus available to a rich, descriptive discussion of its self-evident characteristics. Getting nondesigners to pay attention to the material properties of graphical elements is difficult enough. Undoing the assumptions that support the idea of literal materiality is even harder. Consider the palette and toolbar categories of graphic entities, for instance. These menus reinforce the idea that the appearances of graphical entities are chosen from a finite list of named, discrete elements. The problem is that graphical elements, like anything else material, are defined circumstantially, in relation to the other elements with which they are juxtaposed or surrounded. Even in the seemingly simplest case—black type on a white ground—the letters aren't self-identical things that have the same weight, look, and effect of legibility no matter what. Rather, each assumes a character according to its use. One senses this vividly when working with a line of hand-set lead, fine-tuning placement, juxtaposition, leading, and surrounding space until its weight can be felt most effectively. But it is less tangible in a digital context.

The fallacy of regarding any graphical element as an entity is dramatically demonstrated when we try to name and discuss *the ground*—the page, the material support, or the base—essential to a graphic work, whether in traditional or electronic format. Each of these terms (page etc.) again tends to imply that the ground is a thing to be selected from an inventory and used. Terms like "ground" and "support" also reinforce a hierarchy in which the base is subservient to the presumably more substantive text and graphical elements that will be placed "on" it. I doubt I would have much difficulty convincing readers with the argument I sketched earlier—the value of understanding graphical features in their historical dimensions. But I've set myself a different task here, to dispel the notion of design elements as graphic "entities" and to dislodge the presumptions that carries. I want to rework the conventional approach to the idea of the "page" as an a priori space for graphical construction. In its place, I want to propose an understanding of all graphical elements as dynamic entities in the "quantum field" of a probabilistic system.[21]

Not only are graphical codes the very site and substance of historical meaning, rich and redolent with genealogical traces of origin and use, trailing their vestiges of experience in the counters and serifs of their fine faces. Not only are conventions for the organization of text into textual apparatus and paratextual appendices themselves a set of codes that predispose us to read according to the instructions embedded therein. And not only are both physical materials and the graphically expressive arrangement of verbal materials integral parts of the semantic value of any text. These elements all deserve specific, descriptively analytic attention for the contribution they make to our processes of interpretation. But we also have to understand that the very possibility of interpretive acts occurs within this "quantum system." This field is not a preexisting literal, physical, metrical "space" that underlies the graphical presentation of a text, but a relational, dynamic, dialectically potential *espace* that constitutes it.

To reiterate, I'm suggesting that the specific properties of evident and obvious graphical elements, though frequently unnoticed, make an important contribution to the production of semantic meaning—that the expressivity of these "inflections" is more than superficial, and can and should be understood as integral to textuality. But that only gets us part of the way, and is still within the horizon of an analysis based in literal materiality. I term this mechanistic because it is still premised on the concept of discrete, apparently autonomous, entities. But a radical reconsideration of the process by which these "appearances" are constituted brings about a shift toward a probabilistic approach.

Studying the white space in a page of William Morris's Kelmscott *Canterbury Tales* offers an exemplary opportunity for such a reconsideration. The unprinted area here is not a given, inert and neutral space, but an *espace,* or field, in which forces among mutually constitutive elements make themselves available to be read.[22] The same observation applies to the garden-variety encounters of daily reading. Any page or screen is divided into text blocks and margins, with line space, letterspace, space between page number and margin, and so on. Areas of white space each have their own quality or character, as if they marked variations in atmospheric pressure in different parts of a graphic microclimate.[23] "White" space is thus visually inflected, given a tonal value through relations rather than according to some intrinsic property.

White spaces can be divided into three basic categories depending on their behavior and character: graphic, pictorial, or textual. I define these as follows: (1) graphic—providing framing and structural organization to the supposed ground, with no figural or semantic referent; (2) pictorial—part of an identifiable image or visual meaning in shape or pattern; and (3) textual—keeping characters, lines, and blocks discrete, consistent with organizational convention. On any given page, each area takes on a particular graphic value. By this I mean a tone, or color acquired in relation to the density of other graphic elements in proximity, and also a signifying (if not quite semantic) value.

Typographic elements depend upon the use of white space to sustain the careful articulation that gives them their stylistic specificity. Letterforms are as much an effect of the way the spaces breathe through the lines of type as they are of the character of the strokes. The white space plays a primary role as a supporting medium in guaranteeing the typography its stylistic identity. We see evidence of this in the way the space holds open the counters of letters keeping them to specific degrees of curvature or slant (textual). The incredibly obvious and yet utterly essential space between image and text lines often divides the elements of the graphic universe into word and picture, separating the verbal heavens from the visual earth. This fundamental vocabulary can be subdivided almost indefinitely into the spaces between lines, between the text block and the background, and other distinct margins within the area of the text, each of which has a place within the visual hierarchy that organizes our reading. Similarly, the space around text blocks creates the measured pace for reading while referencing the specific histories of book design and format features (textual). Lower margins keep the text block from slipping off the page while also giving an indication of textual continuity or termination. All of these distinctions could be refined even further, to a surprisingly high degree of granularity and specificity.

The conceptual leap required to move beyond a literal, mechanistic understanding of graphic elements should be easy. We no longer think of the atom as a Tinkertoy model with balls and sticks and rings of wire constraining electrons in fixed orbits. That notion, so charmingly modular, has the scientific validity of Ptolemaic models of the structure of the solar system—or of Newtonian, rather than quantum, physics. Atoms, molecules—the mechanistic understanding of these "entities" were dispelled in the early twentieth century by a theoretical frame that replaced entities with forces and introduced the principle of uncertainty into the account of atomic physics. In quantum physics a phenomenon is produced by the intersection of a set of possibilities and an act of perceptual intervention. At the level of granularity we are used to experiencing, matter appears to operate with a certain consistency according to Newton's laws. But at the atomic and subatomic level, these consistencies dissolve into probabilities, providing contingent, rather than absolute, identities. We should think of letters, words, typefaces, and graphic forms in the same way. Think of the page or screen as a force field, a set of tensions in relation, which assumes a form when intervened in through the productive act of reading. Peculiar? Not really, just unfamiliar as a way to think about "things" as experienced. A slight vertigo can be induced by considering a page as a set of elements in contingent relation, a set of instructions for a potential event. But every reading reinvents a text, and that is a notion we have long felt comfortable invoking. I'm merely shifting our attention from the "produced" nature of signified meaning to the "productive" character of the signifying field.

In historiographic perspective, this approach draws from three areas: (1) work on typography, printing, and graphic design; (2) texts on visual representation, printmaking, and literary, critical, and cultural studies; and (3) speculative work in the realm of documents, cognitive studies, and systems theory.[24] All contribute to the significant shift from a *literal* to an *emergent,* codependent conception of materiality. As literary scholars and design critics engage with graphical aesthetics and material properties of text, I suggest we should not limit ourselves to a literal reading of materiality but consider instead a probabilistic approach to materiality in textual and visual studies. In a digital encounter, the constitutive character of code and expression are all the more compelling because of the generative, iterative, and interactive aspects of display.

3.3 Modeling Functionality: From Codex to e-Book

The material properties of textual artifacts can be modeled, as we've seen, in markup and metadata. They can be described and attended to within mechanistic as well as more probabilistic or constitutive approaches. But considerations of the ways material features are understood should also include attention to their functionality, not just their formal qualities. The peculiar history of the "e-book" shows the ways in which a too-literal misapprehension of what constitute the distinctive features of a material form can give rise to a misconceived model of what it should be when redesigned in another media environment.

The brief career of the e-book has been plagued with fits and starts. In the short time in which personal computers and hand-held devices have come into widespread use, a whole host of surrogates for traditional books has been trotted out with great fanfare and high expectations. In almost every case, these novelties have been accompanied by comparisons between familiar forms and their reinvented electronic shape. That pattern can be discerned in nearly every descriptive title: the expanded book, the superbook, the hyperbook, "the book emulator" (my personal favorite for its touching, underdog sensibility). Such nomenclature seems charged by a need to acknowledge the historical priority of books and to invoke a link with their established cultural identity.

The rhetoric that accompanies these hybrids tends to suggest that all of the advantages are on the electronic side. The copy written in support of them, as new products bidding for market share, contains conspicuous promises of improvement. The idea that electronic "books" will "supercede the limitations" and overcome the "drawbacks" of their paper-based forebears features largely in such promotional claims. Such rhetoric presumes that traditional books are static, fixed, finite forms that can be vastly improved through the addition of so-called interactive features. Testing those claims against the gadgets themselves, however, one encounters a field fraught with contradictions. Electronic presentations often mimic the kitschiest elements of book iconography, while potentially useful features of electronic functionality are excluded. So we see simulacral page drape but little that indicates the capacity for such specifically electronic abilities as rapid refresh, time-stamped updates, or collaborative and aggregated work. E-book "interactivity" has been largely a matter of multiple options within fixed link-and-node hyperstructures.[1]

That e-books have been limited no one doubts. But their limitations have stemmed in part from a flawed understanding of what traditional books are. There has been too much emphasis on formal replication of layout, graphic, and physical features and too little analysis of how those features affect the book's function. Rather than thinking about simulating the way a book *looks,* then, designers might do well to consider extending the ways a book *works.*

A glance at the literature on electronic books shows the persistence of hyperbolic claims spanning more than a decade. Bob Stein's early experiment, Voyager, was adventurous and visionary.[2] Anticipating the design of online formats for hypertext and other new media presentations of experimental works, his company launched its "Expanded Book" in the early 1990s, before the Web was in operation, using CDs and other storage devices. Earlier forms, particularly CDs and the alternative reading practices of hypertext story structures, have not found the large followings their advocates anticipated. Hypertext fiction and the chimera of interactive film have had their vogue and faded. Attempts to develop new reading formats would appear to have reached an impasse if we judge by continuing addictions to traditional fictional forms, or by the persistence of online reading by scrolling through a single text. But during the same decade that hypertext fiction went the way of Kohoutek, the Net has become a fixture in contemporary life. Links and hyperlinks abound, and using these networked structures has become as familiar as turning the pages of a print newspaper. The vision of a reconfigured reading environment has been realized. Geographically dispersed textual, visual, graphic, navigational, and multimedia artifacts can now be aggregated in a single space for study and use, manipulated in ways that traditional means of access don't permit. The telecommunications aspect of new media allows creation of an intersubjective social space—arguably an extension of the social space of traditional scholarly or communicative exchange distinguished mainly by the change in rate, the immediacy, and the capacity to engage simultaneously in shared tasks or common projects.

But what of e-books? The slowness with which new formats have arisen is as much the result of conceptual obstacles as technical ones. The absence of an e-book with the brand-recognition of Kleenex or Xerox isn't due only to the fact that the phrase "electronic document management and information display systems and spaces for intersubjective and associative hyperlinked communication using aggregation, real-time authoring, and participatory editing" doesn't trip off the tongue. The real difficulty is in understanding which aspects of the familiar book have relevance for the design and use of information in an electronic environment. Are they the features that researchers such as IBM's Harold Henke refer to when they identify "metaphors" of book structure?[3] What metaphors does he mean? What does the malleable electronic display of data whose outstanding characteristic is its mutability have to do with the material object familiar to us as the codex book? What, in short, do we mean by the "idea of a book"?

A look at the designs of graphical interfaces for e-books gives some indication of the way conventional answers to this question lead to a conceptual impasse. Ex-libris, Voyager's expanded book, and other "superbook" and "hyperbook" formats have all attempted to simulate in flat-screen space certain obvious physical characteristics of traditional books. IBM's research suggested that readers "prefer features in electronic books that emulate paper book functions." But functions are not the same as formal features. The activity of page turning is not the same as the binary structure of the two-page opening or the recto-verso relations of paper pages. Nonetheless, electronic books have relied heavily on fairly literal simulations of formal features, offering, for instance, a kitschy imitation of page drape from a central gutter. This serves absolutely no purpose, like preserving a coachman's seat on a motor vehicle. Icons that imitate paper clips or book marks, by contrast, allow the reader to place milestones within a large electronic document. As in paper formats, these not only serve navigational purposes but call attention to significant passages. The replacement of pages and volumes with a slider that indicates one's position within the whole reinforces the need to understand information in a gestalt, rather than experiencing it piecemeal. Finally, the reader's urge to annotate, to write into the text with responsive immediacy, has been accommodated as note-taking capabilities for producing e-marginalia have been introduced.

The list of "drawbacks" of traditional books that electronic ones purport to overcome is easy to ridicule. Features like bookmarks, search capabilities, navigation, and spaces for annotation and comments by the author are already fully present in a traditional codex. Indeed, it is very difficult in another medium to simulate their time-tested efficiency. But other features of electronic space do add functionality—live links and real-time or frequent refresh of information. These are unique to digital media; even if linking merely extends the traditional reference function of bibliography or footnotes, it does so in a manner that is radically different. Links don't just indicate a reference route. They either retrieve material or take the reader to that material. And the ideas of rapid refresh, date stamping, and annotating the history of editions materially change the encoded information that constitutes a text in any state. The capacity to materially alter electronic surrogates, customizing actual artifacts, or, at the very least, specifying particular relations among them, presents compelling and unique opportunities.

So what possible function, beyond a nostalgic clue to the reader, do features like gutter and page drape serve in electronic space? The icon of the "book" that casts its long shadow over the production of new electronic instruments is a grotesquely distorted and reductive idea of the codex as a material object. The cover of the book within the video game *Myst* that contains links and clues is a perfect example of the pseudo-gothic, book-as-repository-of-secret-knowledge clichés that abound in the use of the codex as an icon in popular culture.

Let us return to the design of electronic books for one more moment, however. If we ask what is meant by a "metaphor" in Henke's discussion and look at examples of e-book design, we see familiar formats, text/image relations, visual cues that allude to traditional books, and other navigation devices meant to facilitate use by novices. The assumption that familiar forms translate into ease of use may be correct in the first iteration of electronic book-type presentations. But when we look at a table of contents, or an index, or even headers/footers or page numbers—or any of the other structuring elements of book design—it's difficult to imagine how we can consider these metaphors in Henke's sense. These format elements aren't figures of meaning, or presentations of an idea in an unfamiliar form. Quite the contrary, these are instruction sets for cognitive performance.[4] I would argue that as long as visual cues suggest a literal book, our expectations will continue to be constrained by the idea that books are communication devices whose form is static and formal, rather than active and functional. But if we shift our approach we can begin to abstract that functional activity from the familiar iconic presentation. One place to begin this inquiry is by paying attention to the conceptual and intellectual motivations that led to these format features. From there we can extrapolate the design implications that follow for new media.

Instead of reading a book as a formal structure, then, we should understand it in terms of what is known in the architecture profession as a "program," that is, as constituted by the activities that arise from a response to the formal structures. Rather than relying on a literal reading of book "metaphors" grounded in a formal iconography of the codex, we should instead look to scholarly and artistic practices for insight into ways the programmatic function of the traditional codex has been realized. Many aspects of traditional codex books are relevant to the conception and design of virtual books. These depend on the idea of the book as a performative space for the production of reading. This virtual space is created through the dynamic relations that arise from the activity that formal structures make possible. I suggest that the traditional book also produces a virtual space, but this fact tends to be obscured by attention to its iconic and formal properties. The literal has a way with us, its graspable and tractable rhetoric is readily consumed. But concrete conceptions of the performative approach also exist. I shall turn my attention to these in order to sketch a little more fully this idea of the program of the codex.

We should also keep in mind that the traditional codex is as fully engaged with this virtual space as electronic works are. For instance, think of the contrast between the *literal* book—that familiar icon of bound pages in finite, fixed sequence—and the *phenomenal* book—the complex production of meaning and effect that arises from dynamic interaction with the literal work. Here, as elsewhere in my discussion, I base my model of the phenomenal codex on cognitive science, critical theory, and applied aesthetics. The first two set some of the basic parameters for my discussion. Invoking cognitive models suggests that a work is created through interaction with a reader/viewer in a codependent manner. A book (whether thought of as a text or a physical object), is not an inert thing that exists in advance of interaction, but rather is produced anew by the activity of each reading. This idea comports well with the critical legacy of poststructuralism's emphasis on performativity. We make a work through our interaction with it, we don't "receive" a book as a formal structure. Poststructuralist performativity is distinguished from its more constrained meaning in work like that of John Austin, for whom performative language is defined by its instrumental effect. Performativity in a contemporary sense borrows from cognitive science and systems theory, in which entities and actions have codependent relations rather than existing as discrete entities. Performance invokes the kind of constitutive action within a field of constrained possibilities referred to through my argument. Thus, in thinking of a book, whether literal or virtual, we should paraphrase Heinz von Foerster, one of the founding figures of cognitive science, and ask *how a book does* its particular actions, rather than *what a book is.*

With these reference frames in mind, I return to my original question: What features of traditional codex books are relevant to the conception and design of virtual books? My approach can be outlined as follows: (1) start by analyzing how a book works rather than describing what we think it is; (2) describe the program that arises from a book's formal structures; and (3) discard the idea of iconic metaphors of book structure in favor of understanding the way these forms serve as constrained parameters for performance. The literal space of the book thus serves as a field of possibilities, waiting to be "intervened" by a reader. The *espace* of the page arises as a virtual program, interactive, dialogic, dynamic in the fullest sense. Once we see the broader outlines of this program, we can extend it through an understanding of the specific functions that are part of electronic space.

Roger Chartier, tracking the development of book culture, notes several crucial technological and cultural milestones.[5] The shift from scroll to codex in the second to fourth centuries and the invention of printing in the fifteenth century are possibly the two most significant transformations in the technology of book production. Other substantive changes, famously noted by medievalist Malcolm Parkes, came as reading habits were transformed, and when monastic approaches were replaced by scholastic attitudes toward texts in the twelfth through fourteenth centuries, bringing about dramatic changes in format.[6] In earlier usage, books were the basis for linear, silent reading of sacred texts, punctuated by periods of contemplative prayer. These habits gave way to the study and creation of argument as the influence of Aristotle on medieval thought brought about increased attention to rhetoric and the structure of knowledge. Readers began to see the necessity to create metatextual structures for purposes of analysis. To facilitate the creation of arguments, heads and subheads appeared to mark the divisions of a text. Marginal commentary not only added a gloss, an authorial indication of how to read the text, but also outlined and summarized points visually buried in the linear text. Contents pages provided a condensed argument, calling attention to themes and structures and their order within the volume as a whole. The graphic devices that became conventions in this period are aspects of functional activity. They allow for arguments to be abstracted so they can be used, discussed, refuted. These elements are devices for engaging with texts in a manner radically different from that of reflection and prayer. Argument, not reading, is the purpose to which such works are put, and their formal features are designed to provide a reader with both a schematic overview and the means to use the work in rhetorical activity.

Using a book for prayer is clearly an active engagement with the text. But the linear, sequential reading style did not require any extra apparatus as a guide. The development of graphical features that abstract the book's contents thus reflect a radical change in attitudes toward knowledge. Ordered, hierarchical, with an analytical synthesis of contents, the artifact that arose as the instrument of scholastic *lectio* was a new type of book. Readers came to rely on multiple points of access and the search capabilities offered by metatextual apparatus.

The important point here is not just that format features have their origin within specific reading practices but that they are functional, not merely formal. The significant principle is relevant to all reading practices: that the visual hierarchy and use of space and color don't simply reference or reflect an existing hierarchy in a text, they make it, producing the structure through the graphical performance. Such approaches seem self-evident because they are so familiar to us as conventions. But conceptualizing the book in terms of its paratextual apparatus required a leap from literal, linear reading to the spatialized abstraction of an analytic metastructure. Differentiating and identifying various parts of a codex went hand in hand with the recognition of separate functions for these elements. Function gives rise to form, and form sustains functional activity as a program that arises from its structure.

We have inherited that scholastic model but are frequently oblivious to the dynamic agency of its graphic elements. We may find headers a delightful feature on a page, chapter breaks and subheads convenient for our reading in reference materials, but rarely do we step back and recognize them as coded instructions for use. The lines in a modern table of contents, and the accompanying page numbers, function as cognitive cues, pointers into the volume. The information space of a book appears as the structure of its layout. And the analytic synopses in the index and contents are organized to show something in their own right as well as to enable specialized reading tasks.

Various statistical analyses of content appeared as paratextual apparatuses in medieval manuscripts and even their classical predecessors, sometimes motivated by the need to estimate fees (counting of lines) rather than more studious purposes. The habit of creating commentary through marginal notes established a space for conversation within a single page. The palimpsestic nature of such conversations has a particularly rich lineage in commentaries upon sacred texts; an interwoven cultural document like the Talmud is in effect a record of directives for reading. The interpretive gloss is designed to instruct and guide, disposing the reader toward a particular understanding. By contrast, as Anthony Grafton points out, the footnote makes a demonstration of the sources on which a text has been constructed.[7] Justification and verification are the primary purpose of mustering a scholarly bibliography to support one's own work. Thus footnotes may occupy a humbler place, set in smaller type at the bottom of a page or transformed into endnotes at the finish of a section or work, whereas marginalia must be ready to hand, allowing the eye to take in their presence as a visual adjunct if they are to be digested in tandem with the flow of the original text.[8]

The familiarity of conventions causes them to become invisible, and obscures their origin within activity. The figured presentation of meaning in the codex is a condensation of an argument, specific to that form, an argument made in material and graphical structure as well as through textual or visual matter. Recovering the dynamic principles that gave rise to those formats reminds us that graphical elements are not arbitrary or decorative, but serve as functional cognitive guides.

This brief glance at the historical origins of familiar conventions for layout and design should also underscore the fundamental distinction between scroll and codex. The seemingly unified, emphatically linear scroll format, in which navigation depended on markers (protruding ribbons or strips) and a capacity to gauge the volume of the roll on its handle, is striking in contrast to the codex format. When the paratextual features are added, the codex becomes a dynamic knowledge system, organized and structured to allow various routes of access. The replication of such features in electronic space, however, is based on the false premise that they function as well in simulacral form as in their familiar physical instantiation. In thinking toward a design of electronic textual instruments, we would do well to reflect on the *function* that every graphical feature can serve, as well as the informational reference it contains as part of its production or reception history.

Media do matter. The specific properties of electronic technology and digital conditions allow for the continual transformation of artifacts at the most fundamental level of their materiality—their code. The data file of an electronic document can be continually reconfigured. And each intervening act, operating on the field of potentialities, brings a work into being. In digital files we can take advantage of the capacity of electronic instruments to mark such changes rather than merely registering them within the space of interpretation. In addition, two other functions mentioned above are given specific extension within electronic space: the aggregation of documents (as documents and as data) and the creation of an intersubjective exchange. The calling of surrogates through a "portal" in electronic space (as pointed out by Joseph Esposito) allows materials from dispersed collections to be put into proximity for study and analysis.[9] But beyond this, the ability to resize, rescale, alter, or manipulate these documents provides possibilities that traditional paper-based documents simply don't possess. (Looking at a manuscript scanned in raked light, enlarging it until the fibers of the paper show, is a different experience from handling an autographic work in most special collections.) The electronic space engages these technological mediations of the information in a surrogate. But electronic space serves as a site of collaboration and exchange, of generative communication in an intersubjective community that is integral to knowledge production. Information, as Paul Duguid and John Seely Brown have so clearly pointed out, gains its value through social use, not through inherent or abstract properties.[10] The virtual *espace* we envision takes all of these features, themselves present in many aspects of the traditional codex, but often difficult to grasp clearly, and makes them evident. All those traces of reading, of exchange, or of new arrangements and relations of documents, expressions of the shared and social conditions in which a text is produced, altered, and received, can be made visible within an electronic space. These very real and specific features of virtual space can be featured in a graphical interface that acknowledges the codex and traditional document formats as a point of reference but conceives of this new format as quite distinct.

The functions that digital technology accommodates more readily than print media are those of accretion (and processing) of data, aggregation (pulling things together in virtual space that are either separated in physical space or don't exist in physical space), real-time and time-based work, and community interactions in multiauthored environments. But the iterative aspect of digital work fostered by multiple-author environments is also a crucially distinctive feature. Developing a graphical code for representing these functions in an analytic and legible semiotics of new media will still take some time. Ivanhoe is one attempt in this direction, because it is meant to abstract and schematize information in diagrammatic form. Other information visualizations lie ahead, and the conventions for linking functionality and format are emerging.

Writing persists, to this day, with its intimacy and immediacy, while print forms and other mass-production technologies continue to carve up the ecology of communication systems according to an ever more complex division of specialized niches. Books of the future depend very much on how we meet the challenge to understand what a book is and has been.

Frequently, the idea of "the book" guiding design of e-books has been a reductive and unproductive example of inadequate modeling. The multiplicity of physical structures and graphic conventions are manifestations of activity, returned to book form as conventions because of their efficacy in guiding use. An element like a table of contents is not a metaphor, we must recognize, but a program, a set of instructions for performance. By looking to scholarly work for specific understanding of attitudes toward the book as literal space and virtual *espace,* and to artists and poets for evidence of the way the spaces of a book work, we realize that the traditional codex is also, in an important and suggestive way, already virtual. But also, that the format features of virtual spaces of e-space, electronic space, have yet to encode conventions of use within their graphical forms. As that happens, we will witness the conceptual form of virtual spaces for reading, writing, and exchange take shape in the formats that figure their functions in layout and design.

3.4 Aesthetics and New Media

The aesthetics of new media has generated discussion from various perspectives. Some have focused on the specificity of electronic and digital modes, mapping the networked, algorithmic, and procedural character of digital media. Others focus on the rates of transformation, rapid refresh, or mutability of digital media, or its code base and electronic infrastructure. The social immediacy and networked transformation of economies of communication, capital, and currency also come into consideration.[1] All are useful approaches. The resulting insights are accurate and to the point, as are analyses of the "forensics" of new media, the embodied subjects who engage with its practice, and database aesthetics.[2] Other approaches have attempted to frame new media aesthetics in paradigms from the disciplines of literature, film, or semiotics—as a "language" or "montage," or as an extension of narrative or textual practice. These attempts sometimes entail an awkward fit between the conceptual model and terminology derived from older media.

But what happens when instead of pursuing an aesthetics *of* new media, we consider the relation of aesthetics *and* new media? The purpose of this shift would be to bring digital art into dialogue with other artistic practices that are part of a contemporary landscape of imaginative and creative work. Work in new media would then be included among the practices on which contemporary aesthetics is to be based, rather than kept apart as a special case. This puts the fate of *aesthetics* in an era of new media under consideration.

The dialogue between critical theory and aesthetics that descends from modernism was rooted in notions of formalism and autonomy, as well as the innovations and strategic antagonisms of the avant-garde. In the work of Theodor Adorno, this was expressed as an attachment to resistance and its negative capabilities (see chapter 3.5). In a high modern mode, the work of Clement Greenberg, with its emphasis upon media and their specificity, or the literary approaches of new criticism, reinforced the tendency to see works of art as bounded, finite expressions, self-evident and complete. Their highest achievement was to embody a self-referential engagement with the characteristics of their media. In the visual arts, this meant an approach to painting that famously embraced flatness, eschewing illusion and representation, figuration, or literary allusion. High modernism was a last-gasp attempt to salvage utopian beliefs in the face of fascism's rise, the coming of the culture industries as a recognized force, and the period of the Second World War and its chilly aftermath. That modern aesthetic has lost most of its grip. But an attachment to critical negativity still dogs our artistic steps, largely for lack of a better belief system within which to conceptualize art in the current world. If concepts of the avant-garde seem outmoded, so do notions of a language of art.

At the point when high modernism held sway, another approach to media was articulated by Marshall McLuhan.[3] Beginning in the early 1960s his theoretical writings put forward a set of principles that addressed modern media as message and massage. McLuhan and the passing of high modernism shared not only a historical moment but certain assumptions, evident in the ways they defined their objects of study. McLuhan's intensely wrought discussions of the specificity of media as artifacts, modes, and social practices shared high modernism's attachment to formal attributes as essential identities. His intellectual range and his roots in literary criticism and semiotics, as well as the later moment of his intellectual maturity, enriched his approach to media with techniques from cultural and critical studies. And his work became, of course, a defining discourse of mass media as a sphere of activity that could further aesthetic investigation, not be bracketed out of it.

By the 1970s, postmodernism displaced notions of modern autonomy with a model of contingency (and here McLuhan's approach forms another bridge, in part through his connection with Harold Innis and issues of empire, power, and history).[4] Whether inspired by the poststructuralist play of *différance* in it deconstructed mode, or by the theoretical formulations of postcolonial theory and its emphasis on critiques of hegemony and discourses of power, the critical shift in postmodernism was marked by attention to reading works as texts produced across social practices and signs. The writings of Roland Barthes, usually read alongside those of Jean-Francois Lyotard, Jean Baudrillard, and other justly renowned figures of French theory, can also be read in tandem with the shockwaves introduced by McLuhan.

Taken together, these many intellectual frameworks have provided useful descriptive and analytic methods for an aesthetics of digital media, as well as for a rethinking of aesthetics in relation to new media. For the first task, the formal analysis of media specificity remains a valuable and compelling tool. The ontological properties of digital and electronic instruments are key to their material identity and the meanings they thus enable and produce. Critical theory and cultural studies offer useful frames for reading digital works at the macro level of media systems, social practices, and cultural networks of value and control.

But to conceptualize contemporary aesthetics, we have to confront the ways new media push artistic practice into a systems-based, codependent relation with their conditions of *use* and *discourse*, not merely their formal properties or their capacity to function as social signs in a semiotic mode. Aesthetics is transformed, hybridized, by the challenges of *mediation* as a central feature of artistic work. The very situatedness and codependent character of mediation calls forth a host of other terms apt for describing the aesthetic properties of digital media works: embodied, complicit, experiential, participatory. Mediation, as a space between, is registered in digital expressions as an ephemeral but material trace, a time-based inscription, transiently configured, and constituted by and as an experiential field.

To anchor this discussion in concrete examples, I will draw on two artworks: Janet Zweig's *The Medium* and Jim Campbell's *5th Avenue Cutaway,* both from 2002. Zweig's piece models mediation at a fine, micro level of granularity. She tightens our focus, showing us the basic activities through which digital media function within cultural practice. The very concepts of communicative mediation that inform our exchanges within the social sphere, and the practices of representation through which they are embodied, are brought into view in this piece. By contrast, Campbell's work extends the horizon of postmodernity by returning to that quintessential modern figure, the *flâneur,* and bringing it into the paradoxes and ironies of contemporary existence. Zweig's piece is emblematic of the current condition of aesthetic work—its integration into media systems as a fundamental feature of their representational capacity. Campbell's connects to the longer historical axis through which the role and function of the image of contemporary life is figured—and links to our conception of aesthetic practice now. Neither of these pieces can be accommodated within the critical legacy of modernism, postmodernism, or the various didacticisms left over from earlier paradigms.

In *The Medium,* two people sit facing each other in a small alcove, an intimate conversation space carved out of a large, public hall.[5] A screen between them interrupts their direct view. Cameras are installed on each side, and the image of each person is displayed to the other on the screen. Each sees the other, more or less life-size, in what appears to be real time, thus creating a subtle comment on face-to-face communication as pure or unmediated experience. The screen, like the familiar television monitor, offers a talking head. But the programmed algorithms slowly alter the digital video feed. Haloes appear, ghost outlines of the face fade slowly, the image switches from color to black and white, becomes posterized, or is inverted into a stark negative. Occasionally the display shows the viewer's own face, up close, life-size. Less distant than a mirror, the visages are more highly mediated for being recorded live and fed. The time lag between lived and perceived moments is so brief that it barely registers. From time to time a floating frame inserts one image into the other so that the persons see themselves. When both cameras enact this effect, the images form an infinite regress.

Elegant and engaging, *The Medium* reifies its sitters and their interactions, objectifying their presence through representation. The conversational exchange takes place through and as their images. No matter how "natural" those images appear, they are undeniably the product of digital processing, images of mediation. The in-between space of exchange is an entity, and it is precisely that fact that is revealed by this work of art. Without a conversation, *The Medium* doesn't function; it is made anew in every instant and instance. The digital intervention in the conversational exchange embodies and consolidates the intangible and ephemeral trace of face-to-face talk while exposing the embodied materials of digital media as site and instrument of such mediation.

The installation thus embodies several tropes of displacement. We imagine, for instance, that the authentic *real* is always just behind or beyond the screen, where we believe the "actual" person exists. But the screen, cameras, and processors create a relation between the two sitters that is mediated through a "technological imaginary." A space is constituted through mediation and then represented as the work. The deftness of this piece lies in is its conflation of a real place of conversation—the two people are in the physical proximity required for an exchange—and a mediated intervention through a highly technological device, simultaneously an object and an instrument, that both obscures and calls attention to its own existence. This overlap of actual and mediated activity reinforces Zweig's extension of McLuhan's observation that electronic media have subtly pervaded all aspects of our existence. We now *telecommunicate* even in proximity, and all communication falls into a space of "betweenness." The medium is the situation—the ephemeral circumstance, the technological device, and the cultural context of its creation and use as an aesthetic object, one that has no use value except to provoke an experience of wonder, slight strangeness, and awareness.

As a work of art, *The Medium* familiarizes and calls attention to complex networks of technological intermediation that now process communications to an unprecedented degree. McLuhan was not, though he is often misrepresented as such, a technodeterminist. New media don't cause transformations of conception, they participate in them. Technology is opportunistic, not deterministic. New forms of technological infrastructure and media emerge where they can, and thus become the foundation of opportunities and possibilities that they enable. We are better off thinking in terms of an ecology of media than a technology. The imaginary, that place in which we conceive of the world as representation, acts on and receives cultural forms of knowledge and experience. In this age and in response to the technologies of image production that mediate our "imaginary" relation to lived experience, aesthetics has a role in explaining such relations.

The exceptional character of Zweig's piece is that it is less an object than a circumstance for experience. The work constructs conditions of use through which it is in turn constituted. This is a remarkable feature of new media projects. They are not simply experienced, but require participation as a premise of their existence. Understanding the distinctions between Zweig's constructed circumstance of viewing and that of, for instance, a home entertainment system or an arcade is significant. We have to have some grounds on which to distinguish the aesthetic object or experience, ways of knowing how it is marked or defined.

If we refer back, again, to the modern period, we recall that the categories of "the literary" and "visual art" were linked to the differences between the mass production of the culture industries and the practices of high art. Such distinctions depend more on institutional settings and frames rather than on the inherent character of objects. Marcel Duchamp's readymades and Andy Warhol's oeuvre make this very clear indeed. Though brand identity and cultural institutions can still be identified in a digital world—and still carry a validating and legitimating authority—all material appears on the same screen. Digital addresses do not have the same character as Fifth Avenue or Park Avenue ones; .edu or .gov brands something as authoritative in a different way than .com. But the inherent properties of screen display are the same in each case. The quality of a digital image can be just as high in an independent site as in a corporate or institutional one, though information architecture and computational power divide the independent from the powerful.

The distinction between high art and cultural product is increasingly in the value or character of ideas, not the material identity of their expression. Pinning down the "art" status of a digital work depends upon the framework in which it is encountered (UbuWeb vs. Microsoft), but this realization already shows that the strategic position of a work within various cultural categories is part of its identity. Aesthetic objects create a space for reflection, thought, experience. They break the unity of object as product and thing as self-identical that are the hallmarks of a consumerist culture. They do this through their conceptual structure and execution, in the play between idea and expression. An aesthetic object may be simple or complex, but it inserts itself into a historical continuum of ideas in such a way as to register. Aesthetic objects make an argument about the nature of art as expression and experience. They perform that argument about what art is and can be, and what can be expressed and in what ways, at any given moment.[6]

These questions about new media are related to questions about the function of aestheticization in a broader sense. At the end of the 1960s, the cultural authority of information received a boost from important exhibitions like *Information, Software,* and *Cybernetic Serendipity* held in New York and London.[7] These helped familiarize and glamorize new technology. Works of digital art were one of the instruments that helped publicize information and software concepts so that they could circulate outside technical communities. This was long before equipment or software for individual use had been designed, and the advanced wing of information art served more to legitimate the cultural work of computational activity in nonart fields than to shift the realm of art production toward electronic means. Experimentation among early adopters such as Roy Ascott, Melvin Prueitt, and researchers at Bell Labs was often done in the context of scientific or industry sponsorship. Digital productions were regarded as novelties for the most part, even as conceptual art absorbed many of their instruction-based principles for wider dissemination and use.

Both lines of inquiry—the cultural function of aestheticization and the character of contemporary aesthetics—have generated discussion for several decades. Secular literary and artistic traditions have drawn on aleatory and combinatoric practices and an algorithmic or procedural sensibility. Others have engaged mechanistic means of production within programmed constraints, exposing the rule-bound character of art making. Oulipo, the Workshop of Potential Literature, founded by French writer Raymond Queneau in 1960, sought working methods based on constraints that would resist the highly romantic subjectivity of Surrealist automatism. Using mathematics or procedural methods to derive their forms, Oulipian writers worked contemporaneously with the popularization of the concept of "information" within 1960s culture.

A larger procedural turn permeated literary, artistic, and musical production throughout the 1960s. Sol LeWitt articulated this sensibility succinctly: "The idea becomes a machine that makes the art." In the context of 1960s procedural conceptualism, the explicit expression of a (somewhat mechanistic) generative aesthetics was not surprising.

Physicist and poet Max Bense, mentioned in chapter 2.2, put forth a program of what he termed "generative aesthetics" in the early 1960s. His was a fairly literal concept of how such programming would be conceived. Algorithmically generated computations would give rise to data sets capable of being expressed in visual or other output displays. Does such work have the crucial capacity to become emergent at a level beyond the looping processes of its original conception? Or is it limited because in focusing on calculation rather than the symbolic properties of computing it remains mechanistic in conception and execution. Reconceptualizing the mathematical premises of combinatoric and permuatational processes so they work at the level of the symbolic, even the semantic and expressive, is a central tenet of the extension of generative aesthetics to include critical reflection on the role of aesthetic artifacts within an ideological frame.[8]

The mass-culture and material conditions of the development of digital art brought other challenges. Minimalist and fabricated work had established the idea of an industrially produced artwork. Here again the problem arose: How was digital art to be distinguished from other industrial products such as games or information design? The characteristic differences had to reside in some identifiable feature of the work. Works of digital art had to operate with an aesthetic defined as a category of knowledge based in sensory experience. The question remains as to whether aestheticizing information technology and computational activity merely absorbs fine art practice, or whether aesthetic work offers an alternative to instrumentalization and efficient functionality. The answer hinges on the extent to which these works can be shown to resist the smooth functioning, efficient operation, and totalized claims of mathesis.[9]

Digital "things" are highly formalized, obviously, since they exist as data. As I argued in chapter 3.1 and 3.2, the material/immaterial binarism that is so often mindlessly used to distinguish traditional and digital artifacts is simply wrong. The conceptual artists of the 1960s struggled to dematerialize art. In the process they made us aware of a very fundamental principle of art making—the distinction between the idea or algorithmic procedure that instigates a work and the manifestation or execution of a specific iteration. What was clear from that point onward was that the disconnect between provocation and instantiation contributes to the non–self-identicality of all of these elements and to the work of art as a whole. But without a material expression of some kind—an instruction, an utterance, a performance—even the most conceptual of conceptual works did not exist. Every iteration of a digital work is inscribed in the memory trace of the computational system in a highly explicit expression. Aesthetics is a property of experience and knowledge provoked by works structured or situated to maximize that provocation. The mediated character of experience becomes intensified in digital work.

In the 1990s, as digital media came out of the margins and into mainstream galleries and museums, hybridity was a topic that broached identity and border politics and, combined with tropes of genetic mutation, cyborg imaginings and other morphological mutations. The cultural politics surrounding the term "hybridity" deserve their own discussion; the concept was used simultaneously within debates about challenges posed to the fixed categories governing lived conditions and within an artistic domain, where it indicated formal challenges to the traditions of media and classical form-giving. Morphing, figurative distortion, surfaces hinting of impossible combinations of flesh, flora, fauna, and mechanico-robotic engineered organisms proliferated in art even as Monsanto and other major industries branded their own genetic modifications and *Time* magazine flaunted the face of America's future in a Photoshopped blend of racial characteristics whose percentages matched the demographics of our new melting pot.[10]

These thematic concerns, I suggest, all related to a deeper issue. In that moment it seemed that an *aesthetics of hybridity* might acknowledge the broad-based cultural anxiety about the blurring of the category "human."[11] Discussions of hybridity, as well as the popularization of the idea of *posthuman,* risk serving the same normalizing interests. The function of criticism and artistic activity in such a situation is to familiarize the novel concept, rendering it comfortable and acceptable within the cultural sphere. But whose interests were served by a discourse of hybridity? And what administrative decisions can be engineered and authorized on the basis of the category "posthuman," whatever its value in raising critical concerns? The answers to these questions are as complex as our current cultural systems, in which information flows across networks linking very different users. Are we ready to abandon humanness—or the project of humanistic inquiry and beliefs?

The concept of a *hybrid aesthetic* seemed more attractive, since it reflected upon a philosophical tradition and its capacity to address the current cultural condition. The aesthetics bequeathed to us from classical tradition focuses on matters of form, specificity of media, and particulars of expression. In the eighteenth century, aesthetics became concerned with matters of taste and questions of judgment, thus reflecting the shift from objective criteria of value to individual perception, knowledge, and experience. In the twentieth century, concerns with ideology, and anxiety about the role and function of art within the culture, claimed greater attention. But a hybrid aesthetics drawing together various strains of critical and theoretical disciplines is a different concept entirely. These traditions of aesthetic and critical thought might be hybridized in order to address a work like *The Medium.*[12]

Obviously, at the micro level, digital work exists materially as code. At the macro level, we can analyze the repurposing of information by tapping its fungible character, unlocking input and output identities. We can address issues of truth and authority, shifting grounds of cultural power, claims for visual epistemology and for the use of digital media in production of subjectivity, collectivity, and identity. But ultimately, digital artifacts function in relation to contemporary culture. Their properties cannot be discerned through simple analogy to "language" or to a discrete or formalized ontology of "new media." Hybrid aesthetics, unlike an aesthetics of hybridity, induces a self-consciousness into the very practice of critical thought that shifts its ground toward the subjective and nontotalizing. The aesthetics of hybridity, of posthuman and cyborg conditions, especially when posed as *the* language of new media, merely extended the premises of system-building thought. Speculative aesthetics resists this, and the ideas of the post-Cartesian and the metahuman as countermodels.

The ideology of the virtual factors in here, and it appears that we have to do away with yet another false binarism. The distinction between truth and imitation that comes directly from classical thought is still regularly introduced to frame the discussion of simulation. Encounters in virtual worlds, now common in such environments as Second Life, are so many removes from "truth" as to be indicted for aesthetic violations in any Socratic court. Platonic hierarchies, and their negative stigmatization of images as imitations of illusions, are famously entrenched in Western thought. But the ethics of the avatar and the question of whether social rules obtain in a virtual world have brought the symbolic character of lived experience to new levels of discussion.[13] No simple distinction between real and imagined worlds makes sense at the level of symbolic exchange. Physical life exists outside "the machine," but what of emotional, intellectual, and social life? As Zweig's *Medium* makes clear, the line between virtual and its other is unsustainable.

I suggest that the term *metahuman* might be used to more accurately describe this condition. For what such circumstances and artifacts offer is an insight into the nature of our humanity, not a chance to opt out of it or a condition surpassing or rendering it obsolete. The idea of the *metahuman* takes up the tenets of the post-Cartesian sketched in chapter 1.2. Why should we frame our consideration of the virtual as one in which the image must either be a thing-in-itself, with ontological status as a first-order imitation, or a debased mimetic form further removed from those Ideas whose truth we attempt to ascertain? We are not Platonists. The purpose of such an intellectual framework was to create a structure of moral values within a fixed hierarchy. Such approaches reject the positive aspects of creative and imaginative thought. They assume a world in which art and artifice are debased from the outset—as hubris, as deception, as indulgence. Platonic philosophy and its legacy is hardly the place to look for an aesthetic understanding of artifacts of elaborate production.

We might draw on another classical tradition as a foundation for studying the specificity of media, even though the condition of hybridity in artistic practice poses all manner of challenges to its assumptions. Aristotle's poetics lays out the basis for a hierarchical authority of formal structures. Concerned with how things are made, not just how "truthful" they are, Aristotle articulated four causes in works of poetic expression: final, formal, material, and efficient. Aristotelian inquiry sought the properties of each object that are particular to its medium. The "proper" character for poetry, then, is opposed to—or at least distinct from—that of visual forms. These principles were central to the development of modern aesthetics and remain useful, even though "new media" challenge such boundaries.[14] In the late 1960s, for instance, the experimental Computer Technology Group created a painting machine triggered by sensors that responded to a dancer's movements; its 1968 installation-performance thus challenged age-old ideas about the purity of media and specificity of expression. They were hardly alone, and the practices of conceptual art and electronic experimentation in the 1960s share much common ground.[15] Metatechnologies and the intermedia sensibility in art world practices combined conceptual, procedural, and computational practices.

The search for a basis for contemporary aesthetics, however much it may borrow from those discussions of specificity of media, has more to gain from engagement with subjective experience than with objectified forms. Subjectivity must be understood in a contemporary, post-Cartesian mode. In the vernacular, it is understood as the unique experience of a self-identical, intact subject who *has* experience, customized, unique, and consumerist. But the concept of subjectivity that advances contemporary aesthetics is modeled differently, as a codependent configuration. This is a "subject" in the poststructuralist sense, positioned and constituted within discursive and sensorial networks. The concept of subjectivity that arises from this model allows the elaboration of a notion of aesthetics as a particular kind of knowledge formation—partial, situated, experiential, emergent—to be built on its foundation.

High modern and classical approaches to aesthetics, no matter how different in other ways, share one common characteristic: they are *descriptive* systems. Whether concerned with hierarchical, empirical, or formal values, they conceived of their task as the apprehension of an *object* by a perceiving *subject.* Even the eighteenth-century turn to taste and judgment kept these distinctions intact.[16] Note Alexander Baumgarten's statements that the object of aesthetics is "to analyse the faculty of knowledge" or " to investigate the kind of perfection proper to perception which is a lower level of cognition but autonomous and possessed of its own laws."[17]

Because the approaches bequeathed to us by tradition are descriptive, they assume form exists prior to the act of apprehension, that artifacts stand outside the experience of awareness. They assume that stable, static forms of knowledge representation are equally available for perception no matter how hybrid their methods or concerns. By contrast—and the difference is radical, going straight to the ways we conceive the ontology of media—the foundations of new media are not only procedural, generative, iterative, and intersubjective but situated, embodied, and participatory. Though the concept of generative aesthetics provided a crucial turn in consideration of contemporary work because it was conceived very differently from that of formal, rational, empirical, or classical aesthetics, it offered only part of what was necessary. To get beyond the mechanistic aesthetics of earlier models, I turn to a second work of art.

In Jim Campbell's work *5th Avenue Cutaway* (2002), a grid of illuminated spots recalling halftone dots swell and shrink on a video screen, forming fleeting images. Fascinating illusions, the sense they create is of images more nuanced, more detailed than the grid that displays them, as if the granularity of vision and that of the display device were in dialogue. The banal image, produced from a video feed, of a street with figures walking resonates with the memory of the Baudelairian flâneur, whose presence announces modernity. Campbell's screen snatches at modern life with full cognizance of the futility of the attempt, the image is always already gone, mutated and morphed through movement in an illusory continuum. I am seduced by the hypnotic repetition of the work, drawn to its red light tapestry, as if by staring I might recover the lost presence of a past nostalgic for this particular future. Modernity, a modern past, one self-consciously remembering its peculiar relation to temporality and imagery. Digital technology. Beautiful imagery. The modern scene, present and absent, is elusive and passing. Campbell's screen produces a remote relation, in a contemporary mode, to the city of modern life. Its method of display reifies the street in a rendering that is both too crude and too refined for our perceptual apparatus, alien and seductive. In the aesthetics of its display a dialogue between media and culture and traditions of fine art—the red and the black, print modes and reproductive technology, the readout and printout frame of a realistic illusion—all are in play. From autonomy to contingency to complicity and the embedded, entangled condition of all knowledge production, visual or other,

Campbell's work is neither a window on the world nor a mirror. But a produced mediation of the symbolic image of the once real. Integration and streaming virtuality, it can be read as a Virilian vision machine whose subject is production without humanity. Or has it moved beyond such a nihilistic point of view? Is it an argument for recognition of the metahuman, mediated and extended, situated condition of experience I have posited? Campbell's piece embodies a struggle, a compromised optimism. It appears to preserve a humanist striving, trying to recover a purpose for its works of signification and to cover the aporia of "meaning" with a continual stream of information, a pattern of bits passing into aesthetic form through a compromising filter.

If Campbell's work extends the humanistic struggles of modernity into a metacondition, it is because of its contrast to the "cool" antihumanistic relation to form production that characterizes much electronic work made from data display.[18] The difference is striking. Campbell's window on the world, the digital flâneur rooted in individual subjectivity, and—at least the illusion of—a telecommunicated intersubjectivity is not simply information processed by Web crawlers and put up on the screen as a visual pattern. In antihumanist work, information visualization creates form for its own sake in an image no more charged with value than the stars in the night sky. Which is to say, such visualizations are full of potential for meaning—any incidentally produced configuration may be apprehended, rendered meaningful—but they do not display a human intention to communicate. Short of embracing a fully antihumanistic aesthetics, an approach with many advocates in our time, we have to preserve the place of communicative effects within human mediation and our codependent relation to each other and these mediating systems. The distinction is not only in the objects but in the critical conception of aesthetic experience.

Campell's piece embodies an irresolvable distinction between information feed and human perceiver. It comments constantly on the process of mediation. It shows the desperate impossibility of unmediated perception while rendering the activity of looking and seeing within a reification of process and image. Both are figured. The medium is made palpable, perceptible, its production of display inseparable from the message. They are mutually embodied and the viewer's experience of decoding is rendered through a situated engagement, complicit by necessity with the human and electronic mechanisms of contemporary processing and display.

Zweig and Campbell both use dynamic, real-time screen space. The images they make are contingent, circumstantial—embodied and situated but also open-ended. In each instantiation, the images are different, even as the procedures through which they are generated remain the same. Can images produced through such highly "engineered" technological means ever move beyond the mathematics that prescribe them? Can digital representations enable ways of thinking beyond instrumentality?

Aesthetics is the self-conscious attention to that condition of knowledge that returns the knowing mind to its own awareness. But the term *hybrid* aesthetics seems as problematic as that of *generative* aesthetics. Both carry overtones of genetic engineering and rationality. I prefer *speculative* aesthetics, grounded in the language of computational method, but with a recognition that imaginatively stimulated reinvention plays a part in any aesthetic understanding, and that conditions, not objects, are the focus of our investigation.

As human beings, we have the ability to process sensorily and physically. If physical labor and tasks return to greater popularity, then a manual/tactile interface to knowledge can be created that reintegrates what our bodies know more completely into the symbolic processing we are so engaged with. We have to think using a worldview that is sustainable, not just globally but individually. Neither technology and applied knowledge nor capital were partners in the social contract of the Enlightenment. The nonanthropocentric aspect of the world should give us pause and humility, as well as inspiring awe and a need for decisions about responsibility and limits. Human imagination may still preserve and foster a space for human things to be said and for experience to flourish outside the programmatic life of the monoculture and the drives of technocapital. That is the task for imagination expressed in the form of aesthetic works. The machines will not do this for us. Why would they?

3.5 Digital Aesthetics and Critical Opposition

At the turn of the twentieth-first century, discussions of aesthetics are bound to questions of ideology. Artists and critics seem to feel that the task of fine art and imagination is a political one, and so the consideration of how electronic artifacts engage questions of aesthetic opposition arises.

The sheer power ascribed to the idea of digital media gives it an extra potency in our culture. But does the capacity of electronic media to absorb all forms of human expression and experience into data-formats create an inevitability that is ideological as well as technological? In his discussions of aesthetics and ideology, Theodor Adorno continually reiterated the caveat that when positivist logic invades culture to an extreme degreee, representation appears to present a "unitary" truth, a totalizing model of thought which leaves little room for critical action or agency.[1] Pulling this unity apart is essential to critical rationality (as distinct from instrumental rationality) in its struggle to maintain a gap between data and idea, form and experience, the absolute and the lived. In the hybrid condition of the digital the separation necessary to sustain the distinctions between the instrumental and the critical appears to be precluded. The "absolute" nature of the mathematical underpinnings of all digital activity threatens to collapse concept and materiality into a state of identity with an encoded file. As the popular idea of technological truth continues to function as an instrumental force in the increasing rationalization of culture, artwork that renders such "truths" consumable perform in what Adorno would term a reconciliatory manner. Such work seems profoundly insidious—unless it can be qualified within a critique of its assumptions, claims, and premises.

Contemporary technological innovation pushes the boundaries of once discrete areas of cultural activity, subjecting an ever-increasing number of arenas to the managerial bureaucracy of data processing. As it does so, the opposition between two traditions that had been markedly distinct in the visual arts becomes more difficult to sustain—in part because each depends upon identification with contradictory concepts of the role of Reason. The first is the antilyrical, antisubjective, rational tradition of art that aspires to the condition of science. The second is a humanistic, lyrical, subjective romanticism that has opposed emotional, natural, and chaotic forces to those of technologically driven progress.

Hybridity of mechanical and organic entities is a current condition, undermining old oppositions. The machine is now flesh, the body technological, and nature is culture. The cyborg *may* be the sign and actualization of the current lived condition of humanity. Certainly the arts have helped familiarize and legitimate many of these once unthinkable ideas. But the idea that aesthetics regulates boundaries between rational and irrational regimes of technology is limited. As mentioned in chapter 3.4, the notion of a post-Cartesian metahumanism may provide a way to think beyond such oppositions, by shifting the terms of the discussion. *Posthuman*, N. Katherine Hayles's term, suggests that we have passed beyond a condition that fits with humanistic traditions and concerns, even though she frames the term within a synthetic, cyborg sensibility, not a binaristic opposition. But have we? At stake in this conception and its application is nothing short of a commitment to remodel our image of self and world in relation to technology. *Metahumanism* suggests, as McLuhan did, that media are extensions, not negations, of humanity. This concept returns responsibility as well as priority to the culture conceived as a humanistic endeavor. As in any other arena of human endeavor, the long history of the Enlightenment reads as the aggressive attempt to bring nature under control, but the contract that defines humanity is only as strong as the terms on which limits and distinctions are defined. Whether technology has a will and life of its own, like capital, hardly matters, since the process of human beings' defining their own identity as a sentient species begins precisely in our articulation of how to address what is under (however illusory) our control. Metahumanism reengages that enterprise with full recognition of the always mediated and now technologically and digitally extended apparatus of human activity.

These themes could be examined in many areas of contemporary art activity: the imagery of mutation, sculpture and installation work that merges new technology with conventional media, work that extends the human body through technological prostheses or otherwise toys with machine aesthetics in new, synthetic ways.[2] All are interesting manifestations of a profound transformation. But the implications of this change can be brought into focus through a narrower and perhaps more fundamental avenue: an inquiry into the identity of digital technology.

As art's dialogue with technology extends into the digital arena, the political implications of such activity come to the fore. As a touchstone in such debates, the work of Adorno, particularly his conception of aesthetics as potentially resistant (no longer liberatory, given his profound pessimism), provides a starting point and still-relevant reference. At midcentury, Adorno struggled to articulate the capacity of aesthetics to resist the forces and tropes of instrumental reason informing an increasingly commodified and administered consumer society. All cultural production (and reception), he suggested, had come to mirror the processes of capitalism, with their mind-dulling repetitions and formulas, while the capacity of what passed for Reason to perform with destructive force had been made all too vividly clear by the events of the Second World War. Aesthetics could only effectively resist such processes through a refusal of utility and systematicity. Adorno placed considerable weight on formal strategies of artistic production as a means to achieving this goal. These can be identified by the terms *determinate irreconcilability, dissonance,* and *nonidentity.*[3] Are these concepts sustainable within the context of the digital production of works of art? Or does the qualified role of materiality in the digital environment fundamentally alter the way an artwork's identity and self-identity can be conceived?

As we have seen, the notion that an image can be "reduced to" or rendered "equivalent to" a data file, algorithm, program, or any mathematical, quantifiable identity gives rise to a notion of digital identity as absolute and certain. Leibniz's dream is Adorno's nightmare—Adorno and Max Horkheimer were quite clear on this point in *Dialectics of Enlightenment,* when they suggested that Kant's aesthetics of purposelessness is the necessary (and only) antidote and means of resistance to the enslavement of all sections of society and culture by the deceptive forces of mass-culture capitalism (read as an extension of rationalized modes of production into the cultural sphere). In considering how such a technorationality infuses itself into cultural practice, Horkheimer and Adorno take issue with the ways perceptions of representation are themselves subjected to a positivist logic. This occurs, Horkheimer and Adorno suggest, in the approach to language/image in which a literalist, positivist interpretation forces representation into a collusion/elision with the "real" such that one takes the "word for the thing," the "image for the real," the "representation for the referent."[4]

This collapse of the discrete structures of representation into a perceived unity makes it virtually (in the technical sense of the word) impossible to insert any critical distance into an understanding of representation and its social function. The distance between Leibniz and Adorno maps the temporal span that encompasses cultural modernity. What was for Leibniz the glimpsed possibilibity of all-encompassing descriptive and analytic Reason has become for Adorno (and Horkheimer and others) the nightmare image of totalizing control effected by instrumental Rationality. And commodification, it turns out, is the most potent means through which this social and cultural transformation is effected. Capital, as Jean-Francois Lyotard points out in *The Postmodern Condition,* did not sit down at the table at which the terms of the Enlightenment were established.

Artistic engagement with modernity as a phenomenon rapidly transformed every area of cultural production, communication, and administration. Modernity attached a wide range of valences to rationality and technology. A keen awareness of the effects of the radical changes wrought on lived experience produced celebratory as well as critical responses. The utopianism of avant-gardes was premised on the belief that progress promised liberation from oppressive labor and its social constraints, while the nihilistic negation of technorationality attacked these premises and the entire tradition of positivist thought.[5] There was also a tension between the idea that an escape from reason was the means to salvation for the human spirit and a position that invoked rationality as an antidote to injustice and social inequities.

Many reformist and counterculture movements of the late nineteenth century expressed incipient opposition to an unchecked and unquestioned concept of "progress" as the automatic gloss on any feature of modern technology or its effects. By the early twentieth century, aesthetic engagement with the technological manifestations of rationality (and later the irrational manifestations of supposed Reason) covered a considerable range of more sharply defined positions. These included the enthusiastic production of a machine aesthetic—as in Robert Delaunay's renowned paintings of the Eiffel Tower, Fernand Leger's machine motifs, and Futurist works (and rhetoric) in praise of motorcars, trains, and industrially produced objects. The inorganic and technological was clearly privileged over and against the holistic, humanistic, and organic. An ironic retort came in the work of Dada poets and painters, conspicuously nuanced in the oeuvre of Marcel Duchamp. The embrace of chance operations and resolutely asystematic negations of the very premises of rationality eschewed any nostalgic return to the lyrical, personal, or artisanal, which had been held out as antidotes to industrialization and modernity in earlier decades in the Arts and Crafts movement, in Symbolist aesthetics, and among the Pre-Raphaelites.

But mathematical logic, and not a machine aesthetic, is the basis of digital work's relation to the technological as a concept and as a practice. Both the fine arts and the humanities experienced aspirations to the disciplinary rigor of the hard sciences at the turn of the twentieth century. Thus the "machine aesthetic" can be read as a link between the industrial machinery of overproduction and the need to provoke consumption while familiarizing industrial motifs and artifacts. Such aestheticization was also a response to the cultural malaise provoked by increased industrialization, as iconically signaled in Charlie Chaplin's film *Modern Times*.[6] This malaise may not have resulted in a full-blown critically articulated opposition, though sporadic and organized resistances have arisen throughout industrialized nations practically from the moment of the inception of industrial methods of manufacture. The point is not that machines are negatively stigmatized or positively depicted but that their coming causes a radical reconfiguring of attitudes toward work, the body, and aesthetic experience.

Mid-twentieth-century developments in the dialogue of logic and art were as varied as those of the earlier decades, and were tempered considerably by post–World War II reflections on the destructive character of technorationality. Information science, which flourished within the adminstrative and military industries of the war years, expanded out of the restricted domains of esoteric application, extending its reach ever more broadly, until, toward the end of the twentieth century, it had penetrated every aspect of contemporary culture. The rational surfaces of modernist grids, replayed in the minimalist canvases and structures of system-generated art, give ample evidence of the persistence and aesthetic potency of this aesthetic. As mentioned in the preceding chapter, Sol LeWitt's famous dictum "The idea becomes a machine that makes the art" sutures the machinic and rational within his aesthetic—no matter how qualified each of these terms must be in actually assessing his projects.

Computers began to find their way into art projects in the 1950s and 1960s, usually in elaborately conceived works whose conceptual parameters required a considerable amount of input for a relatively piddling material output. The burden of production was such that this work has a high coefficient of conceptualization in proportion to the quality of final product. But such work established the basic paradigm, distinguishing input from output and idea (algorithm, program) from material (print-out, template, form)—so that the "stuff" of a piece is data to be manipulated through process. Fundamental to conceptual art as well as computer-generated work, this paradigm signals a radical break within the aesthetic underpinnning of the fine arts since it renders overt the terms on which idea and material are distinguished—or cease to be distinguishable, depending on the extent to which the identity of a work of digital art is posited to reside within its digital file.

Many works from the 1960s onward, within conceptual art narrowly defined and art broadly understood, take the paradigmatic premises of this distinction between idea as program and form as (incidental) output as their basis. There is a corollary to be drawn between the classical sense of "essence" and "accident" here: the digital file is regarded as having an immutable identity, while individual outputs, though unique, are seen as ever so slightly debased and individuated instances of that original. The question, much discussed in these pages, is whether the essence of the digital work is in fact its file, encoded and encrypted and clearly mathematical, or in a fulfilled, material expression of that file.

Here it seems useful to come back to a framework provided by Adorno: "The Same, which the artworks mean as their what, becomes through how they mean it, an Other."[7] Adorno's aesthetic agenda depends on a separation between subject and object, on a persistent and irremediable difference at all levels of artistic production that allows criticality to function. This "difference" takes the form of keeping idea separate from material—the "same," which is idea or content or thematic problem, is made into an "other" through extrusion into and embodiment in a materiality. The inability of these two to be identical can be maximized in works that promote a dissonant, nonunified, nonunifiable condition in their formal realization. Such work, by its own form and by its disruptive place in the cultural landscape of otherwise too easily consumable objects, serves to keep the critical function alive. Dissonance without closure, difficulty without reconciliation ("determinate irreconcilability")—these functions keep the rational from dominating the material, keep reason from turning nature into its perfect image as some mere imitative form.

William Latham's *Evolution of Form* (1990) consists of algorithmically generated forms that grow, mutate, and adapt within the parameters of an also mutating program. Its operation translates the problems of evolution into a program—the forms evolve according to the various solutions. The images are organic in appearance—tiny units of pale fleshy tissue emerging like some kind of mutant growth. They look more like brains, ovaries, or internal organs than self-sufficient organisms—like clusters of cells organizing in order to perform a specialized rather than an autonomous function. As forms, their identity is entirely linked to—is even, on some conceptual level, isomorphic to—the files generated through the mathematical operation of the algorithms evolving and acting on each other. But though the image cannot exist without the file—except in secondary format as photograph, printout, or other hard-copy manifestation—it *is* in every sense in its material manifestation. The "how" and the "what" of Latham's work are distinct entities—and also merely two aspects of a single, non–self-identical entity. They evolve simultaneously.

To frame it another way, in a paraphrase of the Adorno quote: in certain instances the *how* of the digital environment is never quite precisely the same as the *what*. The means and matter of expression—form/content, idea/expression, essence/accident—only appear to be indissolubly unified in the mathetic condition of digital storage. Difference and distinction remain inscribed within the material expressions of aesthetic artifacts, and with them the possibilities of critical insight. We cannot ask that fine art, or digital aesthetics, answer the need for critical opposition within the sphere of contemporary culture. But we can assert that a theory of knowledge grounded in aesthetic precepts has the possibility of changing the conditions on which cultural authority is ascribed to knowledge. Metahumanism suggests that a language to describe our condition can emerge from our encounter with the extremes of a rationalized instrumentalism. Cognitive studies of codependence and emergence have shifted the foundations of knowledge beyond the mechanistic binarisms on which restrictive thought was instrumentalized and legitimated. The challenge is to foster the awareness to design and imagine alternatives.

4.0

Lessons of SpecLab

Many specific insights came out of our project-based activities at SpecLab and the theoretical and critical investigations that ran in parallel, and sometimes dialogue, with them. Those have been covered in preceding chapters; I won't repeat or elaborately summarize them here. But if any single lesson stands out, for me it would be the value of design as an instrument of humanities work. I mean design in the literal sense, as the task of giving functional form to a project, but also in the most profound sense, as the charge to conceive of the way intellectual undertakings work.

Design is not the usual term used to describe the outline of arguments or research in the humanities, but the meta level of conceptualization and description required for the creation of electronic environments or digital tools is not, I think, different in kind from that required to write a book or structure a scholarly argument. But the tasks focus on attention to models of knowledge and function, rather than on what is being said or presented.

In the more literal sense, the design activity in digital humanities (and for the purpose of these remarks, I include all of our speculative activities under that larger rubric) has followed that in the larger history of graphic design. In an initial phase, problems of display and communication ordered the design of artifacts and determined the rules implemented to achieve these ends. In the world of writing and print, conventions for display, including compositional principles, hierarchies, and orderings, have a history centuries long. In the digital environment, the era in which design was limited to display was relatively short, as issues of navigation soon brought with them requirements to think more systematically about relations among elements (pages, nodes, links, levels). The navigational devices of print culture, combined with the montage sensibility of film and other time-based media, have provided the essential vocabulary for navigational design. The kinds of functionalities built into the codex book, for instance, have found an equivalent standardization within the sidebars, menu bars, and other now familiar features of Web design.

As the Web has matured, the phases of its development can be demarcated as Web 1.0 (static display and navigation), Web.2.0 (interactivity within structured sites), and Web 3.0 (collaborative content development by users, aggregation in real time, and on-the-fly analysis). Web 4.0 will increasingly emphasize the customization of Web resources and require intensified attention to the design of conditions of use.

What will this mean for the future of the humanities and for digital practices that serve scholarship, pedagogy, and professional lie? The design of authoring and editing environments that allow the functionality of peer-review and professional standards is already well under way. The creation of modes of scholarship based on an economy of plenitude and availability of original source materials will continue to foster interpretative innovation. The use of tools for the study of aggregated materials, analysis of patterns of production, distribution, and use, will make use of visualizations and other instruments that work on large bodies of texts and metadata.

But the future of digital scholarship and pedagogy is a social future as much as a technological one. Along with my colleagues in these undertakings and the broader community of digital practitioners, I have said and no doubt will continue to say that the migration of our cultural legacy into digital formats makes this as radical a historical moment as the coming of print. In spite of my aversion to hyperbole, I think this is true, and that the issues of access, preservation, and use will depend upon the way we model and design that legacy. The deepest error is to imagine that the task ahead is a technical one, not an intellectual and cultural one. The theoretical questions that will set the direction for that design are rooted in basic concerns with the interpretative power of models in the creation of any cultural resource.

The other lessons of SpecLab are intellectual points, arguments about the way we were led to an understanding of aesthesis as a foundation for situated, subjective, and partial knowledge through a synthesis of theoretical and critical traditions brought into focus by specific projects and their design. Perhaps, more than anything else, the experiences of SpecLab have provided a way to integrate imagination and intellect, design and theory, individual vision and collaborative work within a variety of professional and institutional settings, into production in ways that demonstrate the rich, possible future for interdisciplinary work within the requirements of digital environments. These projects and their lessons are baby steps in what will turn out to have been the incunabula period of the development of future digital projects. Some of those, I hope, will be inspired by the sensibility that infused our experiments at SpecLab.

NOTES

Introduction

1. For an introduction to digital humanities, see Susan Schreibman, Ray Siemens, and John Unsworth, eds., *A Companion to Digital Humanities* (Oxford: Blackwell, 2004). Jerome McGann's *Radiant Textuality* (New York: Palgrave, 2001) offers critical reflections that trace the development of the field from one perspective. For a philosopher's view on the field of humanities computing, as it is called in a British context, see Willard McCarty's *Humanities Computing* (Hampshire: Palgrave, 2005).

2. For documentation of the field, see the publications of the Association for Literary and Linguistic Computing and the Association for Computers and the Humanities, including *Research in Humanities Computing: Selected Papers from the ALLC/ACH Conferences*, published annually beginning in 1991, and *Literary & Linguistic Computing*, an electronic journal published by ALLC. The online archive of the Institute for Advanced Technology in the Humanities gives a vivid picture of the range of experiments undertaken at Virginia beginning in the 1990s. See the home page for IATH projects: http://www.iath.virginia.edu/IathProjects/projects/homepage.

3. My interests in this area go back to the early 1980s, when digital technology and its mythologies were being introduced at Berkeley, where I was a graduate student. The impact of digital tools on artistic production of print artifacts was particularly striking. In the early 1990s, I served as guest editor for an issue of *Art Journal* that addressed the topic of digital media and arts, at a moment when such activity was still marginal and the community

of theorists and practitioners still relatively small. "Digital Reflections: The Dialogue of Art and Technology" (*Art Journal,* Fall 1997) included articles by Simon Penny, Janet Zweig, Deborah Haynes, Paul Zelevansky, Eduardo Kac, Dew Harrison, Jonathan Harris, and Jon Ippolito.

4. Monroe Beardsley, *Aesthetics from Classical Greece to the Present: A Short History* (Tuscaloosa: University of Alabama Press, 1975), provides a useful introduction to this field. The best reference text is Michael Kelly, ed., *Encyclopedia of Aesthetics* (Oxford: Oxford University Press, 1998).

5. Theories of subjectivity are crucial to this discussion. Mine come from structural linguistics, psychoanalysis, film theory, feminist theory, and cultural studies. Lessons from Claude Levi-Strauss, Ferdinand de Saussure, Sigmund Freud, Jacques Lacan, Julia Kristeva, and Gerard Genette are, for me, foundational. See also the applied and synthetic work of Rosalind Coward and John Ellis (*Language and Materialism: Developments in Semiology and the Theory of the Subject* [London: Routledge and Paul, 1977]), Paul Smith, Jacqueline Rose, Christian Metz, Jean-Louis Comolli, Bertand Augst, Stephen Heath, Elizabeth Grosz, Lisa Tickner, Peter Wollen, Laura Mulvey, Mary Kelly, and other writers in *Screen, Tel Quel, Camera Obscura, Representations,* and *Discourse.*

6. The concept of performativity, derived from the work of John L. Austin (*How to Do Things with Word*s [Cambridge: Harvard University Press, 1967]), has spread across disciplines and fields, resulting in many variations, but the fundamental concept remains the same: that a word, action, or behavior effects change, rather than simply stating, describing, or representing an idea, thought, feeling, or expression.

7. This observation was informed by my background in book arts, printing, letterpress, design, and studio production in the graphic and visual arts, my ongoing work as a practicing artist, and my experience in the 1990s teaching visual art, graphic design history, and theory of visuality and representation.

8. The use of graphs, charts, maps, timelines, and other graphical conventions adopted from statistics are all based in shared assumptions about the transparency of graphical forms. Such premises were apparent in many of the IATH projects created over the years (see URL in note 2 above). The most renowned practitioner in this field is Edward Tufte, whose biases derive from engineering and statistical methods, and whose influence has been enormous even while his assumptions go unexamined. Edward Tufte, *The Visual Display of Quantiative Information* (Chesire, CT: Graphic Press, 1983) and other titles.

9. For an extreme example, see McGann, *Radiant Textuality,* and his thorough discussion of the Rossetti Archive design.

10. TEI Consortium, *Guidelines for Electronic Text Encoding and Interchange* (Humanities Computing Unit, University of Oxford, 2002), http://www.tei-c.org/Guidelines/index.xml.

11. Traditional semiotics and other descriptive methods of analysis point out the difference between the nature of linguistic codes, particularly the double articulation of language, and those of graphical media. No equivalent to the morpheme as a signifying unit exists in graphical systems. Roland Barthes, *Image/Music/Text* (New York: Hill and Wang, 1977); Nelson Goodman, *Languages of Art: An Approach to a Theory of Symbols* (Indianapolis: Bobbs-Merrill, 1968).

12. Rossetti Archive, http://www.rossettiarchive.org; Valley of the Shadow, http://valley.vcdh.virginia.edu.

13. Janet Abbate, *Inventing the Internet* (Cambridge: MIT Press, 1999); Margot Lovejoy, *Postmodern Currents: Art and Artists in the Age of Electronic Media* (Ann Arbor:

UMI Research Press, 1989); Timothy Druckrey, *Iterations* (New York: Institute for Contemporary Photography, 1993) and *Ars Electronica* (Cambridge: MIT Press, 1999); Tom Corby, *Network Art* (New York: Routledge, 2006); Frank Popper, *From Technological to Virtual Art* (Cambridge: MIT Press, 2007); Lynn Hershman-Leeson, *Clicking In: Hot Links to a Digital Culture* (Seattle: Bay Press, 1996); and many articles in *Leonardo* (which has focused in a serious way on art and computers), for a start.

14. For documentation and an introduction to the MOO culture, see http://www.hayseed.net/MOO/, http://ebbs.english.vt.edu/mudmoo.clients.html, and http://personal.georgiasouthern.edu/~jwalker/MOO/index.html to get started (all last accessed August 15, 2007). For a print resource, see Cynthia Haynes and Jan Rune Holmevik, eds., *High Wired: On the Design, Use, and Theory of Educational MOOs* (Ann Arbor: University of Michigan Press, 1998).

15. Daniel Dennett, *Brainchildren* (Cambridge: MIT Press, 1998); "Can Silicon Based Life Exist?" http://www.cmste.uncc.edu/new/papers; Clarence W. De Silva, *Intelligent Machines: Myths and Realities* (Boca Raton: CRC Press, 2000); Robert Reynolds and Thomas Zummer, eds., *CRASH: Nostalgia for the Absence of Cyberspace* (New York: Thread Waxing Space, 1994); Ray Kurzweil, *The Age of Intelligent Machines* (Cambridge: MIT Press, 1990).

16. For detailed and technical debates about electronic scholarship, see Susan Hockey, *Electronic Texts in the Humanities* (Oxford: Oxford University Press, 2000), and Kathryn Sutherland, ed., *Electronic Text* (Oxford: Clarendon Press, 1997). For analysis of the culture of digital technology and critical insight into its workings on the broader contemporary imagination, read Alan Liu, *The Laws of Cool: Knowledge Work and the Culture of Information* (Chicago: University of Chicago Press, 2004). For critical discussions of aesthetics, the concept of control, or matters of textuality, see Mark Hansen, *New Philosophy for New Media* (Cambridge: MIT Press, 2004); Lev Manovich, *The Language of New Media* (Cambridge: MIT Press, 2001); N. Katherine Hayles, *How We Became Posthuman* (Chicago: University of Chicago Press, 1999); Wendy Chun, *Control and Freedom* (Cambridge: MIT Press, 2006); Rita Raley, "Reveal Codes: Hypertext and Performance," http://www.iath.virginia.edu/pmc/text-only/issue.901/12.1raley.txt; and Matthew Kirschenbaum, home page (http://www.otal.umd.edu/~mgk/blog/) and *Mechanisms: New Media and the Forensic Imagination* (Cambridge: MIT Press, 2008). (All URLs last accessed August 15, 2007.)

1.1

1. See the publications of the Association for Computational Linguistics (http://www.aclweb.org). Pioneering projects include the William Blake Archive (http://www.blakearchive.org/blake/main.html), Perseus Digital Library (http://www.perseus.tufts.edu/hopper/), the Rossetti Archive, http://www.rossettiarchive.org, the Canterbury Tales Project (http://www.canterburytalesproject.org/CTPresources.html), the Piers Plowman Electronic Archive (http://www.iath.virginia.edu/piers), and the Brown University Women Writers Project and other work of the Scholarly Technology Group (http://www.stg.brown.edu).

2. Sutherland, *Electronic Text.*

3. See, for instance, William J. Mitchell, *City of Bits: Space, Place, and the Infobahn* (Cambridge: MIT Press, 1995).

4. Bertrand Russell, Alfred North Whitehead, young Ludwig Wittgenstein, Rudolf Carnap, Gottlub Frege, and W. V. Quine are the outstanding figures in this tradition. See Robert J. Stainton, *Philosophical Perspectives on Language* (Peterborough, Ontario:

Broadview Press, 1996); W. G. Lycan, *Philosophy of Language: A Contemporary Introduction.* (New York: Routledge, 2000).

5. Edmund C. Berkeley, *Giant Brains: Or Machines That Think* (New York: Wiley, 1950), is a classic in this genre.

6. For some of the early systematic thinking, see Marvin Minsky, *The Society of Mind* (New York: Simon and Schuster, 1986); Minsky with Seymour Papert, *Artificial Intelligence* (Eugene: Oregon State System of Higher Education, 1973); Herbert Simon, *Representation and Meaning: Experiments with Information Processing Systems* (Englewood Cliffs, NJ: Prentice Hall, 1972); Terry Winograd, *Artificial Intelligence and Language Comprehension* (Washington, DC: U.S. Department of Health, Education, and Welfare, National Institute of Education, 1976); Hubert Dreyfus, *What Computers Can't Do* (New York: Harper and Row, 1979); and even Richard Powers, *Galatea 2.2* (New York: Farrar, Straus, and Giroux, 1995). Daniel Crevier, *AI: The Tumultuous History of the Search for Artificial Intelligence* (New York: Basic Books, 1993), is still a useful introduction to the emergence of crucial fault lines in this area.

7. For a range of imaginings on this topic, see Philip K. Dick, *Do Androids Dream of Electric Sheep* (Garden City, NY: Doubleday, 1968); Rudy Rucker, *Software* (New York: Eos/HarperCollins, 1987); Bill Joy, "Why the Future Doesn't Need Us," http://www.wired.com/wired/archive/8.04/joy.html; Donna Haraway, *Simians, Cyborgs, and Women* (New York: Routledge, 1991); and Chris Habels Gray and Steven Mentor, *The Cyborg Handbook* (New York: Routledge, 1995).

8. For an extreme view, see Arthur Kroker, *Digital Delirium* (New York: St. Martin's Press, 1997); Chun, *Control and Freedom,* offers a more tempered assessment. The writings of Critical Art Ensemble, at http://www.critical-art.net, synthesize critical philosophy and digital culture studies.

9. See Howard Besser's published works and online references, including "Digital Libraries, Standards, Metadata, and Longevity Activities," http://besser.tsoa.nyu.edu/howard/#standards.

10. For an overview of the field from varying perspectives, see Schreibman, Siemens, and Unsworth, *Companion to Digital Humanities.*

11. The EText Center at University of Virginia is a useful example: http://etext.virginia.edu.

12. Barbara Stafford, *Good Looking* (Cambridge: MIT Press, 1996), provides a striking demonstration of the extent to which logocentrism prevails in academic work. Her arguments, from the point of view of artists, art historians, or practitioners of visual knowledge production, seem so obvious as to be unnecessary, and yet, for textual scholars, they seemed to challenge basic assumptions, at least in some quarters. See also Franco Moretti, "Graphs, Maps, and Trees," *New Left Review,* no. 24, November–December 2003 (first of three articles, http://www.newleftreview.net/Issue24.asp?Article=05).

13. On codework, see the exchanges between Rita Raley and John Cayley in Raley, "Interferences: [Net.Writing] and the Practice of Codework" (http://www.electronicbookreview.com/thread/electropoetics/net.writing), and the discussions of digital poetry at http://www.poemsthatgo.com/ideas.htm.

14. Michael Heim, *Electric Language* (New Haven: Yale University Press, 1987); George Landow, *Hypertext* (Baltimore: Johns Hopkins University Press, 1992); J. David Bolter, *Writing Space* (Matwah, NJ: L. Erlbaum Associates, 1991).

15. Hockey, *Electronic Texts,* is an extremely useful, objective introduction to the

field and its history. In addition see: Schreibman, Siemens, and Unsworth, *Companion to Digital Humanities;* McCarty, *Humanities Computing* (Hampshire: Palgrave, 2005); and Elizabeth Bergmann Loizeaux and Neil Fraistat, *Reimagining Textuality* (Madison: University of Wisconsin Press, 2002).

16. For information on metadata standards, see http://dublincore.org; http://www.tei-c.org.

17. Jerome McGann, "Texts in N-Dimensions and Interpretation in a New Key," expands this list in a broader discussion that summarizes our work at SpecLab from his perspective (http://www.humanities.mcmaster.ca/~texttech/pdf/vol12_2_02.pdf).

18. Michael Day, "Metadata for Digital Preservation: A Review of Recent Developments" (http://www.ukoln.ac.uk/metadata/presentations/ecdl2001-day/paper.html), and the discussion of metadata at http://digitalarchive.oclc.org/da/ViewObjectMain.jsp provide useful starting points.

19. Henry Kucera and Nelson Francis of *Computational Analysis of Present-Day American English* in 1967 is considered a turning point for this field. For current work in this area, see *Studies in Corpus Linguistics,* Elena Tognini-Bonelli, general editor, and the professional journal publications *Corpora,* and *Studies in Corpus Linguistics.*

20. Adam Mathes, "Folksonomies—Cooperative Classification and Communication through Shared Metadata," http://www.adammathes.com/academic/computer-mediated-communication/folksonomies.html; overview of search engines, http://jamesthornton.com/search-engine-research; updates on data mining, http://dataminingresearch.blogspot.com.

21. This distinction, more than any other, differentiates natural and formal languages and, not surprisingly, demarcates work in the fields of artificial intelligence from that in cognitive studies, for instance. The shift in the attitudes with which Ludwig Wittgenstein approached the study of language in his *Tractatus* and, later, in *Philosophical Investigations* registers the recognition that the project of totalized, objectified, formalistic approaches to language was a failure and the subsequent realization that only use, situated and specific, could provide insight into the signifying capabilities of language.

22. Jerome McGann, *The Critique of Modern Textual Criticism* (Chicago: University of Chicago Press, 1983) and *Black Riders: The Visible Language of Modernism* (Princeton: Princeton University Press, 1993); Randall McLeod, lectures on "Material Narratives" (http://www.sas.upenn.edu/~traister/pennsem.html); Steve McCaffery and bp nichol, *Rational Geomancy* (Vancouver: Talonbooks, 1992); Johanna Drucker, *The Visible Word: Experimental Typography and Modern Art, 1909–1923* (Chicago: University of Chicago Press, 1994); and essays by McLeod, Nick Frankel, Peter Robinson, Manuel Portela, et al. in *Marking the Text,* ed. Joe Bray, Miriam Handley, and Anne C. Henry (Aldershot: Ashgate, 2000).

23. Joëlle Despeyroux and Robert Harper, "Logical Frameworks and Metalanguages," *Journal of Functional Programming* 13 (2003): 257–60.

24. McGann, *Radiant Textuality,* esp. chap. 5, "Rethinking Textuality," 137–66.

25. Allen Renear, "Out of Praxis: Three (Meta)Theories of Textuality," in Sutherland, *Electronic Text,* 107–26; Allen Renear, Steve De Rose, David G. Durand, and Elli Mylonas, "What Is Text, Really?" *Journal of Computing in Higher Education* 2, no. 1 (Winter 1990): 3–26.

26. Probably the most significant critique of markup, Dino Buzzetti's paper "Text Representation and Textual Models" (http://www.iath.virginia.edu/ach-allc.99/

proceedings/buzzetti.html), brought much of that debate to a close by pointing out another important issue—that the insertion of tags directly into the text created a conflict between the text as a representation at the level of discourse (to use the classic terms of structuralist linguistics) and at the level of reference. Putting content markers into the plane of discourse (a tag that identifies the semantic value of a text relies on reference, even though it is put directly into the character string) *as if* they are marking the plane of reference is a flawed practice. Markup, in its very basis, embodies a contradiction. It collapses two distinct orders of linguistic operation in a confused and messy way. Buzzetti's argument demonstrated that XML itself is built on this confused model of textuality. Thus the foundation of metatextual activity was itself the expression of a model, even as the metatexts embodied in markup schemes model the semantic content of text documents.

27. Holly Shulman, ed., Dolley Madison Digital Edition, University of Virginia Press, http://www.upress.virginia.edu/books/shulman.html.

1.2

1. Heinz von Foerster, *Observing Systems* (Salinas, CA: Intersystems Publications, 1981); Ernst von Glasersfeld, *Radical Constructivism: A Way of Knowing and Learning* (London: Falmer Press, 1995) and "An Introduction to Radical Constructivism," in *The Invented Reality: How Do We Know?*, ed. P. Watzlawick, 17–40 (New York: W. W. Norton, 1984); H. R. Maturana and F. J. Varela, *Autopoiesis and Cognition: The Realization of the Living* (Boston: D. Reidel, 1980).

2. Alan MacEachren, *How Maps Work* (New York: Guilford Press, 1995).

3. Alfred Jarry, *Exploits and Opinions of Dr. Faustroll, Pataphysician* (New York: Exact Change, 1996); Werner Heisenberg, *Philosophical Problems of Quantum Physics* (Woodbridge, CT: Ox Bow Press, 1979); William Charles Price, *The Uncertainty Principle and Foundations of Quantum Mechanics: A Fifty Years' Survey* (New York: Wiley, 1977); McGann, "Texts in N-Dimensions"; Erwin Schrödinger, *Science and Humanism: Physics in Our Time* (Cambridge: Cambridge University Press, 1951) and *My View of the World* (Cambridge: Cambridge University Press, 1964).

4. Mikhail Bakhtin, *The Dialogic Imagination* (Austin: University of Texas Press , 1981).

5. Jacques Bertin, *The Semiology of Graphics* (Madison: University of Wisconsin Press, 1973); Fernande Saint-Martin, *Semiotics of Visual Language* (Bloomington: University of Indiana Press, 1990); Paul Mijksenaar, *Visual Function* (Princeton: Princeton Architectural Press, 1987); Stephen Kosslyn, *Image and Mind* (Cambridge: Harvard University Press, 1980); Richard L. Gregory, *Eye and Brain* (Princeton: Princeton University Press, 1990); James Jerome Gibson, *The Ecological Approach to Visual Perception* (1979; Hillsdale, NJ: Lawrence Erlbaum Associates, 1986); Peter Galison, *Image and Logic: A Material Culture of Microphysics* (Chicago: University of Chicago Press, 1997); Martin Kemp, *Visualizations: The Nature Book of Art and Science* (Berkeley: University of California Press, 2000); Stephen Wilson, *Information Arts: Intersections of Art, Science, and Technology* (Cambridge: MIT Press, 2002); or David Freedberg, *Eye of the Lynx* (Chicago: University of Chicago Press, 2002); James Elkins, *The Domain of Images* (Ithaca: Cornell University Press, 1999); David Marr, *Vision: A Computational Investigation into the Human Representation and Processing of Visual Information* (San Francisco: W. H. Freeman, 1982).

6. Charles Peirce, *The Philosophical Writings of Charles Peirce*, ed. Justus Buchler (New

York: Dover, 1955); Ferdinand de Saussure, *Course in General Linguistics,* ed. Charles Bally and Albert Sechehaye (London: Duckworth, 1983).

7. Jerome McGann, *Dante Gabriel Rossetti and the Game That Must Be Lost* (New Haven: Yale University Press, 2000); Gérard Genette, *Nouveau discours du récit* (Paris: Editions du Seuil, 1983); Paul Smith, *Discerning the Subject* (Minneapolis: University of Minnesota Press, 1988); Emile Benveniste, *Problèmes de linguistique générale* (Paris: Gallimard, 1966–84).

8. *Research in Humanities Computing: Selected Papers from the ALLC/ACH Conference, Association for Literary and Linguistic Computing* (Oxford: Clarendon Press, ongoing from 1991), and Schreibman, Siemens, and Unsworth, *Companion to Digital Humanities,* are excellent starting points for the critical discourse of digital humanities over the last fifteen years.

9. Charles Babbage, *Charles Babbage and His Calculating Engines* (New York: Dover, 1961).

10. Claude Shannon, "A Mathematical Theory of Communication" (Bell Labs, 1947), http://cm.bell-labs.com/cm/ms/what/shannonday/paper.html.

11. Alan Turing, "On Computable Numbers, with an Application to the Entscheidungsproblem," *Proceedings of the London Mathematical Society,* series 2, vol. 42 (1936); Martin Davis, *Universal Computer* (New York: W. W. Norton, 2000).

12. John Comaromi, *Dewey Decimal Classification: History and Current Status* (New Delhi: Sterling, 1989). Wayne Weigand, "The 'Amherst Method': The Origins of the Dewey Decimal Classification Scheme," *Libraries & Culture* 33, no. 2 (Spring 1998), http://www.gslis.utexas.edu/~landc/fulltext/LandC_33_2_Wiegand.pdf; Fritz Machlup, *Information: Interdisciplinary Messages* (New York: Wiley and Sons, 1983).

13. Warren Weaver and Claude Shannon, *The Mathematical Theory of Communication* (Urbana: University of Illinois Press, 1949).

14. Works across disciplines have discussed the dialogue of formal reasoning and cultural issues in representation and knowledge, among them: Liu, *Laws of Cool;* McCarty, *Humanities Computing;* Manovich, *Language of New Media;* Wilson, *Information Arts;* and William J. Mitchell, *The Reconfigured Eye: Visual Truth in the Post-Photographic Era* (Cambridge: MIT Press, 1994). This initial list crosses disciplines.

15. Digital humanities draws heavily on the traditions of René Descartes, Gottfried Leibniz (*mathesis universalis*), and the nineteenth-century algebra of George Boole's tellingly named *Laws of Thought.* It has direct origins in the logical investigations of language in Gottlob Frege and Bertrand Russell, as well as the early Ludwig Wittgenstein (during his youthful, mathematical optimism), and the legacy of formalist approaches as extended to natural language by Noam Chomsky. Mind-as-computer models dominated early cybernetic thought; Norbert Weiner's feedback loops had the aim of making human behavior conform to the perceived perfection of a machine. Systems theory has since become considerably more sophisticated. AI debates have moved into the realms of cognitive studies, away from rule-based notions of programmable intelligence or even bottom-up, neural-net experiential modeling. But complexity theory and chaos models are often only higher-order formalisms, still advancing claims to totalizing explanations and descriptions, not refutations of formal logic's premises. Stephen Wolfram's totalizing explanation of a combinatoric and permutational system is only one extreme manifestation of a pervasive sensibility. Mathesis still undergirds a persistent belief shared by some humanities digerati with closet aspirations to control all knowledge-as-information.

16. Jarry, *Exploits and Opinions;* Warren Motte, *OuLiPo: A Primer of Potential Literature* (Normal, IL: Dalkey Archive Press, 1998); Harry Mathews, *Oulipo Compendium* (London: Atlas Press, 1998).

17. One undergraduate who worked with us created a project in Gnomic, a game that is entirely about rule making. http://www.gnomic.com (accessed September 26, 2006).

18. McGann, "Texts in N-Dimensions,"cites George Spencer-Brown, *The Laws of Form* (New York: Julian Press, 1972).

19. This is garden variety poststructuralism and deconstruction, drawing on Roland Barthes, *S/Z* (Paris: Editions du Seuil, 1970), *Mythologies* (Paris: Editions du Seuil, 1957), and *Image/Music/Text;* Jacques Derrida, *Of Grammatology* (Baltimore: Johns Hopkins University Press, 1976) and *Writing and Difference* (Chicago: University of Chicago Press, 1978); and Paul de Man, *Blindness and Insight* (New York: Oxford University Press, 1971) and *Allegories of Reading* (New Haven: Yale University Press, 1979).

20. Von Glasersfeld, *Radical Constructivism;* von Foerster, *Observing Systems;* Maturana and Varela, *Autopoiesis and Cognition;* Norbert Wiener, *Cybernetics of Control and Communication in the Animal and the Machine* (New York: Wiley, 1948).

21. Cultural theorists of media, from Harold Innis and James Carey through Kaja Silverman, Janice Radway, and Dick Hebdige stress the constitutive rather than instrumental function of mediating systems. James Carey, "Cultural Approach to Communication," chap. 1 in *Communication as Culture* (Boston: Unwin Hyman, 1989).

2.0

1. McGann, *Radiant Textuality,* chap. 9 and appendix (209–48), contains McGann's account of the first game and his moves.

2.1

1. Temporal Modeling is an Intel-sponsored research project of the Speculative Computing Lab and Media Studies at University of Virginia. Demonstrations, work in progress, and research reports are available at http://www.iath.virginia.edu/time.

2. This in fact is what Bruce Robertson's Temporal Markup Scheme attempts to do. The disadvantage is in the assumptions about parent-child relations and the difficulty of accommodating vague information, as well as the persistent difficulty of dealing with overlapping hierarchies within any XML-based scheme.

3. For an overview of some of these issues, see Stuart K. Card, Jock D. Mackinlay, and Ben Shneiderman, *Readings in Information Visualization: Using Vision to Think* (San Francisco: Morgan Kaufmann Publishers, 1999).

4. J. T. Fraser, *Time, the Familiar Stranger* (Cambridge: Massachusetts University Press, 1987); F. A. Schreiber, "Is Time a Real Time? An Overview of Time Ontology in Informatics," in *Real Time Computing,* ed. W. A. Halang and A. D. Stoyenko, 283–307 (Springer Verlag, 1992).

5. Schreiber, "Is Time a Real Time?"

6. Patterns of human activity, even belief systems grounded in cyclic progression toward enlightenment, prove on examination to be temporal arrows "wrapped" in circular loops.

7. J. F. Allen, "Time and Time Again: The Many Ways to Represent Time," *International Journal of Intelligent Systems* 6, no. 4 (July 1991): 341–55.

8. C. S. Jensen et al., "A Glossary of Temporal Database Concepts," *Proceedings of ACM SIGMOD International Conference on Management of Data* 23, no. 1 (March 1994).

9. M. Steedman, "The Productions of Time," draft tutorial notes 2.0, University of Edinburgh, ftp://ftp.cis.upenn.edu/pub/steedman/temporality/.

10. J. Burg, A. Boyle, and S.-D. Lang, "Using Constraint Logic Programming to Analyze the Chronology in *A Rose for Emily*," *Computers and the Humanities* 34, no. 4 (December 2000): 377–92.

11. P. W. Jordan, "Determining the Temporal Ordering of Events in Discourse," master's thesis, Carnegie Mellon Computational Linguistics Program, 1994.

12. H. Bronstein, "Time Schemes, Order, and Chaos: Periodization and Ideology," in *Time, Order, Chaos: The Study of Time IX*, ed. J. T. Fraser, Marlene P. Soulsby, and Alexander J. Argyros (Madison, CT: International Universities Press, 1998).

13. Ira Bashow, Seminar Presentation, University of Virginia, June 2001.

14. H. Price, "The View from Nowhen," in *Time's Arrow and Archimedes' Point* (New York: Oxford University Press, 1996).

15. Teri Reynolds, "Spacetime and Imagetext," *Germanic Review* 73, no. 2 (Spring 1998): 161–74.

16. M. A. O'Toole, "The Theory of Serialism in *The Third Policeman*," *Irish University Review* 18, no. 2 (1988): 215–25.

17. The resulting archive can be found at http://www.iath.virginia.edu/time/time.html.

18. Jerome McGann received a Mellon Lifetime Achievement award in 2002 that brought him a $1.5 million budget. This allowed Ivanhoe to be built but also, shifted the direction of SpecLab's ARP (Applied Research in Patacriticism) into development of Collex, Juxta, and NINES. While this was in every way a positive development, it marked a change in our activities from playful imaginings to serious software creation. My own interests shifted to ABsOnline and Subjective Meteorology, as well as other scholarly projects, and ARP went on under Jerry's direction. Bethany went back to work directly on the ARP projects as well, and our work shifted phase, but she had been the crucial partner in creating Temporal Modeling.

2.2

1. Temporal Modeling was designed to create XML output through a Flash interface. As a result, the kinds of hierarchies and standard metrics that obtain in a Cartesian coordinate system were built into the execution. We struggled over this, but given the budgetary limitations, we lacked the programming muscle to create a discontinuous and/or malleable spatial field. Our goal of making a primary space for creating data through visual means became more important than making sure all the specific characteristics of our design plan were realized, and rather than resort to an environment that would simply, as our doubting peers described it, "make a picture" of a subjective, interpretative space, we opted for making a workable XML platform for creating data on the fly.

2. Subsequent SpecLab projects—Collex (a digital collections tool), Juxta (a collation tool), and NINES, the Networked Infrastructure for Nineteenth-Century Electronic Scholarship—continued in this direction while I became involved in ABsOnline and Subjective Meteorology.

3. See McGann, *Radiant Textuality*, for another discussion of critical issues in their intersection with digital humanities.

4. Barthes, "From Work to Text," in *Image/Music/Text;* also at http://homepage.newschool.edu/~quigleyt/vcs/barthes-wt.html.

5. Colleagues will protest this last assertion, no doubt, pointing to the volumes of scholarly writing that engage seriously with deconstruction and its methods, but anyone observing the daily practices of pedagogy knows all too well how persistently the "mining for meaning" approach to reading continues to hold sway in the classroom.

6. Points and scoring are among the many features of the game that could be toggled on and off. They seem largely unnecessary, though in the *Turn of the Screw,* since we were testing a design with a particular game economy built into it, the points were linked to "inkwells" needed for making moves, thus putting certain constraints into play in the structure of the game.

7. Chandler Sansing's account of this version can be found in the Ivanhoe documents at http://www.speculativecomputing.org.

8. The first designs, as the figures will make clear, were worked out by me in dialogue with Jerry, using what I had learned from storyboarding Temporal Modeling and from talking with Louise Sandhaus and watching her teach her students at Cal Arts, a revelation. Contributions from Bethany Nowviskie, Nathan Piazza, Ben Cummings, Andrea Laue, Steve Ramsey, Worthy Martin, and John Unsworth in the initial rounds transformed the project from an idea into a software development model. We spent at least a year having Ivanhoe lunches during the period when Geoffrey Rockwell and Rune Dalgaard were visiting us. Much of that activity was recorded in notes and minutes, and thanks to Bethany's efforts these are archived on SpecLab. Another major change occurred with the arrival of the Mellon funding and the hiring of Nick Laicona, Lou Foster, Duane Gran, Erik Hatcher, and others who helped build Ivanhoe and other ARP projects.

9. Ben Shneiderman and Catherine Plaisant, *Designing the User Interface: Strategies for Effective Human-Computer Interaction* (Boston: Pearson/Addison Wesley, 2005); Card, Mackinlay, and Shneiderman, *Information Visualization;* Aaron Marcus, Nick Smilonich, and Lynne Thompson, *The Cross-GUI Handbook for Multiplatform User Interface Design* (Boston: Addison-Wesley Longman, 1994). For a sensible introduction and overview, see James Hobart, "Principles of Good GUI Design" (http://www.iie.org.mx/Monitor/v01n03/ar_ihc2.htm), Antionio Drommi, Gregory W. Ulferts, and Dan Shoemaker, "Interface Design: A Focus on Cognitive Science" (http://isedj.org/isecon/2001/02a/ISECON.2001.Drommi.pdf), or Atta Badii and Sylvia Truman, "Cognitive Factors in Interface Design: An E-Learning Environment for Memory Performance and Retention Optimisation," Proceedings of the Eighth European Conference on Information Technology Management: E-Content Management Stream (http://kmi.open.ac.uk/people/sylvia/papers%20pdf/BadiiTruman%202001.pdf).

10. See Jakob Nielson, "Top Ten Web Design Mistakes," http://www.useit.com/alertbox/990530.html; Jeff Johnson, *GUI Bloopers* (Morgan Kaufmann, 2000); Theo Mandel, *The Elements of User Interface Design* (New York: Wiley and Sons, 1997); and, for a collection of the worst errors in design and aesthetics, http://www.webpagesthatsuck.com.

11. Elizabeth Würtz, "A Cross-Cultural Analysis of Websites from High-Context Cultures and Low-Context Cultures," *Journal of Computer-Mediated Communication* 11, no. 1, article 13, has an excellent bibliography (http://jcmc.indiana.edu/vol11/issue1/wuertz.html).

12. Tufte, *Visual Display;* Richard Saul Wurman, *Information Architects* (New York:

Watson-Guptill, 1997); Robert E. Horn, *Visual Language* (Bainbridge Island, WA: Macrovu, 1991).

13. Erwin Panofsky, *Studies in Iconology* (New York: Harper and Row, 1972); Barthes, *Image/Music/Text;* Michel Foucault, *The Order of Things* (New York: Pantheon, 1970) and *The Achaeology of Knowledge* (New York: Harper & Row, 1972); Mieke Bal, *Looking In* (New York: Routledge, 2000); Norman Bryson, Michael Holly, and Keith Moxey, eds., *Visual Culture: Images and Interpretations* (Hanover, NH: University Press of New England, 1994); Laurie Adams, *The Methodologies of Art* (New York: Icon, 1996).

14. Staggeringly little material exists on the history, ideology, and semiotics of diagrams. See Martin Gardner, *Logic Machines and Diagrams* (Chicago: University of Chicago Press, 1982), and Moretti, "Graphs, Maps, and Trees." The study of cultural differences is, however, a major area of Web development now.

15. Graphic design history and theory has a substantial inventory of work at the intersection of critical, cultural studies and visual communication, such as that by Ellen Lupton, Lorraine Wild, Max Gallo, Roland Marchand, Stuart Ewen, Neil Harris, and Michelle Bogart.

16. Tufte is practically synonymous with this view of information.

17. See chapter 3.2, particularly the coda, on going beyond literal materialism.

18. Maturana and Varela, *The Tree of Knowledge* (Boston: Shambala, 1992); Francisco Varela, Evan Thompson, and Eleanor Rosch, *The Embodied Mind: Cognitive Science and Human Experience* (Cambridge: MIT Press, 1991).

19. MacEachren, *How Maps Work.*

20. See Michel Seuphor, *Abstract Painting* (New York: Abrams, 1962); Clement Greenberg, *Art and Culture* (Boston: Beacon, 1961); Theodor Adorno, *Aesthetic Theory* (Minneapolis: University of Minnesota Press, 1997); and my essay on modernism in Kelly, *Encyclopedia of Aesthetics.*

21. Charles Harrison and Paul Wood, eds., *Art in Theory* (Oxford: Blackwell, 1993).

22. Saussure, *Course in General Linguistics.* See also Umberto Eco, *Theory of Semiotics* (Bloomington: University of Indiana Press, 1976); Tveztan Todorov, *Theories of the Symbol* (Oxford: Blackwell, 1982); Louis Hjelmslev, *Prolegomena to a Theory of Language* (Madison: University of Wisconsin Press, 1963); and Thomas Sebeok, *Introduction to Semiotics* (London: Pinter, 1994).

23. Herschell Browning Chipp, *Theories of Modern Art* (Berkeley: University of California Press, 1968).

24. H. G. Wells, *World Brain: The Idea of a Permanent World Encyclopedia* (1937), https://sherlock.sims.berkeley.edu/wells/world_brain.html; J. C. Licklider, *Libraries of the Future* (Cambridge: MIT Press 1965).

25. Max Bense, "The Projects of Generative Aesthetics," in *Cybernetics, Art and Ideas,* ed. Jasia Reichardt, 57–60 (New York: Graphics Society, 1971).

26. Ted Nelson, "Project Xanadu: The Original Hypertext Project," http://www.xanadu.net; James M. Nyce and Paul Kahn, eds., *From Memex to Hypertext: Vannevar Bush and the Mind's Machine* (Boston: Academic Press, 1991).

27. Malcolm B. Parkes, "The Influence of the Concepts of *Ordinatio* and *Compilatio* on the Development of the Book," in *Medieval Learning and Literature,* ed. J. J. G. Alexander and M. T. Gibson, 115–41 (Oxford: Clarendon Press, 1976). See also L. Avrin, *Scribes, Scripts and Books* (Chicago: American Library Association; British Library, 1991); M. M. Smith, "The Design Relationship between the MSS. and the Incunable,"

in *A Millennium of the Book,* ed. R. Meyers and M. Harris (Winchester, England: St. Paul's, 1994); L. Febvre and H.-J. Martin, *The Coming of the Book* (London: Verso, 1997); Douglas McMurtrie, *The Book: The Story of Printing and Bookmaking* (New York: Dorset, 1943); and Robert Stillman, *The New Philosophy and Universal Languages in Seventeenth-Century England: Bacon, Hobbes, and Wilkins* (Lewisburg: Bucknell University Press, 1995).

28. Shneiderman and Plaisant, *Designing the User Interface.*

29. Frances Yates, *Lull and Bruno* (London: Routledge and Kegan Paul, 1982); Gershem Scholem, *Kabbalah* (New York: Meridian, 1978). See also the Oulipo references cited above.

30. See my essay "Graphesis," http://www.noraproject.org/reading.ph.

2.3

1. See Drucker, "Graphesis." See also Donald Ahrens, *Meteorology Today* (Belmont, CA: Thomson/Brooks/Cole, 2007), one of the standard texts in the field. Other works used for reference in this project include Calvin Schmid, *Handbook of Graphic Presentation* (New York: Ronald Press, 1954), and L. Hasse and F. Dobson, *Introductory Physics of the Atmosphere and Ocean* (Dordrecht: D. Reidel, 1986).

2. *Goodman, Languages of Art,* is still useful in describing allographic systems and distinguishing them from other notation forms.

3. *Jarry, Exploits and Opinions.*

4. Christian Bök, *Pataphysics: The Poetics of an Imaginary Science* (Evanston, IL: Northwestern University Press, 2002).

2.4

1. The design of the metadata for ABsOnline takes the form of a Document Type Description (DTD), which is used to generate the XML files that belong to each book/work.

2. Other projects at the University of Virginia use metadata to record the sequence of activities within a constrained space (the space of game play in Ivanhoe) or to enable controlled participation (Jerome McGann's NINES project as a discrete, bounded, but networked environment for academic publishing and scholarship). See http://www.speculativecomputing.org.

2.5

1. "Faustroll defined the universe as that which is the exception to oneself." Jarry, *Exploits and Opinions,* 98. See also Bök, *Pataphysics.*

3.0

1. Gardner, *Logic Machines;* Moretti, "Graphs, Maps, and Trees."

2. See Drucker, *Visible Word* and "Graphesis."

3. Beardsley, *Aesthetics;* Conrad Fiedler, *On Judging Works of Visual Art* (Berkeley: University of California Press, 1978); Clive Bell, *Art* (New York: Capricorn Books, 1958); Roger Fry, *Vision and Design* (Harmondsworth: Penguin, 1937); Michael Fried, *Art and Objecthood* (Chicago: University of Chicago Press, 1998); Greenberg, *Art and Culture;* Seuphor, *Abstract Painting;* Adorno, *Aesthetic Theory.*

4. Derrida, *Grammatology* and *Writing and Difference;* Foucault, *Order of Things* and *Archaeology of Knowledge;* Friedrich Nietzsche, *Beyond Good and Evil* (New York: Vin-

tage, 1966); Martin Heidegger, *Existence and Being* (London: Vision Press, 1956); Gilles Deleuze and Felix Guattari, *Anti-Oedipus: Capitalism and Schizophrenia* (New York: Viking, 1977) and *Rhizome: Introduction* (Paris: Editions de Minuit, 1976); Jean Baudrillard, *For a Critique of the Political Economy of the Sign* (St. Louis: Telos Press, 1981) and *Simulations* (New York: Semiotext(e), 1983).

5. Drucker, "Electronic Media and the Status of Writing," in *Figuring the Word* (New York: Granary Books, 1998), 232–36.

6. The *British Journal of Aesthetics*, I think.

7. Melvin Prueitt, *Art and the Computer* (New York: McGraw Hill, 1984); Roy Ascott, *Telematic Embrace* (Berkeley: University of California Press, 2003); Kenneth Knowlton as described in Jasia Reichardt, *Cybernetics, Art, and Ideas* (Greenwich, CT: New York Graphic Society, 1971) and *Cybernetic Serendipity* (New York: Praeger, 1968).

8. See Paul Binski, *Cambridge Illuminations: Ten Centuries of Book Production in the Medieval West* (London: Harvey Miller, 2005).

9. Druckrey, *Iterations;* Hershman-Leeson, *Clicking In;* Hansen, *New Philosophy;* Manovich, *Language of New Media.* Advocates of the value of code include Eduardo Kac, Alan Sondheim, and Loss Glazier.

10. Kirschenbaum, *Mechanisms,* is the most outstanding and insightful new text in this field.

11. Elkins, *Domain of Images;* Goodman, *Languages of Art.*

12. Roman Jakobson, *Six Lectures on Sound and Meaning* (Cambridge: MIT Press, 1978) and "Closing Statement: Linguistics and Poetics," in *Style in Language*, ed. T. A. Sebeok, 350–77 (New York: Wiley, 1960); Barthes, *Image/Music/Text;* Eco, *Theory of Semiotics;* Bal, *Looking In;* and Bryson, Holly, and Moxey, *Visual Culture.*

3.1

1. This piece took many forms: "Digital Ontologies: The Ideality of Form" (Digital Arts Conference, 1999), "Ontology of the Digital Image" (Wesleyan University, 1997), "Theoretical Informational Aesthetics" (Cal Arts, 1998), and various versions of "Code Storage" (keynote, Mixed Messages Conference, University of North Carolina, 1997; New York University, 1998). It was first published as "Digital Ontologies," *Leonardo* 34, no. 2 (2001): 141–45.

2. Martin Lister, ed., *The Photographic Image in Digital Culture* (London: Routledge, 1995). H. Amelunxen, S. Iglhaut, and F. Rötzer, eds., in collaboration with A. Cassel and N. G. Schneider, *Photography after Photography* (Basel, Switzerland: G&B Arts International, 1996); Mitchell, *Reconfigured Eye;* Druckrey, *Iterations;* Fred Ritchin, *In Our Own Image* (New York: Aperture, 1990).

3. Crevier, *AI.*

4. Jacques Derrida, *Edmund Husserl's Origin of Geometry: An Introduction,* trans. John P. Leavey Jr. (Lincoln: University of Nebraska Press, 1989).

5. Ritchin, *In Our Own Image;* Lister, *Photographic Image;* Mitchell, *Reconfigured Eye;* Kirschenbaum, *Mechanisms.*

6. See Amelunxen et al., *Photography after Photography,* for the specific characterization of the Platonic hierarchy relevant here.

7. Kirschenbaum, *Mechanisms.*

8. Peter Osborne, "Adorno and the Metaphysics of Modernism: The Problem of a Postmodern Art," 23–48, and Peter Dews, "Adorno, Poststructuralism, and the

Critique of Identity," 1–22, both in *The Problems of Modernity: Adorno and Benjamin*, ed. Andrew Benjamin (London: Routledge, 1989); H. Brunkhorst, "Irreconcilable Modernity: Adorno's Aesthetic Experimentalism and the Transgression Theorem," in *The Actuality of Adorno*, ed. M. Pensky (Albany: State Univ. of New York, 1997).

9. Brunkhorst, "Irreconcilable Modernity."

10. Derrida, *Husserl's Origin of Geometry.*

11. For a discussion of Griffiths and Wright, see *Faires: The Cottingley Photographs and their Sequel* (Theosophical Publishing House, 1966). On the ethics of digital manipulation, see http://www.astropix.com/HTML/J_DIGIT/ETHICS.HTM.

12. For Peter Campus images, see http://moma.org and http://www.gravus.net/

13. Lister, *Photographic Image.* Amelunxen et al., *Photography after Photography.*

14. Amelunxen et al., *Photography after Photography.*

15. For a useful resource on digital image manipulation, see http://www.media-awareness.ca/english/resources/educational/teachable_moments/photo_truth.cfm.

16. A. Besant and C. W. Leadbeater, *Thought Forms* (London: Theosophical Publishing Society, 1905).

17. I am thinking of the context in which Wilhelm Worringer's work was produced, for instance, or that of Wassily Kandinsky: that early twentieth-century investment in aesthetic systems of correspondence and universals that came out of late-nineteenth-century symbolism.

18. Herbert W. Franke, *Computer Graphics Computer Art* (New York: Phaidon, 1971).

19. Source for these is Jasia Reichart, *Cybernetic Sensibility* (New York: Praeger, 1968) and *The Computer in Art* (New York: Van Nostrand Reinhold; London: Studio Vista, 1971).

20. Alan Sondheim, *Disorders of the Real* (Barrytown, NY: Station Hill Press, 1988); Loss P. Glazier, *Digital Poetics: The Making of E-Poetries* (Tuscaloosa: University of Alabama Press, 2002); Brian Kim Stefans, *Fashionable Noise* (Berkeley, CA: Atelos, 2003). See online the previously cited Rita Raley references; Jim Rosenberg, essay (http://www.well.com/user/jer/NNHI.html) and poetry (http://www.eastgate.com/people/Rosenberg.html); John Cayley (http://homepage.mac.com/shadoof/net/in/inhome.html); and Jim Andrews, "Vispo, Langu(im)age" (http://www.vispo.com).

21. Prueitt, *Art and the Computer.*

22. Paul Virilio, *The Vision Machine* (Cambridge: MIT Press, 1995).

23. Kirschenbaum, *Mechanisms,* makes this point more strongly and clearly as the basis of a definition of forensic materiality. See chapter 1, "Every Contact Leaves a Trace."

3.2

1. Heim, *Electric Language;* Landow, *Hypertext;* Bolter, *Writing Space.*

2. We could quibble over taking the letter as a starting point. Some would favor an originary inscription of difference as the basis of signification, others a higher-order morphemic-word chunk.

3. Donald Knuth, *Tex and Metafont* (Bedford, MA: American Mathematical Society and Digital Press, 1979).

4. Douglas Hofstadter, *Metamagical Themas* (New York: Basic Books, 1985).

5. For further discussion, see Drucker, "What Is a Letter?" in *The Education of a Typographer,* ed. Steven Heller (New York: Allworth Press, 2004).

6. Walter Ong, *Ramus, Method, and the Decay of Dialogue* (Cambridge: Harvard University Press, 1958).

7. John Wilkins, *An Essay towards a Real Character and Philosophical Language* (London: Printed for Sa. Gellibrand, and for John Martyn, 1668).

8. Crevier, *AI,* provides a useful introduction and overview.

9. Gordon P. Baker, *Wittgenstein, Frege, and the Vienna Circle* (Oxford: Blackwell, 1988); Peter Lewis, *Wittgenstein, Aesthetics and Philosophy* (Burlington, VT: Ashgate, 2004); Noam Chomsky, *Cartesian Linguistics: A Chapter in the History of Rationalist Thought* (Christchurch, New Zealand: Cybereditions, 2002).

10. See Drucker, "Graphesis."

11. Mary J. Carruthers, *The Craft of Thought : Meditation, Rhetoric, and the Making of Images, 400–1200* (Cambridge: Cambridge University Press, 1998)

12. Frances Yates, *The Art of Memory* (Chicago: University of Chicago Press, 1966).

13. Bertin, *Semiology of Graphics.*

14. Charles Bernstein, *Veil* (LaFarge, WI: Xexoxial Editions, 1987).

15. Matthew Kirschenbaum, "Lines for a Virtual T[y/o]pography," http://www.iath.virginia.edu/-mgk3k/dissertation/title.html.

16. Kirschenbaum, *Mechanisms.*

17. René Moreau, *The Computer Comes of Age* (Cambridge: MIT Press, 1984).

18. See the discussion of systems theory, and related references, in chapter 1.2.

19. See, for example, Dick Higgins, *FOEW&OMBWHNW* (New York: Something Else Press, 1969)

20. James Mosely, *Romain du Roi* (Lyon: Musée de l'imprimerie, 2002); Stanley Morison, *Selected Essays on the History of Letter-Forms in Manuscript and Print* (Cambridge: Cambridge University Press, 1981); Jeremy Austen and Christopher Perfect, *The Complete Typographer* (Engelwood Cliffs, NJ: Prentice Hall, 1992); Paul Rand, *Thoughts on Design* (New York: Wittenborn and Company, 1947); Gerald Cinamon, *Rudolf Koch : Letterer, Type Designer, Teacher* (New Castle, DE: Oak Knoll Press, 2000)

21. McGann, "Texts in N-Dimensions."

22. William Morris, *The Works of Geoffrey Chaucer* (Kelmscott Press, 1896).

23. This paper was published in a different form as "Graphical Readings and the Visual Aesthetics of Textuality," *Text, Transactions of the Society for Textual Scholarship* 16 (2006): 267–76.

24. On printing: Daniel B. Updike, *Printing Types* (Cambridge: Harvard University Press/Belknap, 1961); Ellen Lupton and J. Abbott Miller, *Design Writing Research* (New York: Kiosk, 1996); David Pankow, *The Printer's Manual* (Rochester: Cary Graphic Arts Press, 2005); Michael Twyman, *Printing 1770–1970: An Illustrated History of Its Development and Uses in England* (London: British Library, 1998). On visual studies: John Berger, *Ways of Seeing* (New York: Penguin, 1972); Estelle Jussim, *Visual Communication and the Graphic Arts: Photographic Technologies in the Nineteenth Century* (New York: R. R. Bowker, 1974); Kemp, *Visualizations;* Wilson, *Information Arts;* Freedberg, *Eye of the Lynx;* Elkins, *Domain of Images.* William Ivins, *Art and Geometry* (New York: Dover, 1946), makes the point that geometrical figures can be understood and manipulated tangibly as well, and that many geometric proofs are elaborations of physical actions such as turning, layering, or placing shapes in relation to each other. On cultural studies, see von Glasersfeld, *Radical Constructivism;* Maturana and Varela, *Autopoiesis and Cognition;* and the work of Raymond Williams, Stuart Hall, John Tagg, and Francis Frascina.

3.3

1. The iterative aspects of digital processing have now begun to make themselves felt in tools that are genuinely interactive and intersubjective and result in material transformation of the text and knowledge produced through the activity they support. Two authoring and editing environments—Sophie, being prototyped by Bob Stein, and Collex, being developed by Bethany Nowviskie and Jerome McGann at SpecLab at the University of Virginia—are addressing some of the issues that hindered e-spaces from coming into their own. Sophie embodies certain echoes of book structure, particularly in the way it segments or modularizes its spaces and their sequencing, but it also incorporates features of time-based, animated multimedia alongside in software that is accessible enough for classroom use but multipurpose in its applications. Collex is conceived entirely within digital functionalities meant to support electronic publishing and scholarship (collecting, aggregating, making use of folksonomy technology and other networking capabilities). Its interface is strictly functional, with viewing areas for search, display, and notation features rather than a global view of activity. Both projects are so new that issues of scale and sustainability, patterns of use, and graphical navigation have yet to reveal themselves, but both are highly promising. Still, I would argue, these and other electronic environments for reading and authoring expose our indebtedness to print culture at the conceptual level. Understanding the way the basic spatiotemporal structure of the codex undergirds the conceptual organization of reading spaces remains important as we move forward with designing new environments for publication.

2. For discussions of the development of e-books, see Clifford Lynch, "The Battle to Define the Future of the Book in the Digital World," *First Monday* 6, no. 6 (2001), http://www.firstmonday.org/issues/issue6_6/lynch.

3. H. A. Henke, "The Global Impact of eBooks on ePublishing," *Proceedings of the 19th Annual International Conference on Computer Documentation*, 172–80 (New York: ACM, 2001), http://portal.acm.org/citation.cfm?id=501551.

4. One might instead think along the lines of medievalist Mary Carruthers's reassessment of memory theaters, which she views as designs for enacting a cognitive task rather than simply formal structures for information storage and retrieval. Carruthers, *Craft of Thought.*

5. Chartier, R. (1995). *Forms and meanings.* Philadelphia: University of Pennsylvania Press.

6. Parkes, "Influence of the Concepts of *Ordinatio* and *Compilatio.*" See also Avrin, *Scribes, Scripts and Books;* Smith, "Design Relationship"; Febvre and Martin, *Coming of the Book;* McMurtrie, *The Book.*

7. Anthony Grafton, *The Footnote* (Cambridge: Harvard University Press, 1997).

8. Other familiar features of the codex, such as page numbers, are linked to devices like the signature key and register list of first words on sheets. These originally functioned as instructions from printer to binder. The half-title is also an artifact of production history, having come into being with the printing press; sheets already finished, folded, and awaiting binding needed protection on their outer layer. Medieval manuscript scribes, keenly aware of the scarcity and preciousness of their vellum sheets, indicated the start of a text with a simple "Incipit" rather than waste an entire sheet on naming the work, author, or place of production.

9. Joseph Esposito, "The Processed Book," *First Monday* 8, no. 3 (2003), http://www.firstmonday.org/issues/issue8_3/esposito.

10. John Seeley Brown and Paul Duguid, *The Social Life of Information* (Cambridge: Harvard Business School Press, 2000).

3.4

1. Beth E. Kolko, Lisa Nakamura, and Gilbert B. Rodman, *Race in Cyberspace* (New York : Routledge, 2000); Anne Balsamo, *Technologies of the Gendered Body: Reading Cyborg Women* (Durham, NC: Duke University Press, 1996); Mary Flanagan, http://www.maryflanagan.com; Matthew Fuller, *Behind the Blip* (Brooklyn: Autonomedia, 2003)

2. Kirschenbaum, *Mechanisms;* Hansen, *New Philosophy;* Victoria Vesna, ed., *Database Aesthetics: Art in the Age of Information Overflow* (Minneapolis: University of Minnesota Press, 2008); Simon Penny, *Critical Issues in Electronic Media* (Albany: State University of New York Press, 1995); Jay David Bolter and Diane Gromala, *Windows and Mirrors: Experience Design, Digital Art and the Myth of Transparency* (Cambridge: MIT Press, 2005).

3. Marshall McLuhan, *The Medium Is the Massage* (New York: Bantam Books, 1967) and *Understanding Media* (New York: New American Library, 1964).

4. Harold Innis, *Empire and Communications* (Toronto: University of Toronto Press, 1972).

5. *The Medium* was slated for installation in fall 2002 as a permanent public art piece in Murphy Hall, at the School of Journalism and Mass Communication at the University of Minnesota.

6. Industrial and commercial objects often do the same within their own sphere, and the challenge of industrial design is to advance a class of objects through similar self-consciousness. What can a car be now? Or a house? Such questions are resolved not through formal solutions but through conceptual ones. Works of art insert themselves instead into the discourse of art making, an obvious but important difference. Art objects that do not acknowledge this fundamental condition are, in effect, merely well-made products, and are abundant in the art world.

7. *Information,* curated by Kynaston McShine, was held at the Museum of Modern Art in New York in 1970; *Software* was put together by Jack Burnham at the Jewish Museum in New York, also in 1970; and *Cybernetic Serendipity* was curated by Jasia Reichardt at the Institute of Contemporary Arts in London in 1968.

8. Here the key points of reference are not Baumgarten and Kant, Hegel and Arnold, Fry, Bell, and Adorno but the generative morphology of Leibniz, Babbage and Turing, Boole and Simon, Minsky, and the fifth-century-BC Sanskrit grammarian Panini—or the traditions of self-consciously procedural poetics and art: Lautréamont, Duchamp, Cage, Lewitt, Maciunas, Stockhausen, and so on.

9. See Kynaston McShine, *Information* (New York: Museum of Modern Art, 1970), and Reichardt, *Cybernetic Serendipity.*

10. See, for example, http://www.education.mcgill.ca/profs/cartwright/edpe300/mirabel.jpg and Exit Art, *Hybrid State* (New York: 1991). Artworks exemplifying hybridity include Ann Preston's *Twins* (1993), a small sculpture of a head with two faces; Jake and Dino Chapman's potato-headed figures; Alan Rath's machine-sensoria; and Alexis Rockman's fantasy worlds.

11. Hayles, *How We Became Posthuman.* Hayles's important contributions engaged digital technology with enthusiasm, coining this term and calling attention to its implications. But I wonder if, a decade later, we might be in a position to reflect and reconsider our critical agenda.

12. W. J. T. Mitchell, Timothy Druckrey, Hubertus Amelunxen, Fred Ritchin, Jasia

Reichardt, Margaret Morse, Martin Lister, Harold Robins, and more recently Lev Manovich and N. Katherine Hayles, among others, have helped establish some of the frames for description.

13. The classic case is the one reported by Julian Dibbell, "A Rape in Cyberspace," *Village Voice* 38, no. 51, December 21, 1993. Sociological studies abound in this area: Steve Jones, *CyberSociety 2.0: Revisiting Computer-Mediated Communication and Community* (Thousand Oaks: Sage Publications, 1998); Philip N. Howard, *Society Online: The Internet in Context* (Thousand Oaks: Sage Publications, 2004); and the work of Brenda Laurel and others.

14. In *Laocoön* (1766), Gotthold Lessing made distinctions among media that continue to serve as a foundation of disciplinary and critical activity to the present day. Boundaries are still surprisingly well policed. Painting, printmaking, and sculpture departments are frequently defined by media in more ways than one would imagine possible.

15. See my essay "Interactive, Algorithmic, Networked," in *At a Distance: Precursors to Art and Activism on the Internet,* ed. Annmarie Chandler and Norie Neumark (Cambridge: MIT Press, 2005), for an extended discussion and references.

16. As the discussion of taste came to the fore in the eighteenth century, subjective opinion came under scrutiny. The development, marked by contributions of the Earl of Shaftesbury, was well suited to an era of rational cultivation of sensibility; the discussion of taste and refinement builds on the idea of knowledge as expertise, connoisseurship of sorts, created through the systematic accumulation of experience through sampling and refining of sensation.

17. Beardsley, 157. Immanuel Kant's *Critique of Judgment* designated the function of aesthetics as the understanding of design, order, form—"purposiveness without purpose"—design outside of utility—knowledge seeking must be "free," disinterested, without end, aim, or goal. Among the three modes of consciousness, knowledge (governed by pure reason), and desire (subject to practical reason), Kant positioned aesthetics as the bridge between mind and sense, aligning it with feeling and judgment.

18. Lisa Jevbratt and Geri Wittig, "Mapping the Web Informe," http://jevbratt.com/projects.html.

3.5

1. Brunkhorst, "Irreconciable Modernity."

2. Examples include the work of Mark Pauline and Survival Research Lab, performance and robotic artist Stelarc, video artist Alan Rath, sculptor Janet Zweig, and collaborators Heather Schatz and Eric Chan.

3. Osborne, "Adorno and the Metaphysics of Modernism," and Dews, "Adorno, Poststructuralism, and the Critique of Identity," in Benjamin, *Problems of Modernity.*

4. Brunkhorst, "Irreconcilable Modernity," 52.

5. Stephan Bann, ed., *The Tradition of Constructivism* (New York: Da Capo, 1974); Richard Hollis, *Graphic Design: A Concise History* (London: Thames and Hudson, 1994).

6. Max Kozloff, *Cubism/Futurism,* (New York: Harper Icon, 1973) for a general overview of these artists and their work. Stephen Kern, *The Culture of Time and Space* (Cambridge: Harvard University Press, 1983).

7. Brunkhorst, "Irreconcilable Modernity," 52, citing Adorno, "Die Kunst und die Künste," 160.

BIBLIOGRAPHY

Abbate, Janet. *Inventing the Internet*. Cambridge: MIT Press, 1999.

Adams, Laurie. *The Methodologies of Art*. New York: Icon, 1996.

Adorno, Theodor. *Aesthetic Theory*. Minneapolis: University of Minnesota Press, 1997.

Ahrens, Donald. *Meteorology Today*. Belmont, CA : Thomson/Brooks/Cole, 2007.

Allen, J. F. "Time and Time Again: The Many Ways to Represent Time." *International Journal of Intelligent Systems* 6, no. 4 (July 1991): 341–55.

Amelunxen, H. V., S. Iglhaut, and F. Rötzer, eds., in collaboration with A. Cassel and N. G. Schneider. *Photography after Photography*. Munich: G&B Arts, 1996.

Andrews, Jim. "Vispo, Langu(im)age." http://www.vispo.com

Ascott, Roy. *Telematic Embrace*. Berkeley: University of California Press, 2003.

Austen, Jeremy, and Christopher Perfect. *The Complete Typographer*. Engelwood Cliffs, NJ: Prentice Hall, 1992.

Austin, John L. *How to Do Things with Word*s. Cambridge: Harvard University Press, 1967.

Avrin, Leila. *Scribes, Scripts and Books*. Chicago: American Library Association; London: British Library. 1991.

Babbage, Charles. *Charles Babbage and His Calculating Engines*. New York: Dover, 1961.

Baker, Gordon P. *Wittgenstein, Frege, and the Vienna Circle*. Oxford: Blackwell, 1988.

Bakhtin, Mikhail. *The Dialogic Imagination*. Austin: University of Texas Press, 1981.

Bal, Mieke. *Looking In.* New York: Routledge, 2000.
Balsamo, Anne. *Technologies of the Gendered Body: Reading Cyborg Women.* Durham, NC: Duke University Press, 1996.
Barthes, Roland. *S/Z.* Paris: Editions du Seuil, 1970.
———. *Image/Music/Text.* New York: Hill and Wang, 1977.
———. *Mythologies.* Paris: Editions du Seuil, 1957.
Baudrillard, Jean. *For a Critique of the Political Economy of the Sign.* St. Louis: Telos Press, 1981.
———. *Simulations.* New York: Semiotext(e), 1983.
Beardsley, Monroe. *Aesthetics from Classical Greece to the Present: A Short History.* Tuscaloosa: University of Alabama Press, 1966.
Bell, Clive. *Art.* New York: Capricorn Books, 1958.
Benjamin, Andrew, ed. *The Problems of Modernity: Adorno and Benjamin.* London: Routledge, 1989.
Bense, Max. "The Projects of Generative Aesthetics." In *Cybernetics, Art and Ideas,* ed. J. Reichardt, 57–60. New York: Graphics Society Limited, 1971.
Benveniste, Emile. *Problemes de linguistique générale.* Paris: Gallimard, 1966–84.
Berger, John. *Ways of Seeing.* New York: Penguin, 1972.
Bernstein, Charles. *Veil.* LaFarge, WI: Xexoxial Editions, 1987.
Bertin, Jacques. *The Semiology of Graphics.* Madison: University of Wisconsin Press, 1973.
Besser, Howard. "Digital Libraries, Standards, Metadata, and Longevity Activities." http://besser.tsoa.nyu.edu/howard/#standards
Binski, Paul. *Cambridge Illuminations: Ten Centuries of Book Production in the Medieval West.* London: Harvey Miller, 2005.
Bök, Christian. *Pataphysics: The Poetics of an Imaginary Science.* Evanston, IL: Northwestern University Press, 2002.
Bolter, J. David. *Writing Space.* Matwah, NJ: L. Erlbaum Associates, 1991.
Bolter, J. David, with Diane Gromala. *Windows and Mirrors: Experience Design, Digital Art and the Myth of Transparency.* Cambridge: MIT Press, 2005.
Bray, Joe, Miriam Handley, and Anne C. Henry, eds. *Marking the Text.* Aldershot: Ashgate, 2000.
Bronstein, H. "Time Schemes, Order, and Chaos: Periodization and Ideology." In *Time, Order, Chaos: The Study of Time IX,* ed. J. T. Fraser. Madison, CT: International Universities Press, 1998.
Brunkhorst, H. "Irreconcilable Modernity: Adorno's Aesthetic Experimentalism and the Transgression Theorem." In *The Actuality of Adorno,* ed. M. Pensky. Albany: State University of New York Press, 1997.
Bryson, Norman, Michael Holly, and Keith Moxey, eds. *Visual Culture: Images and Interpretations.* Hanover, NH: University Press of New England, 1994.
Burg, J., A. Boyle, and S.-D. Lang. "Using Constraint Logic Programming to Analyze the Chronology in *A Rose for Emily.*" *Computers and the Humanities* 34, no. 4 (December 2000): 377–92.
Buzzetti, Dino. "Text Representation and Textual Models." http://www.iath.virginia.edu/ach-allc.99/proceedings/buzzetti.html
Card, Stuart K., Jock D. Mackinlay, and Ben Shneiderman. *Readings in Information Visualization: Using Vision to Think.* San Francisco: Morgan Kaufmann Publishers, 1999.

Carruthers, Mary J. *The Craft of Thought: Meditation, Rhetoric, and the Making of Images, 400–1200.* Cambridge: Cambridge University Press, 1998.

Carey, James. *Communication as Culture.* Boston: Unwin Hyman, 1989.

Cayley, John. "Indra's Net or Hologography." http://homepage.mac.com/shadoof/net/in/inhome.html

Chipp, Herschell Browning. *Theories of Modern Art.* Berkeley: University of California Press, 1968.

Chomsky, Noam. *Cartesian Linguistics: A Chapter in the History of Rationalist Thought.* Christchurch, New Zealand: Cybereditions, 2002.

Chun, Wendy. *Control and Freedom.* Cambridge: MIT Press, 2006.

Comaromi, John. *Dewey Decimal Classification: History and Current Status.* New Delhi: Sterling, 1989.

Corby, Tom. *Network Art.* New York: Routledge, 2006.

Coward, Rosalind, and John Ellis. *Language and Materialism: Developments in Semiology and the Theory of the Subject.* London: Routledge and Paul, 1977.

Crevier, Daniel. *AI: The Tumultuous History of the Search for Artificial Intelligence.* New York: Basic Books, 1993.

Critical Art Ensemble. Home page. http://www.critical-art.net

Davis, Martin. *Universal Computer.* New York: W. W. Norton, 2000.

Day, Michael. "Metadata for Digital Preservation: A Review of Recent Developments." http://www.ukoln.ac.uk/metadata/presentations/ecdl2001-day/paper.html

Deleuze, Gilles, and Felix Guattari. *Rhizome: Introduction.* Paris: Les Editions de Minuit, 1976.

———. *Anti-Oedipus: Capitalism and Schizophrenia.* New York: Viking, 1977.

de Man, Paul. *Allegories of Reading.* New Haven: Yale University Press, 1979.

———. *Blindness and Insight.* New York: Oxford University Press, 1971.

Dennett, Daniel. *Brainchildren.* Cambridge: MIT Press, 1998.

De Silva, Clarence W. *Intelligent Machines: Myths and Realities.* Boca Raton: CRC Press, 2000.

Derrida, Jacques. *Edmund Husserl's Origin of Geometry: An Introduction.* Trans. John P. Leavey Jr. Lincoln: University of Nebraska Press, 1989.

———. *Of Grammatology.* Baltimore: Johns Hopkins University Press, 1976.

———. *Writing and Difference.* Chicago: University of Chicago Press, 1978.

Despeyroux, Joëlle, and Robert Harper. "Logical Frameworks and Metalanguages." *Journal of Functional Programming* 13 (2003):257–60.

Dibbell, Julian. "A Rape in Cyberspace." *Village Voice* 38, no. 51, December 21, 1993.

Dick, Philip K. *Do Androids Dream of Electric Sheep?* Garden City, NY: Doubleday, 1968.

Dreyfus, Hubert. *What Computers Can't Do.* New York: Harper and Row, 1979.

Drucker, Johanna. "Digital Ontologies." *Leonardo* 34, no. 2 (2001): 141–45.

———. *Figuring the Word* (New York: Granary Books, 1998).

———. "Graphesis." http://www.noraproject.org/reading.ph.

———. "Graphical Readings and the Visual Aesthetics of Textuality." *Text, Transactions of the Society for Textual Scholarship* 16 (2006): 267–76.

———. "Interactive, Algorithmic, Networked." In *At a Distance: Precursors to Art and Activism on the Internet,* ed. Annmarie Chandler and Norie Neumark, 34–59. Cambridge: MIT Press, 2005.

———. *The Visible Word: Experimental Typography and Modern Art, 1909–1923.* Chicago: University of Chicago Press, 1994.

———. "What Is a Letter?" In *The Education of a Typographer,* ed. Steven Heller, 78–90. New York: Allworth Press, 2004.
———, ed. "Digital Reflections: The Dialogue of Art and Technology." Comment introducing special issue of the same title. *Art Journal,* Fall 1997, 2.
Druckrey, Timothy. *Iterations.* New York: International Center of Photography; Cambridge: MIT Press, 1993.
———. *Ars Electronica.* Cambridge: MIT Press, 1999.
Duggan, Hoyt. "The Piers Plowman Electronic Archive." http://www.iath.virginia.edu/piers/
Dunn, D., ed. *Pioneers of Electronic Art.* Santa Fe: Ars Electronica and the Vasulkas, 1992.
Eco, Umberto. *Theory of Semiotics.* Bloomington: University of Indiana Press, 1976.
Elkins, James. *The Domain of Images.* Ithaca: Cornell University Press, 1999.
Esposito, Joseph. "The Processed Book." *First Monday* 8, no. 3 (2003). http://www.firstmonday.org/issues/issue8_3/esposito
Febvre, Lucien, and H.-J. Martin. *The Coming of the Book.* London: Verso, 1997.
Fiedler, Conrad. *On Judging Works of Visual Art.* Berkeley: University of California Press, 1978.
Foucault, Michel. *The Order of Things.* New York: Pantheon, 1970.
———. *The Achaeology of Knowledge.* New York: Harper & Row, 1972.
Franke, Herbert W. *Computer Graphics Computer Art.* New York: Phaidon, 1971.
Franke, Herbert W., and Horst S. Helbig. "Generative Mathematics: Mathematically Described and Calculated Visual Art." In *The Visual Mind,* ed. M. Emmer, 101–4. Cambridge: MIT Press, 1993.
Fraser, James T. "From Chaos to Conflict." In *Time, Order, Chaos: The Study of Time IX,* ed. J. T. Fraser, Marlene P. Soulsby, and Alexander J. Argyros, 3–19. Madison, CT: International Universities Press, 1998.
———. *Time, the Familiar Stranger.* Cambridge: Massachusetts University Press, 1987.
Freedberg, David. *Eye of the Lynx.* Chicago: University of Chicago Press, 2002.
Fried, Michael. *Art and Objecthood.* Chicago: University of Chicago Press, 1998.
Fry, Roger. *Vision and Design.* Harmondsworth: Penguin, 1937.
Fuller, Matthew. *Behind the Blip.* Brooklyn: Autonomedia, 2003.
Galison, Peter. *Image and Logic: A Material Culture of Microphysics.* Chicago: University of Chicago Press, 1997.
Gardner, Martin. *Logic Machines and Diagrams.* Chicago: University of Chicago Press, 1982.
Genette, Gérard. *Nouveau discours du récit.* Paris: Editions du Seuil, 1983.
Gibson, James Jerome. *The Ecological Approach to Visual Perception.* 1979; Hillsdale, NJ: Lawrence Erlbaum Associates, 1986.
Glazier, Loss P. *Digital Poetics: The Making of E-Poetries.* Tuscaloosa: University of Alabama Press, 2002.
Goodman, Nelson. *Languages of Art: An Approach to a Theory of Symbols.* Indianapolis: Bobbs-Merrill, 1968.
Grafton, Anthony. *The Footnote.* Cambridge: Harvard University Press, 1997.
Gray, Chris Habels, and Steven Mentor. *The Cyborg Handbook.* New York: Routledge, 1995.
Greenberg, Clement. *Art and Culture.* Boston: Beacon, 1961.
Gregory, Richard L. *Eye and Brain.* Princeton: Princeton University Press, 1990.

Hansen, Mark. *New Philosophy for New Media.* Cambridge: MIT Press, 2004.
Haraway, Donna. *Simians, Cyborgs, and Women.* New York: Routledge, 1991.
Harrison, Charles, and Paul Wood, eds. *Art in Theory.* Oxford: Blackwell, 1993.
Hasse, L., and F. Dobson. *Introductory Physics of the Atmosphere and Ocean.* Dordrecht: D. Reidel, 1986.
Hayles, N. Katherine. *How We Became Posthuman.* Chicago: University of Chicago Press, 1999.
Haynes, Cynthia, and Jan Rune Holmevik, eds. *High Wired: On the Design, Use, and Theory of Educational MOOs.* Ann Arbor: University of Michigan Press, 1998.
Heim, Michael. *Electric Language.* New Haven: Yale University Press, 1987.
Heisenberg, Werner. *Philosophical Problems of Quantum Physics.* Woodbridge, CT: Ox Bow Press, 1979.
Henke, H. A. "The Global Impact of eBooks on ePublishing." *Proceedings of the 19th Annual International Conference on Computer Documentation, 172–80.* New York: ACM, 2001. http://portal.acm.org/citation.cfm?id=501551
Hershman-Leeson, Lynn. *Clicking In: Hot Links to a Digital Culture.* Seattle: Bay Press, 1996.
Higgins, Dick. *FOEW&OMBWHNW.* New York: Something Else Press, 1969.
Hjelmslev, Louis. *Prolegomena to a Theory of Language.* Madison: University of Wisconsin Press, 1963.
Hobart, James. "Principles of Good GUI Design." http://www.iie.org.mx/Monitor/v01n03/ar_ihc2.htm
Hockey, Susan. *Electronic Texts in the Humanities.* Oxford: Oxford University Press, 2000.
Hofstadter, Douglas. *Metamagical Themas.* New York: Basic Books, 1985.
Hollis, Richard. *Graphic Design: A Concise History.* London: Thames and Hudson, 1994.
Horn, Robert E. *Visual Language.* Bainbridge Island, WA: Macrovu, 1991.
Howard, Philip N. *Society Online: The Internet in Context.* Thousand Oaks: Sage Publications, 2004.
Innis, Harold. *Empire and Communications.* Toronto: University of Toronto Press, 1972.
Ivins, William. *Art and Geometry.* New York: Dover, 1946.
Jarry, Alfred. *Exploits and Opinions of Dr. Faustroll, Pataphysician.* Boston: Exact Change, 1996.
Jensen, C. S., J. Clifford, S. K. Gadia, A. Segev, and R. T. Snodgrass. "A Glossary of Temporal Database Concepts." *Proceedings of ACM SIGMOD International Conference on Management of Data* 23, no. 1, March 1994.
Jones, Steve. *CyberSociety 2.0: Revisiting Computer-Mediated Communication and Community.* Thousand Oaks: Sage Publications, 1998.
Jordan, P. W. "Determining the Temporal Ordering of Events in Discourse." Masters thesis, Carnegie Mellon Computational Linguistics Program, 1994.
Joy, Bill. "Why the Future Doesn't Need Us." http://www.wired.com/wired/archive/8.04/joy.html
Jussim, Estelle. *Visual Communication and the Graphic Arts: Photographic Technologies in the Nineteenth Century.* New York: R. R. Bowker, 1974.
Kelly, Michael, ed. *Encyclopedia of Aesthetics.* Oxford: Oxford University Press, 1998.
Kemp, Martin. *Visualizations: The Nature Book of Art and Science.* Berkeley: University of California Press, 2000.

Kern, Stephen. *The Culture of Time and Space.* Cambridge: Harvard University Press, 1983.

Kirschenbaum, Matthew. Home page. http://www.otal.umd.edu/-mgk/blog/

———. "Lines for a Virtual T[y/o]pography," http://www.iath.virginia.edu/-mgk3k/dissertation/title.html

———. *Mechanisms: New Media and the Forensic Imagination.* Cambridge: MIT Press, 2008.

Knuth, Donald. *Tex and Metafont.* Bedford, MA: American Mathematical Society and Digital Press, 1979.

Kolko, Beth E., Lisa Nakamura, and Gilbert B. Rodman. *Race in Cyberspace.* New York: Routledge, 2000.

Kosslyn, Stephen. *Image and Mind.* Cambridge: Harvard University Press, 1980.

Kroker, Arthur. *Digital Delirium.* New York: St. Martin's Press, 1997.

Kurzweil, Ray. *The Age of Intelligent Machines.* Cambridge: MIT Press, 1990.

Landow, George. *Hypertext.* Baltimore: Johns Hopkins University Press, 1992.

Lewis, Peter. *Wittgenstein: Aesthetics and Philosophy.* Burlington, VT: Ashgate, 2004.

LeWitt, Sol. "Paragraphs on Conceptual Art." *Artforum,* June 1967.

Licklider, J. C. *Libraries of the Future.* Cambridge: MIT Press, 1965.

Lister, Martin, ed. *The Photographic Image in Digital Culture.* London: Routledge, 1995.

Liu, Alan. *The Laws of Cool: Knowledge Work and the Culture of Information.* Chicago: University of Chicago Press, 2004.

"Logical Frameworks and Metalanguages." Special issue of *Journal of Functional Programming.* 13 (2003): 257–60.

Loizeaux, Elizabeth Bergmann, and Neil Fraistat. *Reimagining Textuality.* Madison: University of Wisconsin Press, 2002.

Lovejoy, Margot. *Postmodern Currents: Art and Artists in the Age of Electronic Media.* Ann Arbor: UMI Research Press, 1989.

Lupton, Ellen, and J. Abbott Miller. *Design Writing Research.* New York: Kiosk, 1996.

Lycan, W. G. *Philosophy of Language: A Contemporary Introduction.* New York: Routledge, 2000.

Lynch, Clifford. "The Battle to Define the Future of the Book in the Digital World." *First Monday* 6, no. 6 (2001). http://www.firstmonday.org/issues/issue6_6/lynch

MacEachren, Alan. *How Maps Work.* New York: Guilford Press, 1995.

Machlup, Fritz. *Information: Interdisciplinary Messages.* New York: Wiley and Sons, 1983.

Mandel, Theo. *The Elements of User Interface Design.* New York: Wiley and Sons, 1997.

Manovich, Lev. *The Language of New Media.* Cambridge: MIT Press, 2001.

Marcus, Aaron, Nick Smilonich, and Lynne Thompson. *The Cross-GUI Handbook for Multiplatform User Interface Design.* Boston: Addison-Wesley Longman, 1994.

Marr, D. *Vision: A Computational Investigation into the Human Representation and Processing of Visual Information.* San Francisco: W. H. Freeman, 1982.

Mathes, Adam. "Folksonomies—Cooperative Classification and Communication through Shared Metadata." http://www.adammathes.com/academic/computer-mediated-communication/folksonomies.html

Mathews, Harry. *Oulipo Compendium.* London: Atlas Press, 1998.

Maturana, H. R., and F. J. Varela. *Autopoiesis and Cognition: The Realization of the Living.* Boston: D. Reidel, 1980.

McCaffery, Steve, and bp nichol. *Rational Geomancy.* Vancouver: Talonbooks, 1992.

McCarty, Willard. *Humanities Computing*. Hampshire: Palgrave, 2005.
McGann, Jerome. *Black Riders: The Visible Language of Modernism*. Princeton: Princeton University Press, 1993.
———. *The Critique of Modern Textual Criticism*. Chicago: University of Chicago Press, 1983.
———. *Dante Gabriel Rossetti and the Game That Must Be Lost*. New Haven: Yale University Press, 2000.
———. *Radiant Textuality*. New York: Palgrave, 2001.
———. "Texts in N-Dimensions and Interpretation in a New Key." http://texttechnology.mcmaster.ca/pdf/vol12_2_02.pdf
———. *The Textual Condition*. Princeton: Princeton University Press, 1991.
McLeod, Randall. Lectures on "Material Narratives." http://www.sas.upenn.edu/~traister/pennsem.html
McLuhan, Marshall. *The Medium Is the Massage*. New York: Bantam Books, 1967.
———. *Understanding Media*. New York: New American Library, 1964.
McMurtrie, Douglas. *The Book: The Story of Printing and Bookmaking*. New York: Dorset, 1943.
McShine, Kynaston. *Information*. New York: Museum of Modern Art, 1970.
Minsky, Marvin. *The Society of Mind*. New York: Simon and Schuster, 1986.
Minsky, Marvin, with Seymour Papert. *Artificial Intelligence*. Eugene: Oregon State System of Higher Education, 1973.
Mijksenaar, Paul. *Visual Function*. Princeton: Princeton Architectural Press, 1987.
Mitchell, William J. *City of Bits: Space, Place, and the Infobahn*. Cambridge: MIT Press, 1995.
———. *The Reconfigured Eye: Visual Truth in the Post-Photographic Era*. Cambridge: MIT Press, 1994.
Moreau, René. *The Computer Comes of Age*. Cambridge: MIT Press, 1984.
Moretti, Franco. "Graphs, Maps, and Trees." *New Left Review*, no. 24, November–December 2003.
Morison, Stanley. *Selected Essays on the History of Letter-Forms in Manuscript and Print*. Cambridge: Cambridge University Press, 1981.
Morse, Margaret. *Virtualities*. Bloomington: Indiana University Press, 1998.
Moser, Mary Anne, ed., with Douglas MacLeod. *Immersed in Technology*. Cambridge: MIT Press, 1996.
Mosely, James. *Romain du Roi*. Lyons: Musée de l'imprimerie, 2002.
Motte, Warren. *OuLiPo: A Primer of Potential Literature*. Normal, IL: Dalkey Archive Press, 1998.
Nelson, Ted. "Project Xanadu." http://www.xanadu.net
Nielson, Jakob. "Top Ten Web Design Mistakes." May 30, 1999. http://www.useit.com/alertbox/990530.html
Nyce, James M., and Paul Kahn, eds. *From Memex to Hypertext: Vannevar Bush and the Mind's Machine*. Boston: Academic Press, 1991.
Ong, Walter. *Ramus, Method, and the Decay of Dialogue*. Cambridge: Harvard University Press, 1958.
O'Toole, M. A. "The Theory of Serialism in *The Third Policeman*." *Irish University Review* 18, no. 2 (1988): 215–25. 1988.
Panofsky, Erwin. *Studies in Iconology*. New York: Harper and Row, 1972.
Parkes, Malcolm B. "The Influence of the Concepts of *Ordinatio* and *Compilatio* on the

Development of the Book." In *Medieval Learning and Literature,* ed. J. J. G. Alexander and M. T. Gibson, 115–41. Oxford: Clarendon Press, 1976.

Peirce, Charles. *The Philosophical Writings of Charles Peirce.* Ed. Justus Buchler. New York: Dover, 1955. (Republication of *The Philosophy of Peirce: Selected Writings.* London: Routledge and Kegan Paul, 1940.)

Penny, Simon. *Critical Issues in Electronic Media.* Albany: State University of New York Press, 1995.

Popper, Frank. *From Technological to Virtual Art.* Cambridge: MIT Press, 2007.

Powers, Richard. *Galatea 2.2.* New York: Farrar, Straus, and Giroux, 1995.

Price, H. "The View from Nowhen." In *Time's Arrow and Archimedes' Point.* New York: Oxford University Press, 1996.

Price, William Charles. *The Uncertainty Principle and Foundations of Quantum Mechanics: A Fifty Years' Survey.* New York: Wiley, 1977.

Prueitt, Melvin. *Art and the Computer.* New York: McGraw-Hill, 1984.

Raley, Rita. "Interferences: [Net.Writing] and the Practice of Codework." http://www.electronicbookreview.com/thread/electropoetics/net.writing

———. "Reveal Codes: Hypertext and Performance." http://www.iath.virginia.edu/pmc/text-only/issue.901/12.1raley.txt

Reichardt, Jasia. *The Computer in Art.* New York: Van Nostrand Reinhold; London: Studio Vista, 1971.

———. *Cybernetics, Art, and Ideas.* Greenwich, CT: New York Graphic Society, 1971.

———. *Cybernetic Serendipity.* New York: Praeger, 1968.

Renear, Allen. "Out of Praxis: Three (Meta) Theories of Textuality." In *Electronic Text,* ed. K. Sutherland, 107–26. Oxford: Clarendon Press, 1997.

Renear, Allen, with Steve De Rose, David G. Durand, and Elli Mylonas. "What Is Text, Really?" *Journal of Computing in Higher Education* 1, no.2 (Winter 1990): 3–26.

Reynolds, Robert, and Thomas Zummer, eds. *CRASH: Nostalgia for the Absence of Cyberspace.* New York: Thread Waxing Space, 1994.

Reynolds, Teri. "Spacetime and Imagetext." *Germanic Review* 73, no. 2 (Spring 1998): 161–74.

Ritchin, Fred. *In Our Own Image.* New York: Aperture, 1990.

Rucker, Rudy. *Software.* New York: Eos/HarperCollins, 1987.

Saint-Martin, Fernande. *Semiotics of Visual Language.* Bloomington: University of Indiana Press, 1990.

Saussure, Ferdinand de. *Course in General Linguistics.* Ed. Charles Bally and Albert Sechehaye. London: Duckworth, 1983.

Schmid, Calvin. *Handbook of Graphic Presentation.* New York: Ronald Press, 1954.

Scholem, Gershem. *Kabbalah.* New York: Meridian, 1978.

Schreiber, F. A. "Is Time a Real Time? An Overview of Time Ontology in Informatics." In *Real Time Computing,* ed. W. A. Halang and A. D. Stoyenko, 283–307 (Springer Verlag, 1992).

Schreibman, Susan, Ray Siemens, and John Unsworth, eds. *A Companion to Digital Humanities.* Malden, MA: Blackwell, 2004.

Schrödinger, Erwin. *My View of the World.* Cambridge: Cambridge Uniersity Press, 1964.

———. *Science and Humanism: Physics in Our Time.* Cambridge: Cambridge University Press, 1951.

Sebeok, Thomas. *Introduction to Semiotics.* London: Pinter, 1994.

Seuphor, Michel. *Abstract Painting.* New York: Abrams, 1962.
Shanken, E. "The House That Jack Built." http://www.duke.edu/-giftwrap/, http://mitpress.mit.edu/e-journals/LEA/ARTICLES/jack.html
Shannon, Claude. "A Mathematical Theory of Communication." Bell Labs, 1947. http://cm.bell-labs.com/cm/ms/what/shannonday/paper.html
Shneiderman, Ben, and Catherine Plaisant. *Designing the User Interface: Strategies for Effective Human-Computer Interaction.* Boston: Pearson/Addison Wesley, 2005.
Shulman, Holly, ed. Dolley Madison Digital Edition. University of Virginia Press. http://www.upress.virginia.edu/books/shulman.html
Simon, Herbert. *Representation and Meaning: Experiments with Information Processing Systems.* Englewood Cliffs, NJ: Prentice Hall, 1972.
Smith, M. M. "The Design Relationship between the MSS. and the Incunable." In *A Millennium of the Book,* ed. R. Meyers and M. Harris. Winchester, England: St. Paul's, 1994.
Smith, Paul. *Discerning the Subject.* Minneapolis: University of Minnesota Press, 1988.
Sondheim, Alan. *Disorders of the Real.* Barrytown, NY: Station Hill Press, 1988.
Spencer-Brown, George. *The Laws of Form.* New York: Julian Press, 1972.
Stafford, Barbara. *Good Looking.* Cambridge: MIT Press, 1996.
Stainton, Robert J. *Philosophical Perspectives on Language.* Peterborough, Ontario: Broadview Press, 1996.
Steedman, M. "The Productions of Time." Draft tutorial notes 2.0. University of Edinburgh. ftp://ftp.cis.upenn.edu/pub/steedman/temporality/
Stefans, Brian Kim. *Fashionable Noise.* Berkeley, CA: Atelos, 2003.
Stillman, Robert. *The New Philosophy and Universal Languages in Seventeenth-Century England: Bacon, Hobbes, and Wilkins.* Lewisburg: Bucknell University Press, 1995.
Sutherland, Kathryn, ed. *Electronic Text.* Oxford: Clarendon Press, 1997.
TEI Consortium. *Guidelines for Electronic Text Encoding and Interchange.* Humanities Computing Unit, University of Oxford, 2002. http://www.tei-c.org/Guidelines/index.xml
Todorov, Tveztan. *Theories of the Symbol.* Oxford: Blackwell, 1982.
Tufte, Edward. *The Visual Display of Quantiative Information.* Cheshire, CT: Graphic Press, 1983.
Turing, Alan. "On Computable Numbers, with an Application to the Entscheidungsproblem." *Proceedings of the London Mathematical Society,* series 2, vol. 42 (1936).
Twyman, Michael. *Printing 1770–1970: An Illustrated History of Its Development and Uses in England.* London: British Library, 1998.
Updike, Daniel B. *Printing Types.* Cambridge: Harvard University Press/Belknap, 1961.
Varela, Francisco J., Evan Thompson, and Eleanor Rosch. *The Embodied Mind: Cognitive Science and Human Experience.* Cambridge: MIT Press, 1991.
Vesna, Victoria, ed. *Database Aesthetics: Art in the Age of Information Overflow.* Minneapolis: University of Minnesota Press, 2008.
Virilio, Paul. *The Vision Machine.* Cambridge: MIT Press, 1995.
von Foerster, Heinz. *Observing Systems.* Salinas, CA: Intersystems Publications, 1981.
von Glasersfeld, Ernst. "An Introduction to Radical Constructivism." In *The Invented Reality: How Do We Know?,* ed. P. Watzlawick, 17–40. New York: W. W. Norton, 1984.
———. *Radical Constructivism: A Way of Knowing and Learning.* London: Falmer Press, 1995.

Weaver, Warren, and Claude Shannon. *The Mathematical Theory of Communication.* Urbana: University of Illinois Press, 1949.
Weigand, Wayne. "The 'Amherst Method': The Origins of the Dewey Decimal Classification Scheme." *Libraries & Culture* 33, no. 2, Spring 1998. http://www.gslis.utexas.edu/~landc/fulltext/LandC_33_2_Wiegand.pdf
Wells, H. G. *World Brain: The Idea of a Permanent World Encyclopedia.* 1937. https://sherlock.sims.berkeley.edu/wells/world_brain.html
Wiener, Norbert. *Cybernetics of Control and Communication in the Animal and the Machine.* New York: Wiley, 1948.
Wilkins, John. *An Essay towards a Real Character and Philosophical Language.* London: Printed for Sa. Gellibrand, and for John Martyn, 1668.
Wilson, Stephen. *Information Arts: Intersections of Art, Science, and Technology.* Cambridge: MIT Press, 2002.
Winograd, Terry. *Artificial Intelligence and Language Comprehension.* Washington, DC: U.S. Department of Health, Education, and Welfare, National Institute of Education, 1976.
Wurman, Richard Saul. *Information Architects.* New York: Watson-Guptill, 1997.
Würtz, Elizabeth. "A Cross-Cultural Analysis of Websites from High-Context Cultures and Low-Context Cultures." *Journal of Computer-Mediated Communication* 11, no. 1, article 13. http://jcmc.indiana.edu/vol11/issue1/wuertz.html
Yates, Frances. *The Art of Memory.* Chicago: University of Chicago Press, 1966.
———. *Lull and Bruno.* London: Routledge and Kegan Paul, 1982.

INDEX

List of Figures

1. Introduction

1.1 Nonlinear Control Design: Motivation and Overview

Since all physical systems are inherently nonlinear with system nonlinearities arising from numerous sources including, for example, friction (e.g., Coulomb, hysteresis), gyroscopic effects (e.g., rotational motion), kinematic effects (e.g., backlash), input constraints (e.g., saturation, deadband), and geometric constraints, plant nonlinearities must be accounted for in the control-system design process. However, since nonlinear systems can exhibit multiple equilibria, limit cycles, bifurcations, jump resonance phenomena, and chaos, general nonlinear system stabilization is notoriously difficult and remains an open problem. Control system designers have usually resorted to Lyapunov methods [137, 89, 73] in order to obtain stabilizing controllers for nonlinear systems. In particular, for smooth feedback, Lyapunov-based methods were inspired by Jurdjevic and Quinn [70] who give sufficient conditions for smooth stabilization based on the ability of constructing a Lyapunov function for the closed-loop system [134]. Unfortunately, however, there does not exist a unified procedure for finding a Lyapunov function candidate that will stabilize the closed-loop system for general nonlinear systems. This is further exacerbated when addressing robustness in uncertain nonlinear systems.

Recent work involving differential geometric methods [25, 66, 26, 117, 28, 27] has made the design of controllers for certain classes of nonlinear systems more methodical. Such frameworks include the concepts of zero dynamics and feedback linearization and require that the system zero dynamics are asymptotically stable assuring the existence of globally defined diffeomorphisms to transform the nonlinear system into a normal form [66]. For this class of systems, feedback linearization techniques usually rely on cancelling out system nonlinearities using feedback and may therefore lead to inefficient designs since feedback linearizing controllers may generate unnecessarily large control effort to cancel beneficial system nonlinearities [47]. Furthermore, a serious drawback of all feedback linearization techniques is the failure to account for system uncertainty since exact cancellation of the nonlinear dynamics via feedback is required and hence an exact knowledge of the dynamics is as-

sumed resulting in non-robust designs. Even though robustness frameworks to parametric uncertainty via feedback linearization techniques involving a two stage design consisting of nominal feedback linearization followed by additional state feedback designed to guarantee robustness have been developed [118, 126, 130, 131], the fact that such approaches do not *directly* account for system uncertainty can result in severe robustness problems with respect to nonlinear errors internal to the system dynamics. Furthermore, restrictive matching conditions are imposed to the structure of the uncertainty in order to address general feedback linearizable systems [129].

Backstepping control has also recently received a great deal of attention in the nonlinear control literature [72, 76, 77]. The popularity of this control methodology can be explained in a large part due to the fact that it provides a systematic procedure for finding a Lyapunov function for nonlinear closed-loop cascade systems. Furthermore, the controller is obtained in such a way that the nonlinearities of the dynamical system, which may be useful in attaining performance objectives, do not need to be cancelled as in state or output feedback linearization techniques. However, once again this approach is limited to strict-feedback systems [77, 119].

To address optimality issues within nonlinear control-system design, one may consider an optimal control problem in which a performance functional is minimized along the closed loop system trajectories. This problem leads to the maximum principle [24, 74] which usually provides open-loop optimal control laws, characterized via nonlinear two-point boundary-value problems requiring iterative solution schemes. Alternatively, to obtain feedback optimal controllers, one may formulate the optimal control problem as a dynamic programming problem [21] by considering a value function [127], or return function [24], that minimizes a cost functional among all possible system trajectories. In this case, the value function is given by the solution to the Hamilton-Jacobi-Bellman equation [24, 74]. Computational methods for solving the Hamilton-Jacobi-Bellman partial differential equation have not been developed to a level comparable to solving Riccati equations arising in linear optimal control problems and, in general, it is very difficult to solve. In fact, solutions may not even exist unless one allows a generalized notion of viscosity solutions [19]. In order to avoid the complexity in solving the Hamilton-Jacobi-Bellman equation one may consider an inverse optimal control problem [105, 47, 119] where one does not attempt to minimize a given cost functional, but rather, minimizes a derived cost functional. The basic underlying idea of inverse optimality is the fact that the steady-state solution to the Hamilton-Jacobi-Bellman equation is a Lyapunov function for the nonlinear controlled system [22, 67]. However, in this case the complexity of solving the Hamilton-Jacobi-Bellman equation is shifted to finding a Lyapunov

function for the closed-loop system limiting this approach to strict-feedback systems. Furthermore, the performance of inverse optimal controllers can be arbitrarily poor when compared to the optimal performance as measured by a designer specified cost functional.

If the operating range of the control system is small and if the system nonlinearities are smooth, then the control system can be locally approximated by a linearized system about a given operating condition and linear multivariable control theory can be used to maintain local stability and performance. However, in high performance engineering applications such as advanced tactical fighter aircraft and variable-cycle gas turbine aeroengines, the locally approximated linearized system does not cover the operating range of the system dynamics. In this case, gain scheduled controllers can be designed over several fixed operating points covering the system's operating range and controller gains interpolated over this range [132, 115]. However, due to approximation linearization errors and neglected operating point transitions, the resulting gain scheduled system does not have any guarantees of performance or stability. Even though stability properties of gain scheduled controllers are analyzed in [122, 84] and stability guarantees are provided for plant output scheduling, a ***design*** framework for gain scheduling control guaranteeing system stability over an operating range of the nonlinear plant dynamics has not been addressed in the literature.

In an attempt to develop a design framework for gain scheduling control, linear parameter-varying system theory has been developed [123, 124, 121, 108]. Since gain scheduling control involves a linear parameter-dependent plant, linear parameter-varying methods for gain scheduling seem natural. However, even though this is indeed the case for linear dynamical systems involving exogenous parameters, this is not the case for nonlinear dynamical systems. This is due to the fact that a nonlinear system cannot be represented as a true linear parameter-varying system since the varying system parameters are endogenous, that is, functions of the system state. Hence, stability and performance guarantees of linear parameter-varying systems do ***not*** extend to the nonlinear system. Of course, in the case where the magnitude and rate of the endogenous parameters are constrained such that the linear parameter-varying system ***hopefully*** behaves closely to the actual nonlinear system, then stable controllers can be designed using quasi-linear parameter-varying representations [80]. However, in the case of unexpectedly large amplitude uncertain exogenous disturbances and/or system parametric uncertainty, ***a priori*** assumptions on magnitude and rate constraints on endogenous parameters are unverifiable.

For nonlinear controlled dynamical systems null controllability does not in general imply continuous or smooth stabilizability [10, 113]. Variable struc-

ture control is perhaps the quintessential discontinuous nonlinear stabilization approach for controlling nonlinear systems (see [135, 39] and the numerous references therein). In particular, variable structure control utilizes a high-speed switching control law to drive the system state trajectories onto an invariant sliding manifold. In order to establish the existence of a sliding mode, the controller is constructed such that the system state velocity vector is directed towards the switching surface. In an ideal case, this results in infinitely fast switching. However, in practical implementation, infinitely fast switching leads to high frequency chattering which may excite unmodeled high frequency system dynamics leading to system instabilities. In an attempt to overcome this problem, boundary layer controllers which continuously approximate the discontinuous control action in a neighborhood of the switching surface have been proposed [39]. However, in this case the resulting controller does not guarantee asymptotic stability but rather uniform ultimate boundedness [39]. Finally, variable structure controllers are generally limited to nonlinear affine systems.

1.2 Brief Outline of the Monograph

The main objective of this monograph is to develop a general nonlinear control design methodology for nonlinear uncertain systems with input saturation constraints. The results are then applied to the control of rotating stall and surge in jet engine compression systems. The main contents of the monograph are as follows. In Chapter 2 we develop generalized Lyapunov and invariant set theorems for nonlinear dynamical systems wherein all regularity assumptions on the Lyapunov function and the system dynamics are removed. In particular, local and global stability theorems are presented using generalized Lyapunov functions that are lower semicontinuous. In the case where the generalized Lyapunov function is taken to be a C^1 function, our results collapse to the standard Lyapunov stability and invariant set theorems. The present formulation provides *new* invariant set stability theorem generalizations by explicitly characterizing system limit sets in terms of lower semicontinuous Lyapunov functions.

The generalized Lyapunov and invariant set theorems developed in Chapter 2 are used in Chapter 3 to develop a nonlinear control design framework predicated on a hierarchical switching controller architecture. Specifically, using ***equilibria-dependent Lyapunov functions***, or instantaneous (with respect to a given parameterized equilibrium manifold) Lyapunov functions, a hierarchical nonlinear control strategy is developed that stabilizes a given nonlinear system using a supervisory switching controller that coordinates lower-level

stabilizing subcontrollers. The hierarchical switching nonlinear controller architecture is developed based on a *generalized* lower semicontinuous Lyapunov function obtained by minimizing a potential function, associated with each domain of attraction, over a given switching set induced by the parameterized system equilibria. In Chapter 4 the hierarchical switching nonlinear controller architecture is extended to address the problem of robust stabilization for nonlinear *uncertain* systems.

In Chapter 5 we develop a self-contained first principles derivation of a multi-mode model for rotating stall and surge in axial flow compression systems, that is accessible to control-system designers requiring state space models for modern nonlinear control design. Then, we apply the hierarchical switching nonlinear control framework to mitigate the aerodynamics instabilities of rotating stall and surge in multi-mode axial compressor models while accounting for the effects of system parametric uncertainty and a rate saturation constraint on the system actuator throttle opening.

In Chapter 6 we address the problem of nonlinear stabilization for centrifugal compression systems. First, we obtain a three-state lumped parameter model for surge in centrifugal flow compression systems. The low-order centrifugal compression system model presented involves pressure and mass flow compression system dynamics using principles of conservation of mass and momentum. Furthermore, in order to account for the influence of speed transients on the compression surge dynamics, turbocharger spool dynamics are also considered. Next, we develop globally stabilizing control laws for the lumped parameter centrifugal compressor surge model with spool dynamics using the nonlinear switching control framework. Inverse optimal nonlinear switching controllers are also developed. Both switching nonlinear control frameworks are directly applicable to centrifugal compression systems with amplitude and rate saturation constraints.

Finally, in Chapter 7 conclusions are discussed.

2. Generalized Lyapunov and Invariant Set Theorems for Nonlinear Dynamical Systems

2.1 Introduction

One of the most basic issues in system theory is stability of dynamical systems. The most complete contribution to the stability analysis of nonlinear dynamical systems is due to Lyapunov [95]. Lyapunov's results, along with the Barbashin-Krasovskii-LaSalle invariance principle [18, 75, 82], provide a powerful framework for analyzing the stability of nonlinear dynamical systems. In particular, Lyapunov's direct method can provide local and global stability conclusions of an equilibrium point of a nonlinear dynamical system if a smooth (at least C^1) positive definite function of the nonlinear system states (Lyapunov function) can be constructed for which its time rate of change due to perturbations in a neighborhood of the system's equilibrium is always negative or zero, with strict negative definiteness ensuring asymptotic stability. Alternatively, using the Barbashin-Krasovskii-LaSalle invariance principle [18, 75, 82] the strict negative definiteness condition on the Lyapunov derivative can be relaxed while assuring asymptotic stability. In particular, if a smooth function defined on a compact invariant set with respect to the nonlinear dynamical system can be constructed whose derivative along the system's trajectories is negative semidefinite and no system trajectories can stay indefinitely at points where the function's derivative identically vanishes, then the system's equilibrium is asymptotically stable.

Most Lyapunov stability and invariant set theorems presented in the literature require that the Lyapunov function candidate for a nonlinear dynamical system be a C^1 function with a negative-definite derivative (see [59, 71, 73, 83, 137, 141] and the numerous references therein). This is due to the fact that the majority of the dynamical systems considered are systems possessing continuous dynamics and hence Lyapunov theorems provide stability conditions that do not require knowledge of the system trajectories [59, 71, 73, 83, 137, 141]. However, in light of the increasingly complex nature of dynamical systems such as biological systems [90], hybrid systems [140], sampled-data systems [58], discrete-event systems [110], gain scheduled systems [111, 96, 86], and constrained mechanical systems [11], system

discontinuities arise naturally. Even though standard Lyapunov theory is applicable for systems with discontinuous system dynamics and continuous motions, it might be simpler to construct discontinuous "Lyapunov" functions to establish system stability. For example, in gain scheduled control it is not uncommon to use several different controllers designed over several fixed operating points covering the system's operating range and to switch between them over this range. Even though for each operating range one can construct a C^1 Lyapunov function, to show closed-loop system stability over the whole system operating envelope for a given switching control strategy, a *generalized* Lyapunov function involving combinations of the Lyapunov functions for each operating range can be constructed [86, 96, 111]. However, in this case, as shown in Chapter 3, the generalized Lyapunov function is non-smooth and non-continuous [86, 96, 111].

In this chapter we develop generalized Lyapunov and invariant set theorems for nonlinear dynamical systems wherein all regularity assumptions on the Lyapunov function and the system dynamics are removed. In particular, local and global stability theorems are presented using generalized Lyapunov functions that are lower semicontinuous. Furthermore, generalized invariant set theorems are derived wherein system trajectories converge to a union of largest invariant sets contained on the boundary of the intersections over finite intervals of the closure of generalized Lyapunov level surfaces. In the case where the generalized Lyapunov function is taken to be a C^1 function, our results collapse to the standard Lyapunov stability and invariant set theorems. Finally, we note that nondifferentiable Lyapunov functions have been considered in the literature. Specifically, continuous and lower semicontinuous Lyapunov functions have been considered in [23, 8, 140, 6, 7, 9], with [8, 7, 9] focusing on viability theory and differential inclusions. Furthermore, significant extensions of LaSalle's invariance principle for continuous Lyapunov functions are developed in [29, 116]. However, the present formulation provides *new* invariant set stability theorem generalizations not considered in [8, 29, 116] by explicitly characterizing system limit sets in terms of lower semicontinuous Lyapunov functions.

2.2 Mathematical Preliminaries

In this section we establish definitions, notation, and several key results used later in the chapter. Let $\mathbb{R}$ denote the set of real numbers, let $\mathbb{R}^n$ denote the set of $n \times 1$ real column vectors, and let $(\cdot)^{\mathrm{T}}$ denote transpose. Furthermore, let $\partial\mathcal{S}$, $\overset{\circ}{\mathcal{S}}$, and $\bar{\mathcal{S}}$ denote the boundary, the interior, and the closure of the set $\mathcal{S} \subset \mathbb{R}^n$, respectively. A set $\mathcal{S} \subseteq \mathbb{R}^n$ is *connected* if there does not exist open sets $\mathcal{O}_1$ and $\mathcal{O}_2$ in $\mathbb{R}^n$ such that $\mathcal{S} \subset \mathcal{O}_1 \cup \mathcal{O}_2$, $\mathcal{S} \cap \mathcal{O}_1 \neq \emptyset$, $\mathcal{S} \cap \mathcal{O}_2 \neq \emptyset$,

and $\mathcal{S} \cap \mathcal{O}_1 \cap \mathcal{O}_2 = \emptyset$. Recall that $\mathcal{S}$ is a connected subset of $\mathbb{R}$ if, and only if, $\mathcal{S}$ is either an interval or a single point. Let $\|\cdot\|$ denote the Euclidean vector norm, let $V'(x)$ denote the gradient of V at x, and let $D^+V(x)$ denote the lower Dini derivative of V at x [114]. Finally, let C^0 denote the set of continuous functions and C^r denote the set of functions with r-continuous derivatives.

In this chapter we consider the general nonlinear dynamical system

$$\dot{x}(t) = f(x(t)), \qquad x(0) = x_0, \qquad t \in \mathcal{I}_{x_0}, \tag{2.1}$$

where $x(t) \in \mathcal{D} \subseteq \mathbb{R}^n$, $t \in \mathcal{I}_{x_0}$, is the system state vector, $\mathcal{I}_{x_0} \subseteq \mathbb{R}$ is the maximal interval of existence of a solution $x(\cdot)$ of (2.1), $\mathcal{D}$ is an open set, $0 \in \mathcal{D}$, and $f : \mathcal{D} \to \mathbb{R}^n$. A function $x : \mathcal{I}_{x_0} \to \mathcal{D}$ is said to be a solution to (2.1) on the interval $\mathcal{I}_{x_0} \subseteq \mathbb{R}$ with initial condition $x(0) = x_0$, if $x(t)$ satisfies (2.1) for all $t \in \mathcal{I}_{x_0}$. Note that we do not assume any regularity condition on the function $f(\cdot)$. However, we do assume that for every $y \in \mathcal{D}$ there exists a unique solution $x(\cdot)$ of (2.1) defined on $\mathcal{I}_y$ satisfying $x(0) = y$. Furthermore, we assume that all the solutions $x(t)$, $t \in \mathcal{I}_{x_0}$, to (2.1) are continuous functions of the system initial conditions $x_0 \in \mathcal{D}$ which, with the assumption of uniqueness of solutions, implies continuity of solutions $x(t)$, $t \in \mathcal{I}_{x_0}$, to (2.1) [60, p. 24].

If $f(\cdot)$ is Lipschitz continuous on $\mathcal{D}$ then there exists a unique solution to (2.1). In this case, the semi-group property $s(t+\tau, x_0) = s(t, s(\tau, x_0))$, $t, \tau \in \mathcal{I}_{x_0}$, and the continuity of $s(t, \cdot)$ on $\mathcal{D}$, $t \in \mathcal{I}_{x_0}$, hold, where $s(\cdot, x_0)$ denotes the solution of the nonlinear dynamical system (2.1) with initial condition $x(0) = x_0$. Alternatively, uniqueness of solutions in time along with the continuity of $f(\cdot)$ ensure that the solutions to (2.1) satisfy the semi-group property and are continuous functions of the initial condition $x_0 \in \mathcal{D}$ even when $f(\cdot)$ is not Lipschitz continuous on $\mathcal{D}$ (see [35, Theorem 4.3, p. 59]). More generally, $f(\cdot)$ need not be continuous. In particular, if $f(\cdot)$ is discontinuous but bounded and $x(\cdot)$ is the unique solution to (2.1) in the sense of Filippov [44], then the semi-group property along with the continuous dependence of solutions on initial conditions hold [44].

Next, we introduce several definitions and two key results that are necessary for the main results of this chapter.

Definition 2.1. *Let $\mathcal{D}_c \subseteq \mathcal{D}$ and let $V : \mathcal{D}_c \to \mathbb{R}$. For $\alpha \in \mathbb{R}$, the set $V^{-1}(\alpha) \triangleq \{x \in \mathcal{D}_c : V(x) = \alpha\}$ is called the* α-level set. *For $\alpha, \beta \in \mathbb{R}$, $\alpha \leq \beta$, the set $V^{-1}([\alpha, \beta]) \triangleq \{x \in \mathcal{D}_c : \alpha \leq V(x) \leq \beta\}$ is called the* $[\alpha, \beta]$-sublevel set.

Definition 2.2. *The* trajectory $x(t) \in \mathcal{D} \subseteq \mathbb{R}^n$, $t \in \mathcal{I}_{x_0}$, *of (2.1) denotes the solution to (2.1) corresponding to the initial condition $x(0) = x_0$ evaluated*

at time t. *The trajectory* $x(t)$, $t \in \mathcal{I}_{x_0}$, *is* bounded *on* $\mathcal{I}_{x_0} \subseteq \mathbb{R}$ *if there exists* $\gamma > 0$ *such that* $\|x(t)\| < \gamma$, $t \in \mathcal{I}_{x_0}$.

Definition 2.3. *A set* $\mathcal{M}^+ \subseteq \mathcal{D} \subseteq \mathbb{R}^n$ *(resp.,* $\mathcal{M}^-$*) is a* positively *(resp.,* negatively*)* invariant set *for the nonlinear dynamical system (2.1) if* $x_0 \in \mathcal{M}^+$ *(resp.,* $\mathcal{M}^-$*) implies that* $[0, +\infty) \subseteq \mathcal{I}_{x_0}$ *(resp.,* $(-\infty, 0] \subseteq \mathcal{I}_{x_0}$*) and* $x(t) \in \mathcal{M}^+$ *(resp.,* $\mathcal{M}^-$*) for all* $t \geq 0$ *(resp.,* $t \leq 0$*). A set* $\mathcal{M} \subseteq \mathcal{D} \subseteq \mathbb{R}^n$ *is* an invariant set *for the nonlinear dynamical system (2.1) if* $x_0 \in \mathcal{M}$ *implies that* $\mathcal{I}_{x_0} = \mathbb{R}$ *and* $x(t) \in \mathcal{M}$ *for all* $t \in \mathbb{R}$.

Definition 2.4. $p \in \overline{\mathcal{D}} \subseteq \mathbb{R}^n$ *is a* positive limit point *of the trajectory* $x(t)$, $t \in \mathcal{I}_{x_0}$, *if* $[0, +\infty) \subseteq \mathcal{I}_{x_0}$ *and for all* $\varepsilon > 0$ *and finite time* $T > 0$ *there exists* $t > T$ *such that* $\|x(t) - p\| < \varepsilon$. *The set of all positive limit points of* $x(t)$, $t \in \mathcal{I}_{x_0}$, *is the* positive limit set $\mathcal{P}^+_{x_0}$ *of* $x(t)$, $t \in \mathcal{I}_{x_0}$.

Note that $\|x(t) - p\| < \varepsilon$ for all $\varepsilon > 0$ and $t > T > 0$ is equivalent to the existence of a sequence $\{t_n\}_{n=0}^{\infty}$, with $t_n \to \infty$ as $n \to \infty$, such that $x(t_n) \to p$ as $n \to \infty$. The following result on positive limit sets is fundamental and forms the basis for all later developments.

Lemma 2.1 ([73]). *Suppose the forward solution* $x(t)$, $t \geq 0$, *to (2.1) corresponding to an initial condition* $x(0) = x_0$ *exists and is bounded. Then the positive limit set* $\mathcal{P}^+_{x_0}$ *of* $x(t)$, $t \in \mathcal{I}_{x_0}$, *is a nonempty, compact, connected invariant set. Furthermore,* $x(t) \to \mathcal{P}^+_{x_0}$ *as* $t \to \infty$.

It is important to note that Lemma 2.1 holds for time-invariant nonlinear dynamical systems (2.1) possessing unique solutions with solutions being continuous functions of the system initial conditions. More generally, Lemma 2.1 holds if $s(t + \tau, x_0) = s(t, s(\tau, x_0))$, $t, \tau \in \mathcal{I}_{x_0}$, and $s(\cdot, x_0)$ is a continuous function of $x_0 \in \mathcal{D}$. Finally, Lemma 2.1 also holds for all discrete-time, time-invariant nonlinear dynamical systems that have (unique) solutions which are continuously dependent on the system initial conditions. Of course, in this case, $\mathcal{P}^+_{x_0}$ is not connected.

The following definition introduces three types of stability as well as attraction of (2.1) with respect to a compact positively invariant set.

Definition 2.5. *Let* $\mathcal{D}_0 \subset \mathcal{D}$ *be a compact positively invariant set for the nonlinear dynamical system (2.1).* $\mathcal{D}_0$ *is* Lyapunov stable *if for every open neighborhood* $\mathcal{O}_1 \subseteq \mathcal{D}$ *of* $\mathcal{D}_0$, *there exists an open neighborhood* $\mathcal{O}_2 \subseteq \mathcal{O}_1$ *of* $\mathcal{D}_0$ *such that* $x(t) \in \mathcal{O}_1$, $t \geq 0$, *for all* $x_0 \in \mathcal{O}_2$. $\mathcal{D}_0$ *is* attractive *if there exists an open neighborhood* $\mathcal{O}_3 \subseteq \mathcal{D}$ *of* $\mathcal{D}_0$ *such that* $\mathcal{P}^+_{x_0} \subseteq \mathcal{D}_0$ *for all* $x_0 \in \mathcal{O}_3$. $\mathcal{D}_0$ *is* asymptotically stable *if it is Lyapunov stable and attractive.* $\mathcal{D}_0$ *is* globally asymptotically stable *if it is Lyapunov stable and* $\mathcal{P}^+_{x_0} \subseteq \mathcal{D}_0$ *for all* $x_0 \in \mathbb{R}^n$. *Finally,* $\mathcal{D}_0$ *is* unstable *if it is not Lyapunov stable.*

Next, we give a set theoretic definition involving the domain, or region, of attraction of the compact positively invariant set $\mathcal{D}_c$ of (2.1).

Definition 2.6. *Suppose the compact positively invariant set* $\mathcal{D}_0 \subset \mathcal{D}$ *of (2.1) is attractive. Then the* domain of attraction $\mathcal{D}_A$ *of* $\mathcal{D}_0$ *is defined as*

$$\mathcal{D}_A \triangleq \{x_0 \in \mathcal{D} : \mathcal{P}^+_{x_0} \subseteq \mathcal{D}_0\}. \tag{2.2}$$

Recall that $\mathcal{D}_A$ is an open, connected invariant set [23, Proposition 4.15, p. 88].

Next, we present a key theorem due to Weierstrass involving lower semicontinuous functions on compact sets. For the statement of this result the following definition is needed.

Definition 2.7. *Let* $\mathcal{D}_c \subset \mathcal{D}$. *A function* $V : \mathcal{D}_c \to \mathbb{R}$ *is* lower semicontinuous *on* $\mathcal{D}_c$ *if for every sequence* $\{x_n\}_{n=0}^{\infty} \subset \mathcal{D}_c$ *such that* $\lim_{n\to\infty} x_n = x$, $V(x) \leq \liminf_{n\to\infty} V(x_n)$.

Note that a bounded function $V : \mathcal{D}_c \to \mathbb{R}$ is lower semicontinuous at $x \in \mathcal{D}_c$ if, and only if, for each $\epsilon > 0$ there exists $\delta > 0$ such that $\|x - y\| < \delta$, $y \in \mathcal{D}_c$, implies $V(x) - V(y) \leq \epsilon$. Furthermore, if $\mathcal{D}_c \subset \mathcal{D}$ is compact and $V : \mathcal{D}_c \to \mathbb{R}$ is lower semicontinuous at $x \in \mathcal{D}_c$, then for each $\alpha \in \mathbb{R}$ the set $\{x \in \mathcal{D}_c : V(x) \leq \alpha\}$ is compact.

Theorem 2.1 ([114]). *Suppose* $\mathcal{D}_c \subset \mathcal{D}$ *is compact and* $V : \mathcal{D}_c \to \mathbb{R}$ *is lower semicontinuous. Then there exists* $x^* \in \mathcal{D}_c$ *such that* $V(x^*) \leq V(x)$, $x \in \mathcal{D}_c$.

2.3 Generalized Stability Theorems

As discussed in Section 2.1, most Lyapunov stability theorems presented in the literature require that the Lyapunov function candidate for a nonlinear dynamical system be a C^1 function with a negative-definite derivative along the system trajectories. In this section, we present several generalized stability theorems where we relax both of these assumptions while guaranteeing local and global stability of a nonlinear dynamical system. The following result gives sufficient conditions for Lyapunov stability of a compact positively invariant set with respect to a nonlinear dynamical system.

Theorem 2.2. *Consider the nonlinear dynamical system (2.1), let* $\mathcal{D}_0$ *be a compact positively invariant set with respect to (2.1) such that* $\mathcal{D}_0 \subset \mathcal{D}$, *and let* $x(t)$, $t \in \mathcal{I}_{x_0}$, *denote the solution to (2.1) with* $x_0 \in \mathcal{D}$. *Assume that*

there exists a lower semicontinuous function $V : \mathcal{D} \to \mathbb{R}$ *such that* $V(\cdot)$ *is continuous on* $\mathcal{D}_0$ *and*

$$V(x) = 0, \qquad x \in \mathcal{D}_0, \tag{2.3}$$

$$V(x) > 0, \qquad x \in \mathcal{D}, \qquad x \notin \mathcal{D}_0, \tag{2.4}$$

$$V(x(t)) \le V(x(\tau)), \qquad 0 \le \tau \le t. \tag{2.5}$$

Then $\mathcal{D}_0$ *is Lyapunov stable.*

Proof. Let $\mathcal{O}_1 \subseteq \mathcal{D}$ be an open neighborhood of $\mathcal{D}_0$. Since $\partial\mathcal{O}_1$ is compact and $V(x)$, $x \in \mathcal{D}$, is lower semicontinuous, it follows from Theorem 2.1 that there exists $\alpha = \min_{x\in\partial\mathcal{O}_1} V(x)$. Note that $\alpha > 0$ since $\mathcal{D}_0 \cap \partial\mathcal{O}_1 = \emptyset$ and $V(x) > 0$, $x \in \mathcal{D}$, $x \notin \mathcal{D}_0$. Next, using the facts that $V(x) = 0$, $x \in \mathcal{D}_0$, and $V(\cdot)$ is continuous on $\mathcal{D}_0$, it follows that the set $\mathcal{O}_2 \triangleq \{x \in \mathcal{O}_1 : V(x) < \alpha\}^\circ$ is not empty. Now, it follows from (2.5) that for all $x(0) \in \mathcal{O}_2$,

$$V(x(t)) \le V(x(0)) < \alpha, \qquad t \ge 0,$$

which, since $V(x) \ge \alpha$, $x \in \partial\mathcal{O}_1$, implies that $x(t) \notin \partial\mathcal{O}_1$, $t \ge 0$. Hence, for every open neighborhood $\mathcal{O}_1 \subseteq \mathcal{D}$ of $\mathcal{D}_0$, there exists an open neighborhood $\mathcal{O}_2 \subseteq \mathcal{O}_1$ of $\mathcal{D}_0$ such that, if $x(0) \in \mathcal{O}_2$, then $x(t) \in \mathcal{O}_1$, $t \ge 0$, which proves Lyapunov stability of the compact positively invariant set $\mathcal{D}_0$ of (2.1). ■

A lower semicontinuous function $V(\cdot)$, with $V(\cdot)$ being continuous on $\mathcal{D}_0$, satisfying (2.3) and (2.4) is called a *generalized Lyapunov function candidate* for the nonlinear dynamical system (2.1). If, additionally, $V(\cdot)$ satisfies (2.5), $V(\cdot)$ is called a *generalized Lyapunov function* for the nonlinear dynamical system (2.1). Note that in the case where the function $V(\cdot)$ is C^1 on $\mathcal{D}$ in Theorem 2.2, it follows that $V(x(t)) \le V(x(\tau))$, for all $t \ge \tau \ge 0$, is equivalent to $\dot{V}(x) \triangleq V'(x)f(x) \le 0$, $x \in \mathcal{D}$. In this case conditions (2.3)–(2.5) in Theorem 2.2 specialize to the standard Lyapunov stability conditions [59, 73, 83].

Next, we generalize the Barbashin-Krasovskii-LaSalle invariant set theorems [18, 75, 82, 83, 73] to the case in which the function $V(\cdot)$ is lower semicontinuous. For the remainder of the results of this chapter define the notation

$$\mathcal{R}_\gamma \triangleq \bigcap_{c>\gamma} \overline{V^{-1}([\gamma, c])}, \tag{2.6}$$

for arbitrary $V : \mathcal{D} \subseteq \mathbb{R}^n \to \mathbb{R}$ and $\gamma \in \mathbb{R}$, and let $\mathcal{M}_\gamma$ denote the largest invariant set (with respect to (2.1)) contained in $\mathcal{R}_\gamma$.

Theorem 2.3. *Consider the nonlinear dynamical system (2.1), let* $x(t)$, $t \in \mathcal{I}_{x_0}$, *denote the solution to (2.1), and let* $\mathcal{D}_c \subset \mathcal{D}$ *be a compact positively invariant set with respect to (2.1). Assume that there exists a lower*

semicontinuous function $V : \mathcal{D}_c \to \mathbb{R}$ *such that* $V(x(t)) \leq V(x(\tau))$, *for all* $\tau \in [0,t]$ *and* $x_0 \in \mathcal{D}_c$. *If* $x_0 \in \mathcal{D}_c$, *then* $x(t) \to \mathcal{M} \triangleq \cup_{\gamma \in \mathbb{R}} \mathcal{M}_\gamma$ *as* $t \to \infty$.

Proof. Let $x(t)$, $t \in \mathcal{I}_{x_0}$, be the solution to (2.1) with $x_0 \in \mathcal{D}_c$ so that $[0,+\infty) \subseteq \mathcal{I}_{x_0}$. Since $V(\cdot)$ is lower semicontinuous on the compact set $\mathcal{D}_c$, there exists $\beta \in \mathbb{R}$ such that $V(x) \geq \beta$, $x \in \mathcal{D}_c$. Hence, since $V(x(t))$, $t \geq 0$, is nonincreasing, $\gamma_{x_0} \triangleq \lim_{t\to\infty} V(x(t))$, $x_0 \in \mathcal{D}_c$, exists. Now, for all $p \in \mathcal{P}^+_{x_0}$ there exists an increasing unbounded sequence $\{t_n\}_{n=0}^\infty$, with $t_0 = 0$, such that $\lim_{n\to\infty} x(t_n) = p$. Next, since $V(x(t_n))$, $n \geq 0$, is nonincreasing it follows that for all $n \geq 0$, $\gamma_{x_0} \leq V(x(t_n)) \leq V(x(t_N))$, $n \geq N$, or, equivalently, since $\mathcal{D}_c$ is positively invariant, $x(t_n) \in V^{-1}([\gamma_{x_0}, V(x(t_N))])$, $n \geq N$. Now, since $\lim_{n\to\infty} x(t_n) = p$ it follows that $p \in \overline{V^{-1}([\gamma_{x_0}, V(x(t_n))])}$, $n \geq 0$. Furthermore, since $\lim_{n\to\infty} V(x(t_n)) = \gamma_{x_0}$ it follows that for every $c > \gamma_{x_0}$, there exists $n \geq 0$ such that $\gamma_{x_0} \leq V(x(t_n)) \leq c$ which implies that for every $c > \gamma_{x_0}$, $p \in \overline{V^{-1}([\gamma_{x_0}, c])}$. Hence, $p \in \mathcal{R}_{\gamma_{x_0}}$ which implies that $\mathcal{P}^+_{x_0} \subseteq \mathcal{R}_{\gamma_{x_0}}$. Now, since $\mathcal{D}_c$ is compact and positively invariant it follows that the forward solution $x(t)$, $t \geq 0$, to (2.1) is bounded for all $x_0 \in \mathcal{D}_c$ and hence it follows from Lemma 2.1 that $\mathcal{P}^+_{x_0}$ is a nonempty compact connected invariant set which further implies that $\mathcal{P}^+_{x_0}$ is a subset of the largest invariant set contained in $\mathcal{R}_{\gamma_{x_0}}$, that is, $\mathcal{P}^+_{x_0} \subseteq \mathcal{M}_{\gamma_{x_0}}$. Hence, for all $x_0 \in \mathcal{D}_c$, $\mathcal{P}^+_{x_0} \subseteq \mathcal{M}$. Finally, since $x(t) \to \mathcal{P}^+_{x_0}$ as $t \to \infty$ it follows that $x(t) \to \mathcal{M}$ as $t \to \infty$. ■

If, in Theorem 2.3, $\mathcal{M}$ contains no other trajectory other than the trivial trajectory $x(t) \equiv 0$, then the zero solution $x(t) \equiv 0$ to (2.1) is attractive and $\mathcal{D}_c$ is a subset of the domain of attraction. Furthermore, note that if $V : \mathcal{D}_c \to \mathbb{R}$ is a lower semicontinuous function such that all the conditions of Theorem 2.3 are satisfied, then for every $x_0 \in \mathcal{D}_c$ there exists $\gamma_{x_0} \leq V(x_0)$ such that $\mathcal{P}^+_{x_0} \subseteq \mathcal{M}_{\gamma_{x_0}} \subseteq \mathcal{M}$. Finally, since $V^{-1}([\gamma, c]) = \{x \in \mathcal{D}_c : V(x) \geq \gamma\} \cap \{x \in \mathcal{D}_c : V(x) \leq c\}$ and $\{x \in \mathcal{D}_c : V(x) \leq c\}$ is a closed set, it follows that $\hat{\mathcal{R}}_{\gamma,c} \subset \{x \in \mathcal{D}_c : V(x) < \gamma\}$, where $\hat{\mathcal{R}}_{\gamma,c} \triangleq \overline{V^{-1}([\gamma,c])} \setminus V^{-1}([\gamma,c])$, $c > \gamma$, for a fixed $\gamma \in \mathbb{R}$. Hence,

$$\mathcal{R}_\gamma = \bigcap_{c>\gamma} \left(V^{-1}([\gamma,c]) \cup \hat{\mathcal{R}}_{\gamma,c}\right) = V^{-1}(\gamma) \cup \hat{\mathcal{R}}_\gamma,$$

where $\hat{\mathcal{R}}_\gamma \triangleq \cap_{c>\gamma} \hat{\mathcal{R}}_{\gamma,c}$, is such that $V(x) < \gamma$, $x \in \hat{\mathcal{R}}_\gamma$. Finally, if $V(\cdot)$ is C^0 then $\hat{\mathcal{R}}_{\gamma,c} = \emptyset$, $\gamma \in \mathbb{R}$, $c > \gamma$, and hence $\mathcal{R}_\gamma = V^{-1}(\gamma)$.

It is important to note that even though the stability conditions appearing in Theorem 2.3 are system trajectory dependent, in Chapter 3 a hierarchical switching nonlinear control strategy is developed using Theorem 2.3 *without* requiring knowledge of the system trajectories. Furthermore, note that as in standard Lyapunov and invariant set theorems involving C^1 functions, Theorem 2.3 allows one to characterize the invariant set $\mathcal{M}$ *without* knowledge

of the system trajectories $x(t)$, $t \in \mathcal{I}_{x_0}$. Similar remarks hold for the rest of the theorems in this section. To illustrate the utility of Theorem 2.3 consider the simple scalar nonlinear dynamical system given by

$$\dot{x}(t) = -x(t)(x(t)-1)(x(t)+2), \quad x(0) = x_0, \quad t \geq 0, \tag{2.7}$$

with generalized Lyapunov function candidate $V(x)$ given by

$$V(x) = \begin{cases} (x+2)^2, x < 0, \\ (x-1)^2, x \geq 0. \end{cases}$$

Now, note that

$$\begin{aligned} \dot{V}(x) &\triangleq D^+V(x)[-x(x-1)(x+2)] \\ &= \begin{cases} -2x(x-1)(x+2)^2, x < 0, \\ -2x(x-1)^2(x+2), x \geq 0, \end{cases} \leq 0, \qquad x \in \mathbb{R}, \end{aligned}$$

which implies that $V(x(t))$, $t \geq 0$, is nonincreasing along the system trajectories. Next, note that $\mathcal{R}_\gamma = V^{-1}(\gamma)$, $\gamma \in \mathbb{R}\backslash\{4\}$, and $\mathcal{R}_4 = V^{-1}(4) \cup \{0\}$. Since the only invariant sets contained in $\mathcal{R}_\gamma$ are the equilibrium points $x_{e1} = -2$, $x_{e2} = 0$, $x_{e3} = 1$, it follows that $\mathcal{M}_\gamma = \emptyset$, $\gamma \notin \{0,\ 1,\ 4\}$, $\mathcal{M}_0 = \{-2,\ 1\}$, $\mathcal{M}_1 = \{0\}$, and $\mathcal{M}_4 = \{0\}$ which implies that $\mathcal{M} = \{-2,\ 0,\ 1\}$. Hence, it follows from Theorem 2.3 that for every $x_0 \in \mathbb{R}$ the solution to (2.7) approaches the invariant set $\mathcal{M} = \{-2,\ 0,\ 1\}$ as $t \to \infty$ which can be easily verified. As shown by the above example, Theorem 2.3 allows for a systematic way of constructing system Lyapunov functions by piecing together a collection of functions.

The following corollary to Theorem 2.3 presents sufficient conditions that guarantee local asymptotic stability of a compact positively invariant set with respect to the nonlinear dynamical system (2.1).

Corollary 2.1. *Consider the nonlinear dynamical system (2.1), let $\mathcal{D}_c$ and $\mathcal{D}_0$ be compact positively invariant sets with respect to (2.1) such that $\mathcal{D}_0 \subset \overset{\circ}{\mathcal{D}}_c$, $\mathcal{D}_c \subset \mathcal{D}$, and let $x(t)$, $t \in \mathcal{I}_{x_0}$, denote the solution to (2.1) corresponding to $x_0 \in \mathcal{D}_c$. Assume that there exists a lower semicontinuous function $V : \mathcal{D}_c \to \mathbb{R}$ such that $V(\cdot)$ is continuous on $\mathcal{D}_0$ and*

$$V(x) = 0, \qquad x \in \mathcal{D}_0, \tag{2.8}$$

$$V(x) > 0, \qquad x \in \mathcal{D}_c, \qquad x \notin \mathcal{D}_0, \tag{2.9}$$

$$V(x(t)) \leq V(x(\tau)), \qquad 0 \leq \tau \leq t, \tag{2.10}$$

and assume that $\mathcal{M} \triangleq \cap_{\gamma \geq 0} \mathcal{M}_\gamma \subseteq \mathcal{D}_0$. Then $\mathcal{D}_0$ is locally asymptotically stable and $\mathcal{D}_c$ is a subset of the domain of attraction.

Proof. The result is a direct consequence of Theorems 2.2 and 2.3. ■

Next, we specialize Theorem 2.3 to the Barbashin-Krasovskii-LaSalle invariant set theorem wherein $V(\cdot)$ is a C^1 function.

Corollary 2.2. *Consider the nonlinear dynamical system (2.1), let $\mathcal{D}_c \subset \mathcal{D}$ be a compact positively invariant set with respect to (2.1) and let $x(t)$, $t \in \mathcal{I}_{x_0}$, denote the solution to (2.1) corresponding to $x_0 \in \mathcal{D}_c$. Assume that there exists a C^1 function $V : \mathcal{D}_c \to \mathbb{R}$ such that $V'(x)f(x) \leq 0$, $x \in \mathcal{D}_c$. Let $\mathcal{R} \triangleq \{x \in \mathcal{D}_c : V'(x)f(x) = 0\}$ and let $\mathcal{M}$ be the largest invariant set contained in $\mathcal{R}$. Then $x(t) \to \mathcal{M}$ as $t \to \infty$.*

Proof. The result follows from Theorem 2.3. Specifically, since $V'(x)f(x) \leq 0$, $x \in \mathcal{D}_c$, it follows that

$$V(x(t)) - V(x(\tau)) = \int_\tau^t V'(x(s))f(x(s))\mathrm{d}s \leq 0, \qquad t \geq \tau,$$

and hence $V(x(t)) \leq V(x(\tau))$, $t \geq \tau$. Now, since $V(\cdot)$ is C^1 it follows that $\mathcal{R}_\gamma = V^{-1}(\gamma)$, $\gamma \in \mathbb{R}$. In this case, it follows from Theorem 2.3 that for every $x_0 \in \mathcal{D}_c$ there exists $\gamma_{x_0} \in \mathbb{R}$ such that $\mathcal{P}^+_{x_0} \subseteq \mathcal{M}_{\gamma_{x_0}}$, where $\mathcal{M}_{\gamma_{x_0}}$ is the largest invariant set contained in $\mathcal{R}_{\gamma_{x_0}} = V^{-1}(\gamma_{x_0})$ which implies that $V(x) = \gamma_{x_0}$, $x \in \mathcal{P}^+_{x_0}$. Hence, since $\mathcal{M}_{\gamma_{x_0}}$ is an invariant set it follows that for all $x(0) \in \mathcal{M}_{\gamma_{x_0}}$, $x(t) \in \mathcal{M}_{\gamma_{x_0}}$, $t \geq 0$, and thus $\dot{V}(x(0)) \triangleq \frac{\mathrm{d}V(x(t))}{\mathrm{d}t}\Big|_{t=0} = V'(x(0))f(x(0)) = 0$, which implies the $\mathcal{M}_{\gamma_{x_0}}$ is contained in $\mathcal{M}$ which is the largest invariant set contained in $\mathcal{R}$. Hence, since $x(t) \to \mathcal{P}^+_{x_0} \subseteq \mathcal{M}$ as $t \to \infty$, it follows that $x(t) \to \mathcal{M}$ as $t \to \infty$. ■

Next, we sharpen the results of Theorem 2.3 by providing a refined construction of the invariant set $\mathcal{M}$. In particular, we show that the system trajectories converge to a union of largest invariant sets contained on the boundary of the intersections over *finite* intervals of the closure of generalized Lyapunov level surfaces.

Theorem 2.4. *Consider the nonlinear dynamical system (2.1), let $\mathcal{D}_c$ and $\mathcal{D}_0$ be compact positively invariant sets with respect to (2.1) such that $\mathcal{D}_0 \subset \mathcal{D}_c \subset \mathcal{D}$, and let $x(t)$, $t \in \mathcal{I}_{x_0}$, denote the solution to (2.1) corresponding to $x_0 \in \mathcal{D}_c$. Assume that there exists a lower semicontinuous function $V : \mathcal{D}_c \to \mathbb{R}$ such that*

$$V(x) = 0, \qquad x \in \mathcal{D}_0, \tag{2.11}$$

$$V(x) > 0, \qquad x \in \mathcal{D}_c, \quad x \notin \mathcal{D}_0, \tag{2.12}$$

$$V(x(t)) \leq V(x(\tau)), \qquad 0 \leq \tau \leq t. \tag{2.13}$$

Furthermore, assume that for all $x_0 \in \mathcal{D}_c$, $x_0 \notin \mathcal{D}_0$, there exists an increasing unbounded sequence $\{t_n\}_{n=0}^{\infty}$, with $t_0 = 0$, such that

$$V(x(t_{n+1})) < V(x(t_n)), \qquad n = 0, 1, \ldots. \tag{2.14}$$

Then, either $\mathcal{M}_\gamma \subset \hat{\mathcal{R}}_\gamma \triangleq \mathcal{R}_\gamma \setminus V^{-1}(\gamma)$, or $\mathcal{M}_\gamma = \emptyset$, $\gamma > 0$. Furthermore, if $x_0 \in \mathcal{D}_c$, then $x(t) \to \hat{\mathcal{M}} \triangleq \cup_{\gamma \in \mathcal{G}} \mathcal{M}_\gamma$ as $t \to \infty$, where $\mathcal{G} \triangleq \{\gamma \geq 0 : \mathcal{R}_\gamma \cap \mathcal{D}_0 \neq \emptyset\}$. If, in addition, $\mathcal{D}_0 \subset \overset{\circ}{\mathcal{D}}_c$ and $V(\cdot)$ is continuous on $\mathcal{D}_0$ then $\mathcal{D}_0$ is locally asymptotically stable and $\mathcal{D}_c$ is a subset of the domain of attraction.

Proof. Since $\mathcal{D}_c$ is a compact positively invariant set, it follows that for all $x_0 \in \mathcal{D}_c$, the forward solution $x(t)$, $t \geq 0$, to (2.1) is bounded. Hence, it follows from Lemma 2.1 that, for all $x_0 \in \mathcal{D}_c$, $\mathcal{P}_{x_0}^+$ is a nonempty, compact, connected invariant set. Next, it follows from Theorem 2.3 and the fact that $V(\cdot)$ is positive definite (with respect to $\mathcal{D}_c \setminus \mathcal{D}_0$), that for every $x_0 \in \mathcal{D}_c$ there exists $\gamma_{x_0} \geq 0$ such that $\mathcal{P}_{x_0}^+ \subseteq \mathcal{M}_{\gamma_{x_0}} \subseteq \mathcal{R}_{\gamma_{x_0}}$. Now, given $x(0) \in V^{-1}(\gamma_{x_0})$, $\gamma_{x_0} > 0$, (2.14) implies that there exists $t_1 > 0$ such that $V(x(t_1)) < \gamma_{x_0}$ and $x(t_1) \notin V^{-1}(\gamma_{x_0})$. Hence, $V^{-1}(\gamma_{x_0}) \subset \mathcal{R}_{\gamma_{x_0}}$ does not contain any invariant set. Alternatively, if $x(0) \in \hat{\mathcal{R}}_{\gamma_{x_0}}$ then $V(x(0)) < \gamma_{x_0}$ and (2.14) implies that $x(t) \notin V^{-1}(\gamma_{x_0})$, $t \geq 0$. Hence, any invariant set contained in $\mathcal{R}_{\gamma_{x_0}}$ is a subset of $\hat{\mathcal{R}}_{\gamma_{x_0}}$, which implies that $\mathcal{M}_{\gamma_{x_0}} \subset \hat{\mathcal{R}}_{\gamma_{x_0}}$, $\gamma_{x_0} > 0$. If $\hat{\gamma} > 0$ is such that $\hat{\gamma} \neq \gamma_{x_0}$, for all $x_0 \in \mathcal{D}_c$, then there does not exist $x_0 \in \mathcal{D}_c$ such that $\mathcal{P}_{x_0}^+ \subseteq \mathcal{R}_{\hat{\gamma}}$ and hence $\mathcal{M}_{\hat{\gamma}} = \emptyset$. Now, *ad absurdum*, suppose $\mathcal{D}_0 \cap \mathcal{P}_{x_0}^+ = \emptyset$. Since $V(\cdot)$ is lower semicontinuous it follows from Theorem 2.1 that there exists $\hat{x} \in \mathcal{P}_{x_0}^+$ such that $\alpha = V(\hat{x}) \leq V(x)$, $x \in \mathcal{P}_{x_0}^+$. Now, with $x(0) = \hat{x} \notin \mathcal{D}_0$ it follows from (2.14) that there exists an increasing unbounded sequence $\{t_n\}_{n=0}^{\infty}$, with $t_0 = 0$, such that $V(x(t_{n+1})) < V(x(t_n))$, $n = 0, 1, \ldots$, which implies that there exists $t > 0$ such that $V(x(t)) < \alpha$ which further implies that $x(t) \notin \mathcal{P}_{x_0}^+$ contradicting the fact that $\mathcal{P}_{x_0}^+$ is an invariant set. Hence, there exists $q \in \mathcal{D}_0$ such that $q \in \mathcal{P}_{x_0}^+ \subseteq \mathcal{R}_{\gamma_{x_0}}$ which implies that $\mathcal{R}_{\gamma_{x_0}} \cap \mathcal{D}_0 \neq \emptyset$. Thus, $\gamma_{x_0} \in \mathcal{G}$ for all $x_0 \in \mathcal{D}_c$ which further implies that $\mathcal{P}_{x_0}^+ \subseteq \hat{\mathcal{M}}$. Now, since $x(t) \to \mathcal{P}_{x_0}^+ \subseteq \hat{\mathcal{M}}$ as $t \to \infty$ it follows that $x(t) \to \hat{\mathcal{M}}$ as $t \to \infty$.

If $V(\cdot)$ is continuous on $\mathcal{D}_0 \subset \overset{\circ}{\mathcal{D}}_c$, then Lyapunov stability of the compact positively invariant set $\mathcal{D}_0$ follows from Theorem 2.2. Furthermore, from the continuity of $V(\cdot)$ on $\mathcal{D}_0$ and the fact that $V(x) = 0$ for all $x \in \mathcal{D}_0$, it follows that $\mathcal{G} = \{0\}$ and $\hat{\mathcal{M}} \equiv \mathcal{M}_0$. Hence, $\mathcal{P}_{x_0}^+ \subseteq \mathcal{D}_0$ for all $x_0 \in \mathcal{D}_c$ establishing local asymptotic stability of the compact positively invariant set $\mathcal{D}_0$ of (2.1) with a subset of the domain of attraction given by $\mathcal{D}_c$. ■

In all of the above results we explicitly assumed that there exists a compact positively invariant set $\mathcal{D}_c \subset \mathcal{D}$ with respect to (2.1). Next, we provide a result that does not require the existence of such $\mathcal{D}_c$.

Theorem 2.5. *Consider the nonlinear dynamical system (2.1), let $\mathcal{D}_0$ be a compact positively invariant set with respect to (2.1) such that $\mathcal{D}_0 \subset \mathcal{D}$, and let $x(t)$, $t \in \mathcal{I}_{x_0}$, denote the solution to (2.1) corresponding to $x_0 \in \mathcal{D}$. Assume that there exists a lower semicontinuous function $V : \mathcal{D} \to \mathbb{R}$ such that*

$$V(x) = 0, \quad x \in \mathcal{D}_0, \tag{2.15}$$

$$V(x) > 0, \quad x \in \mathcal{D}, \quad x \notin \mathcal{D}_0, \tag{2.16}$$

$$V(x(t)) \leq V(x(\tau)), \quad 0 \leq \tau \leq t. \tag{2.17}$$

Then all forward solutions $x(t)$, $t \geq 0$, to (2.1) that are bounded approach $\mathcal{M} \triangleq \cup_{\gamma \geq 0} \mathcal{M}_\gamma$ as $t \to \infty$. If, in addition, for all $x_0 \in \mathcal{D}$, $x_0 \notin \mathcal{D}_0$, there exists an increasing unbounded sequence $\{t_n\}_{n=0}^{\infty}$, with $t_0 = 0$, such that

$$V(x(t_{n+1})) < V(x(t_n)), \quad n = 0, 1, \ldots, \tag{2.18}$$

then, either $\mathcal{M}_\gamma \subset \hat{\mathcal{R}}_\gamma \triangleq \mathcal{R}_\gamma \setminus V^{-1}(\gamma)$, or $\mathcal{M}_\gamma = \emptyset$, $\gamma > 0$. Furthermore, all forward solutions $x(t)$, $t \geq 0$, to (2.1) that are bounded approach $\hat{\mathcal{M}} \triangleq \cup_{\gamma \in \mathcal{G}} \mathcal{M}_\gamma$ as $t \to \infty$, where $\mathcal{G} \triangleq \{\gamma \geq 0 : \mathcal{R}_\gamma \cap \mathcal{D}_0 \neq \emptyset\}$.

Proof. The proof is a direct consequence of Theorems 2.3 and 2.4 with $\mathcal{D}_c$ given by the union of all bounded trajectories of (2.1). ■

Next, we present a generalized global invariant set theorem for guaranteeing global attraction and global asymptotic stability of a compact positively invariant set of a nonlinear dynamical system.

Theorem 2.6. *Consider the nonlinear dynamical system (2.1) with $\mathcal{D} = \mathbb{R}^n$ and let $x(t)$, $t \in \mathcal{I}_{x_0}$, denote the solution to (2.1) corresponding to $x_0 \in \mathbb{R}^n$. Assume that there exists a compact positively invariant set $\mathcal{D}_0$ with respect to (2.1) and a lower semicontinuous function $V : \mathbb{R}^n \to \mathbb{R}$ such that*

$$V(x) = 0, \quad x \in \mathcal{D}_0, \tag{2.19}$$

$$V(x) > 0, \quad x \in \mathbb{R}^n, \quad x \notin \mathcal{D}_0, \tag{2.20}$$

$$V(x(t)) \leq V(x(\tau)), \quad 0 \leq \tau \leq t, \tag{2.21}$$

$$V(x) \to \infty \text{ as } \|x\| \to \infty. \tag{2.22}$$

Then for all $x_0 \in \mathbb{R}^n$, $x(t) \to \mathcal{M} \triangleq \cup_{\gamma \geq 0} \mathcal{M}_\gamma$, as $t \to \infty$. If, in addition, for all $x_0 \in \mathbb{R}^n$, $x_0 \notin \mathcal{D}_0$, there exists an increasing unbounded sequence $\{t_n\}_{n=0}^{\infty}$, with $t_0 = 0$, such that

$$V(x(t_{n+1})) < V(x(t_n)), \quad n = 0, 1, \ldots, \tag{2.23}$$

then, either $\mathcal{M}_\gamma \subset \hat{\mathcal{R}}_\gamma \triangleq \mathcal{R}_\gamma \setminus V^{-1}(\gamma)$, or $\mathcal{M}_\gamma = \emptyset$, $\gamma > 0$. Furthermore, $x(t) \to \hat{\mathcal{M}} \triangleq \cup_{\gamma \in \mathcal{G}} \mathcal{M}_\gamma$ as $t \to \infty$, where $\mathcal{G} \triangleq \{\gamma \geq 0 : \mathcal{R}_\gamma \cap \mathcal{D}_0 \neq \emptyset\}$.

Finally, if $V(\cdot)$ is continuous on $\mathcal{D}_0$ then the compact positively invariant set $\mathcal{D}_0$ of (2.1) is globally asymptotically stable.

Proof. Note that since $V(x) \to \infty$ as $\|x\| \to \infty$ it follows that for every $\beta > 0$ there exists $r > 0$ such that $V(x) > \beta$ for all $\|x\| > r$, or, equivalently, $V^{-1}([0,\beta]) \subseteq \{x : \|x\| \leq r\}$ which implies that $V^{-1}([0,\beta])$ is bounded for all $\beta > 0$. Hence, for all $x_0 \in \mathbb{R}^n$, $V^{-1}([0,\beta_{x_0}])$ is bounded, where $\beta_{x_0} \triangleq V(x_0)$. Furthermore, since $V(\cdot)$ is a positive-definite lower semicontinuous function, it follows that $V^{-1}([0,\beta_{x_0}])$ is closed and, since $V(x(t))$, $t \geq 0$, is nonincreasing, $V^{-1}([0,\beta_{x_0}])$ is positively invariant. Hence, for every $x_0 \in \mathbb{R}^n$, $V^{-1}([0,\beta_{x_0}])$ is a compact positively invariant set. Now, with $\mathcal{D}_c = V^{-1}([0,\beta_{x_0}])$ it follows from Theorem 2.3 that there exists $\gamma_{x_0} \in [0,\beta_{x_0}]$ such that $\mathcal{P}^+_{x_0} \subseteq \mathcal{M}_{\gamma_{x_0}} \subset \hat{\mathcal{R}}_{\gamma_{x_0}}$ which implies that $x(t) \to \mathcal{M}$ as $t \to \infty$. If, in addition, for all $x_0 \in \mathbb{R}^n$, $x_0 \notin \mathcal{D}_0$, there exists an increasing unbounded sequence $\{t_n\}_{n=0}^{\infty}$, with $t_0 = 0$, such that (2.23) holds then it follows from Theorem 2.4 that $x(t) \to \hat{\mathcal{M}}$ as $t \to \infty$.

Finally, if $V(\cdot)$ is continuous on $\mathcal{D}_0$ then Lyapunov stability follows as in the proof of Theorem 2.4. Furthermore, in this case, $\mathcal{G} = \{0\}$ which implies that $\hat{\mathcal{M}} = \mathcal{M}_0$. Hence, $\mathcal{P}^+_{x_0} \subseteq \mathcal{D}_0$ establishing global asymptotic stability of the compact positively invariant set $\mathcal{D}_0$ of (2.1). ■

If in Theorem 2.4 (resp., Theorem 2.6) the function $V(\cdot)$ is C^1 on $\mathcal{D}_c$ (resp., $\mathbb{R}^n$), $\mathcal{D}_0 \equiv \{0\}$, and $V'(x)f(x) < 0$, $x \in \mathcal{D}_c$ (resp., $x \in \mathbb{R}^n$), $x \neq 0$, then every increasing unbounded sequence $\{t_n\}_{n=0}^{\infty}$, with $t_0 = 0$, is such that $V(x(t_{n+1})) < V(x(t_n))$, $n = 0, 1, \ldots$. In this case, Theorems 2.4 and 2.6 specialize to the standard Lyapunov stability theorems for local and global asymptotic stability, respectively.

Note that the results in this chapter also hold for nonlinear discrete-time dynamical systems described by time-invariant difference equations whose (unique) solutions are continuous functions of the initial conditions. Specifically, in this case all of the above results and proofs proceed exactly as in the continuous-time case by replacing $t \in \mathbb{R}$ with $k \in \mathbb{Z}$, where $\mathbb{Z}$ is the set of integers.

It is important to note that even though the stability conditions appearing in Theorems 2.3–2.6 are system trajectory dependent, in Chapter 3 we present a hierarchical switching nonlinear controller guaranteeing nonlinear system stabilization *without* requiring knowledge of the closed-loop system trajectories. Finally, we note that the concept of lower semicontinuous Lyapunov functions has been considered in the literature. Specifically, lower semicontinuous Lyapunov functions have been considered in [23, 8], with [8] focusing on viability theory and differential inclusions. However, the present formulation provides *new* invariant set stability theorem generalizations characterizing

system limit sets in terms of lower semicontinuous Lyapunov functions not considered in [23, 8].

2.4 Conclusion

Generalized Lyapunov and invariant set stability theorems for nonlinear dynamical systems were developed. In particular, local and global stability theorems were presented using generalized lower semicontinuous Lyapunov functions providing a transparent generalization of standard Lyapunov and invariant set theorems.

3. Nonlinear System Stabilization via Hierarchical Switching Controllers

3.1 Introduction

In this chapter a nonlinear control design framework predicated on a hierarchical switching controller architecture parameterized over a set of moving system equilibria is developed. Specifically, using ***equilibria-dependent Lyapunov functions***, or instantaneous (with respect to a given parameterized equilibrium manifold) Lyapunov functions, a hierarchical nonlinear control strategy is developed that stabilizes a given nonlinear system using a supervisory switching controller that coordinates lower-level stabilizing subcontrollers (see Figure 3.1) [68, 65, 30, 36]. Each subcontroller can be nonlinear

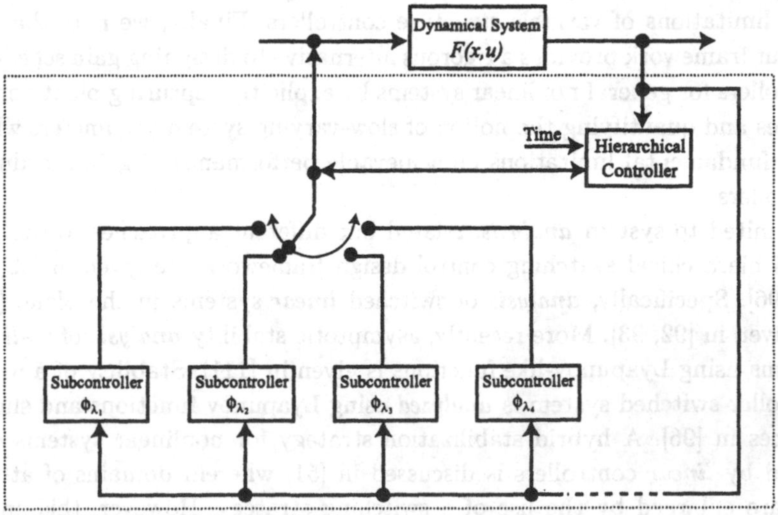

Fig. 3.1. Switching controller architecture

and thus local set point designs can be nonlinear. Furthermore, for each parameterized equilibrium manifold, the collection of subcontrollers provide guaranteed domains of attraction with nonempty intersections that cover the

region of operation of the nonlinear system in the state space. A hierarchical switching nonlinear controller architecture is developed based on a *generalized* lower semicontinuous Lyapunov function obtained by minimizing a potential function, associated with each domain of attraction, over a given switching set induced by the parameterized system equilibria. The switching set specifies the subcontroller to be activated at the point of switching, which occurs within the intersections of the domains of attraction. The hierarchical switching nonlinear controller guarantees that the generalized Lyapunov function is nonincreasing along the closed-loop system trajectories with strictly decreasing values at the switching points, establishing asymptotic stability. In the case where one of the parameterized system equilibrium points is globally asymptotically stable, the proposed nonlinear stabilization framework guarantees global attraction to any given system invariant set. If, in addition, a structural topological constraint is enforced on the switching set, then the proposed framework guarantees global asymptotic stability of any given system equilibrium on the parameterized system equilibrium manifold. Furthermore, since the proposed switching nonlinear control strategy is predicted on a generalized Lyapunov function framework with strictly decreasing values at the switching points, the possibility of a sliding mode is precluded. Hence, the proposed nonlinear stabilization framework avoids the undesirable effects of high-speed switching onto an invariant sliding manifold which is one of the main limitations of variable structure controllers. Finally, we note that the present framework provides a rigorous alternative to designing gain scheduled controllers for general nonlinear systems by explicitly capturing plant nonlinearities and quantifying the notion of slow-varying system parameters which place fundamental limitations on achievable performance of gain scheduling controllers.

Limited to system *analysis*, related but different approaches to the proposed hierarchical switching control design framework are given in [92, 93, 111, 96]. Specifically, *analysis* of switched *linear* systems in the plane ($\mathbb{R}^2$) are given in [92, 93]. More recently, asymptotic stability *analysis* of *m-linear* systems using Lyapunov-like functions is given in [111]. Stability of a multi-controller switched system is *analyzed* using Lyapunov functions and sliding surfaces in [96]. A hybrid stabilization strategy for nonlinear systems controlled by *linear* controllers is discussed in [51] wherein domains of attraction are enlarged by the use of a switching strategy. However, this analysis is limited to linearly controlled systems in the plane. Even though the approach can be extended to higher-order systems, the computational complexity needed to analyze the direction of the closed-loop system flows render the approach impractical. The special issues on hybrid control systems [2, 3] present an excellent analysis expansion on switching systems with controller

design methods limited to specific applications. We note that the concept of equilibria-dependent Lyapunov functions was first introduced by the authors in [85] where a globally stabilizing control design framework for controlling rotating stall and surge in axial flow compressors was developed. In parallel research to [85] a related but different approach was introduced in [101] wherein control Lyapunov functions for nonlinear systems *linearized* about a *finite* number of "trim" points to guarantee stability of a range of operating conditions were given. In the case where we specialize the switching set to a finite number of equilibrium points, we recover the results of [101]. Furthermore, we note that the authors in [101] mainly focus on how to choose the switching points so as to construct a domain of attraction that covers an *a priori* designated region in the state space. Hence, in the special case where the switching set contains a finite number of equilibrium points, the results of [101] can be used within the present framework to construct feasible switching sets. Finally, we emphasize that our approach is constructive in nature rather than existential. In particular, we provide an explicit construction for a hierarchical switching controller for nonlinear system stabilization and in this case our constructive conditions are complementary to existential results on asymptotic controllability via discontinuous feedback [33].

3.2 Mathematical Preliminaries

In this chapter we consider nonlinear controlled dynamical systems of the form

$$\dot{x}(t) = F(x(t), u(t)), \qquad x(0) = x_0, \qquad t \in \mathcal{I}_{x_0}, \tag{3.1}$$

where $x(t) \in \mathcal{D} \subseteq \mathbb{R}^n$, $t \in \mathcal{I}_{x_0}$, is the system state vector, $\mathcal{I}_{x_0} \subseteq \mathbb{R}$ is the maximal interval of existence of a solution $x(\cdot)$ of (3.1), $\mathcal{D}$ is an open set, $0 \in \mathcal{D}$, $u(t) \in \mathcal{U} \subseteq \mathbb{R}^m$, $t \in \mathcal{I}_{x_0}$, is the control input, $\mathcal{U}$ is the set of all admissible controls such that $u(\cdot)$ is a measurable function with $0 \in \mathcal{U}$, and $F : \mathcal{D} \times \mathcal{U} \to \mathbb{R}^n$ is continuous on $\mathcal{D} \times \mathcal{U}$.

Definition 3.1. *The point $\bar{x} \in \mathcal{D}$ is an* equilibrium point *of (3.1) if there exists $\bar{u} \in \mathcal{U}$ such that $F(\bar{x}, \bar{u}) = 0$.*

Here we assume that given an equilibrium point $\bar{x} \in \mathcal{D}$ corresponding to $\bar{u} \in \mathcal{U}$ and a mapping $\varphi : \mathcal{D} \times \Lambda \to \mathcal{U}$, $\Lambda \subseteq \mathbb{R}^q$, $0 \in \Lambda$, such that $\varphi(\bar{x}, 0) = \bar{u}$, there exist neighborhoods $\mathcal{D}_o \subset \mathcal{D}$ of $\bar{x}$ and $\Lambda_o \subset \Lambda$ of 0, and a continuous function $\psi : \Lambda_o \to \mathcal{D}_o$ such that $\bar{x} = \psi(0)$, and, for every $\lambda \in \Lambda_o$, $x_\lambda = \psi(\lambda)$ is an equilibrium point; that is, $F(\psi(\lambda), \varphi(\psi(\lambda), \lambda)) = 0$, $\lambda \in \Lambda_o$. This is

a necessary condition for parametric stability with respect to Λ_o as defined in [64, 107]. Note that the connected set $\Lambda \subseteq \mathbb{R}^q$ corresponds to a parameterization set with the function $\psi(\cdot)$ parameterizing the system equilibria. In the special case where $q = m$ and $\varphi(x, \lambda) = \lambda$, it follows that the parameterized system equilibria are given by the constant control $u(t) \equiv \lambda$. A parameterization that provides a local characterization of an equilibrium manifold, including in neighborhoods of bifurcation points, is given in [81]. Alternatively, the well known Implicit Function Theorem provides *sufficient* conditions for guaranteeing the existence of such a parameterization under the more restrictive condition of continuous differentiability of the mapping $\psi(\cdot)$.

Theorem 3.1 ([60]). *Assume the function $\tilde{F}(x,\lambda) \triangleq F(x,\varphi(x,\lambda))$, $x \in \mathcal{D}$, $\lambda \in \Lambda$, is C^1 at each point $(x,\lambda) \in \mathcal{D} \times \Lambda$. Suppose $\tilde{F}(\bar{x},0) = 0$ for $(\bar{x},0) \in \mathcal{D} \times \Lambda$ and the Jacobian matrix $\frac{\partial \tilde{F}}{\partial x}(\bar{x},0)$ is full rank. Then there exist open neighborhoods $\mathcal{D}_o \subset \mathcal{D}$ of $\bar{x}$ and $\Lambda_o \subset \Lambda$ of 0 such that $\tilde{F}(x,\lambda) = 0$, $\lambda \in \Lambda_o$, has a unique solution $x_\lambda \in \mathcal{D}_o$. In particular, there exists a unique C^1 mapping $\psi : \Lambda_o \to \mathcal{D}_o$ such that $x_\lambda = \psi(\lambda)$, $\lambda \in \Lambda_o$, and $\bar{x} = \psi(0)$.*

Next, we consider nonlinear feedback controlled dynamical systems. A measurable mapping $\phi : \mathcal{D} \to \mathcal{U}$ satisfying $\phi(\bar{x}) = \bar{u}$ is called a *control law*. Furthermore, if $u(t) = \phi(x(t))$, where $\psi(\cdot)$ is a control law and $x(t)$, $t \subset \mathcal{I}_{x_0}$, satisfies (3.1), then $u(\cdot)$ is called a *feedback control law*. Here, we consider nonlinear closed-loop dynamical systems of the form

$$\dot{x}(t) = F(x(t), \phi(x(t))), \qquad x(0) = x_0, \qquad t \in \mathcal{I}_{x_0}. \tag{3.2}$$

A function $x : \mathcal{I}_{x_0} \to \mathcal{D}$ is said to be a solution to (3.2) on the interval $\mathcal{I}_{x_0} \subseteq \mathbb{R}$ with initial condition $x(0) = x_0$, if $x(t)$ satisfies (3.2) for all $t \in \mathcal{I}_{x_0}$. Note that we do not assume any regularity condition on the function $\phi(\cdot)$. However, we do assume that for every $y \in \mathcal{D}$ there exists a unique solution $x(\cdot)$ of (3.2) defined on $\mathcal{I}_y$ satisfying $x(0) = y$. Furthermore, we assume that all the solutions $x(t)$, $t \in \mathcal{I}_{x_0}$, to (3.2) are continuous functions of the system initial conditions $x_0 \in \mathcal{D}$ which, with the assumption of uniqueness of solutions, implies continuity of solutions $x(t)$, $t \in \mathcal{I}_{x_0}$, to (3.2) [60, p. 24].

3.3 Parameterized System Equilibria and Domains of Attraction

The nonlinear control design framework developed in this chapter is predicated on a hierarchical switching nonlinear controller architecture parameterized over a set of system equilibria. It is important to note that both

the dynamical system and the controller for each parameterized equilibrium can be nonlinear and thus local set point designs are in general nonlinear. Hence, the nonlinear controlled system can be viewed as a collection of controlled subsystems with a hierarchical switching controller architecture. In this section we concentrate on nonlinear stabilization of the local set points parameterized in $\mathcal{D}$. Specifically, we consider the nonlinear controlled dynamical system (3.1) with the origin as an equilibrium point corresponding to the control $u = 0$, that is, $F(0,0) = 0$. Furthermore, we assume that given a mapping $\varphi : \mathcal{D} \times \Lambda \to \mathcal{U}$, $\varphi(0,0) = 0$, there exists a continuous function $\psi : \Lambda_o \to \mathcal{D}_o$, where $\mathcal{D}_o \subseteq \mathcal{D}$, $0 \in \mathcal{D}_o$, and $\Lambda_o \subseteq \Lambda$, $0 \in \Lambda_o$, such that $F(x_\lambda, \varphi(x_\lambda, \lambda)) = 0$ with $x_\lambda = \psi(\lambda) \in \mathcal{D}_o$ for all $\lambda \in \Lambda_o$. As discussed in Section 3.2, this is a necessary condition for parametric stability with respect to Λ_o as defined in [64, 107], while Theorem 3.1 provides sufficient conditions for guaranteeing the existence of such a parameterization.

Next, we consider a family of stabilizing feedback control laws given by

$$\Phi \triangleq \{\phi_\lambda : \mathcal{D} \to \mathcal{U} : \phi_\lambda \in \mathrm{C}^0, \phi_\lambda(x_\lambda) = \varphi(x_\lambda, \lambda), \lambda \in \Lambda_s\}, \quad \Lambda_s \subseteq \Lambda_o, \tag{3.3}$$

such that, for $\phi_\lambda(\cdot) \in \Phi$, $\lambda \in \Lambda_s$, the closed-loop nonlinear feedback system

$$\dot{x}(t) = F(x(t), \phi_\lambda(x(t))), \qquad x(0) = x_0, \qquad t \in \mathcal{I}_{x_0}, \tag{3.4}$$

has an asymptotically stable equilibrium point $x_\lambda \in \mathcal{D}_o \subseteq \mathcal{D}$. Hence, in the terminology of [64, 107], (3.4) is parametrically asymptotically stable with respect to $\Lambda_s \subseteq \Lambda_o$. Here, we assume that for each $\lambda \in \Lambda_s$, the linear or nonlinear feedback controllers $\phi_\lambda(\cdot)$ are given. In particular, these controllers correspond to local set point designs and can be obtained using any appropriate standard linear or nonlinear stabilization scheme with a domain of attraction for each $\lambda \in \Lambda_s$. For example, appropriate nonlinear stabilization techniques such as feedback linearization [66], nonlinear H_∞ control [145], constructive nonlinear control [119], optimal nonlinear control [24], and nonlinear regulation via state-dependent Riccati techniques [34], as well as linear-quadratic stabilization schemes based on locally approximated linearizations, can be used to design the controllers $\phi_\lambda(\cdot)$ for each $\lambda \in \Lambda_s$. Furthermore, for an asymptotically stable equilibrium point $x_\lambda \in \mathcal{D}_o \subseteq \mathcal{D}$, $\lambda \in \Lambda_s$, the domain of attraction $\mathcal{D}_\lambda$ of x_λ is given by

$$\mathcal{D}_\lambda \triangleq \{x_0 \in \mathcal{D} : \text{ if } x(0) = x_0 \text{ then } \lim_{t\to\infty} x(t) = x_\lambda\}. \tag{3.5}$$

Next, given a stabilizing feedback controller $\phi_\lambda(\cdot)$ for each $\lambda \in \Lambda_s$, we provide a guaranteed subset of the domain of attraction $\mathcal{D}_\lambda$ of x_λ using classical Lyapunov stability theory.

Theorem 3.2 **([59]).** *Let $\lambda \in \Lambda_s$. Consider the closed-loop nonlinear system (3.4) with $\phi_\lambda(\cdot) \in \Phi$ and let $x_\lambda \in \mathcal{D}_o \subseteq \mathcal{D}$ be an equilibrium point of (3.4). Furthermore, let $\mathcal{X}_\lambda \subset \mathcal{D}$ be a compact neighborhood of x_λ. Then $x_\lambda \in \mathcal{D}_o \subseteq \mathcal{D}$ is locally asymptotically stable if, and only if, there exists a C^1 function $V_\lambda : \mathcal{X}_\lambda \to \mathbb{R}$ such that*

$$V_\lambda(x_\lambda) = 0, \tag{3.6}$$

$$V_\lambda(x) > 0, \quad x \in \mathcal{X}_\lambda, \quad x \neq x_\lambda, \tag{3.7}$$

$$\dot{V}_\lambda(x) \triangleq V'_\lambda(x)F(x, \phi_\lambda(x)) < 0, \quad x \in \mathcal{X}_\lambda, \quad x \neq x_\lambda. \tag{3.8}$$

In addition, a subset of the domain of attraction of x_λ is given by

$$\mathcal{D}_\lambda \triangleq V_\lambda^{-1}([0, c_\lambda]), \tag{3.9}$$

where $c_\lambda \triangleq \max\{\beta > 0 : V_\lambda^{-1}([0, \beta]) \subseteq \mathcal{X}_\lambda\}$.

It follows from Theorem 3.2 that for all $x_0 \in \mathcal{D}_\lambda$, $\lim_{t\to\infty} V_\lambda(x(t)) = 0$ or, equivalently, for each $\delta > 0$ there exists a finite time $T > 0$ such that $V_\lambda(x(t)) \leq \delta$, $t \geq T$. Hence, given the initial condition $x_0 \in \mathcal{D}_\lambda$, it follows that for every $\delta > 0$ there exists a finite time $T > 0$ such that $x(t) \in V_\lambda^{-1}([0, \delta])$, $t \geq T$. We stress that the aim of Theorem 3.2 is not to make direct comparisons with existing methods for estimating domains of attraction, but rather in helping to provide a streamlined presentation of the main results of Section 3.4 requiring estimates of domains of attraction for local set point designs. Since $\mathcal{D}_\lambda$ given in Theorem 3.2 gives an estimate of the domain of attraction using closed Lyapunov sublevel sets, it may be conservative. To reduce conservatism in estimating a subset of the domain of attraction several alternative methods can be used. For example, maximal Lyapunov functions [136], Zubov's method [142, 59], ellipsoidal estimate mappings [37], Carlemann linearizations [94], computer generated Lyapunov functions [102], iterative Lyapunov function constructions [32], trajectory-reversing methods [73], and open Lyapunov sublevel sets [56], can be used to construct less conservative estimates of the domain of attraction.

3.4 Nonlinear System Stabilization via a Hierarchical Switching Controller Architecture

In this section we develop a nonlinear stabilization framework predicated on a hierarchical switching controller architecture parameterized over a set of moving system equilibria. Specifically, using ***equilibria-dependent Lyapunov functions***, or instantaneous (with respect to a given parameterized equilibrium manifold) Lyapunov functions, a hierarchical nonlinear control strategy

is developed that stabilizes a given nonlinear system by stabilizing a collection of nonlinear controlled subsystems while providing an explicit expression for a guaranteed domain of attraction. A switching nonlinear controller architecture is developed based on a *generalized* lower semicontinuous Lyapunov function obtained by minimizing a potential function, associated with each domain of attraction, over a given switching set induced by the parameterized system equilibria. In the case where one of the parameterized equilibrium points is globally asymptotically stable with a given subcontroller and a structural topological constraint is enforced on the switching set, the proposed nonlinear stabilization framework guarantees global asymptotic stability of any given system equilibrium on the parameterized equilibrium manifold.

To state the main results of this section several definitions and a key assumption are needed. Recall that the set $\Lambda_s \subseteq \Lambda_o$, $0 \in \Lambda_s$, is such that for every $\lambda \in \Lambda_s$ there exists a feedback control law $\phi_\lambda(\cdot) \in \Phi$ such that the equilibrium point $x_\lambda \in \mathcal{D}_o \subseteq \mathcal{D}$ of (3.4) is asymptotically stable with an estimate of the domain of attraction given by $\mathcal{D}_\lambda$. Since x_λ, $\lambda \in \Lambda_s$, is an asymptotically stable equilibrium point of (3.4), it follows from Theorem 3.2 that there exists a Lyapunov function $V_\lambda(\cdot)$ satisfying (3.6)–(3.8), and hence, without loss of generality, we can take $\mathcal{D}_\lambda$, $\lambda \in \Lambda_s$, given by (3.9). Furthermore, we assume that the set-valued map $\Psi : \Lambda_s \rightsquigarrow 2^{\mathcal{D}}$, where $2^{\mathcal{D}}$ denotes the collection of all subsets of $\mathcal{D}$, is such that $\mathcal{D}_\lambda = \Psi(\lambda)$, $\lambda \in \Lambda_s$, is continuous. Here, continuity of a set-valued map is defined in the sense of [8, p. 56] and has the property that the limit of a sequence of a continuous set-valued map is the value of the map at the limit of the sequence. In particular, since $\mathcal{D}_\lambda$, $\lambda \in \Lambda_s$, is given by (3.9), the continuity of the set-valued map $\Psi(\cdot)$ is guaranteed provided that $V_\lambda(x)$, $x \in \mathcal{D}_\lambda$, and c_λ are continuous functions of the parameter $\lambda \in \Lambda_s$. Next, let $\mathcal{S} \subseteq \Lambda_s$, $0 \in \mathcal{S}$, denote a *switching set* such that the following key assumption is satisfied.

Assumption 3.1. *The switching set $\mathcal{S} \subseteq \Lambda_s$ is such that the following properties hold:*

i) There exists a continuous positive-definite function $p : \mathcal{S} \to \mathbb{R}$ such that for all $\lambda \in \mathcal{S}$, $\lambda \neq 0$, there exists $\lambda_1 \in \mathcal{S}$ such that

$$p(\lambda_1) < p(\lambda), \qquad x_\lambda \in \mathring{\mathcal{D}}_{\lambda_1}. \tag{3.10}$$

ii) If $\lambda, \lambda_1 \in \mathcal{S}$, $\lambda \neq \lambda_1$, is such that $p(\lambda) = p(\lambda_1)$, then $\mathcal{D}_\lambda \cap \mathcal{D}_{\lambda_1} = \emptyset$.

Note that Assumption 3.1 assumes the existence of a positive-definite *potential function* $p(\lambda)$, for all λ in the switching set $\mathcal{S}$. It follows that, for each $\lambda \in \mathcal{S}$, there exists an equilibrium point x_λ with an associated domain of attraction $\mathcal{D}_\lambda$, and potential value $p(\lambda)$. Hence, every domain of attraction

has an associated value of the potential function such that, by *ii)*, domains of attraction corresponding to different local set point designs intersect each other only if their corresponding potentials are different. Furthermore, given $\mathcal{D}_\lambda$, $\lambda \in \mathcal{S} \setminus \{0\}$, *i)* implies that there exists at least one intersecting domain of attraction $\mathcal{D}_{\lambda_1}$, $\lambda_1 \in \mathcal{S}$, such that the potential function decreases and $\mathcal{D}_{\lambda_1}$ contains x_λ as an internal point. This guarantees that if a forward trajectory $x(t)$, $t \geq 0$, of the controlled system approaches x_λ, then there exists a finite time $T > 0$ such that the trajectory enters $\mathcal{D}_{\lambda_1}$. Finally, it is important to note that the switching set $\mathcal{S}$ is arbitrary. In particular, we do not assume that $\mathcal{S}$ is countable or countably infinite. For example, the switching set $\mathcal{S}$ can have a hybrid topological structure involving isolated points and closed sets homeomorphic to intervals on the real line.

Next, we show that Assumption 3.1 implies that every level set of the potential function $p(\cdot)$ is either empty or consists of only isolated points. Furthermore, in a neighborhood of the origin every level set of $p(\cdot)$ consists of at most *one* isolated point. For the statement of this result, let $\mathcal{B}_\lambda(r)$, $\lambda \in \mathcal{U}_o$, $r > 0$, denote the open ball centered at x_λ with radius r, that is, $\mathcal{B}_\lambda(r) \triangleq \{x \in \mathcal{D} : \|x - x_\lambda\| < r\}$.

Proposition 3.1. *Let $\mathcal{S} \subseteq \Lambda_s$ be such that Assumption 3.1 holds. Then for every $\alpha > 0$, $p^{-1}(\alpha)$ is either empty or consists of only isolated points. Furthermore, there exists $\beta > 0$ such that for every $\alpha < \beta$, $p^{-1}(\alpha)$ consists of at most one isolated point.*

Proof. Suppose, *ad absurdum*, that there exists $\hat{\lambda} \in p^{-1}(\alpha)$, $\alpha > 0$, such that $\hat{\lambda}$ is not an isolated point in $p^{-1}(\alpha)$. Now, let $\hat{\mathcal{N}} \subset p^{-1}(\alpha)$ be a neighborhood of $\hat{\lambda}$ and note that, by continuity of $\psi(\cdot)$ and the fact that $\hat{\lambda} \in p^{-1}(\alpha)$ is not an isolated point, for every $\epsilon > 0$, there exist $\lambda_1, \lambda_2 \in \overset{\circ}{\hat{\mathcal{N}}}$ such that $\|x_{\lambda_1} - x_{\lambda_2}\| < \epsilon$, and $p(\lambda_1) = p(\lambda_2) = \alpha$. However, since $x_{\lambda_1} \in \mathcal{D}_{\lambda_1}$, it follows that there exists $r > 0$ such that $\mathcal{B}_{\lambda_1}(r) \subseteq \mathcal{D}_{\lambda_1}$. Now, choosing $\epsilon < r$ yields $x_{\lambda_2} \in \mathcal{B}_{\lambda_1}(r) \subseteq \mathcal{D}_{\lambda_1}$ and $x_{\lambda_2} \in \mathcal{D}_{\lambda_2}$ which implies that $\mathcal{D}_{\lambda_1} \cap \mathcal{D}_{\lambda_2} \neq \emptyset$ contradicting *ii)* of Assumption 3.1. Hence, if $p^{-1}(\alpha)$, $\alpha > 0$, is non-empty, it must consist of only isolated points.

Next, suppose, *ad absurdum*, that for all $\delta > 0$ there exist two isolated points $\lambda_1, \lambda_2 \in \hat{\mathcal{N}}_\delta \triangleq \{\lambda \in \mathcal{S} : \|\lambda\| < \delta\}$ such that $p(\lambda_1) = p(\lambda_2)$. Now, repeating the above arguments leads to a contradiction. Hence, there exists $\hat{\delta} > 0$ such that if $\lambda_1 \in \hat{\mathcal{N}}_{\hat{\delta}}$, then $\hat{\mathcal{N}}_{\hat{\delta}} \cap p^{-1}(p(\lambda_1)) = \{\lambda_1\}$. Now, since $p(\cdot)$ is continuous and positive definite, it follows that there exists $\beta > 0$ such that $p^{-1}(\alpha) \subseteq \hat{\mathcal{N}}_{\hat{\delta}}$, $\alpha < \beta$, and hence $p^{-1}(\alpha)$, $\alpha < \beta$, consists of at most one isolated point. ■

Note that Proposition 3.1 implies that, if $p^{-1}(\alpha)$, $\alpha > 0$, is bounded, then there exists a finite distance between isolated points contained in $p^{-1}(\alpha)$

which consists of at most a *finite* number of isolated points. Finally, since in a neighborhood of the origin every level set of $p(\cdot)$ consists of at most one isolated point, a particular topology for $\mathcal{S}$, in a neighborhood of the origin, is homeomorphic to the interval $[0, a]$, $a > 0$, with $0 \in \mathcal{S}$ corresponding to $0 \in \mathbb{R}$.

Now, for every $x \in \mathcal{D}_c \triangleq \cup_{\lambda \in \mathcal{S}} \mathcal{D}_\lambda$, define the *viable switching set* $\mathcal{V}_s(x) \triangleq \{\lambda \in \mathcal{S} : x \in \mathcal{D}_\lambda\}$, which contains all $\lambda \in \mathcal{S}$ such that $x \in \mathcal{D}_\lambda$. Note that if we consider a sequence $\{\lambda_n\}_{n=1}^{\infty} \subset \mathcal{V}_s(x)$, that is, $x \in \mathcal{D}_{\lambda_n}$, such that $\lim_{n\to\infty} \lambda_n = \bar{\lambda}$, it follows from the continuity of the set-valued map $\Psi(\cdot)$ that $x \in \mathcal{D}_{\bar{\lambda}}$. Thus, $\bar{\lambda} \in \mathcal{V}_s(x)$ which implies that $\mathcal{V}_s(x)$ is a non-empty closed set since it contains all of its accumulation points.

Next, we introduce the *switching function* $\lambda_S(x)$, $x \in \mathcal{D}_c$, such that the following definitions hold

$$V(x) \triangleq p(\lambda_S(x)), \quad \lambda_S(x) \triangleq \operatorname{argmin}\{p(\lambda) : \lambda \in \mathcal{V}_s(x)\}, \quad x \in \mathcal{D}_c. \quad (3.11)$$

In particular, $\lambda_S(x)$, $x \in \mathcal{D}_c$, corresponds to the value at which $p(\lambda)$ is minimized wherein λ belongs to the viable switching set. The following proposition shows that "min" in (3.11) is attained and hence $V(x)$ is well defined.

Proposition 3.2. *Let $\mathcal{S} \subseteq \Lambda_s$ and let $p : \mathcal{S} \to \mathbb{R}$ be a continuous positive-definite function such that Assumption 3.1 holds. Then, for all $x \in \mathcal{D}_c$, there exists a unique $\lambda_S(x) \in \mathcal{V}_s(x)$ such that $p(\lambda_S(x)) = \min\{p(\lambda) : \lambda \in \mathcal{V}_s(x)\}$.*

Proof. Existence follows from the fact that $p(\cdot)$ is lower bounded and $\mathcal{V}_s(x)$, $x \in \mathcal{D}_c$, is a non-empty closed set. Now, to prove uniqueness suppose, *ad absurdum*, $\lambda_S(x)$ is not unique. In this case, there exist $\lambda_1, \lambda_2 \in \mathcal{S}$, $\lambda_1 \neq \lambda_2$, such that $p(\lambda_1) = p(\lambda_2)$ and $x \in \mathcal{D}_{\lambda_1} \cap \mathcal{D}_{\lambda_2} \neq \emptyset$ which contradicts *ii)* in Assumption 3.1. ■

The next result shows that $V(\cdot)$ given by (3.11) is a generalized Lyapunov function candidate, that is, $V(\cdot)$ is lower semicontinuous on $\mathcal{D}_c$.

Theorem 3.3. *Let $\mathcal{S} \subseteq \Lambda_s$ be such that Assumption 3.1 holds. Then the function $V(\overset{\circ}{x}) = p(\lambda_S(x))$, $x \in \mathcal{D}_c$, is lower semicontinuous on $\mathcal{D}_c$ and continuous on $\mathcal{D}_{\lambda_S(x)}$.*

Proof. Let the sequence $\{x_n\}_{n=0}^{\infty} \subset \mathcal{D}_c$ be such that $\lim_{n\to\infty} x_n = \hat{x}$ and define $\hat{\lambda} \triangleq \liminf_{n\to\infty} \lambda_S(x_n)$. Here we assume without loss of generality that $\{\lambda_S(x_n)\}_{n=0}^{\infty}$ converges to $\hat{\lambda}$; if this is not the case, it is always possible to construct a subsequence having this property. Since $p(\cdot)$ is continuous (and hence $p(\lim_{n\to\infty} \lambda_S(x_n)) = \lim_{n\to\infty} p(\lambda_S(x_n)))$, it suffices to show that $V(\hat{x}) \leq p(\hat{\lambda})$. Suppose, *ad absurdum*, that $V(\hat{x}) > p(\hat{\lambda})$. In this case, there exists a positive integer n_0 such that $V(\hat{x}) > V(x_n)$, $n \geq n_0$. Now, since by

definition $\lambda_S(\hat{x})$ minimizes $p(\lambda)$ for $\lambda \in \mathcal{V}_s(\hat{x})$, it follows that $V(\hat{x}) \leq p(\lambda)$, $\lambda \in \mathcal{V}_s(\hat{x})$. Hence, since $V(\hat{x}) > V(x_n) = p(\lambda_S(x_n))$, $n \geq n_0$, it follows that $\lambda_S(x_n) \notin \mathcal{V}_s(\hat{x})$ and $\hat{x} \notin \mathcal{D}_{\lambda_S(x_n)}$, $n \geq n_0$. Now, define the closed set $\hat{\mathcal{D}} \triangleq \cup_{n=n_0}^{\infty} \mathcal{D}_{\lambda_S(x_n)}$ such that $\{x_n\}_{n=n_0}^{\infty} \subset \hat{\mathcal{D}}$. Since $\hat{\mathcal{D}}$ is closed, it follows that $\hat{x} \in \hat{\mathcal{D}}$ which implies that there exist $n_1 \geq n_0$ such that $\hat{x} \in \mathcal{D}_{\lambda_S(x_{n_1})}$ which is a contradiction.

To show that $V(x)$ is continuous on $\overset{\circ}{\mathcal{D}}_{\lambda_S(\hat{x})}$ it need only be shown that $V(\hat{x})$ is upper semicontinuous on $\overset{\circ}{\mathcal{D}}_{\lambda_S(\hat{x})}$, or, equivalently, $V(\hat{x}) \geq p(\hat{\lambda})$. Since $\lim_{n\to\infty} x_n = \hat{x}$ and $\hat{x} \in \overset{\circ}{\mathcal{D}}_{\lambda_S(\hat{x})}$, there exists a positive integer n_2 such that $x_n \in \mathcal{D}_{\lambda_S(\hat{x})}$, $n \geq n_2$. Hence, $\lambda_S(\hat{x}) \in \mathcal{V}_s(x_n)$ and $V(\hat{x}) \geq V(x_n)$, $n \geq n_2$, which implies that $V(\hat{x}) \geq p(\hat{\lambda})$. ■

Next, we show that with the hierarchical nonlinear feedback control strategy $u = \phi_{\lambda_S(x)}(x)$, $x \in \mathcal{D}_c$, $V(\cdot)$ given by (3.11) is a generalized Lyapunov function for the nonlinear feedback controlled dynamical system (3.2). The controller notation $\phi_{\lambda_S(x)}(x)$ denotes a switching nonlinear feedback controller where the switching function $\lambda_S(x)$, $x \in \mathcal{D}_c$, is such that definition (3.11) of the generalized Lyapunov function $V(x)$, $x \in \mathcal{D}_c$, holds for a given potential function $p(\cdot)$ and switching set $\mathcal{S}$ satisfying Assumption 3.1. Furthermore, note that since $\phi_{\lambda_S(x)}(x)$ is defined for $x \in \mathcal{D}_c$, it follows that the solution $x(\cdot)$ to (3.2) with $x_0 \in \mathcal{D}_c$ and $u = \phi_{\lambda_S(x)}(x)$ is defined for all values of $t \in \mathcal{I}_{x_0}$ such that $x(t) \in \mathcal{D}_c$. However, as will be shown, since $\mathcal{D}_c$ is a positively invariant set, $[0, +\infty) \subseteq \mathcal{I}_{x_0}$, while if $x_0 \in \mathcal{D}_c$ is such that $x(t)$, $t < 0$, is always contained in $\mathcal{D}_c$, then $\mathcal{I}_{x_0} = \mathbb{R}$. Finally, note that since the solution $x(t)$, $t \in \mathcal{I}_{x_0}$, to (3.2) with $x_0 \in \mathcal{D}_c$ and $u = \phi_{\lambda_S(x)}(x)$ is continuous, it follows from Theorem 3.3 that $V(x(t))$, $t \in \mathcal{I}_{x_0}$, is right continuous. Hence, using the continuity of $p(\cdot)$ and the definition of $V(x)$, $x \in \mathcal{D}_c$, it follows that $\lambda_S(x(t))$, $t \in \mathcal{I}_{x_0}$, is also right continuous. Now, the continuity of $F(\cdot,\cdot)$ and $\phi_\lambda(\cdot)$, $\lambda \in \Lambda_s$, imply that $F(x(t), \phi_{\lambda_S(x(t))}(x(t)))$, $t \in \mathcal{I}_{x_0}$, is right continuous.

Theorem 3.4. *Consider the nonlinear controlled dynamical system given by (3.1) with $F(0,0) = 0$ and assume there exists a continuous function $\psi : \Lambda_o \to \mathcal{D}_o$, $0 \in \Lambda_o$, parameterizing an equilibrium manifold of (3.1), such that $x_\lambda = \psi(\lambda)$, $\lambda \in \Lambda_o$. Furthermore, assume that there exists a C^0 feedback control law $\phi_\lambda(\cdot)$, $\lambda \in \Lambda_s \subseteq \Lambda_o$ with $0 \in \Lambda_s$, that locally stabilizes x_λ with a domain of attraction estimate $\mathcal{D}_\lambda$ and let $\mathcal{S} \subseteq \Lambda_s$, $0 \in \mathcal{S}$, be such that Assumption 3.1 holds. If $\lambda_S(x)$, $x \in \mathcal{D}_c$, is such that $V(x)$, $x \in \mathcal{D}_c$, given by (3.11) holds and $x(t)$, $t \in \mathcal{I}_{x_0}$, is the solution to (3.1) with $x(0) = x_0 \in \mathcal{D}_c$ and feedback control law*

$$u = \phi_{\lambda_S(x)}(x), \qquad x \in \mathcal{D}_c, \tag{3.12}$$

then $\mathcal{D}_c$ is positively invariant and $V(x(t))$, $t \in \mathcal{I}_{x_0}$, is nonincreasing. Furthermore, for all $t_1, t_2 \in \mathcal{I}_{x_0}$, $V(x(t)) = V(x(t_1))$, $t \in [t_1, t_2]$, if, and only if, $\lambda_S(x(t)) = \lambda_S(x(t_1))$, $t \in [t_1, t_2]$. Finally, for all $t \in \mathcal{I}_{x_0}$ such that $\lambda_S(x(t)) \neq 0$, there exists a finite time $T > 0$ such that $V(x(t+T)) < V(x(t))$.

Proof. First, note that $x \in \partial\mathcal{D}_c$ implies $x \in \partial\mathcal{D}_{\lambda_S(x)}$. Since $\phi_{\lambda_S(x)}$ stabilizes $x_{\lambda_S(x)}$ with domain of attraction $\mathcal{D}_{\lambda_S(x)}$, it follows that, for all $x \in \partial\mathcal{D}_c$, the flow of $F(x, \phi_{\lambda_S(x)})$ is directed towards the interior of $\mathcal{D}_{\lambda_S(x)}$ and consequently towards the interior of $\mathcal{D}_c$, which proves positive invariance of $\mathcal{D}_c$. Next, let $x(t)$, $t \in \mathcal{I}_{x_0}$, satisfy (3.1) with $u(t) = \phi_{\lambda_t}(x(t))$, where $\lambda_t \triangleq \lambda_S(x(t))$, and let, for an arbitrary time $t_1 \in \mathcal{I}_{x_0}$, the feedback control law $u = \phi_{\lambda_{t_1}}(x)$ asymptotically stabilize the equilibrium point $x_{\lambda_{t_1}}$ of (3.1) with domain of attraction $\mathcal{D}_{\lambda_{t_1}}$. Since $x(t_1) \in \mathcal{D}_{\lambda_{t_1}}$, it follows from Theorem 3.2 that there exists a C^1 Lyapunov function $V_{\lambda_{t_1}}(\cdot)$ such that $\dot{V}_{\lambda_{t_1}}(x(t)) \triangleq V'_{\lambda_{t_1}}(x(t))F(x(t), \phi_{\lambda_S(x(t))}(x(t)))$, $t \in \mathcal{I}_{x_0}$, and $\dot{V}_{\lambda_{t_1}}(x(t_1)) = V'_{\lambda_{t_1}}(x(t_1))F_{\lambda_{t_1}}(x(t_1)) < 0$. Next, since $F(x(t), \phi_{\lambda_S(x(t))}(x(t)))$, $t \in \mathcal{I}_{x_0}$, is right continuous, it follows that there exists $\delta > 0$ such that $\dot{V}_{\lambda_{t_1}}(x(t)) < 0$, $t \in [t_1, t_1 + \delta]$, which implies that $V_{\lambda_{t_1}}(x(t)) < V_{\lambda_{t_1}}(x(t_1))$, $t \in [t_1, t_1 + \delta]$. Hence, $x(t) \in \mathcal{D}_{\lambda_{t_1}}$, $t \in [t_1, t_1 + \delta]$, and $\lambda_{t_1} \in \mathcal{V}_S(x(t))$, $t \in [t_1, t_1 + \delta]$, which implies that $V(x(t)) \leq V(x(t_1)) = p(\lambda_{t_1})$, $t \in [t_1, t_1 + \delta]$. Now, since $t_1 \in \mathcal{I}_{x_0}$ is arbitrary, it follows that $V(x(t))$, $t \in \mathcal{I}_{x_0}$, is a nonincreasing function along the trajectories $x(t)$, $t \in \mathcal{I}_{x_0}$, of (3.1) with $u(t) = \phi_{\lambda_t}(x(t))$.

Next, assume that $x(t) \in \mathcal{D}_{\lambda_{t_1}}$ and $V(x(t)) = V(x(t_1))$, $t \in [t_1, t_1 + \hat{\delta}]$. Now, suppose, *ad absurdum*, that λ_t, $t \in [t_1, t_1 + \hat{\delta}]$, is not constant, that is, there exists $t_2 \in (t_1, t_1 + \hat{\delta}]$ such that $\lambda_{t_1} \neq \lambda_{t_2}$. In this case $x(t_2) \in \mathcal{D}_{\lambda_{t_1}} \cap \mathcal{D}_{\lambda_{t_2}}$ and $p(\lambda_{t_1}) = p(\lambda_{t_2})$, which contradicts *ii)* of Assumption 3.1. Hence, it follows that if $V(x(t)) = V(x(t_2))$, $t \in [t_1, t_1 + \hat{\delta}]$, then $\lambda_t = \lambda_{t_1}$, $t \in [t_1, t_1 + \hat{\delta}]$. Conversely, if for $t_1, t_2 \in \mathcal{I}_{x_0}$, $\lambda_t = \lambda_{t_1}$, $t \in [t_1, t_2]$, then $V(x(t)) = V(x(t_1))$, $t \in [t_1, t_2]$, is immediate.

Finally, for an arbitrary $t_1 \in \mathcal{I}_{x_0}$, suppose, *ad absurdum*, that $V(x(t)) = V(x(t_1)) \neq 0$, $t \geq t_1$, or, equivalently, $\lambda_t = \lambda_{t_1} \in \mathcal{S} \setminus \{0\}$, $t \geq t_1$. Then the feedback control law $\phi_{\lambda_t}(\cdot) = \phi_{\lambda_{t_1}}(\cdot)$ stabilizes the equilibrium point $x_{\lambda_{t_1}}$. In this case, it follows from Assumption 3.1 that there exists $\lambda_1 \neq \lambda_{t_1}$ such that $p(\lambda_1) < V(x(t_1))$ and $x_{\lambda_{t_1}} \in \mathring{\mathcal{D}}_{\lambda_1}$, which implies that there exists $\alpha > 0$ such that $x_{\lambda_{t_1}} \in V_{\lambda_{t_1}}^{-1}([0, \alpha]) \subseteq \mathcal{D}_{\lambda_1}$. Hence, it follows that $x(t)$ approaches the level set $V_{\lambda_{t_1}}^{-1}(\alpha)$ in a finite time $T > 0$ so that $V(x(t_1 + T)) \leq p(\lambda_1) < V(x(t_1))$, which contradicts the original supposition. ∎

Next, we show that the hierarchical switching nonlinear controller (3.12) guarantees that the generalized Lyapunov function (3.11) is nonincreasing along the closed-loop system trajectories with strictly decreasing values *only*

at the switching times which occur when the closed-loop system trajectory enters a new domain of attraction with an associated lower potential value.

Corollary 3.1. *Consider the nonlinear controlled dynamical system given by (3.1) with* $F(0,0) = 0$ *and assume the hypothesis of Theorem 3.4 hold. Then* $V(x(t))$, $t \geq 0$, *is strictly decreasing only at the switching times which occur when the trajectory* $x(t)$, $t \in \mathcal{I}_{x_0}$, *enters a new domain of attraction with an associated lower potential value.*

Proof. First, we consider the case where $x(t_2) \in \overset{\circ}{\mathcal{D}}_{\lambda_{t_2}}$, with $\lambda_{t_2} \triangleq \lambda_S(x(t_2))$ and $t_2 > 0$. It follows from the continuity of the closed-loop system trajectories $x(\cdot)$ that there exists $t_1 < t_2$ such that $x(t_1) \in \mathcal{D}_{\lambda_{t_2}}$, which implies that $\lambda_{t_2} \in \mathcal{V}_S(x(t_1))$ and $V(x(t_1)) \leq V(x(t_2))$. Since $V(x(t))$, $t \geq 0$, is a nonincreasing function of time, it follows that $V(x(t)) = V(x(t_2))$, $t \in [t_1, t_2]$. Alternatively, assume that $x(t_2) \in \partial\mathcal{D}_{\lambda_{t_2}}$, and suppose, *ad absurdum*, that there exists $t_1 < t_2$ such that $x(t_1) \in \mathcal{D}_{\lambda_{t_2}}$. Then $\lambda_t = \lambda_{t_2}$, $t \in [t_1, t_2]$, and, since $V_{\lambda_{t_2}}(x)$, $x \in \mathcal{D}_{\lambda_{t_2}}$, attains its maximum at $x(t_2) \in \partial\mathcal{D}_{\lambda_{t_2}}$, it follows that $V_{\lambda_{t_2}}(x(t)) \leq V_{\lambda_{t_2}}(x(t_2))$, $t \in [t_1, t_2]$, which contradicts the fact that $V_{\lambda_{t_2}}(x(t))$, $t \geq 0$, is a decreasing function of time. Hence, $x(t) \notin \mathcal{D}_{\lambda_{t_2}}$ and $V(x(t)) < V(x(t_2))$, for all $t < t_2$. ■

Finally, we present the main result of this section. Specifically, we show that the hierarchical switching nonlinear controller given by (3.12) guarantees that the closed-loop system trajectories converge to a union of largest invariant sets contained on the boundary of intersections over finite intervals of the closure of the generalized Lyapunov level surfaces. In addition, if the switching set $\mathcal{S}$ is homeomorphic to an interval on the real line and/or consists of only isolated points, then the hierarchical switching nonlinear controller establishes asymptotic stability of the origin.

Theorem 3.5. *Consider the nonlinear controlled dynamical system given by (3.1) with* $F(0,0) = 0$ *and assume there exists a continuous function* $\psi : \Lambda_o \to \mathcal{D}_o$, $0 \in \Lambda_o$, *parameterizing an equilibrium manifold of (3.1), such that* $x_\lambda = \psi(\lambda)$, $\lambda \in \Lambda_o$. *Furthermore, assume that there exists a* C^0 *feedback control law* $\phi_\lambda(\cdot)$, $\lambda \in \Lambda_s \subseteq \Lambda_o$ *with* $0 \in \Lambda_s$, *that locally stabilizes* x_λ *with a domain of attraction estimate* $\mathcal{D}_\lambda$ *and let* $\mathcal{S} \subseteq \Lambda_s$, $0 \in \mathcal{S}$, *be such that Assumption 3.1 holds. In addition, assume* $\lambda_S(x)$, $x \in \mathcal{D}_c$, *is such that* $V(x)$, $x \in \mathcal{D}_c$, *given by (3.11) holds, and, for* $x_0 \in \mathcal{D}_c$, $x(t)$, $t \in \mathcal{I}_{x_0}$, *is the solution to (3.2) with the feedback control law*

$$u = \phi_{\lambda_S(x)}(x), \qquad x \in \mathcal{D}_c. \tag{3.13}$$

If $x_0 \in \mathcal{D}_c$, *then* $x(t) \to \hat{\mathcal{M}} \triangleq \cup_{\gamma \in \mathcal{G}} \mathcal{M}_\gamma$ *as* $t \to \infty$, *where* $\mathcal{G} \triangleq \{\gamma \geq 0 : \mathcal{R}_\gamma \cap \mathcal{D}_0 \neq \emptyset\}$. *If, in addition,* $\mathcal{S}_0 \triangleq \{\lambda \in \mathcal{S} : \mathcal{D}_\lambda \cap \mathcal{D}_0 \neq \emptyset\}$ *is homeomorphic*

to $[0, a]$, $a > 0$, *with* $0 \in \mathcal{S}_0$ *corresponding to* $0 \in \mathbb{R}$, *or* $\mathcal{S}_0$ *consists of only isolated points, then the zero solution* $x(t) \equiv 0$ *to (3.2) is locally asymptotically stable with an estimate of domain of attraction given by* $\mathcal{D}_c$. *Finally, if* $\mathcal{D} = \mathbb{R}^n$ *and there exists* $\hat{\lambda} \in \mathcal{S}$ *such that the feedback control law* $\phi_{\hat{\lambda}}(\cdot)$ *globally stabilizes* $x_{\hat{\lambda}}$, *then the above results are global.*

Proof. The result follows from Theorems 2.4, 3.3, and 3.4. Specifically, Theorem 3.4 implies that if $x(\hat{t}) \in \mathcal{D}_0$ for an arbitrary $\hat{t} \geq 0$, then $V(x(t)) = V(x(\hat{t})) = 0$, $t \geq \hat{t}$. Hence, $\lambda_S(x(t)) = 0$, $t \geq \hat{t}$, and the feedback control law $u = \phi_0(x)$ asymptotically stabilizes the origin with an estimate of the domain of attraction given by $\mathcal{D}_0$. In this case, $\mathcal{D}_0$ is a compact positively invariant set of (3.2) with the feedback control law (3.13). Next, it follows from Theorems 3.3 and 3.4 that $V(\cdot)$ is a generalized Lyapunov function defined on $\mathcal{D}_c$. Now, it follows from Theorem 2.4 that, for all $x_0 \in \mathcal{D}_c$, $x(t) \to \hat{\mathcal{M}}$ as $t \to \infty$.

Next, if $\mathcal{S}_0$ is homeomorphic to $[0, a]$, $a > 0$, with $0 \in \mathcal{S}_0$ corresponding to $0 \in \mathbb{R}$, so that Assumption 3.1 is satisfied, it follows from the continuity of the set-valued map $\Psi(\cdot)$ restricted to $\mathcal{S}_0$ that $V(\cdot)$ is continuous on $\mathcal{D}_0$. Now, it follows from Theorem 2.4 and the fact that the origin is the largest invariant set contained in $\mathcal{R}_0$ that $\hat{\mathcal{M}} \equiv \mathcal{M}_0 \equiv \{0\}$. Hence, $x(t) \to 0$ as $t \to \infty$ establishing local asymptotic stability with an estimate of domain of attraction given by $\mathcal{D}_c$. Alternatively, if $\mathcal{S}_0$ consists of only isolated points with finite pairwise distance, it follows that $\mathcal{G}$ consists of the isolated values of $p(\cdot)$ evaluated on the elements of $\mathcal{S}_0$. Hence, since $\hat{\mathcal{R}}_\gamma \triangleq \mathcal{R}_\gamma \setminus V^{-1}(\gamma)$, $\gamma \in \mathcal{G}$, is bounded, $V(\cdot)$ can only assume a finite number of distinct values on $\hat{\mathcal{R}}_\gamma$, $\gamma \in \mathcal{G}$, including the zero value. Now, it follows from Theorems 2.4 and 3.4, respectively, that $\mathcal{M}_\gamma \subset \hat{\mathcal{R}}_\gamma$ and, for all $x_0 \in \mathcal{D}_c \setminus \mathcal{D}_0$, there exists an increasing unbounded sequence $\{t_n\}_{n=0}^{\infty}$, with $t_0 = 0$, such that $V(x(t_{n+1})) < V(x(t))$, $n = 0, 1, \ldots$. Thus, no forward trajectories can be entirely contained in $\mathcal{R}_\gamma$, $\gamma \in \mathcal{G} \setminus \{0\}$. Hence, $\mathcal{M}_\gamma = \emptyset$, $\gamma \in \mathcal{G} \setminus \{0\}$, and $\hat{\mathcal{M}} \equiv \mathcal{M}_0 \equiv \{0\}$ establishing local asymptotic stability with an estimate of domain of attraction given by $\mathcal{D}_c$.

Finally, let $\mathcal{D} = \mathbb{R}^n$ and assume $\mathcal{S}_g \triangleq \{\lambda \in \mathcal{S} : \mathcal{D}_\lambda \equiv \mathbb{R}^n\}$ is not empty. In particular, if $\hat{\lambda} \in \mathcal{S}_g$, then the feedback control law $\phi_{\hat{\lambda}}(\cdot)$ globally stabilizes $x_{\hat{\lambda}}$. Now, Assumption 3.1 implies that if $\hat{\lambda}_1, \hat{\lambda}_2 \in \mathcal{S}_g$, $\hat{\lambda}_1 \neq \hat{\lambda}_2$, then $p(\hat{\lambda}_1) \neq p(\hat{\lambda}_2)$. Furthermore, since $\mathcal{D}_{\hat{\lambda}_1} \equiv \mathcal{D}_{\hat{\lambda}_2} \equiv \mathbb{R}^n$, we obtain that $\hat{\lambda}_1, \hat{\lambda}_2 \in \mathcal{V}_s(x)$ for all $x \in \mathbb{R}^n$. Next, assume without loss of generality that $p(\hat{\lambda}_1) < p(\hat{\lambda}_2)$, and note that since $\lambda_S(x)$ minimizes $p(\cdot)$ over $\mathcal{V}_s(x)$, we obtain that $\lambda_S(x) \neq \lambda_2$, $x \in \mathbb{R}^n$. It follows that $V(x) \leq \min\{p(\hat{\lambda}) : \hat{\lambda} \in \mathcal{S}_g\}$ for all $x \in \mathbb{R}^n$ and only the (unique) value $\lambda \in \mathcal{S}_g$ that minimizes $p(\cdot)$ over $\mathcal{S}_g$ is assumed by the switching function $\lambda_S(\cdot)$, so that all the other elements of $\mathcal{S}_g$ can be discarded from $\mathcal{S}$. Hence, without loss of generality, assume that there exists a unique $\hat{\lambda} \in \mathcal{S}$ such that $\phi_{\hat{\lambda}}(\cdot)$ globally stabilizes $x_{\hat{\lambda}}$. Now, define $\hat{\mathcal{S}} \triangleq \{\lambda \in \mathcal{S} : p(\lambda) < p(\hat{\lambda})\}$

and $\hat{\mathcal{D}}_c \triangleq \cup_{\lambda \in \hat{\mathcal{S}}} \mathcal{D}_\lambda$ which is a compact positively invariant set. Hence, if $x_0 \in \hat{\mathcal{D}}_c$, it follows from the first part of the theorem that $\hat{\mathcal{M}}$ is a local attractor and, if $\mathcal{S}_0$ is homeomorphic to an interval on $\mathbb{R}$ or consists of only isolated points, the origin is asymptotically stable, with (in both cases) an estimate of the domain of attraction given by $\hat{\mathcal{D}}_c$. Now, global attraction to $\hat{\mathcal{M}}$ as well as global asymptotic stability of the origin is immediate by noting that if $x_0 \notin \hat{\mathcal{D}}_c$, then the forward trajectory of (3.2) approaches $\hat{\mathcal{D}}_c$ in a finite time. If, in fact, $x \notin \hat{\mathcal{D}}_c$, then $\lambda_{\mathcal{S}}(x) = \hat{\lambda}$ which implies that for all $x \notin \hat{\mathcal{D}}_c$ the feedback control law (3.13) stabilizes $x_{\hat{\lambda}}$ and, by Assumption 3.1, $x_{\hat{\lambda}} \in \mathring{\mathcal{D}}_c$. In this case, it follows that for all $x_0 \notin \hat{\mathcal{D}}_c$ there exists a finite time $T > 0$ such that $x(T) \in \hat{\mathcal{D}}_c$. Hence, global attraction as well as global asymptotic stability of the origin is established for the respective cases. ■

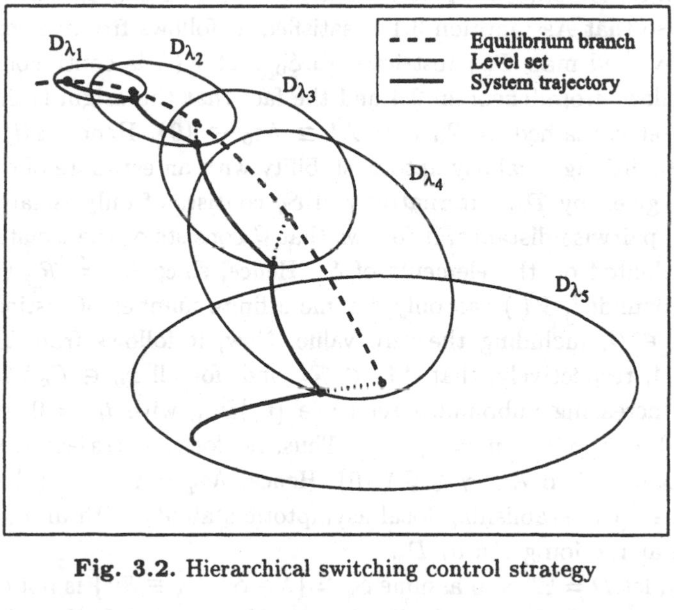

Fig. 3.2. Hierarchical switching control strategy

The switching set $\mathcal{S}$ is quite general in the sense that it can have a hybrid topological structure involving isolated points and closed sets homeomorphic to intervals on the real line. In the special case where the switching set $\mathcal{S}$ consists of only isolated points, the hierarchical switching control strategy given by (3.13) is piecewise continuous (see Figure 3.2). Alternatively, in the special case where the switching set $\mathcal{S}$ is homeomorphic to an interval on $\mathbb{R}$, the hierarchical switching control strategy given by (3.13) is *not* necessarily continuous. The continuous control case will be discussed in Section 3.5. In the case where the switching set $\mathcal{S}$ is homeomorphic to an interval on $\mathbb{R}$ and

a stabilizing controller $\phi_0(\cdot)$ for the origin cannot be obtained, that is, $c_0 = 0$, $0 \in \mathcal{D}_c$ still holds. Hence, Theorem 3.5 guarantees attraction of the origin if $0 \in \partial\mathcal{D}_c$. Alternatively, if $0 \in \mathring{\mathcal{D}}_c$, then the origin is asymptotically stable.

Finally, it is important to note that since the hierarchical switching nonlinear controller $u = \phi_{\lambda_S(x)}(x)$, $x \in \mathcal{D}_c$, is constructed such that the switching function $\lambda_S(x)$, $x \in \mathcal{D}_c$, assures that $V(x)$, $x \in \mathcal{D}_c$, defined by (3.11) is a generalized Lyapunov function with strictly decreasing values at the switching points, the possibility of a sliding mode is precluded with the proposed control scheme. In particular, Theorem 3.4 guarantees that the closed-loop state trajectories cross the boundary of adjacent regions of attraction in the state space in a inward direction. Thus, the closed-loop state trajectories *enter* the lower potential-valued domain of attraction before subsequent switching can occur. Hence, the proposed nonlinear stabilization framework avoids the undesirable effects of high-speed switching onto an invariant sliding manifold.

3.5 Extensions to Nonlinear Dynamic Compensation

In this section we provide an online procedure for computing the switching function $\lambda_S(x)$, $x \in \mathcal{D}_c$, such that (3.11) holds, by constructing an initial value problem for $\lambda_S(x)$ having a fixed-order dynamic compensator structure. Specifically, we consider a switching set $\mathcal{S}$ diffeomorphic to an interval on the real line $\mathbb{R}$ and assume that the switching function $\lambda_S(x)$, $x \in \mathcal{D}_c$, is C^1. To present this result, consider the nonlinear controlled dynamical system given by (3.1) with a nonlinear feedback dynamic controller of the form

$$\dot{x}_c(t) = F_c(x(t), x_c(t)), \qquad x_c(0) = x_{c0}, \qquad t \in \mathcal{I}_{x_0}, \tag{3.14}$$

$$u(t) = \phi_c(x(t), x_c(t)), \tag{3.15}$$

where $x_c(t) \in \mathcal{C} \subseteq \mathbb{R}^{n_c}$, $t \in \mathcal{I}_{x_0}$, is the controller state vector, $\mathcal{C}$ is an open set, $F_c : \mathcal{D} \times \mathcal{C} \to \mathbb{R}^{n_c}$, and $\phi_c : \mathcal{D} \times \mathcal{C} \to \mathcal{U}$. Note that we do not assume any regularity condition on the mappings $F_c(\cdot,\cdot)$ and $\phi_c(\cdot,\cdot)$. However, we do assume that the nonlinear feedback controlled dynamical system given by (3.1), (3.14), and (3.15) is such that the solutions of the closed-loop system on $\mathcal{I}_{x_0}$ are unique and continuously dependent on the closed-loop system initial conditions.

To construct dynamic controllers of the form (3.14), (3.15) we assume that the switching set $\mathcal{S}$ is such that there exists a closed interval $[0, a]$, $a > 0$, and a diffeomorphism $\sigma : [0, a] \to \Lambda_s$, such that $\sigma(s) \in \mathcal{S}$, $s \in [0, a]$, and $\sigma(0) = 0$. Furthermore, we assume that $V_\lambda(\cdot)$ and c_λ are C^1 functions of $\lambda \in \mathcal{S}$. Next, recall that Theorem 3.5 guarantees that the feedback control law $u(x) = \phi_{\lambda_S(x)}(x)$, $x \in \mathcal{D}_c$, where $\lambda_S(x)$, $x \in \mathcal{D}_c$, is given by (3.11),

locally asymptotically stabilizes the origin of the nonlinear feedback controlled dynamical system (3.2), and $\mathcal{D}_c$ is a subset of the domain of attraction. Furthermore, note that if $x_0 \in \mathcal{D}_c \setminus \mathcal{D}_0$ it follows by continuity of the closed-loop trajectories $x(\cdot)$ that there exists a finite time $T_{x_0} > 0$ such that $x(T_{x_0}) \in \partial\mathcal{D}_0$ and $\lambda_S(x(T_{x_0})) = 0$. Now, define $\lambda_t \triangleq \lambda_S(x(t))$ and note that since $\mathcal{S}$ is connected and the set-valued map $\Psi(\cdot)$ is continuous, it follows that $\mathcal{R}_\gamma = V^{-1}(\gamma)$, $\gamma \geq 0$, which implies that $V(x(t))$, $t \geq 0$, can be constant on the interval $[t_1, t_2] \subseteq [0, T_{x_0}]$ only if $x(t) \in \partial\mathcal{D}_{\lambda_t}$, $t \in [t_1, t_2]$, which contradicts the fact that $\dot{V}_{\lambda_t}(x(t)) < 0$. Hence, it follows that $V(x(t))$, $t \in [0, T_{x_0}]$, is a strictly decreasing function and Theorem 3.4 further implies that $x(t) \in \partial\mathcal{D}_{\lambda_t} = V_{\lambda_t}^{-1}(c_{\lambda_t})$, $t \in [0, T_{x_0}]$. Thus,

$$V_{\lambda_t}(x(t)) = c_{\lambda_t}, \qquad t \in [0, T_{x_0}], \tag{3.16}$$

relates the state trajectories $x(\cdot)$ to the switching function $\lambda_S(x(\cdot))$. For the statement of the main result of this section the following definitions and proposition are needed. Define

$$v_\lambda(x) \triangleq \frac{\partial(c_\lambda - V_\lambda(x))}{\partial\lambda}, \quad w_\lambda \triangleq \left.\frac{\mathrm{d}\sigma(s)}{\mathrm{d}s}\right|_{s=\sigma^{-1}(\lambda)}, \quad (x,\lambda) \in \mathcal{D}_c \times \mathcal{S}, \tag{3.17}$$

and $Q_\lambda(x) \triangleq \frac{w_\lambda}{v_\lambda(x) w_\lambda} V'_\lambda(x)$ which, as shown in Proposition 3.3 below, is well defined for all $x \in \mathcal{D}_c \setminus \mathcal{D}_0$ and $\lambda = \lambda_S(x)$. In the case where $x \in \mathcal{D}_0$, define $Q_\lambda(x) \equiv 0$.

Proposition 3.3. *Assume $\sigma : [0, a] \to \mathcal{S}$ is a diffeomorphism, $\lambda_S : \mathcal{D}_c \to \mathcal{S}$ is a C^1 function, and $V_\lambda(\cdot)$ and c_λ are C^1 functions of $\lambda \in \mathcal{S}$. Then $v_\lambda(x)$ and w_λ, $(x,\lambda) \in \mathcal{D}_c \times \mathcal{S}$, defined in (3.17), are such that $v_{\lambda_S(x)}(x) w_{\lambda_S(x)} \neq 0$ for all $x \in \mathcal{D}_c \setminus \mathcal{D}_0$.*

Proof. The result follows by differentiating both sides of (3.16) and noting that for all $x_0 \in \mathcal{D}_c$, $\dot{V}_{\lambda_0}(x_0) < 0$, $\lambda_0 = \lambda_S(x_0)$. Specifically, differentiating both sides of (3.16) yields

$$v_{\lambda_t}(x(t))\dot{\lambda}_t = V'_{\lambda_t}(x(t))\dot{x}(t), \qquad t \in [0, T_{x_0}]. \tag{3.18}$$

Next, let $s(t)$, $t \in [0, T_{x_0}]$, be such that $\lambda_t = \sigma(s(t))$, $t \in [0, T_{x_0}]$, and note that (3.18) evaluated at $t = 0$ yields

$$v_{\lambda_0}(x_0) w_{\lambda_0} \dot{s}(0) = V'_{\lambda_0}(x_0)\dot{x}(0) = \dot{V}_{\lambda_0}(x_0) < 0. \tag{3.19}$$

Now, noting that $v_{\lambda_0}(x_0) w_{\lambda_0}$ and $\dot{s}(0)$ are scalars, it follows that $v_{\lambda_0}(x_0) w_{\lambda_0} \neq 0$, $x_0 \in \mathcal{D}_c \setminus \mathcal{D}_0$. ■

Next, we present the main result of this section which provides an online procedure for computing the switching function $\lambda_S(x(t))$, $t \in \mathcal{I}_{x_0}$. Specifically, differentiating both sides of (3.16) with respect to time yields a Davidenko-type differential equation that defines an initial value problem for the switching function and hence the function $\lambda_S(x(t))$, $t \in \mathcal{I}_{x_0}$, is characterized via a homotopy map.

Theorem 3.6. *Assume $\sigma : [0,a] \to \mathcal{S}$ is a diffeomorphism, $\lambda_S : \mathcal{D}_c \to \mathcal{S}$ is a C^1 function, and $V_\lambda(\cdot)$ and c_λ are C^1 functions of $\lambda \in \mathcal{S}$. Then, the solutions $x(t)$ and $\lambda(t)$, $t \in \mathcal{I}_{x_0}$, of the closed-loop nonlinear feedback controlled dynamical system*

$$\dot{x}(t) = F(x(t), \phi_{\lambda(t)}(x(t))), \qquad x(0) = x_0, \qquad t \in \mathcal{I}_{x_0}, \quad (3.20)$$

$$\dot{\lambda}(t) = Q_{\lambda(t)}(x(t))F(x(t), \phi_{\lambda(t)}(x(t))), \qquad \lambda_0 = \lambda_S(x_0), \quad (3.21)$$

are such that $\lambda(t) = \lambda_S(x(t))$, $t \in \mathcal{I}_{x_0}$, or, equivalently, $V_{\lambda(t)}(x(t)) = c_{\lambda(t)}$, $t \in \mathcal{I}_{x_0}$.

Proof. The result follows by differentiating both sides of (3.16) with respect to time and noting that $\dot{\lambda}_t = w_{\lambda_t}\dot{s}(t)$. ■

Note that the switching function dynamics characterized by (3.21) defines a fixed-order dynamic compensator of the form given by (3.14), (3.15). Specifically, defining the compensator state as $x_c(t) \triangleq \lambda(t)$ so that $n_c = q$, the dynamic compensator structure is given by

$$\dot{x}_c(t) = Q_{x_c(t)}(x(t))F(x(t), \phi_{x_c(t)}(x(t))), \quad x_c(0) = \lambda_S(x_0), \quad t \in \mathcal{I}_{x_0}, \quad (3.22)$$

$$u(t) = \phi_{x_c(t)}(x(t)). \quad (3.23)$$

Now, it follows from Theorems 3.5 and 3.6 that for all $x_0 \in \mathcal{D}_c$ the nonlinear feedback controlled dynamical system given by (3.1), (3.22), and (3.23), drives $x_c(t)$ to $0 \in \mathcal{S}$ in a finite time $T_{x_0} > 0$. Note that this result does not violate the assumption of uniqueness of solutions of (3.22) since $x(T_{x_0}) \in \partial\mathcal{D}_0$ at the finite time $T_{x_0} > 0$ does not correspond to a system equilibrium point. Next, since $\lambda(t) = 0$, $t \geq T_{x_0}$, $x(t)$ reaches $0 \in \mathcal{D}$ asymptotically which guarantees asymptotic stability of the origin with an estimate of the domain of attraction given by $\mathcal{D}_c$. Similar arguments hold for global asymptotic stability in the case where $\mathcal{D} \equiv \mathcal{D}_c \equiv \mathbb{R}^n$.

The compensator dynamics (3.21) characterize the fastest admissible rate of change of the switching function for which the feedback control (3.23) maintains stability of the closed-loop system. As discussed in the Introduction, this quantifies the notion of slow-varying system parameters which has been one of the major shortcomings of gain scheduling practice. It is not

surprising to note that this rate is an explicit function of the gradient of the equilibria-dependent Lyapunov functions and the gradient of the domains of attraction estimates with respect to the parameterized equilibrium manifold. Furthermore, the rate of change of the switching function also depends on the gradient of the diffeomorphism evaluated along the switching set $\mathcal{S}$; such a dependence can be used to enforce desirable structural properties of the switching set. In particular, in Section 3.6, a hierarchical inverse optimal control framework for computing a diffeomorphism online will be presented. Finally, as noted in the Introduction, the proposed framework can be used to address practical actuator limitations such as control rate saturation constraints. Specifically, constraining the rate at which the dynamics of the switching function $\lambda(t)$ can evolve on the equilibrium manifold, it is possible to address input rate saturation constraints. In particular, since the parameterization of the equilibrium manifold introduced in Section 3.2 is obtained by setting $u = \varphi(\psi(\lambda), \lambda)$, $\lambda \in \Lambda_{\rm o}$, it follows that, differentiating both sides of this parameterization with respect to time,

$$\dot{u}(t) = S(\lambda(t))\dot{\lambda}(t), \qquad t \in \mathcal{I}_{x_0}, \tag{3.24}$$

where $S(\lambda(t)) \in \mathbb{R}^{m \times q}$ is given by

$$S(\lambda) \triangleq \left[\frac{\partial \varphi(x,\lambda)}{\partial x}\frac{\mathrm{d}\psi(\lambda)}{\mathrm{d}\lambda} + \frac{\partial \varphi(x,\lambda)}{\partial \lambda}\right]_{x=\psi(\lambda)}.$$

Now, choosing $\varphi(\cdot,\cdot)$ so that $q = m$, that is, the control and parameter spaces have the same dimensions, and constructing the switching set $\mathcal{S}$ so that $S(\lambda)$, $\lambda \in \mathcal{S}$, is not singular, it follows from (3.24) that a constraint on the rate at which the dynamics of the parameter $\lambda(t)$ can evolve on the equilibrium manifold can be enforced effectively placing a rate constraint on the control $u(t)$.

Finally, to elucidate the hierarchical switching nonlinear controller presented in this section and Section 3.4, we present an algorithm that outlines the key steps in constructing the hierarchical switching feedback controller.

Algorithm 3.1. *To construct the hierarchical switching feedback control $\phi_{\lambda_{\mathcal{S}}(x(t))}(x(t))$, $t \geq 0$, perform the following steps:*

Step 1. Construct the equilibrium manifold of (3.1) using $u = \varphi(x,\lambda)$, where $\varphi(\cdot,\cdot)$ is an arbitrary function of $\lambda \in \Lambda_{\rm o}$. Use $F(x,\varphi(x,\lambda)) = 0$ to explicitly define the mapping $\psi(\cdot)$ such that $x_\lambda = \psi(\lambda)$, $\lambda \in \Lambda_{\rm o}$, is an equilibrium point of (3.1) corresponding to the parameter value λ. We note that the above parameterization can be constructed using the approaches given in [81, 64, 107].

Step 2. Construct the set $\Lambda_s \subseteq \Lambda_o$ *such that, for each equilibrium point* x_λ, $\lambda \in \Lambda_s$, *there exists a stabilizing controller* $\phi_\lambda(\cdot)$ *and an associated domain of attraction* $\mathcal{D}_\lambda$ *corresponding to the level set* c_λ *and Lyapunov function* $V_\lambda(\cdot)$. *Here, the controllers* $\phi_\lambda(\cdot)$, $\lambda \in \Lambda_s$, *can be obtained using any appropriate standard linear or nonlinear stabilization scheme.*

Step 3. Construct the switching set $\mathcal{S} \subseteq \Lambda_s$ *and a potential function* $p : \mathcal{S} \to \mathbb{R}^+$ *such that Assumption 3.1 is satisfied. In particular:*

3a. If $\lambda \in \mathcal{S}$ *is an isolated point of* $\mathcal{S}$ *with corresponding equilibrium point* x_λ, *then there exists* $\lambda_1 \in \mathcal{S}$ *such that* $p(\lambda_1) < p(\lambda)$, $x_\lambda \in \mathring{\mathcal{D}}_{\lambda_1}$.

3b. If $\lambda \in \mathcal{S}$ *is an accumulation point of* $\mathcal{S}$ *then Step 3a is automatically satisfied if* $p(\cdot)$ *does not achieve a local minimum at* λ.

3c. If $\lambda, \lambda_1 \in \mathcal{S}$, $\lambda \neq \lambda_1$, *is such that* $p(\lambda) = p(\lambda_1)$, *then* $\mathcal{D}_\lambda \cap \mathcal{D}_{\lambda_1} = \emptyset$.

Step 4. Given the state space point $x(t)$ *at* $t \geq 0$, *search for solutions to* $V_\lambda(x(t)) = c_\lambda$, $\lambda \in \mathcal{S}$.

4a. If no solution exists, $\lambda_{\mathcal{S}}(x(t))$ *is unchanged.*

4b. If one solution λ_1 *exists and* $p(\lambda_1) < p(\lambda)$ *then switch* $\lambda_{\mathcal{S}}(x(t))$ *to* λ_1.

4c. If more than one solution exists, repeat Step 4b with λ_1 *replaced by the solution that minimizes* $p(\cdot)$. *Note that multiple solutions can be avoided by modifying the* c_λ*'s.*

Step 5. Construct the hierarchical switching feedback controller $\phi_{\lambda_{\mathcal{S}}(x(t))}(x(t))$, *where* $\lambda_{\mathcal{S}}(x(t))$, $x \in \mathcal{D}_c$, *constructed in Step 4 is such that (3.11) holds.*

Note that the existence of a switching set $\mathcal{S}$ and a potential function $p(\cdot)$ such that Step 3 is satisfied, can be guaranteed by modifying the first part Step 4 as follows:

Step 4'. Given the state space point $x(t)$ *at* $t = t_k \triangleq k\Delta T$, *where* $\Delta T > 0$ *and* $k = 0, 1, \ldots$, *search for the solutions of* $V_\lambda(x(t_k)) \leq c_\lambda$, $\lambda \in \Lambda_s$.

In this case, the switching set $\mathcal{S} \subseteq \Lambda_s$ need not be explicitly defined and is computed online. Furthermore, the case where $\Delta T \to 0$ recovers the continuous framework described in this section.

3.6 Inverse Optimal Nonlinear Switching Control

In this section we develop optimality notions for the hierarchical switching controller architecture developed in Section 3.4. Specifically, in the case where the switching set induced by the parameterized system equilibria is diffeomorphic to a closed interval on the real line, we construct an inverse

optimal hierarchical switching strategy that is characterized by a Davidenko-type differential equation. The inverse optimal hierarchical control framework developed herein is quite different from the quasivariational inequality methods for optimal switching systems developed in the literature (e.g., [41, 19]). Specifically, quasivariational methods do not guarantee asymptotic stability via Lyapunov functions and do not necessarily yield feedback controllers. In contrast, our results provide hierarchical homotopic *feedback* controllers guaranteeing closed-loop stability via an underlying Lyapunov function.

To provide an optimal online procedure for computing the switching function $\lambda_{\mathcal{S}}(x)$, $x \in \mathcal{D}_{\rm c}$, such that (3.11) holds, we extend the results of Section 3.5 by constructing an initial value problem for the switching function $\lambda_{\mathcal{S}}(x)$ having a fixed-order dynamic compensator structure that minimizes a *derived* cost functional. Specifically, to address optimality notions within our nonlinear switching control framework we consider the following minimization problem.

Optimal Switching Control Problem. Consider the nonlinear controlled dynamical system given by (3.1) with $u(x) = \phi_{\lambda_{\mathcal{S}}(x)}(x)$, $x \in \mathcal{D}_{\rm c}$, where $\lambda_{\mathcal{S}}(x)$, $x \in \mathcal{D}_{\rm c}$, is given by (3.11), and let $\mathcal{S}$ be diffeomorphic to a closed interval on the real line. Then, determine $\lambda_{\mathcal{S}}(x)$, $x \in \mathcal{D}_{\rm c}$, by solving

$$p(\lambda_t) = \min p(\lambda), \qquad \lambda \in \mathcal{V}_s(x(t)), \qquad t \in [0, T_{x_0}], \tag{3.25}$$

subject to (3.16).

It follows from (3.11) and Assumption 3.1 that λ_t is the ***unique*** solution of the Optimal Switching Control Problem. Furthermore, since $\{\lambda : V_\lambda(x) = c_\lambda, \lambda \in \mathcal{S}, x \in \mathcal{D}_{\rm c}\} \subseteq \mathcal{V}_s(x) \subseteq \mathcal{S}$, it follows that $\mathcal{V}_s(x(t))$ in (3.25) can be equivalently replaced by the switching set $\mathcal{S}$. Since the Optimal Switching Control Problem requires that the switching set $\mathcal{S}$ be given, we define an extended minimization problem wherein the switching set is computed online.

Extended Optimal Switching Control Problem. Consider the nonlinear controlled dynamical system given by (3.1) with $u(x) = \phi_{\lambda_{\mathcal{S}}(x)}(x)$, $x \in \mathcal{D}_{\rm c}$, where $\lambda_{\mathcal{S}}(x)$, $x \in \mathcal{D}_{\rm c}$, is given by (3.11). Then, determine $\lambda_{\mathcal{S}}(x)$, $x \in \mathcal{D}_{\rm c}$, where $\mathcal{S} \triangleq \cup_{t \in [0, T_{x_0}]} \lambda_t$, by solving

$$p(\lambda_t) = \min p(\lambda), \qquad \lambda \in \Lambda_{\rm s}, \qquad t \in [0, T_{x_0}], \tag{3.26}$$

subject to (3.16). Note that since λ_t, $t \in [0, T_{x_0}]$, is a C^1 function, the switching set $\mathcal{S}$ is diffeomorphic to an interval on the real line.

Next, we present the main result of this section which provides an inverse optimal online procedure for computing the switching function $\lambda_{\mathcal{S}}(x(t))$, $t \in [0, T_{x_0}]$. For the statement of this result define

$$v_\lambda(x) \triangleq \frac{\mathrm{d}c_\lambda}{\mathrm{d}\lambda} - \frac{\partial V_\lambda(x)}{\partial \lambda}, \tag{3.27}$$

$$M_\lambda(x,q) \triangleq \frac{\mathrm{d}^2 p(\lambda)}{\mathrm{d}\lambda^2} - q\frac{\mathrm{d}^2 c_\lambda}{\mathrm{d}\lambda^2} + q\frac{\partial^2 V_\lambda(x)}{\partial\lambda^2}, \tag{3.28}$$

for $x \in \mathcal{D}_c$, $\lambda \in \mathcal{S}$, and $q \in \mathbb{R}$. Note that since the Hessian of the potential function $p(\cdot)$ explicitly appears in the definition of $M_\lambda(\cdot,\cdot)$, we assume, without loss of generality, that $p(\cdot)$ is such that $M_\lambda(x,q)$ is nonsingular for all $x \in \mathcal{D}_c$, $\lambda \in \mathcal{S}$, and $q \in \mathbb{R}$. Furthermore, define

$$R_\lambda(x,q) \triangleq \frac{V_\lambda'(x) + q\, v_\lambda(x) M_\lambda^{-1}(x,q)\dfrac{\partial^2 V_\lambda(x)}{\partial\lambda\partial x}}{v_\lambda(x) M_\lambda^{-1}(x,q) v_\lambda^{\mathrm{T}}(x)}, \tag{3.29}$$

$$Q_\lambda(x,q) \triangleq M_\lambda^{-1}(x,q)\left(v_\lambda^{\mathrm{T}}(x) R_\lambda(x,q) - q\frac{\partial^2 V_\lambda(x)}{\partial\lambda\partial x}\right), \tag{3.30}$$

where $v_\lambda(x)$ and $Q_\lambda(x,q)$ are such that $v_\lambda(x)Q_\lambda(x,q) = V_\lambda'(x)$.

Theorem 3.7. *For a fixed $x \in \mathcal{D}_c$, assume that $V_\lambda(x) : \mathcal{S} \to \mathbb{R}^+$, $c_\lambda : \mathcal{S} \to \mathbb{R}^+$, and $p : \mathcal{S} \to \mathbb{R}^+$ are C^2 functions of $\lambda \in \mathcal{S}$. Then the solutions $x(t)$, $\lambda(t)$, and $q(t)$, $t \in [0, T_{x_0}]$, of the nonlinear feedback controlled dynamical system*

$$\dot{x}(t) = F(x(t), \phi_{\lambda(t)}(x(t))), \qquad x(0) = x_0, \tag{3.31}$$

$$\dot{\lambda}(t) = Q_{\lambda(t)}(x(t), q(t)) F(x(t), \phi_{\lambda(t)}(x(t))), \qquad \lambda(0) = \lambda_0, \tag{3.32}$$

$$\dot{q}(t) = R_{\lambda(t)}(x(t), q(t)) F(x(t), \phi_{\lambda(t)}(x(t))), \qquad q(0) = q_0, \tag{3.33}$$

where $\lambda(0) = \lambda_0$ and $q(0) = q_0$ satisfy the Extended Optimal Switching Control Problem at $t = 0$, are such that $\lambda(t)$, $t \in [0, T_{x_0}]$, solves the Extended Optimal Switching Control Problem.

Proof. To solve the Extended Optimal Switching Control Problem, form the Lagrangian $\mathcal{L}(\lambda, q(x)) \triangleq p(\lambda) - q(x)[c_\lambda - V_\lambda(x)]$, $\lambda \in \Lambda_s$, $x \in \mathcal{D}_c$, where $q : \mathcal{D}_c \to \mathbb{R}$ is a Lagrange multiplier. If λ_t, $t \in [0, T_{x_0}]$, solves the Extended Optimal Switching Control Problem it follows that

$$0 = \left.\frac{\partial \mathcal{L}(\lambda, q(x(t)))}{\partial\lambda}\right|_{\lambda=\lambda_t} = \left[\frac{\mathrm{d}p(\lambda)}{\mathrm{d}\lambda}\right]^{\mathrm{T}}_{\lambda=\lambda_t} - q(x(t)) v_{\lambda_t}^{\mathrm{T}}(x(t)), \tag{3.34}$$

$$0 = \left.\frac{\partial \mathcal{L}(\lambda_t, q)}{\partial q}\right|_{q=q(x(t))} = c_{\lambda_t} - V_{\lambda_t}(x(t)), \qquad t \in [0, T_{x_0}]. \tag{3.35}$$

Differentiating both sides of (3.34) and (3.35) with respect to time and denoting $q(x(t))$ and λ_t by $q(t)$ and $\lambda(t)$, respectively, it follows that

$$\begin{aligned} 0 = {} & M_{\lambda(t)}(x(t), q(t))\dot{\lambda}(t) - v_{\lambda(t)}^{\mathrm{T}}(x(t))\dot{q}(t) \\ & + q(t)\left.\frac{\partial^2 V_\lambda(x)}{\partial\lambda\partial x}\right|_{\substack{\lambda=\lambda(t)\\ x=x(t)}} F(x(t), \phi_{\lambda(t)}(x(t))), \end{aligned} \tag{3.36}$$

$$0 = v_{\lambda(t)}(x(t))\dot{\lambda}(x(t)) - \left.\frac{\partial V_{\lambda(t)}(x)}{\partial x}\right|_{x=x(t)} F(x(t), \phi_{\lambda(t)}(x(t))), \qquad t \in [0, T_{x_0}]. \tag{3.37}$$

Now, since $M_{\lambda(t)}(x(t), q(t))$ is invertible, (3.32), (3.33) are a direct consequence of (3.36), (3.37), respectively. Finally, the initial conditions $\lambda(0) = \lambda_0$ and $q(0) = q_0$ are computed by solving the algebraic system of equations given by (3.34) and (3.35) for $t = 0$. ■

The update parameters $\lambda(t)$ and $q(t)$, $t \in [0, T_{x_0}]$, in (3.31)–(3.33) should be interpreted as $\lambda(x(t))$ and $q(x(t))$, $t \in [0, T_{x_0}]$, since they are implicit functions of time parameterized via the system trajectory $x(t)$, $t \in [0, T_{x_0}]$. This minor abuse of notation considerably simplifies the presentation.

Note that the switching function and Lagrange multiplier dynamics characterized by (3.32) and (3.33), respectively, define a fixed-order compensator of the form given by (3.14) and (3.15). Specifically, defining the compensator state $x_c(t) = [x_{c1}^{\mathrm{T}}(t)\, x_{c2}(t)]^{\mathrm{T}}$, where $x_{c1}(t) \triangleq \lambda(t)$ and $x_{c2}(t) \triangleq q(t)$ so that $n_c = q + 1$, the dynamic compensator structure is given by

$$\dot{x}_{c1}(t) = Q_{x_{c1}(t)}(x(t), x_{c2}(t))F(x(t), \phi_{x_{c1}(t)}(x(t))), \qquad x_{c1}(0) = \lambda_0, \tag{3.38}$$

$$\dot{x}_{c2}(t) = R_{x_{c1}(t)}(x(t), x_{c2}(t))F(x(t), \phi_{x_{c1}(t)}(x(t))), \qquad x_{c2}(0) = q_0, \tag{3.39}$$

$$u(t) = \phi_{x_{c1}(t)}(x(t)). \tag{3.40}$$

The next result gives an explicit expression for the time derivative of the Lyapunov function $V(x)$, $x \in \mathcal{D}_c$.

Proposition 3.4. *Assume the switching set $\mathcal{S}$ is diffeomorphic to an interval on the real line and, for a fixed $x \in \mathcal{D}_c$, assume $V_\lambda(x) : \mathcal{S} \to \mathbb{R}^+$, $c_\lambda : \mathcal{S} \to \mathbb{R}^+$, and $p : \mathcal{S} \to \mathbb{R}^+$ are C^2 functions of $\lambda \in \mathcal{S}$. Let $q : \mathcal{D}_c \to \mathbb{R}$ be the Lagrange multiplier for the Extended Optimal Switching Control Problem. Then*

$$\dot{V}(x) = q(x)\left[V_\lambda'(x)F(x, \phi_\lambda(x))\right]_{\lambda=\lambda_S(x)}. \tag{3.41}$$

Proof. Since $V(x) = p(\lambda_S(x))$ and, by (3.34), $\left.\frac{\mathrm{d}p(\lambda)}{\mathrm{d}\lambda}\right|_{\lambda=\lambda_S(x)} = q(x)v_{\lambda_S(x)}(x)$, it follows that

$$\dot{V}(x) = \left.\frac{\mathrm{d}p(\lambda)}{\mathrm{d}\lambda}\right|_{\lambda=\lambda_S(x)} \dot{\lambda}_S(x) = q(x)v_{\lambda_S(x)}\dot{\lambda}_S(x). \tag{3.42}$$

Now, using $\dot{\lambda}_S(x) = Q_{\lambda_S(x)}(x, q(x))F(x, \phi_{\lambda_S(x)}(x))$ and noting that $v_\lambda(x) \cdot Q_\lambda(x, q) = V_\lambda'(x)$, (3.41) is immediate. ■

The following proposition gives an implicit characterization for the Lagrange multiplier $q(x) \in \mathbb{R}$, $x \in \mathcal{D}_c$, such that its dynamics satisfy (3.33).

Proposition 3.5. *The Lagrange multiplier $q : \mathcal{D}_{\mathrm{c}} \to \mathbb{R}$ for the Extended Optimal Switching Control Problem is given by*

$$q(x) = \frac{c(x)\left[\frac{\mathrm{d}p(\lambda)}{\mathrm{d}\lambda}\right]^{\mathrm{T}}_{\lambda=\lambda_S(x)}}{c(x)v^{\mathrm{T}}_{\lambda_S(x)}(x)}, \qquad x \in \mathcal{D}_{\mathrm{c}}, \tag{3.43}$$

where $c : \mathcal{D}_{\mathrm{c}} \to \mathbb{R}^{1\times q}$ is an arbitrary row vector such that $c(x)v^{\mathrm{T}}_{\lambda}(x) \neq 0$, $x \in \mathcal{D}_{\mathrm{c}}$, $\lambda \in \mathcal{S}$. For $c(x) = v_{\lambda}(x)M^{-1}_{\lambda}(x, q(x))$, $x \in \mathcal{D}_{\mathrm{c}}$, (3.43) implicitly defines the solution $q(x)$ of (3.33).

Proof. Forming $c(x)$(3.34) and solving for $q(x)$ yields (3.43). Next, differentiating (3.43) with respect to time gives

$$\dot{q}(x) = \frac{c(x)M_{\lambda_S(x)}(x, q(x))\dot{\lambda}_S(x)}{c(x)v^{\mathrm{T}}_{\lambda_S(x)}(x)} + q(x)\frac{c(x)\left.\frac{\partial^2 V_{\lambda}(x)}{\partial\lambda\partial x}\right|_{\lambda=\lambda_S(x)}}{c(x)v^{\mathrm{T}}_{\lambda_S(x)}(x)} F(x, \phi_{\lambda_S(x)}(x)). \tag{3.44}$$

Now, substituting $v_{\lambda_S(x)}(x)M^{-1}_{\lambda_S(x)}(x, q(x))$ for $c(x)$ into (3.44) and using (3.37), yields (3.33). ■

It is important to note that (3.43), with $c(x) = v_{\lambda_S(x)}(x)M^{-1}_{\lambda_S(x)}(x, q(x))$, $x \in \mathcal{D}_{\mathrm{c}}$, *implicitly* characterizes the Lagrange multiplier $q(x)$, $x \in \mathcal{D}_{\mathrm{c}}$, since $q(x)$ appears in $M^{-1}_{\lambda_S(x)}(x, q(x))$. The next result provides an *explicit* characterization for the Lagrange multiplier $q(x)$, $x \in \mathcal{D}_{\mathrm{c}}$.

Proposition 3.6. *Consider the polynomial in $q(x)$, $x \in \mathcal{D}_{\mathrm{c}}$, of degree p given by*

$$0 = \sum_{k=1}^{p}\left[c_k(x, q(x))\left(q(x)v_{\lambda}(x) - \frac{\mathrm{d}p(\lambda)}{\mathrm{d}\lambda}\right)M^{k-1}_{\lambda}(x, q(x))v^{\mathrm{T}}_{\lambda}(x)\right]_{\lambda=\lambda_S(x)}, \tag{3.45}$$

where $c_k(x, q(x))$, $x \in \mathcal{D}_{\mathrm{c}}$, $k = 0, 1, \ldots, p$, are the coefficients of the characteristic polynomial associated with $M_{\lambda_S(x)}(x, q(x))$, $x \in \mathcal{D}_{\mathrm{c}}$. Then, the Lagrange multiplier $q : \mathcal{D}_{\mathrm{c}} \to \mathbb{R}$ for the Extended Optimal Switching Control Problem is the root of (3.45) such that $q(x_0) = q_0$.

Proof. With $c(x) = v_{\lambda_S(x)}(x)M^{-1}_{\lambda_S(x)}(x, q(x))$ it follows from (3.43) that

$$0 = \left[\left(q(x)v_{\lambda}(x) - \frac{\mathrm{d}p(\lambda)}{\mathrm{d}\lambda}\right)M^{-1}_{\lambda}(x, q(x))v^{\mathrm{T}}_{\lambda}(x)\right]_{\lambda=\lambda_S(x)}. \tag{3.46}$$

Next, using the Cayley-Hamilton theorem [31] it follows that

$$M^{-1}_{\lambda_S(x)}(x, q(x)) = -\frac{1}{c_0(x, q(x))} \sum_{k=1}^{p} c_k(x, q(x)) M^{k-1}_{\lambda_S(x)}(x, q(x)), \qquad x \in \mathcal{D}_c, \tag{3.47}$$

where $c_0(x, q(x)) = \det(M_{\lambda_S(x)}(x, q(x))) \neq 0$. Substituting (3.47) into (3.46) yields (3.45). Now, noting that $M_\lambda(x, q(x))$ is affine in $q(x)$, $x \in \mathcal{D}_c$, it follows that $c_k(x, q(x))$, $x \in \mathcal{D}_c$, $k = 0, 1, \ldots, p$, is a polynomial in $q(x)$, $x \in \mathcal{D}_c$, of degree $p - k$ and hence (3.45) is a polynomial in $q(x)$, $x \in \mathcal{D}_c$, of degree p. ■

In the case where $p = 2$ it follows that

$$c_1(x, q(x)) = -\mathrm{tr}\, M_{\lambda_S(x)}(x, q(x)), \qquad c_2(x, q(x)) = 1, \qquad x \in \mathcal{D}_c, \tag{3.48}$$

where "tr" is the trace operator. Now, defining

$$A_\lambda \triangleq \mathrm{tr}\, \frac{\mathrm{d}^2 p(\lambda)}{\mathrm{d}\lambda^2} I_2 - \frac{\mathrm{d}^2 p(\lambda)}{\mathrm{d}\lambda^2}, \tag{3.49}$$

$$B_\lambda(x) \triangleq \mathrm{tr}\left(\frac{\mathrm{d}^2 c_\lambda}{\mathrm{d}\lambda^2} - \frac{\partial^2 V_\lambda(x)}{\partial\lambda^2}\right) I_2 - \left(\frac{\mathrm{d}^2 c_\lambda}{\mathrm{d}\lambda^2} - \frac{\partial^2 V_\lambda(x)}{\partial\lambda^2}\right), \quad x \in \mathcal{D}_c, \tag{3.50}$$

where I_2 is the 2×2 identity matrix, it follows from (3.45) that $q(x)$, $x \in \mathcal{D}_c$, satisfies

$$0 = q(x) v_\lambda(x)(A_\lambda - q(x) B_\lambda(x)) v_\lambda^{\mathrm{T}}(x) - \frac{\mathrm{d}p(\lambda)}{\mathrm{d}\lambda}(A_\lambda - q(x) B_\lambda(x)) v_\lambda^{\mathrm{T}}(x) \Big|_{\lambda = \lambda_S(x)}. \tag{3.51}$$

Next, we show that the controller $\phi_{\lambda_S(x)}(x)$, $x \in \mathcal{D}_c$, minimizes a derived nonlinear-nonquadratic cost functional that explicitly depends on the Lagrange multiplier $q(x)$, the nonlinear closed-loop system dynamics, and the gradient of the equilibria-dependent Lyapunov functions evaluated at the switching function. For the statement of this result define the set of asymptotically stabilizing controllers by

$$\mathcal{S}(x_0) \triangleq \{u(\cdot) : u(\cdot) \text{ is admissible and } \mathcal{D}_0 \text{ is an asymptotically stable positively invariant set of (3.1)}\}, \tag{3.52}$$

and consider the performance functional

$$J(x_0, u(\cdot)) = \int_0^\infty L(x(t), u(t))\, \mathrm{d}t, \tag{3.53}$$

where $L : \mathcal{D}_c \times \mathcal{U} \to \mathbb{R}$.

Theorem 3.8 ([22]). *Consider the nonlinear controlled dynamical system given by (3.1) with* $F(0,0) = 0$ *and performance functional (3.53). Assume* $\mathcal{S}$ *is diffeomorphic to an interval on the real line, let* $\lambda_{\mathcal{S}}(x)$, $x \in \mathcal{D}_c$, *be a* C^1 *function satisfying (3.11), and assume that there exists a* C^1 *function* $V : \mathcal{D}_c \to \mathbb{R}$ *such that*

$$V(x) = 0, \quad x \in \mathcal{D}_0, \tag{3.54}$$

$$V(x) > 0, \quad x \in \mathcal{D}_c \setminus \mathcal{D}_0 \tag{3.55}$$

$$\dot{V}(x) \triangleq V'(x)F(x, \phi_{\lambda_S(x)}(x)) < 0, \quad x \in \mathcal{D}_c \setminus \mathcal{D}_0. \tag{3.56}$$

Then the positively invariant set $\mathcal{D}_0$ *of the closed-loop system*

$$\dot{x}(t) = F(x(t), \phi_{\lambda_S(x(t))}(x(t))), \quad x(0) = x_0, \quad t \in \mathcal{I}_{x_0}, \tag{3.57}$$

is locally asymptotically stable with an estimate of the domain of attraction given by $\mathcal{D}_c$, *and the performance functional (3.53), with* $L(x, \phi_{\lambda_S(x)}(x)) = -\dot{V}(x)$, *is minimized in the sense that*

$$J(x_0, \phi_{\lambda_S(x(\cdot))}(x(\cdot))) = \min_{u(\cdot) \in \mathcal{S}(x_0)} J(x_0, u(\cdot)), \quad x_0 \in \mathcal{D}_c. \tag{3.58}$$

Finally,

$$J(x_0, \phi_{\lambda_S(x(\cdot))}(x(\cdot))) = V(x_0), \quad x_0 \in \mathcal{D}_c. \tag{3.59}$$

Proof. The proof of closed-loop stability is a direct consequence of Lyapunov's theorem as applied to the closed-loop system (3.57). Optimality follows from Theorem 4.1 of [22] with Hamiltonian $H(x,u) = L(x,u) + V'(x)F(x,u)$, $x \in \mathcal{D}_c$, $u \in \mathcal{U}$. ■

It follows from Proposition 3.4 and Theorem 3.8 that the dynamic compensator (3.38)–(3.40) guarantees *inverse optimality* with respect to the performance functional (3.53) with $L(x,u)$ given by $L(x, \phi_{\lambda_S(x)}(x)) = -q(x)V'_{\lambda_S(x)}(x)F(x, \phi_{\lambda_S(x)}(x))$, which by (3.41) and (3.56), is positive.

Finally, we note that Algorithm 3.1 can be used to construct inverse optimal hierarchical controllers as presented in this section. However, in this case Step 3 in the algorithm is unnecessary if we substitute Step 4 with:

Step 4". Given the state space point $x(t)$, *the parameter value* $\lambda(t)$, *and the Lagrange multiplier* $q(t)$ *at* $t \geq 0$, *update* $\lambda(t)$ *and* $q(t)$ *using (3.32) and (3.33).*

In this case, the switching set $\mathcal{S} \subseteq \Lambda_s$ need not be explicitly defined and is computed online.

3.7 Conclusion

A nonlinear control-system design framework predicated on a hierarchical switching controller architecture parameterized over a set of system equilibria was developed. Specifically, a hierarchical switching nonlinear control strategy is constructed to stabilize a given nonlinear system by stabilizing a collection of nonlinear controlled subsystems. The switching nonlinear controller architecture is designed based on a generalized Lyapunov function obtained by minimizing a potential function over a given switching set induced by the parameterized system equilibria. An online procedure for computing the switching scheme was proposed by constructing an initial value problem having a fixed-order dynamic compensator structure. Furthermore, an inverse optimal control strategy was obtained by constructing a hierarchical controller parameterized with respect to a given system equilibrium manifold wherein an inverse optimal morphing strategy is developed to coordinate the hierarchical switching.

Finally, we note that the results presented in this chapter also hold for nonlinear discrete-time dynamical systems described by time-invariant difference equations with (unique) solutions being continuous functions of the initial conditions. Specifically, in this case all of the results proceed exactly as in the continuous-time case by replacing $t \in \mathbb{R}$ with $k \in \mathbb{Z}$, where $\mathbb{Z}$ denotes the set of nonnegative integers. Of course, in this case, the topology of the switching set $\mathcal{S}$ is such that it only consists of countable or countably infinite isolated points.

4. Nonlinear Robust Switching Controllers for Nonlinear Uncertain Systems

4.1 Introduction

In Chapter 3, a nonlinear control design framework predicated on a hierarchical switching controller architecture parameterized over a set of moving system equilibria was developed. In this chapter we extend these results to address the problem of ***robust*** stabilization for nonlinear ***uncertain*** systems. Specifically, using ***equilibria-dependent Lyapunov functions***, or instantaneous (with respect to a given ***nominal*** parameterized equilibrium manifold) Lyapunov functions, a hierarchical ***robust*** nonlinear control strategy is developed that stabilizes a compact positively invariant set of a nonlinear uncertain system using a supervisory robust switching controller that coordinates lower-level stabilizing subcontrollers (see Figure 4.1). Each robust subcontroller can be nonlinear and thus local set point designs can be nonlinear. Furthermore, for each nominally parameterized equilibrium manifold, the collection of the robust subcontrollers provide guaranteed domains of attraction with nonempty intersections that cover the region of operation over the prescribed range of system uncertainty of the nonlinear uncertain system in the state space. A hierarchical robust switching nonlinear controller architecture is developed based on a *generalized* lower semicontinuous Lyapunov function obtained by minimizing a potential function, associated with each domain of attraction, over a given switching set induced by the parameterized ***nominal*** system equilibria. The hierarchical robust switching nonlinear controller guarantees that the generalized Lyapunov function is nonincreasing along the closed-loop system trajectories for all parametric system uncertainty with strictly decreasing values at the switching points, establishing robust asymptotic stability of a compact positively invariant set. Furthermore, as in Chapter 3, since the proposed robust switching nonlinear control strategy is predicated on a generalized Lyapunov framework with strictly decreasing values at the switching points, the possibility of a sliding mode is precluded. Hence, the proposed nonlinear robust stabilization framework avoids the undesirable effects of high-speed switching onto an invariant sliding manifold which is one of the main limitations of variable structure controllers. Finally,

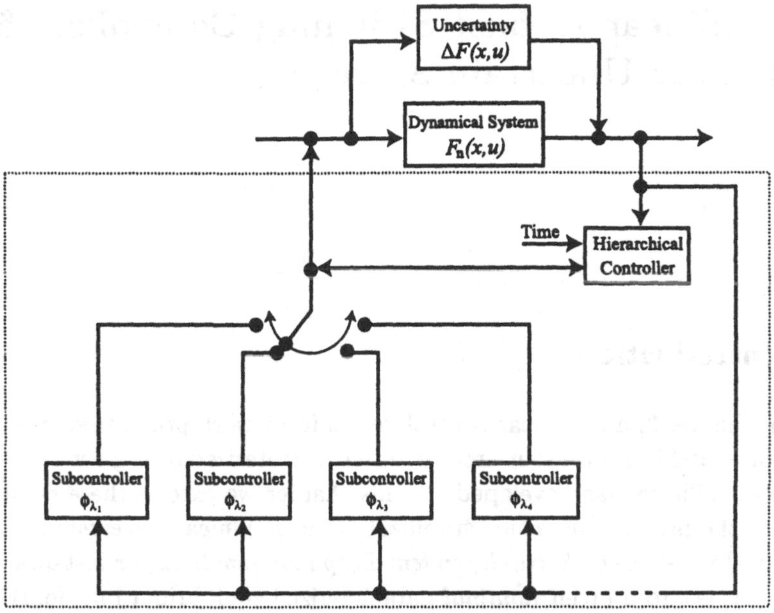

Fig. 4.1. Robust switching controller architecture

since the theory for the robust switching controller framework very closely parallels the theory for the switching control framework developed in Chapter 3, many of the results are similar. Hence, the comments in this chapter are brief and the proofs are omitted.

4.2 Mathematical Preliminaries

In this section we establish definitions and notation used later in the chapter. Specifically, in this chapter we consider nonlinear controlled *uncertain* dynamical systems of the form

$$\dot{x}(t) = F(x(t), u(t)), \qquad x(0) = x_0, \qquad F(\cdot,\cdot) \in \mathcal{F}, \qquad t \in \mathcal{I}_{x_0}, \quad (4.1)$$

where $x(t) \in \mathcal{D} \subseteq \mathbb{R}^n$, $t \in \mathcal{I}_{x_0}$, is the system state vector, $\mathcal{I}_{x_0} \subseteq \mathbb{R}$ is the maximal interval of existence of a solution $x(\cdot)$ of (4.1), $\mathcal{D}$ is an open set, $0 \in \mathcal{D}$, $u(t) \in \mathcal{U} \subseteq \mathbb{R}^m$, $t \in \mathcal{I}_{x_0}$, is the control input, $\mathcal{U}$ is the set of all admissible controls such that $u(\cdot)$ is a measurable function with $0 \in \mathcal{U}$, and $\mathcal{F} \subset \{F : \mathcal{D}\times\mathcal{U} \to \mathbb{R}^n : F(\cdot,\cdot) \in \mathrm{C}^0\}$ denotes the class of uncertain nonlinear dynamics. Furthermore, we introduce the nominal controlled dynamical system

$$\dot{x}(t) = F_{\mathrm{n}}(x(t), u(t)), \qquad x(0) = x_0, \qquad t \in \mathcal{I}_{x_0}, \quad (4.2)$$

where $F_{\rm n}(\cdot,\cdot) \in \mathcal{F}$ represents the nominal system dynamics. Here, we consider nonlinear closed-loop uncertain dynamical systems of the form

$$\dot{x}(t) = F(x(t), \phi(x(t))), \quad x(0) = x_0, \quad F(\cdot,\cdot) \in \mathcal{F}, \quad t \in \mathcal{I}_{x_0}. \tag{4.3}$$

The following definition introduces three types of robust stability as well as attraction of (4.3) with respect to a compact positively invariant set.

Definition 4.1. *Let $\mathcal{D}_0 \subset \mathcal{D}$ be a compact positively invariant set for the nonlinear feedback controlled uncertain dynamical system (4.3). $\mathcal{D}_0$ is* robustly Lyapunov stable *if for every open neighborhood $\mathcal{O}_1 \subseteq \mathcal{D}$ of $\mathcal{D}_0$, there exists an open neighborhood $\mathcal{O}_2 \subseteq \mathcal{O}_1$ of $\mathcal{D}_0$ such that $x(t) \in \mathcal{O}_1$, $t \geq 0$, for all $x_0 \in \mathcal{O}_2$ and $F(\cdot,\cdot) \in \mathcal{F}$. $\mathcal{D}_0$ is* robustly attractive *if there exists an open neighborhood $\mathcal{O}_3 \subseteq \mathcal{D}$ of $\mathcal{D}_0$ such that $\mathcal{P}^+_{x_0} \subseteq \mathcal{D}_0$ for all $x_0 \in \mathcal{O}_3$ and $F(\cdot,\cdot) \in \mathcal{F}$. $\mathcal{D}_0$ is* robustly asymptotically stable *if it is robustly Lyapunov stable and robustly attractive. $\mathcal{D}_0$ is* robust globally asymptotically stable *if it is robustly Lyapunov stable and $\mathcal{P}^+_{x_0} \subseteq \mathcal{D}_0$ for all $x_0 \in \mathbb{R}^n$ and $F(\cdot,\cdot) \in \mathcal{F}$. Finally, $\mathcal{D}_0$ is* unstable *if it is not robustly Lyapunov stable.*

4.3 Parameterized Nominal System Equilibria, System Attractors, and Domains of Attraction

The nonlinear robust control design framework developed in this chapter is predicated on a hierarchical robust switching nonlinear controller architecture parameterized over a set of *nominal* system equilibria. It is important to note that both the nominal dynamical system and the robust controller for each parameterized nominal equilibrium can be nonlinear and thus local set point designs are in general nonlinear. Hence, the nonlinear controlled uncertain dynamical system can be viewed as a collection of controlled uncertain subsystems with a hierarchical robust switching controller architecture. In this section we concentrate on robust nonlinear stabilization of compact positively invariant sets, parameterized in $\mathcal{D}$, of the nonlinear closed-loop uncertain subsystems. Specifically, we consider the nonlinear controlled uncertain dynamical system (4.1) with the origin being an equilibrium point of the *nominal* system corresponding to the control $u = 0$, that is, $F_{\rm n}(0,0) = 0$. Furthermore, we assume that given a mapping $\varphi : \mathcal{D} \times \Lambda \to \mathcal{U}$, $\varphi(0,0) = 0$, there exists a continuous function $\psi : \Lambda_{\rm o} \to \mathcal{D}_{\rm o}$, where $\mathcal{D}_{\rm o} \subseteq \mathcal{D}$, $0 \in \mathcal{D}_{\rm o}$, and $\Lambda_{\rm o} \subseteq \Lambda$, $0 \in \Lambda_{\rm o}$, such that $F_{\rm n}(x_\lambda, \varphi(x_\lambda, \lambda)) = 0$ with $x_\lambda = \psi(\lambda) \in \mathcal{D}_{\rm o}$ for all $\lambda \in \Lambda_{\rm o}$. As discussed in Chapter 3, this is a necessary condition for nominal parametric stability with respect to $\Lambda_{\rm o}$ as defined in [64, 107] while Theorem 3.1 provides sufficient conditions for guaranteeing the existence of such a parameterization for the nominal system.

Next, we consider a family of stabilizing feedback control laws for the *nominal* system given by

$$\Phi \triangleq \{\phi_\lambda : \mathcal{D} \to \mathcal{U} : \phi_\lambda \in \mathrm{C}^0,\, \phi_\lambda(x_\lambda) = \varphi(x_\lambda, \lambda),\, \lambda \in \Lambda_{\mathrm{s}}\}, \quad \Lambda_{\mathrm{s}} \subseteq \Lambda_{\mathrm{o}}, \quad (4.4)$$

such that, for $\phi_\lambda(\cdot) \in \Phi$, $\lambda \in \Lambda_{\mathrm{s}}$, the nonlinear closed-loop *nominal* system

$$\dot{x}(t) = F_{\mathrm{n}}(x(t), \phi_\lambda(x(t))), \qquad x(0) = x_0, \qquad t \in \mathcal{I}_{x_0}, \qquad (4.5)$$

has an asymptotically stable equilibrium point $x_\lambda \in \mathcal{D}_{\mathrm{o}} \subseteq \mathcal{D}$ with a corresponding Lyapunov function $V_\lambda(\cdot)$. Hence, in the terminology of [64, 107], (4.5) is (nominally) parametrically asymptotically stable with respect to $\Lambda_{\mathrm{s}} \subseteq \Lambda_{\mathrm{o}}$. Here, we assume that for each $\lambda \in \Lambda_{\mathrm{s}}$, the linear or nonlinear feedback controllers $\phi_\lambda(\cdot)$ are given. In particular, these controllers correspond to local set point designs and can be obtained using any appropriate standard linear or nonlinear stabilization scheme with a domain of attraction for each $\lambda \in \Lambda_{\mathrm{s}}$. It is important to note that even though x_λ, $\lambda \in \Lambda_{\mathrm{s}}$, is an equilibrium point of the nominal system (4.2), in general, x_λ, $\lambda \in \Lambda_{\mathrm{s}}$, is *not* an equilibrium point for the uncertain system (4.1). Hence, $V_\lambda(\cdot)$ is *not* a standard Lyapunov function for the nonlinear closed-loop uncertain system

$$\dot{x}(t) = F(x(t), \phi_\lambda(x(t))), \qquad x(0) = x_0, \quad F(\cdot,\cdot) \in \mathcal{F}, \qquad t \in \mathcal{I}_{x_0}. \quad (4.6)$$

However, under an additional assumption on the structure of the system uncertainty, it can be shown that $u = \phi_\lambda(x)$ is a robust control law that ***robustly*** asymptotically stabilizes a compact positively invariant set $\mathcal{N}_\lambda$, containing the nominal equilibrium point x_λ, $\lambda \in \Lambda_{\mathrm{s}}$, with domain of attraction $\mathcal{D}_\lambda$. In this case, $V_\lambda(\cdot)$ serves as a Lyapunov function of the uncertain system guaranteeing stability with respect to a compact positively invariant set. In particular, defining $\Delta F(x,u) \triangleq F(x,u) - F_{\mathrm{n}}(x,u)$ and assuming that $V_\lambda'(x)\Delta F(x, \phi_\lambda(x)) < -V_\lambda'(x)F_{\mathrm{n}}(x, \phi_\lambda(x))$ for all $x \in \mathcal{D}_\lambda$ such that $\|x - x_\lambda\| > r$, $r > 0$, it follows that $\phi_\lambda(\cdot)$ is a robustly stabilizing feedback controller of a compact positively invariant set $\mathcal{N}_\lambda$ of (4.6).

Next, given a stabilizing feedback robust controller $\phi_\lambda(\cdot)$ for each $\lambda \in \Lambda_{\mathrm{s}}$, we provide a guaranteed subset of the domain of attraction $\mathcal{D}_\lambda$ of a compact positively invariant set $\mathcal{N}_\lambda$ for the nonlinear closed-loop *uncertain* system using Lyapunov stability theory.

Theorem 4.1 ([59]). *Let $\lambda \in \Lambda_{\mathrm{s}}$. Consider the nonlinear uncertain closed-loop system (4.6) with $\phi_\lambda(\cdot) \in \Phi$ and let $\mathcal{N}_\lambda$ be a compact positively invariant set of (4.6). Furthermore, let $\mathcal{X}_\lambda \subset \mathcal{D}$ be a compact neighborhood of $\mathcal{N}_\lambda$. Then $\mathcal{N}_\lambda$ is a robustly asymptotically stable set of (4.6) for all $F(\cdot,\cdot) \in \mathcal{F}$, if, and only if, there exists a C^0 function $V_\lambda : \mathcal{X}_\lambda \to \mathbb{R}$, with V_λ C^1 on $\mathcal{X}_\lambda \setminus \mathcal{N}_\lambda$, such that*

$$V_\lambda(x) = 0, \qquad x \in \mathcal{N}_\lambda, \tag{4.7}$$
$$V_\lambda(x) > 0, \qquad x \in \mathcal{X}_\lambda \setminus \mathcal{N}_\lambda, \tag{4.8}$$
$$\dot{V}_\lambda(x) \triangleq V_\lambda'(x)F(x, \phi_\lambda(x)) < 0, \qquad x \in \mathcal{X}_\lambda \setminus \mathcal{N}_\lambda, \qquad F(\cdot,\cdot) \in \mathcal{F}. \tag{4.9}$$

In addition, a subset of the domain of attraction of $\mathcal{N}_\lambda$ is given by

$$\mathcal{D}_\lambda \triangleq V_\lambda^{-1}([0, c_\lambda]), \tag{4.10}$$

where $c_\lambda \triangleq \max\{\beta > 0 : V_\lambda^{-1}([0,\beta]) \subseteq \mathcal{X}_\lambda\}$.

It follows from Theorem 4.1 that for all $x_0 \in \mathcal{D}_\lambda$ and each open set $\mathcal{O}$ such that $\mathcal{N}_\lambda \subset \mathcal{O} \subset \mathcal{D}_\lambda$, there exists a finite time $T > 0$ such that $x(t) \in \mathcal{O}$ for all $t \geq T$ and $F(\cdot,\cdot) \in \mathcal{F}$. Alternatively, Theorem 4.1 can be restated by requiring $V_\lambda(\cdot)$ to be a C^1 function on $\mathcal{X}_\lambda$ such that Conditions (4.8) and (4.9) hold and $V_\lambda(x) \geq 0$, $x \in \mathcal{N}_\lambda$. In this case the compact positively invariant set $\mathcal{N}_\lambda$ is defined by $\mathcal{N}_\lambda \triangleq V_\lambda^{-1}([0, b_\lambda])$, where $b_\lambda \triangleq \inf\{\beta \geq 0 : \dot{V}_\lambda(x) < 0,\ x \in V_\lambda^{-1}([\beta, c_\lambda])\}$.

Note that Conditions (4.7)–(4.9) imply that $V_\lambda(x)$ is a Lyapunov function guaranteeing robust stability of a compact positively invariant set $\mathcal{N}_\lambda$ of the closed-loop uncertain system (4.6). However, Condition (4.9) is unverifiable since it is dependent on the uncertain system dynamics $F(\cdot,\cdot) \in \mathcal{F}$. This condition is implied by the conditions

$$V_\lambda'(x)F(x, \phi_\lambda(x)) \leq V_\lambda'(x)F_{\rm n}(x, \phi_\lambda(x)) + \Gamma_\lambda(x, \phi_\lambda(x)), \quad x \in \mathcal{X}_\lambda \setminus \mathcal{N}_\lambda, \tag{4.11}$$
$$V_\lambda'(x)F_{\rm n}(x, \phi_\lambda(x)) + \Gamma_\lambda(x, \phi_\lambda(x)) < 0, \quad x \in \mathcal{X}_\lambda \setminus \mathcal{N}_\lambda, \tag{4.12}$$

where $F(\cdot,\cdot) \in \mathcal{F}$ and $\Gamma_\lambda : \mathcal{D}_\lambda \times \mathcal{U} \longrightarrow \mathbb{R}$, $\lambda \in \Lambda_{\rm s}$. It is important to note that Condition (4.12) is a verifiable condition since it is independent of the uncertain system dynamics $F(\cdot,\cdot) \in \mathcal{F}$. To apply Theorem 4.1, we specify a bounding function $\Gamma_\lambda(\cdot,\cdot)$ for an uncertainty set $\mathcal{F}$ such that $\Gamma_\lambda(\cdot,\cdot)$ bounds $\mathcal{F}$. In this case Conditions (4.11) and (4.12) are satisfied. Hence, if the $V_\lambda(x)$ satisfying (4.7), (4.8), and (4.12) can be determined, then robust stability of a compact positively invariant set $\mathcal{N}_\lambda$ of (4.6) is guaranteed. For further details see Section 5.7 and [53]. For the remainder of the chapter we assume that the structure of the system uncertainty is such that there exists $\Gamma_\lambda(\cdot,\cdot)$ such that (4.11) and (4.12) hold.

4.4 Robust Nonlinear System Stabilization via a Hierarchical Switching Controller Architecture

In this section we develop a nonlinear robust stabilization framework predicated on a hierarchical switching controller architecture parameterized

over a set of moving *nominal* system equilibria. Specifically, using ***equilibria-dependent Lyapunov functions***, or instantaneous (with respect to a given parameterized nominal equilibrium manifold) Lyapunov functions, a hierarchical nonlinear robust control strategy is developed that stabilizes a compact positively invariant set by robustly stabilizing a collection of nonlinear uncertain closed-loop subsystems while providing an explicit expression for a guaranteed domain of attraction. A switching nonlinear robust controller architecture is developed based on a *generalized* lower semicontinuous Lyapunov function obtained by minimizing a potential function, associated with the domain of attraction of each controlled uncertain subsystem, over a given switching set induced by the parameterized nominal system equilibria. In the case where one of the compact positively invariant sets parameterized by the nominal system equilibria is globally robustly asymptotically stable with a given robust subcontroller and a structural topological constraint is enforced on the switching set, the proposed nonlinear robust stabilization framework guarantees global asymptotic robust stability of a compact positively invariant set associated to any given parameterized nominal system equilibrium.

To state the main results of this section several definitions and a key assumption are needed. Recall that the set $\Lambda_{\rm s} \subseteq \Lambda_{\rm o}$, $0 \in \Lambda_{\rm s}$, is such that for every $\lambda \in \Lambda_{\rm s}$ there exists a robust feedback control law $\phi_\lambda(\cdot) \in \Phi$ such that a compact positively invariant neighborhood $\mathcal{N}_\lambda \subseteq \mathcal{D}$ of the nominal equilibrium point $x_\lambda \in \mathcal{D}_{\rm o}$ of (4.5) is robustly asymptotically stable with an estimate of the domain of attraction given by $\mathcal{D}_\lambda$. Since $\mathcal{N}_\lambda$, $\lambda \in \Lambda_{\rm s}$, is a positively invariant set, it follows from Theorem 4.1 that there exists a Lyapunov function $V_\lambda(\cdot)$ satisfying (4.7)–(4.9), and hence, without loss of generality, we can take $\mathcal{D}_\lambda$, $\lambda \in \Lambda_{\rm s}$, given by (4.10). Furthermore, we assume that the set-valued map $\Psi : \Lambda_{\rm s} \rightsquigarrow 2^{\mathcal{D}}$, where $2^{\mathcal{D}}$ denotes the collection of all subsets of $\mathcal{D}$, is such that $\mathcal{D}_\lambda = \Psi(\lambda)$, $\lambda \in \Lambda_{\rm s}$, is continuous. In particular, since $\mathcal{D}_\lambda$, $\lambda \in \Lambda_{\rm s}$, is given by (4.10), the continuity of the set-valued map $\Psi(\cdot)$ is guaranteed provided that $V_\lambda(x)$, $x \in \mathcal{D}_\lambda$, and c_λ are continuous functions of the parameter $\lambda \in \Lambda_{\rm s}$. Next, let $\mathcal{S} \subseteq \Lambda_{\rm s}$, $0 \in \mathcal{S}$, denote a *switching set* such that the following key assumption is satisfied.

Assumption 4.1. *The switching set $\mathcal{S} \subseteq \Lambda_{\rm s}$ is such that the following properties hold:*

i) There exists a continuous positive-definite function $p : \mathcal{S} \to \mathbb{R}$ such that for all $\lambda \in \mathcal{S}$, $\lambda \neq 0$, there exists $\lambda_1 \in \mathcal{S}$ such that

$$p(\lambda_1) < p(\lambda), \qquad \mathcal{N}_\lambda \subset \mathring{\mathcal{D}}_{\lambda_1}. \tag{4.13}$$

ii) If $\lambda, \lambda_1 \in \mathcal{S}$, $\lambda \neq \lambda_1$, is such that $p(\lambda) = p(\lambda_1)$, then $\mathcal{D}_\lambda \cap \mathcal{D}_{\lambda_1} = \emptyset$.

Note that Assumption 4.1 assumes the existence of a positive-definite *potential function* $p(\lambda)$, for all λ in the switching set $\mathcal{S}$. It follows that, for each $\lambda \in \mathcal{S}$, there exists an equilibrium point x_λ with an associated domain of attraction $\mathcal{D}_\lambda$, and potential value $p(\lambda)$. Hence, every domain of attraction has an associated value of the potential function such that, by *ii)*, domains of attraction corresponding to different local set point designs intersect each other only if their corresponding potentials are different. Furthermore, given $\mathcal{D}_\lambda$, $\lambda \in \mathcal{S} \setminus \{0\}$, *i)* implies that there exists at least one intersecting domain of attraction $\mathcal{D}_{\lambda_1}$, $\lambda_1 \in \mathcal{S}$, such that the potential function decreases and $\overset{\circ}{\mathcal{D}}_{\lambda_1}$ contains $\mathcal{N}_\lambda$. This guarantees that if a forward trajectory $x(t)$, $t \geq 0$, of the controlled uncertain system approaches $\mathcal{N}_\lambda$, then there exists a finite time $T > 0$ such that the trajectory enters $\mathcal{D}_{\lambda_1}$. Finally, as in Chapter 3, it is important to note that the switching set $\mathcal{S}$ is arbitrary. In particular, we do not assume that $\mathcal{S}$ is countable or countably infinite.

Next, we note that Assumption 4.1 implies that every level set of the potential function $p(\cdot)$ is either empty or consists of only isolated points. Furthermore, in a neighborhood of $\lambda = 0$ every level set of $p(\cdot)$ consists of at most *one* isolated point. For the statement of this result, let $\mathcal{B}_\lambda$, $\lambda \in \mathcal{U}_o$, denote the largest open ball centered at x_λ and contained in $\mathcal{D}_\lambda$, that is $\mathcal{B}_\lambda \triangleq \{x \in \mathcal{D} : \|x - x_\lambda\| < r_\lambda\}$, where $r_\lambda \triangleq \min_{x \in \partial\mathcal{D}_\lambda} \|x - x_\lambda\|$.

Proposition 4.1. *Let $\mathcal{S} \subseteq \Lambda_s$ be such that Assumption 4.1 holds. Then for every $\alpha > 0$, $p^{-1}(\alpha)$ is either empty or consists of only isolated points. Furthermore, there exists $\beta > 0$ such that for every $\alpha < \beta$, $p^{-1}(\alpha)$ consists of at most one isolated point.*

Note that Proposition 4.1 implies that, if $p^{-1}(\alpha)$, $\alpha > 0$, is bounded, then there exists a finite distance between isolated points contained in $p^{-1}(\alpha)$ which consists of at most a *finite* number of isolated points. Finally, since in a neighborhood of $\lambda = 0$ every level set of $p(\cdot)$ consists of at most one isolated point, a particular topology for $\mathcal{S}$, in a neighborhood of the $\lambda = 0$, is homeomorphic to the interval $[0, a]$, $a > 0$, with $0 \in \mathcal{S}$ corresponding to $0 \in \mathbb{R}$.

Now, for every $x \in \mathcal{D}_c \triangleq \cup_{\lambda \in \mathcal{S}} \mathcal{D}_\lambda$, define the *viable switching set* $\mathcal{V}_s(x) \triangleq \{\lambda \in \mathcal{S} : x \in \mathcal{D}_\lambda\}$, which contains all $\lambda \in \mathcal{S}$ such that $x \in \mathcal{D}_\lambda$. Note that if we consider a sequence $\{\lambda_n\}_{n=1}^\infty \subset \mathcal{V}_s(x)$, that is, $x \in \mathcal{D}_{\lambda_n}$, such that $\lim_{n\to\infty} \lambda_n = \bar{\lambda}$, it follows from the continuity of the set-valued map $\Psi(\cdot)$ that $x \in \mathcal{D}_{\bar{\lambda}}$. Thus, $\bar{\lambda} \in \mathcal{V}_s(x)$ which implies that $\mathcal{V}_s(x)$ is a non-empty closed set since it contains all of its accumulation points.

Next, we introduce the *switching function* $\lambda_{\mathcal{S}}(x)$, $x \in \mathcal{D}_c$, such that the following definitions hold

$$V(x) \triangleq p(\lambda_{\mathcal{S}}(x)), \quad \lambda_{\mathcal{S}}(x) \triangleq \operatorname{argmin}\{p(\lambda) : \lambda \in \mathcal{V}_s(x)\}, \quad x \in \mathcal{D}_c. \quad (4.14)$$

In particular, $\lambda_S(x)$, $x \in \mathcal{D}_c$, corresponds to the value at which $p(\lambda)$ is minimized wherein λ belongs to the viable switching set. The following proposition shows that "min" in (4.14) is attained and hence $V(x)$ is well defined.

Proposition 4.2. *Let $\mathcal{S} \subseteq \Lambda_s$ and let $p : \mathcal{S} \to \mathbb{R}$ be a continuous positive-definite function such that Assumption 4.1 holds. Then, for all $x \in \mathcal{D}_c$, there exists a unique $\lambda_S(x) \in \mathcal{V}_s(x)$ such that $p(\lambda_S(x)) = \min\{p(\lambda) : \lambda \in \mathcal{V}_s(x)\}$.*

The next result shows that $V(\cdot)$ given by (4.14) is a generalized Lyapunov function candidate, that is, $V(\cdot)$ is lower semicontinuous on $\mathcal{D}_c$.

Theorem 4.2. *Let $\mathcal{S} \subseteq \Lambda_s$ be such that Assumption 4.1 holds. Then the function $V(x) = p(\lambda_S(x))$, $x \in \mathcal{D}_c$, is lower semicontinuous on $\mathcal{D}_c$ and continuous on $\mathcal{D}_{\lambda_S(x)}$.*

Next, we show that with the hierarchical nonlinear robust feedback control strategy $u = \phi_{\lambda_S(x)}(x)$, $x \in \mathcal{D}_c$, $V(\cdot)$ given by (4.14) is a generalized Lyapunov function for the nonlinear feedback controlled uncertain dynamical system (4.3). The controller notation $\phi_{\lambda_S(x)}(x)$ denotes a switching nonlinear robust feedback controller where the switching function $\lambda_S(x)$, $x \in \mathcal{D}_c$, is such that definition (4.14) of the generalized Lyapunov function $V(x)$, $x \in \mathcal{D}_c$, holds for a given potential function $p(\cdot)$ and switching set $\mathcal{S}$ satisfying Assumption 4.1. Furthermore, note that since $\phi_{\lambda_S(x)}(x)$ is defined for $x \in \mathcal{D}_c$, it follows that the solution $x(\cdot)$ to (4.3) with $x_0 \in \mathcal{D}_c$ and $u = \phi_{\lambda_S(x)}(x)$ is defined for all values of $t \in \mathcal{I}_{x_0}$ such that $x(t) \in \mathcal{D}_c$. However, as will be shown, since $\mathcal{D}_c$ is a positively invariant set, $[0, +\infty) \subseteq \mathcal{I}_{x_0}$, while if $x_0 \in \mathcal{D}_c$ is such that $x(t)$, $t < 0$, is always contained in $\mathcal{D}_c$, then $\mathcal{I}_{x_0} = \mathbb{R}$. Finally, note that since the solution $x(t)$, $t \in \mathcal{I}_{x_0}$, to (4.3) with $x_0 \in \mathcal{D}_c$ and $u = \phi_{\lambda_S(x)}(x)$ is continuous, it follows from Theorem 4.2 that $V(x(t))$, $t \in \mathcal{I}_{x_0}$, is right continuous. Hence, using the continuity of $p(\cdot)$ and the definition of $V(x)$, $x \in \mathcal{D}_c$, it follows that $\lambda_S(x(t))$, $t \in \mathcal{I}_{x_0}$, is also right continuous. Now, the continuity of $F(\cdot,\cdot) \in \mathcal{F}$ and $\phi_\lambda(\cdot)$, $\lambda \in \Lambda_s$, imply that $F(x(t), \phi_{\lambda_S(x(t))}(x(t)))$, $F(\cdot,\cdot) \in \mathcal{F}$, $t \in \mathcal{I}_{x_0}$, is right continuous.

Theorem 4.3. *Consider the nonlinear controlled uncertain dynamical system given by (4.1) with $F_n(0,0) = 0$, and assume there exists a continuous function $\psi : \Lambda_o \to \mathcal{D}_o$, $0 \in \Lambda_o$, parameterizing a nominal equilibrium manifold of (4.2), such that $x_\lambda = \psi(\lambda)$, $\lambda \in \Lambda_o$. Furthermore, assume that there exists a C^0 feedback control law $\phi_\lambda(\cdot)$, $\lambda \in \Lambda_s \subseteq \Lambda_o$ with $0 \in \Lambda_s$, that locally stabilizes a compact positively invariant neighborhood $\mathcal{N}_\lambda$ of x_λ for all $F(\cdot,\cdot) \in \mathcal{F}$ with a domain of attraction estimate $\mathcal{D}_\lambda$ and let $\mathcal{S} \subseteq \Lambda_s$, $0 \in \mathcal{S}$, be such that Assumption 4.1 holds. If $\lambda_S(x)$, $x \in \mathcal{D}_c$, is such that $V(x)$, $x \in \mathcal{D}_c$, given by (4.14) holds and $x(t)$, $t \in \mathcal{I}_{x_0}$, is the solution to (4.1) with $x(0) = x_0 \in \mathcal{D}_c$ and robust feedback control law*

$$u = \phi_{\lambda_{\mathcal{S}}(x)}(x), \qquad x \in \mathcal{D}_{\mathrm{c}}, \tag{4.15}$$

then $\mathcal{D}_{\mathrm{c}}$ is robustly positively invariant and $V(x(t))$, $t \geq 0$, is nonincreasing. Furthermore, for all $t_1, t_2 \geq 0$, $V(x(t)) = V(x(t_1))$, $t \in [t_1, t_2]$, if, and only if, $\lambda_{\mathcal{S}}(x(t)) = \lambda_{\mathcal{S}}(x(t_1))$, $t \in [t_1, t_2]$. Finally, for all $t \in \mathcal{I}_{x_0}$ such that $\lambda_{\mathcal{S}}(x(t)) \neq 0$, there exists a finite time $T > 0$ such that $V(x(t+T)) < V(x(t))$.

Finally, we present the main result of this chapter. Specifically, we show that the hierarchical robust switching nonlinear controller given by (4.15) guarantees that the closed-loop system trajectories converge to a union of largest invariant sets contained on the boundary of intersections over finite intervals of the closure of the generalized Lyapunov level surfaces. In addition, if the scheduling set $\mathcal{S}$ is homeomorphic to an interval on the real line and/or consists of only isolated points, then the hierarchical switching nonlinear controller establishes robust asymptotic stability of the compact positively invariant set $\mathcal{N}_0$.

Theorem 4.4. *Consider the nonlinear controlled uncertain dynamical system given by (4.1) with $F_{\mathrm{n}}(0,0) = 0$ and assume there exists a continuous function $\psi : \Lambda_{\mathrm{o}} \to \mathcal{D}_{\mathrm{o}}$, $0 \in \Lambda_{\mathrm{o}}$, parameterizing a nominal equilibrium manifold of (4.2), such that $x_\lambda = \psi(\lambda)$, $\lambda \in \Lambda_{\mathrm{o}}$. Furthermore, assume that there exists a C^0 feedback control law $\phi_\lambda(\cdot)$, $\lambda \in \Lambda_{\mathrm{s}} \subseteq \Lambda_{\mathrm{o}}$ with $0 \in \Lambda_{\mathrm{s}}$, that locally stabilizes a compact positively invariant neighborhood $\mathcal{N}_\lambda$ of x_λ for all $F(\cdot,\cdot) \in \mathcal{F}$ with a domain of attraction estimate $\mathcal{D}_\lambda$ and let $\mathcal{S} \subseteq \Lambda_{\mathrm{s}}$, $0 \in \mathcal{S}$, be such that Assumption 4.1 holds. In addition, assume $\lambda_{\mathcal{S}}(x)$, $x \in \mathcal{D}_{\mathrm{c}}$, is such that $V(x)$, $x \in \mathcal{D}_{\mathrm{c}}$, given by (4.14) holds, and, for $x_0 \in \mathcal{D}_{\mathrm{c}}$, $x(t)$, $t \in \mathcal{I}_{x_0}$, is the solution to (4.3) with the robust feedback control law*

$$u = \phi_{\lambda_{\mathcal{S}}(x)}(x), \qquad x \in \mathcal{D}_{\mathrm{c}}. \tag{4.16}$$

If $x_0 \in \mathcal{D}_{\mathrm{c}}$, then $x(t) \to \hat{\mathcal{M}} \triangleq \cup_{\gamma \in \mathcal{G}} \mathcal{M}_\gamma$ as $t \to \infty$ for all $F(\cdot,\cdot) \in \mathcal{F}$, where $\mathcal{G} \triangleq \{\gamma \geq 0 : \mathcal{R}_\gamma \cap \mathcal{D}_0 \neq \emptyset\}$. If, in addition, $\mathcal{S}_0 \triangleq \{\lambda \in \mathcal{S} : \mathcal{D}_\lambda \cap \mathcal{D}_0 \neq \emptyset\}$ is homeomorphic to $[0,a]$, $a > 0$, with $0 \in \mathcal{S}_0$ corresponding to $0 \in \mathbb{R}$, or $\mathcal{S}_0$ consists of only isolated points, then the compact positively invariant set $\mathcal{N}_0$ is locally robustly asymptotically stable with an estimate of domain of attraction given by $\mathcal{D}_{\mathrm{c}}$. Finally, if $\mathcal{D} = \mathbb{R}^n$ and there exists $\hat{\lambda} \in \mathcal{S}$ such that the feedback control law $\phi_{\hat{\lambda}}(\cdot)$ globally robustly asymptotically stabilizes the compact positively invariant set $\mathcal{N}_{\hat{\lambda}}$, then the above results are global.

Proof. The proof follows from Theorems 2.6, 4.2, and 4.3. The details of the proof are similar to those given in Theorem 3.5 and hence are omitted. ■

In the case where the switching set $\mathcal{S}$ is homeomorphic to an interval on $\mathbb{R}$ and a robust stabilizing controller $\phi_0(\cdot)$ for $\mathcal{N}_0$ cannot be obtained, that

is, $c_0 = \beta_0$, $\mathcal{N}_0 \subset \mathcal{D}_c$ still holds. Hence, Theorem 4.4 guarantees attraction to $\mathcal{N}_0$ if $\partial\mathcal{N}_0 \cap \partial\mathcal{D}_c \neq \emptyset$. Alternatively, if $\mathcal{N}_0 \subset \overset{\circ}{\mathcal{D}}_c$, then $\mathcal{N}_0$ is robustly asymptotically stable.

As in the nominal case discussed in Chapter 3, since the hierarchical robust switching nonlinear controller $u = \phi_{\lambda_S(x)}(x)$, $x \in \mathcal{D}_c$, is constructed such that the switching function $\lambda_S(x)$, $x \in \mathcal{D}_c$, assures that $V(x)$, $x \in \mathcal{D}_c$, defined by (4.14) is a generalized Lyapunov function with strictly decreasing values at the switching points, the possibility of a sliding mode is precluded with the proposed robust control scheme. In particular, Theorem 4.3 guarantees that the closed-loop state trajectories cross the boundary of adjacent regions of attraction in the state space in a inward direction. Thus, the closed-loop state trajectories *enter* the lower potential-valued domain of attraction before subsequent switching can occur. Hence, the proposed robust nonlinear stabilization framework avoids the undesirable effects of high-speed switching onto an invariant sliding manifold.

Finally, to elucidate the hierarchical ***robust*** switching nonlinear controller presented in this chapter, we present an algorithm that outlines the key steps in constructing the robust hierarchical switching feedback controller.

Algorithm 4.1. *To construct the robust hierarchical switching feedback control $\phi_{\lambda_S(x(t))}(x(t))$, $t \geq 0$, perform the following steps:*

Step 1. Construct the nominal equilibrium manifold of (4.2) using $u = \varphi(x, \lambda)$, where $\varphi(\cdot, \cdot)$ is an arbitrary function of $\lambda \in \Lambda_o$. Use $F_n(x, \varphi(x, \lambda)) = 0$ to explicitly define the mapping $\psi(\cdot)$ such that $x_\lambda = \psi(\lambda)$, $\lambda \in \Lambda_o$, is a nominal equilibrium point of (4.2) corresponding to the parameter value λ. We note that the above parameterization can be constructed using the approaches given in [81, 64, 107].

Step 2. Construct the set $\Lambda_s \subseteq \Lambda_o$ such that, for each $\lambda \in \Lambda_s$, there exists a compact positively invariant set $\mathcal{N}_\lambda$ containing the nominal equilibrium point x_λ. Furthermore, for each $\lambda \in \Lambda_s$, construct an asymptotically stabilizing controller $\phi_\lambda(\cdot)$ for the positively invariant set $\mathcal{N}_\lambda$ with an associated domain of attraction $\mathcal{D}_\lambda$ corresponding to the level set c_λ and Lyapunov function $V_\lambda(\cdot)$. Here, the controllers $\phi_\lambda(\cdot)$, $\lambda \in \Lambda_s$, can be obtained using any appropriate standard linear or nonlinear stabilization scheme.

Step 3. Construct the switching set $S \subseteq \Lambda_s$ and a potential function $p : S \to \mathbb{R}^+$ such that Assumption 4.1 is satisfied. In particular:

3a. If $\lambda \in S$ is an isolated point of S with corresponding compact positively invariant set $\mathcal{N}_\lambda$, then there exists $\lambda_1 \in S$ such that $p(\lambda_1) < p(\lambda)$, $\mathcal{N}_\lambda \subset \overset{\circ}{\mathcal{D}}_{\lambda_1}$.

3b. If $\lambda \in \mathcal{S}$ is an accumulation point of $\mathcal{S}$ then Step 3a is automatically satisfied if $p(\cdot)$ does not achieve a local minimum at λ.

3c. If $\lambda, \lambda_1 \in \mathcal{S}$, $\lambda \neq \lambda_1$, is such that $p(\lambda) = p(\lambda_1)$, then $\mathcal{D}_\lambda \cap \mathcal{D}_{\lambda_1} = \emptyset$.

Step 4. Given the state space point $x(t)$ at $t \geq 0$, search for solutions to $V_\lambda(x(t)) = c_\lambda$, $\lambda \in \mathcal{S}$.

4a. If no solution exists, $\lambda_S(x(t))$ is unchanged.

4b. If one *solution λ_1 exists and $p(\lambda_1) < p(\lambda)$ then switch $\lambda_S(x(t))$ to λ_1.*

4c. If more than one solution exists, repeat Step 4b with λ_1 replaced by the solution that minimizes $p(\cdot)$. Note that multiple solutions can be avoided by modifying the c_λ's.

Step 5. Construct the hierarchical robust switching feedback controller $\phi_{\lambda_S(x(t))}(x(t))$ where $\lambda_S(x(t))$, $x \in \mathcal{D}_c$, constructed in Step 4 is such that (4.14) holds.

Note that the existence of a switching set $\mathcal{S}$ and a potential function $p(\cdot)$ such that Step 3 is satisfied, can be guaranteed by modifying the first part Step 4 as follows:

Step 4′. Given the state space point $x(t)$ at $t = t_k \triangleq k\Delta T$, where $\Delta T > 0$ and $k = 0, 1, \ldots$, search for the solutions of $V_\lambda(x(t_k)) \leq c_\lambda$, $\lambda \in \Lambda_S$.

In this case, the switching set $\mathcal{S} \subseteq \Lambda_s$ need not be explicitly defined and is computed online.

4.5 Conclusion

A nonlinear robust control-system design framework predicated on a hierarchical switching controller architecture parameterized over a set of nominal system equilibria was developed. Specifically, a hierarchical robust switching nonlinear control strategy is constructed to stabilize a given uncertain nonlinear system by robustly stabilizing a collection of nonlinear controlled subsystems. The switching nonlinear controller architecture is designed based on a generalized Lyapunov function obtained by minimizing a potential function over a given switching set induced by the parameterized nominal system equilibria.

5. Hierarchical Switching Control for Multi-Mode Axial Flow Compressor Models

5.1 Introduction

The desire for developing an integrated control system-design methodology for advanced propulsion systems has led to significant activity in modeling and control of flow compression systems in recent years (see, for example, [49, 103, 104, 88, 5, 52, 109, 138, 20, 79, 78, 143, 14, 50] and the numerous references therein). However, unavoidable discrepancies between compression system models and real-world compression systems can result in degradation of control-system performance including instability. In particular, jet engine compression systems with poorly modeled dynamics and exogenous disturbances can severely limit jet engine compression system performance by inducing the compressor aerodynamic instabilities of rotating stall and surge. Rotating stall is an inherently three-dimensional[1] local compression system oscillation which is characterized by regions of flow that rotate at a fraction of the compressor rotor speed while surge is a one-dimensional axisymmetric global compression system oscillation which involves axial flow oscillations and in some cases even axial flow reversal which can damage engine components and cause flameout to occur.

Rotating stall and surge arise due to perturbations in stable system operating conditions involving steady, axisymmetric flow and can severely limit compressor performance. The transition from stable compressor operating conditions to rotating stall and surge is shown in Figure 5.1 representing a schematic of a compressor characteristic map where the abscissa corresponds to the circumferentially averaged mass flow through the compressor and the ordinate corresponds to the normalized total-to-static pressure rise in the compressor. For maximum compressor performance, operating conditions require that the pressure rise in the compressor correspond to the maximum pressure operating point on the stable axisymmetric branch for a given throttle opening. Here, we distinguish between compressor performance (pressure rise) and compressor efficiency (specific power consumption) where, depend-

[1] When analyzing high hub-to-tip ratio compressors, rotating stall can be approximated as a two-dimensional local compression system oscillation.

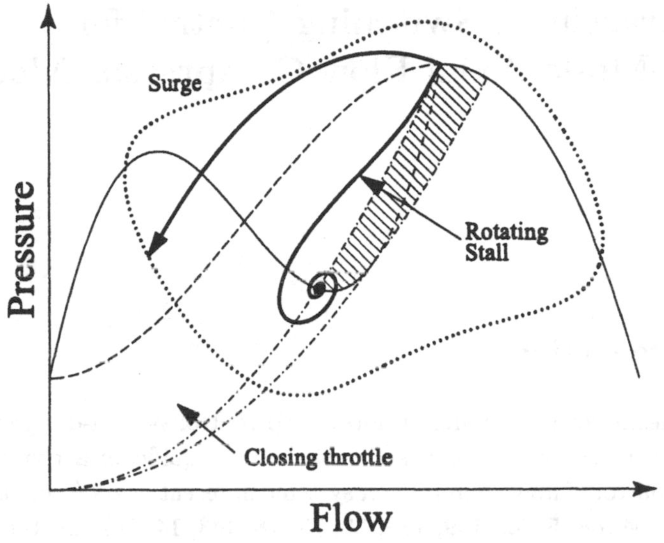

Fig. 5.1. Schematic of compressor characteristic map for a typical compression system (—— stable equilibria, – – – unstable equilibria)

ing on how the compressor is designed, the most *efficient* operating point may be to the right of the peak of the compressor characteristic map.

In practice, however, compression system uncertainty and compression system disturbances can perturb the operating point into an unstable region driving the system to a stalled stable equilibrium, a stable limit cycle (surge), or both. In the case of rotating stall, an attempt to recover to a high pressure operating point by increasing the flow through the throttle traps the system within a flow range corresponding to two stable operating conditions involving steady axisymmetric flow and rotating stall resulting in severe hysteresis.

To avoid rotating stall and surge, traditionally system designers allow for a safety margin (rotating stall or surge margin) in compression system operation. However, to account for compression system uncertainty such as system modeling errors, in-service changes due to aging, etc., and compression system disturbances such as compressor speed fluctuations, combustion noise, etc., operating at or below the rotating stall/surge margin significantly reduces the efficiency of the compression system. In contrast, active control can enhance stable compression system operation to achieve peak compressor performance. However, compression system uncertainty and compression system disturbances are often significant and the need for robust disturbance rejection control is severe.

In order to develop robust control-system design methodologies for compression systems, reliable models capturing the intricate physical phenom-

ena of rotating stall and surge are necessary. A fundamental development in compression system modeling for low speed axial compressors is the Moore-Greitzer model given in [103, 104]. Specifically, utilizing a one-mode expansion of the disturbance velocity potential in the compression system and assuming a nonlinear (cubic) characteristic for the compressor performance map the authors in [104] develop a low-order three-state nonlinear model involving the mean flow in the compressor, the pressure rise, and the amplitude of the rotating stall. Starting from infinitesimal perturbations in the flow field the model captures the development of rotating stall and surge. In particular, the model predicts the experimentally verified pitchfork bifurcation at the onset of rotating stall [100]. Extensions to the Moore-Greitzer model that include blade row time lags and viscous transport terms have been reported in [62] and [4, 5, 138], respectively.

Using the Moore-Greitzer model a bifurcation-based control methodology for rotating stall and surge is developed in [4, 16]. The bifurcation-based controllers guarantee local asymptotic stability with guaranteed domains of attraction. This approach has been successfully implemented in industrial turbomachinery by Nett and co-workers [15, 12]. Alternatively, the authors in [79, 78, 77, 55] develop globally stabilizing controllers for controlling rotating stall and surge. In particular, a Lyapunov-based recursive backstepping globally stabilizing controller is given in [79, 78, 77] while an optimality-based nonlinear globally stabilizing controller is given in [55]. In both cases the controllers are predicated on the Moore-Greitzer model.

A fundamental shortcoming of the low-order three-state Moore-Greitzer model and, as a consequence, the control design methodologies based on the model, is the fact that only a one-mode expansion of the disturbance velocity potential in the compression system is considered. Since the second and higher-order disturbance velocity potential harmonics strongly interact with the first harmonic during stall inception they must be accounted for in the control-system design process. A notable exception to the low-order three-state model is given in [98] where a discrete Fourier transform formulation is used to obtain a distributed nonlinear model for axial compression systems.

In this chapter we first develop a self-contained first principles derivation of a multi-mode model for rotating stall and surge in axial flow compression systems that is accessible to control-system designers requiring state space models for modern nonlinear control design. Specifically, the formulation is based on a generalized multi-mode expansion of the disturbance velocity potential in the flow field which accounts for the coupling between higher-order system harmonics and the pressure rise and mean flow through the compressor. Then, we apply the hierarchical switching nonlinear control framework developed in Chapters 3 and 4 to mitigate the aerodynamics instabilities of

rotating stall and surge in a two-mode axial compressor model. To reflect a more realistic design we account for uncertainty in the pressure-flow compressor performance characteristic map as well as impose a rate saturation constraint on the system actuator throttle opening.

5.2 Governing Fluid Dynamic Equations for Axial Flow Compression Systems

In this section we develop a first principles multi-mode model for rotating stall and surge in axial flow compressors. Specifically, we consider the basic compression system shown in Figure 5.2, consisting of an inlet duct, a compressor, an outlet duct, a plenum, and a control throttle. We assume that the plenum dimensions are large as compared to the compressor-duct dimensions so that the fluid velocity and acceleration in the plenum are negligible. In this case, the pressure in the plenum is spatially uniform. Furthermore, we assume that the flow is controlled by a throttle at the plenum exit. Finally, we assume a low speed compression system with oscillation frequencies much lower than the acoustic resonance frequencies so that the flow can be considered incompressible. However, we do assume that the gas in the plenum is compressible and hence acts as a gas spring.

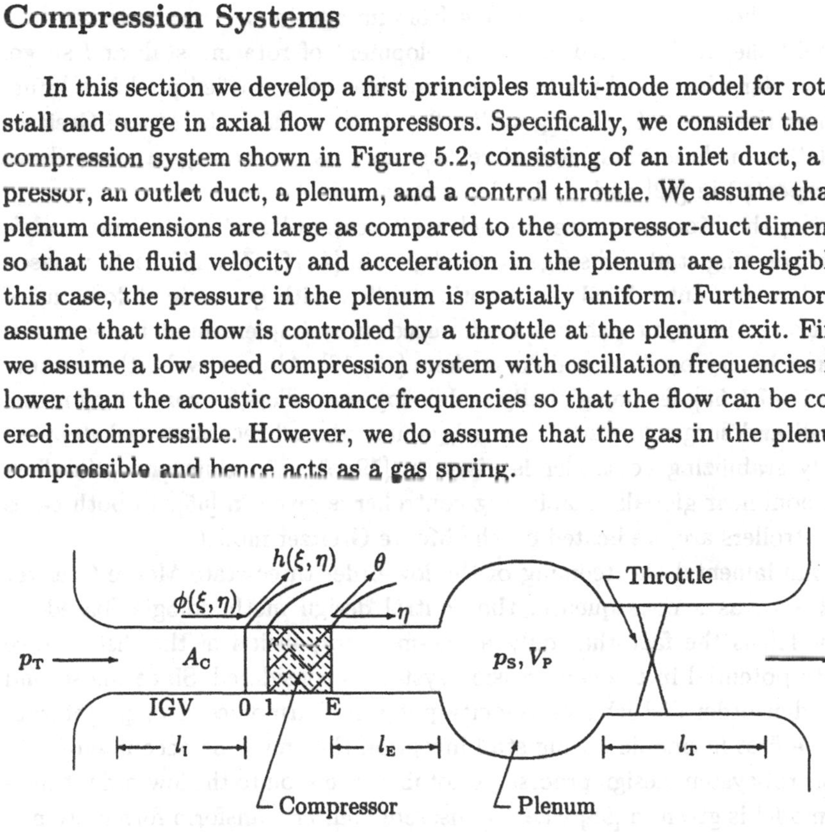

Fig. 5.2. Compressor system geometry

5.2.1 Entrance Duct and Inlet Guide Vane Entrance

The fundamental principle used for modeling the pressure increase in the entrance duct for inviscid, irrotational, and incompressible flow is the

Bernouilli equation applied between the upstream (atmospheric) reservoir and the inlet guide vane entrance. Since the irrotational flow assumption implies the absence of any radial flow variation we utilize a two-dimensional reference frame involving circumferential and axial axis with s, y, u, and v denoting circumferential and axial positions and velocities, respectively. The assumption of inviscid, irrotational, and incompressible flow is valid for low speed compressors with high hub-to-tip ratio. Finally, we also model the pressure difference associated with the circumferential velocity component just ahead of the inlet guide vane entrance.

The first fundamental principle used for modeling the increase of pressure in the entrance duct is the conservation of mass which states that, for any arbitrary material volume $\mathcal{V}_m$ within a continuum, the rate of change of mass in $\mathcal{V}_m$ is zero. Hence, the mass balance for an arbitrary material volume $\mathcal{V}_m$ is given by

$$0 = \frac{d}{dt}\int_{\mathcal{V}_m} \rho \, d\mathcal{V}, \tag{5.1}$$

where ρ is the local density and $d\mathcal{V}$ is the infinitesimal volume element. Using the Reynolds transport theorem [120], (5.1) can be re-written as

$$0 = \int_{\mathcal{V}_m} \left[\frac{\partial \rho}{\partial t} + \nabla \cdot (\rho V)\right] d\mathcal{V}, \tag{5.2}$$

where V is the local velocity field, ∇ is the "nabla" operator and "$\cdot$" denotes the usual dot product. Since (5.2) is valid for any arbitrary material volume $\mathcal{V}_m$, it follows that

$$0 = \frac{\partial \rho}{\partial t} + \nabla \cdot (\rho V). \tag{5.3}$$

The second fundamental principle used for modeling the increase of pressure in the entrance duct is the conservation of momentum which states that, for any arbitrary material volume $\mathcal{V}_m$ within a continuum, the rate of change of momentum in $\mathcal{V}_m$ is equal to the resultant of all internal stresses and volume forces. Hence, the momentum balance for an arbitrary material volume $\mathcal{V}_m$ is given by

$$\frac{d}{dt}\int_{\mathcal{V}_m} \rho V \, d\mathcal{V} = \oint_{\mathcal{S}_m} \hat{t} \, d\mathcal{S} + \int_{\mathcal{V}_m} \rho f \, d\mathcal{V}, \tag{5.4}$$

where $\mathcal{S}_m$ is the surface that encloses the arbitrary material volume $\mathcal{V}_m$, $d\mathcal{S}$ is the infinitesimal surface element, $\hat{t}$ is the stress vector, and f is the body force per unit mass. Denoting the outward normal vector to the infinitesimal surface element $d\mathcal{S}$ by $\hat{n}$, the stress vector $\hat{t}$ is defined by

$$\hat{t} \triangleq T\hat{n}, \tag{5.5}$$

where T is the Cauchy stress tensor [120].

Next, using (5.5) and the divergence theorem [133], $\oint_{\mathcal{S}_{\mathrm{m}}} \hat{t}\,\mathrm{d}\mathcal{S}$ in (5.4) can be replaced by $\int_{\mathcal{V}_{\mathrm{m}}} \nabla \cdot T\,\mathrm{d}\mathcal{V}$. Now, using the Reynolds transport theorem and the fact that (5.4) is valid for any arbitrary material volume $\mathcal{V}_{\mathrm{m}}$, it follows that

$$\frac{\partial(\rho V)}{\partial t} + \nabla \cdot (\rho V V) = \nabla \cdot T + \rho f. \tag{5.6}$$

Under the assumption of inviscid flow, the stress tensor T is defined by

$$T \triangleq -pI_3, \tag{5.7}$$

where p is the local pressure and I_3 denotes the 3×3 identity matrix. Now, neglecting the body force f and using (5.3) and (5.7) it follows from (5.6) that

$$0 = \rho\frac{\partial V}{\partial t} + \rho V \cdot \nabla V + \nabla p. \tag{5.8}$$

Furthermore, introducing the vorticity vector ζ defined by $\zeta \triangleq \nabla \times v$, where "$\times$" denotes the usual cross product, (5.8) simplifies to

$$0 = \rho\left(\frac{\partial V}{\partial t} + \tfrac{1}{2}\nabla V^2 + \zeta \times V\right) + \nabla p. \tag{5.9}$$

Finally, in the case of incompressible ($\rho =$ constant) and irrotational ($\zeta = 0$) flow, (5.9) yields

$$0 = \nabla\left(\tfrac{1}{2}\rho V^2 + p\right) + \frac{\partial(\rho V)}{\partial t}. \tag{5.10}$$

Under the assumption of irrotational flow it follows that the velocity field can be represented by $V = \nabla\varphi$, where $\varphi(t, s, y)$ is the velocity potential function. Substituting the potential representation of the velocity into (5.10) yields

$$0 = \nabla\left(\tfrac{1}{2}\rho V^2 + p + \rho\frac{\partial\varphi}{\partial t}\right). \tag{5.11}$$

Next, integrating (5.11) along an arbitrary contour inside the arbitrary material volume $\mathcal{V}_{\mathrm{m}}$ we obtain the well known Bernouilli equation

$$p + \tfrac{1}{2}\rho V^2 + \rho\frac{\partial\varphi}{\partial t} = f(t), \tag{5.12}$$

which states that the sum of the static pressure, the dynamic pressure, and the time rate of change of the velocity potential is a function of time. Now, writing (5.12) at the reservoir (subscript "T") and at the inlet guide vane entrance (subscript "0") for a fixed time $t = t^*$ yields

$$p_{\mathrm{T}} + \tfrac{1}{2}\rho V_{\mathrm{T}}^2 + \rho\frac{\partial \varphi_{\mathrm{T}}}{\partial t} = p_0 + \tfrac{1}{2}\rho V_0^2 + \rho\frac{\partial \varphi_0}{\partial t}. \tag{5.13}$$

Finally, appropriately choosing the spatially constant term in the representation of the velocity potential to ensure that there is no explicit dependence on time of the velocity potential at the reservoir ($\frac{\partial \varphi_{\mathrm{T}}}{\partial t} = 0$) and considering zero initial velocity ($V_{\mathrm{T}} = 0$) or, equivalently, redefining p_{T} to be the total pressure rather than the static pressure, we obtain

$$p_{\mathrm{T}} - p_0 = \tfrac{1}{2}\rho V_0^2 + \rho\frac{\partial \varphi_0}{\partial t}. \tag{5.14}$$

Next, we present the nondimensional form of (5.14). Specifically, we define the nondimensional velocity potential function

$$\hat{\varphi}(\xi,\theta,\eta) \triangleq \frac{1}{UR}\varphi(\frac{\xi R}{U},\theta R,\eta R), \tag{5.15}$$

where U is the circumferential blade speed at mean diameter, R is the mean compressor radius, and η, θ, and ξ are the nondimensional axial coordinate, circumferential coordinate, and time, respectively, defined by

$$\eta \triangleq \frac{y}{R}, \qquad \theta \triangleq \frac{s}{R}, \qquad \xi \triangleq \frac{Ut}{R}. \tag{5.16}$$

Now, using (5.15) it follows that

$$\hat{\nabla}\hat{\varphi} = R\nabla\frac{\varphi}{UR} = \frac{\nabla\varphi}{U} = \frac{V}{U}, \tag{5.17}$$

$$\frac{\partial\hat{\varphi}}{\partial\xi} = \frac{1}{UR}\frac{\partial\varphi}{\partial\xi} = \frac{1}{U^2}\frac{\partial\varphi}{\partial t}, \tag{5.18}$$

where $\hat{\nabla} \triangleq R\nabla$ is the nabla operator defined with respect to the nondimensional spatial variables. Substituting (5.18) into (5.14) and setting $\eta = 0$ at the inlet guide vane entrance we obtain

$$p_{\mathrm{T}} - p_0 = \tfrac{1}{2}\rho(u_0^2 + v_0^2) + \rho U^2 \left.\frac{\partial\hat{\varphi}}{\partial\xi}\right|_{\eta=0}, \tag{5.19}$$

where $u_0^2 + v_0^2 = V_0^2$, $u_0(\xi,\theta) \triangleq u(\xi,\theta,0)$, and $v_0(\xi,\theta) \triangleq v(\xi,\theta,0)$. Finally, dividing both sides of (5.19) by ρU^2 we obtain

$$\frac{p_{\mathrm{T}} - p_0}{\rho U^2} = \tfrac{1}{2}(\phi^2 + h^2) + \left.\frac{\partial\hat{\varphi}}{\partial\xi}\right|_{\eta=0}, \tag{5.20}$$

where

$$h(\xi,\theta) \triangleq \frac{u_0(\xi,\theta)}{U} = \left.\frac{\partial\hat{\varphi}}{\partial\theta}\right|_{\eta=0}, \tag{5.21}$$

$$\phi(\xi,\theta) \triangleq \frac{v_0(\xi,\theta)}{U} = \left.\frac{\partial\hat{\varphi}}{\partial\eta}\right|_{\eta=0}, \tag{5.22}$$

correspond to the nondimensional components of the velocity field at the inlet guide vane entrance. Note that ϕ also corresponds to the axial flow coefficient at $\eta = 0$ since

$$\phi = \frac{\rho v_0 A_c}{\rho U A_c} = \frac{v_0}{U}, \tag{5.23}$$

where A_c is the cross sectional area at the inlet guide vane entrance.

The complexity of the unsteady contribution in (5.20) can be reduced by considering a straight inlet duct, of dimensionless length l_1, preceded by a much shorter contracting passage. In this case the circumferential averaged flow is constant and hence we decompose the general velocity field into a circumferential averaged part and a perturbed part. Furthermore, even though the reservoir pressure is constant, ϕ can depend on ξ and θ. If ϕ is a function of θ it follows that the inlet guide vane entrance will involve a circumferential velocity component $h(\xi,\theta)$. However, assuming that any circumferential nonuniformity of axial velocity within the compressor must, by continuity arguments, be sucked through the compression system, it further follows that no circulation can arise in the entrance duct and hence $h(\xi,\theta)$ must have a vanishing circumferential average; that is,

$$0 = \int_0^{2\pi} h(\xi,\theta)\,d\theta. \tag{5.24}$$

Next, we decompose the axial flow coefficient ϕ as

$$\phi(\xi,\theta) = \Phi(\xi) + g(\xi,\theta), \tag{5.25}$$

where

$$\Phi(\xi) \triangleq \frac{1}{2\pi}\int_0^{2\pi} \phi(\xi,\theta)\,d\theta, \tag{5.26}$$

represents the circumferential averaged component and $g(\xi,\theta)$ is the perturbed component such that, by definition,

$$0 = \int_0^{2\pi} g(\xi,\theta)\,d\theta. \tag{5.27}$$

Now, letting

$$\hat{\varphi}(\xi,\theta,\eta) = (\eta + l_{\text{I}})\Phi(\xi) + \tilde{\varphi}(\xi,\theta,\eta), \tag{5.28}$$

where $\tilde{\varphi}(\xi,\theta,\eta)$ represents the disturbance velocity potential, we seek $\hat{\varphi}$ and $\tilde{\varphi}$ such that the Laplace equations $\nabla^2\hat{\varphi} = 0$ and $\nabla^2\tilde{\varphi} = 0$ are satisfied. Since the averaged flow is assumed to be constant it follows that the boundary condition associated with $\nabla^2\hat{\varphi} = 0$ is

$$\left.\frac{\partial\hat{\varphi}}{\partial\eta}\right|_{\eta=-l_{\text{I}}} = \Phi(\xi), \tag{5.29}$$

which implies that the circumferential averaged flow at the reservoir must be equal to the circumferential averaged flow at the inlet guide vane entrance. The boundary condition for the disturbance velocity potential $\tilde{\varphi}(\xi,\theta,\eta)$ follows directly from (5.29) using (5.28) and is given by

$$\left.\frac{\partial\tilde{\varphi}}{\partial\eta}\right|_{\eta=-l_{\text{I}}} = 0. \tag{5.30}$$

Using (5.21), (5.22), and (5.25) the disturbance velocity potential at the inlet guide vane entrance satisfies

$$g(\xi,\theta) = \left.\frac{\partial\tilde{\varphi}}{\partial\eta}\right|_{\eta=0}, \qquad h(\xi,\theta) = \left.\frac{\partial\tilde{\varphi}}{\partial\theta}\right|_{\eta=0}. \tag{5.31}$$

Now, differentiating (5.28) with respect to the nondimensional time ξ we obtain

$$\left.\frac{\partial\hat{\varphi}}{\partial\xi}\right|_{\eta=0} = l_{\text{I}}\frac{\mathrm{d}\Phi}{\mathrm{d}\xi} + \left.\frac{\partial\tilde{\varphi}}{\partial\xi}\right|_{\eta=0}, \tag{5.32}$$

which, substituted into (5.20), gives

$$\frac{p_{\text{T}} - p_0}{\rho U^2} = \tfrac{1}{2}(\phi^2 + h^2) + l_{\text{I}}\frac{\mathrm{d}\Phi}{\mathrm{d}\xi} + \left.\frac{\partial\tilde{\varphi}}{\partial\xi}\right|_{\eta=0}. \tag{5.33}$$

Finally, the overall pressure rise from the compressor inlet to the compressor exit involves the pressure difference associated with the circumferential velocity component ahead of the inlet guide vane entrance. In particular [104],

$$\frac{p_1 - p_0}{\rho U^2} = \tfrac{1}{2}K_{\text{G}}h^2, \tag{5.34}$$

where the entrance recovery coefficient $K_{\text{G}} = 1$ if the inlet guide vane is lossless and $K_{\text{G}} < 1$ if the inlet guide vane is dissipative.

5.2.2 Compressor

We consider an N-stage compressor model with a static pressure rise through each row given by [103]

$$\frac{\Delta p}{\frac{1}{2}\rho U^2} = F(\phi) - \tau(\phi)\frac{\mathrm{d}\phi}{\mathrm{d}t}, \tag{5.35}$$

where $F(\phi)$ represents the axisymmetric steady performance of either a stator or a rotor blade row and $\tau(\phi)\frac{\mathrm{d}\phi}{\mathrm{d}t}$ represents hysteresis due to flow acceleration, flow separation, and viscous effects in the blade passage. We assume that $\tau(\phi)$ is constant and $F(\phi)$ and $\tau(\phi)$ do not change across each stator and rotor row. The latter assumption is rigorously valid only for symmetric blade configurations. A reasonable value for τ, which can be viewed as a time constant associated with the internal lags in the compressor, can be obtained by analyzing the inertia of the fluid in the passage [103].

Next, we evaluate $\frac{\mathrm{d}\phi}{\mathrm{d}t}$ for one rotor-stator stage and compute the increase in pressure through that stage. The overall pressure rise in the compressor is computed by simply adding the pressure rise across each rotor-stator stage. For an ith stator we have

$$\frac{\mathrm{d}\phi}{\mathrm{d}t} = \frac{\partial \phi}{\partial t} = \frac{U}{R}\frac{\partial \phi}{\partial \xi}, \tag{5.36}$$

while for an ith rotor we also need to account for flow unsteadiness due to the rotor blades moving (with velocity U) through a circumferentially non-uniform flow so that

$$\frac{\mathrm{d}\phi}{\mathrm{d}t} = \frac{\partial \phi}{\partial t} + U\frac{\partial \phi}{\partial s} = \frac{U}{R}\left(\frac{\partial \phi}{\partial \xi} + \frac{\partial \phi}{\partial \theta}\right). \tag{5.37}$$

Hence, for one rotor-stator stage we obtain the pressure rise in the compressor by

$$\begin{aligned}\frac{\Delta p_1}{\frac{1}{2}\rho U^2} &= \left[F(\phi) - \tau\frac{U}{R}\frac{\partial \phi}{\partial \xi}\right] + \left[F(\phi) - \tau\frac{U}{R}\left(\frac{\partial \phi}{\partial \xi} + \frac{\partial \phi}{\partial \theta}\right)\right] \\ &= 2F(\phi) - \tau\frac{U}{R}\left(2\frac{\partial \phi}{\partial \xi} + \frac{\partial \phi}{\partial \theta}\right). \end{aligned}\tag{5.38}$$

Applying (5.38) to an N-stage compressor, the pressure rise across the compressor is given by

$$\frac{p_{\mathrm{E}} - p_1}{\rho U^2} = NF(\phi) - \frac{1}{2a}\left(2\frac{\partial \phi}{\partial \xi} + \frac{\partial \phi}{\partial \theta}\right), \tag{5.39}$$

where $a \triangleq \frac{R}{\tau N U}$. Finally, substituting (5.25) and (5.31) into (5.39) yields

$$\frac{p_{\mathrm{E}} - p_1}{\rho U^2} = NF(\phi) - \frac{1}{a}\frac{\mathrm{d}\Phi}{\mathrm{d}\xi} - \frac{1}{2a}\left[2\frac{\partial^2 \tilde{\varphi}}{\partial \eta \partial \xi} + \frac{\partial^2 \tilde{\varphi}}{\partial \eta \partial \theta}\right]_{\eta=0}. \tag{5.40}$$

5.2.3 Exit Duct

To compute the pressure increase across the exit duct we use the momentum conservation equations written in terms of a stream function formulation. We begin by defining the stream function ψ satisfying

$$\frac{u}{U} = \frac{\partial \psi}{\partial \eta}, \qquad \frac{v}{U} = -\frac{\partial \psi}{\partial \theta}. \tag{5.41}$$

Next, for notational convenience, we introduce a dimensionless form for pressure given by

$$\pi(\xi, \theta, \eta) \triangleq \frac{p_S(\xi) - p(\xi, \theta, \eta)}{\rho U^2}, \tag{5.42}$$

where $p_S(\xi)$ denotes the discharge pressure. In this case the momentum conservation equation yields

$$0 = \frac{\partial^2 \psi}{\partial \eta \partial \xi} + \frac{\partial \psi}{\partial \eta}\frac{\partial^2 \psi}{\partial \eta \partial \theta} - \frac{\partial \psi}{\partial \theta}\frac{\partial^2 \psi}{\partial \eta^2} - \frac{\partial \pi}{\partial \theta}, \tag{5.43}$$

$$0 = -\frac{\partial^2 \psi}{\partial \theta \partial \xi} - \frac{\partial \psi}{\partial \eta}\frac{\partial^2 \psi}{\partial \theta^2} + \frac{\partial \psi}{\partial \theta}\frac{\partial^2 \psi}{\partial \theta \partial \eta} - \frac{\partial \pi}{\partial \eta}, \tag{5.44}$$

which further imply

$$\tfrac{1}{2}\nabla^2 \pi = -\frac{\partial^2 \psi}{\partial \theta^2}\frac{\partial^2 \psi}{\partial \eta^2} + \left(\frac{\partial^2 \psi}{\partial \eta \partial \theta}\right)^2. \tag{5.45}$$

If the pressure in the exit duct slightly differs from the static discharge pressure $p_S(\xi)$, then the disturbance velocity field is small which implies that, due to circumferential flow non-uniformities, the right-hand side of (5.45) is negligible and hence

$$0 = \nabla^2 \pi. \tag{5.46}$$

Since the pressure is uniform downstream (subscript "S") of the exit duct it follows that

$$0 = \pi_S(\xi, \theta). \tag{5.47}$$

Finally, comparing the potential problem for π with the inlet disturbance velocity potential problem, it follows that [104]

$$\pi = (\eta - l_E)\frac{d\Phi}{d\xi} - \frac{\partial \tilde{\varphi}}{\partial \xi}. \tag{5.48}$$

In the above analysis we assumed that l_E is greater than or equal to the distance at which the entrance disturbance velocity potential vanishes. If this

were not the case, the second term in (5.48) should be omitted. Hence, in general

$$\frac{p_{\mathrm{S}} - p_{\mathrm{E}}}{\rho U^2} = -l_{\mathrm{E}} \frac{\mathrm{d}\Phi}{\mathrm{d}\xi} - (m-1) \left. \frac{\partial \tilde{\varphi}}{\partial \xi} \right|_{\eta=0}, \tag{5.49}$$

where $m = 1$ for a very short exit duct and $m = 2$ otherwise.

5.2.4 Governing System Flow Equations

In this subsection we combine the results of the previous subsections to obtain the pressure rise between the upstream reservoir and the exit duct discharge. Since the plenum and the throttle are subject to axisymmetric disturbances only, they are considered separately.

Combining equations (5.33), (5.34), (5.40), and (5.49), it follows that

$$\begin{aligned} \frac{p_{\mathrm{S}} - p_{\mathrm{T}}}{\rho U^2} &= \frac{p_{\mathrm{S}} - p_{\mathrm{E}}}{\rho U^2} + \frac{p_{\mathrm{E}} - p_1}{\rho U^2} + \frac{p_1 - p_0}{\rho U^2} + \frac{p_0 - p_{\mathrm{T}}}{\rho U^2} \\ &= -l_{\mathrm{E}} \frac{\mathrm{d}\Phi}{\mathrm{d}\xi} - (m-1) \left. \frac{\partial \tilde{\varphi}}{\partial \xi} \right|_{\eta=0} - \frac{1}{2a} \left[2\frac{\partial^2 \tilde{\varphi}}{\partial \eta \partial \xi} + \frac{\partial^2 \tilde{\varphi}}{\partial \eta \partial \theta} \right]_{\eta=0} + NF(\phi) \\ &\quad - \frac{1}{a} \frac{\mathrm{d}\Phi}{\mathrm{d}\zeta} + \tfrac{1}{2} K_{\mathrm{G}} h^2 - \tfrac{1}{2}(\phi^2 + h^2) - l_{\mathrm{I}} \frac{\mathrm{d}\Phi}{\mathrm{d}\xi} - \left. \frac{\partial \tilde{\varphi}}{\partial \xi} \right|_{\eta=0}, \end{aligned} \tag{5.50}$$

which, assuming $K_{\mathrm{G}} = 1$, can be re-written as

$$\Psi = \psi_{\mathrm{C}}(\phi) - l_{\mathrm{C}} \frac{\mathrm{d}\Phi}{\mathrm{d}\xi} - m \left. \frac{\partial \tilde{\varphi}}{\partial \xi} \right|_{\eta=0} - \frac{1}{2a} \left[2\frac{\partial^2 \tilde{\varphi}}{\partial \eta \partial \xi} + \frac{\partial^2 \tilde{\varphi}}{\partial \eta \partial \theta} \right]_{\eta=0}, \tag{5.51}$$

where

$$\Psi \triangleq \frac{p_{\mathrm{S}} - p_{\mathrm{T}}}{\rho U^2}, \qquad \psi_{\mathrm{C}}(\phi) \triangleq NF(\phi) - \tfrac{1}{2}\phi^2, \qquad l_{\mathrm{C}} \triangleq l_{\mathrm{E}} + l_{\mathrm{I}} + \frac{1}{a}, \tag{5.52}$$

are the total-to-static pressure rise coefficient, the quasi-steady axisymmetric compressor characteristic, and the effective flow-passage length through the compressor and ducts measured in radii of the compressor wheel, respectively. $\psi_{\mathrm{C}}(\phi)$ is the compressor performance map in the case where the flow through the compressor is circumferentially uniform and steady, even in a stalled condition.

Next, recalling that the perturbation velocity potential satisfies

$$\nabla^2 \tilde{\varphi} = 0, \qquad \left. \frac{\partial \tilde{\varphi}}{\partial \eta} \right|_{\eta=-l_{\mathrm{I}}} = 0, \tag{5.53}$$

it follows that

$$\tilde{\varphi}(\xi,\theta,\eta) = \sum_{k=1}^{\infty}[a_k(\xi)\sin(k\theta) + b_k(\xi)\cos(k\theta)]\frac{\cosh[k(\eta+l_I)]}{k\sinh(kl_I)}, \quad \eta \le 0. \quad (5.54)$$

Now, substituting (5.54) into (5.51) and computing the circumferential mean of the resulting equation we obtain

$$l_C\frac{d\Phi}{d\xi} = -\Psi + \frac{1}{2\pi}\int_0^{2\pi}\psi_C(\phi)\,d\theta, \quad (5.55)$$

which relates the change of mass flow through the compressor to the total pressure rise. The circumferential averaged flow coefficient Φ changes to balance the difference between the circumferential averaged pressure rise that is generated by the compressor in quasi-steady conditions and the pressure rise Ψ that actually exists across the compressor.

5.2.5 Plenum and Throttle Discharge

Since the plenum dimensions are large as compared to the compressor-duct dimensions, the pressure in the plenum is spatially uniform. We assume that the mass flow rate at the plenum entrance is $\frac{dm_C}{dt}$ and the mass flow rate at the plenum is $\frac{dm_T}{dt}$. In general, $\frac{dm_C}{dt} \neq \frac{dm_T}{dt}$. Furthermore, the cross-sectional areas at the plenum entrance and exit are assumed to be different from the compressor cross sectional area A_C. By continuity we have

$$\frac{dm_C}{dt} - \frac{dm_T}{dt} = V_P\frac{d\rho_S}{dt}, \quad (5.56)$$

where V_P denotes the plenum volume. Now, assuming that the flow in the plenum is isentropic it follows that

$$\frac{d\rho_S}{dt} = \frac{1}{a_S^2}\frac{dp_S}{dt}, \quad (5.57)$$

where a_S is the sound velocity in the plenum. Substituting (5.57) into (5.56) we obtain

$$\frac{dm_C}{dt} - \frac{dm_T}{dt} = \frac{V_P}{a_S^2}\frac{dp_S}{dt}, \quad (5.58)$$

which can be written in nondimensional form as

$$\rho U A_C\Phi(\xi) - \rho U A_C\Phi_T(\xi) = \frac{V_P}{a_S^2}\frac{U}{R}\frac{d}{d\xi}[\rho U^2\Psi(\xi)], \quad (5.59)$$

or, equivalently,

$$l_C\frac{d\Psi}{d\xi} = \frac{1}{4B^2}[\Phi(\xi) - \Phi_T(\xi)], \quad (5.60)$$

where

$$\Phi_{\mathrm{T}}(\xi) \triangleq \frac{1}{\rho U A_{\mathrm{C}}} \frac{\mathrm{d}m_{\mathrm{T}}}{\mathrm{d}t}, \qquad B \triangleq \frac{U}{2a_{\mathrm{S}}} \sqrt{\frac{V_{\mathrm{P}}}{A_{\mathrm{C}} L_{\mathrm{C}}}}, \qquad L_{\mathrm{C}} \triangleq R l_{\mathrm{C}}. \tag{5.61}$$

The compliance parameter B is a function of the compressor rotor speed and the system plenum size. For large values of B a surge limit cycle can occur while rotating stall can occur for any value of B.

Next, assuming that the throttle discharges to an infinite reservoir with pressure p_{T} it follows that the pressure difference $p_{\mathrm{S}} - p_{\mathrm{T}}$ must balance both the throttle pressure loss and the net difference in pressure due to the flow acceleration through the throttle duct. Hence,

$$\Psi(\xi) = F_{\mathrm{T}}(\Phi_{\mathrm{T}}) + l_{\mathrm{T}} \frac{\mathrm{d}\Phi_{\mathrm{T}}}{\mathrm{d}\xi}, \tag{5.62}$$

where $F_{\mathrm{T}}(\Phi_{\mathrm{T}})$ represents the throttle pressure loss and $l_{\mathrm{T}} \frac{\mathrm{d}\Phi_{\mathrm{T}}}{\mathrm{d}\xi}$ represent the change of pressure due to flow acceleration. Here, we assume that the throttle duct is short enough so that l_{T} can be neglected. Furthermore, we consider a quadratic throttle characteristic given by

$$F_{\mathrm{T}}(\Phi_{\mathrm{T}}) = \tfrac{1}{2} K_{\mathrm{T}} \Phi_{\mathrm{T}}^2, \tag{5.63}$$

where K_{T} is a constant throttle coefficient. In this case it follows that

$$\Phi_{\mathrm{T}} = F_{\mathrm{T}}^{-1}(\Psi) = \gamma_{\mathrm{th}} \sqrt{\Psi}, \tag{5.64}$$

where the parameter γ_{th} is proportional to the throttle opening. Finally, substituting (5.64) into (5.60) yields

$$l_{\mathrm{C}} \frac{\mathrm{d}\Psi}{\mathrm{d}\xi} = \frac{1}{4B^2} [\Phi(\xi) - F_{\mathrm{T}}^{-1}(\Psi)]. \tag{5.65}$$

5.3 Multi-Mode State Space Model

In this section we develop a multi-mode state space model for the axial compression system addressed in Section 5.2. The governing system flow equations for the axial flow compressor model are given by (5.51), (5.55), and (5.65). Using (5.55), (5.51) can be re-written as

$$\psi_{\mathrm{C}}(\phi) - \frac{1}{2\pi} \int_0^{2\pi} \psi_{\mathrm{C}}(\phi)\, \mathrm{d}\theta = m \left. \frac{\partial \tilde{\varphi}}{\partial \xi} \right|_{\eta=0} + \frac{1}{2a} \left[2 \frac{\partial^2 \tilde{\varphi}}{\partial \eta \partial \xi} + \frac{\partial^2 \tilde{\varphi}}{\partial \eta \partial \theta} \right]_{\eta=0}. \tag{5.66}$$

Substituting (5.54) for $\tilde{\varphi}(\xi, \theta, \eta)$ into (5.66) we obtain

$$\psi_{\mathrm{c}}(\phi) - \frac{1}{2\pi}\int_0^{2\pi} \psi_{\mathrm{c}}(\phi)\,\mathrm{d}\theta = \sum_{k=1}^{\infty}\Bigg[\left(\alpha_k \frac{\mathrm{d}a_k}{\mathrm{d}\xi} - \beta_k b_k\right)\sin(k\theta) + \left(\alpha_k \frac{\mathrm{d}b_k}{\mathrm{d}\xi} + \beta_k a_k\right)\cos(k\theta)\Bigg], \quad (5.67)$$

where

$$\alpha_k \triangleq \frac{m}{k}\frac{\cosh(kl_{\mathrm{I}})}{\sinh(kl_{\mathrm{I}})} + \frac{1}{a}, \qquad \beta_k \triangleq \frac{k}{2a}, \quad (5.68)$$

and where, using (5.25) and (5.31),

$$\phi = \Phi + \left.\frac{\partial\tilde{\varphi}}{\partial\eta}\right|_{\eta=0} = \Phi + \sum_{k=1}^{\infty}[a_k(\xi)\sin(k\theta) + b_k(\xi)\cos(k\theta)]. \quad (5.69)$$

Note that if the length of the inlet duct is large, that is, $l_{\mathrm{I}} \to \infty$, the expression for α_k, $k = 1, 2, \ldots$, becomes

$$\alpha_k = \frac{m}{k} + \frac{1}{a}. \quad (5.70)$$

Applying a Galerkin formulation to (5.67) and using $\sin(k\theta)$ and $\cos(k\theta)$, $k = 1, \ldots, n_{\mathrm{m}}$, as projection functions, we obtain the n_{m}-mode model given by

$$\alpha_k \frac{\mathrm{d}a_k}{\mathrm{d}\xi} - \beta_k b_k = \hat{\psi}^{\sin}_{\mathrm{c},k}, \qquad k = 1, \ldots, n_{\mathrm{m}}, \quad (5.71)$$

$$\alpha_k \frac{\mathrm{d}b_k}{\mathrm{d}\xi} + \beta_k a_k = \hat{\psi}^{\cos}_{\mathrm{c},k}, \quad (5.72)$$

where

$$\hat{\psi}^{\sin}_{\mathrm{c},k}(\xi) \triangleq \frac{1}{\pi}\int_0^{2\pi} \psi_{\mathrm{c}}(\phi)\sin(k\theta)\,\mathrm{d}\theta, \qquad k = 1, \ldots, n_{\mathrm{m}}, \quad (5.73)$$

$$\hat{\psi}^{\cos}_{\mathrm{c},k}(\xi) \triangleq \frac{1}{\pi}\int_0^{2\pi} \psi_{\mathrm{c}}(\phi)\cos(k\theta)\,\mathrm{d}\theta, \quad (5.74)$$

and

$$\hat{\psi}_{\mathrm{c},0}(\xi) \triangleq \frac{1}{2\pi}\int_0^{2\pi} \psi_{\mathrm{c}}(\phi)\,\mathrm{d}\theta. \quad (5.75)$$

In the above formulation (5.71) and (5.72) were obtained using the fact that

$$0 = \int_0^{2\pi} \hat{\psi}_{\mathrm{c},0}(\phi)\sin(k\theta)\,\mathrm{d}\theta = \int_0^{2\pi} \hat{\psi}_{\mathrm{c},0}(\phi)\cos(k\theta)\,\mathrm{d}\theta, \qquad k = 1, \ldots, n_{\mathrm{m}}.$$

Next, combining (5.71) and (5.72) and re-writing (5.55) and (5.65) we obtain

$$\frac{\mathrm{d}\hat{x}}{\mathrm{d}\xi} = \hat{A}\hat{x} + \hat{f}(\hat{x}, \Phi), \tag{5.76}$$

$$\frac{\mathrm{d}\Phi}{\mathrm{d}\xi} = \frac{1}{l_{\mathrm{c}}}\left(\hat{\psi}_{\mathrm{c},0} - \Psi\right), \tag{5.77}$$

$$\frac{\mathrm{d}\Psi}{\mathrm{d}\xi} = \frac{1}{4B^2 l_{\mathrm{c}}}[\Phi - \gamma_{\mathrm{th}}\sqrt{\Psi}], \tag{5.78}$$

where

$$\hat{x} \triangleq \begin{bmatrix} a_1 \\ b_1 \\ \vdots \\ a_{n_{\mathrm{m}}} \\ b_{n_{\mathrm{m}}} \end{bmatrix}, \quad D_\alpha \triangleq \underset{k=1,\ldots,n_{\mathrm{m}}}{\text{block-diag}} \begin{bmatrix} \alpha_k & 0 \\ 0 & \alpha_k \end{bmatrix}, \quad \hat{f}(\Phi, \hat{x}) \triangleq D_\alpha^{-1} \begin{bmatrix} \hat{\psi}^{\sin}_{\mathrm{c},1} \\ \hat{\psi}^{\cos}_{\mathrm{c},1} \\ \vdots \\ \hat{\psi}^{\sin}_{\mathrm{c},n_{\mathrm{m}}} \\ \hat{\psi}^{\cos}_{\mathrm{c},n_{\mathrm{m}}} \end{bmatrix}, \tag{5.79}$$

$$\hat{A} \triangleq D_\alpha^{-1} \underset{k=1,\ldots,n_{\mathrm{m}}}{\text{block-diag}} \begin{bmatrix} 0 & \beta_k \\ -\beta_k & 0 \end{bmatrix}, \tag{5.80}$$

Equations (5.76)–(5.78) give a $2n_{\mathrm{m}} + 2$ order state space model for the compression problem with state variables $\hat{x}$, Φ, and Ψ and control variable γ_{th}.

The quasi-steady, axisymmetric compressor characteristic map $\psi_{\mathrm{C}}(\phi)$ considered in the literature [104] is

$$\psi_{\mathrm{C}}(\phi) = \psi_{\mathrm{C}_0} + H\left[1 + \tfrac{3}{2}\left(\frac{\phi}{W} - 1\right) - \tfrac{1}{2}\left(\frac{\phi}{W} - 1\right)^3\right], \tag{5.81}$$

where ψ_{C_0}, H, and W are parameters that can be used to shape the compressor characteristic map. In the case where $n_{\mathrm{m}} = 1$, it follows from (5.69) that

$$\phi = \Phi + a_1(\xi)\sin(\theta) + b_1(\xi)\cos(\theta). \tag{5.82}$$

In this case (5.73)-(5.75) yield

$$\begin{aligned}\hat{\psi}_{\mathrm{c},0} &= \frac{H}{2}\left[2\frac{\psi_{\mathrm{C}_0}}{H} - \frac{\Phi^3}{W^3} + 3\frac{\Phi^2}{W^2} + \tfrac{3}{2}\left(1 - \frac{\Phi}{W}\right)\frac{a_1^2 + b_1^2}{W^2}\right] \\ &= \psi_{\mathrm{C}}(\Phi) + 3H\left(1 - \frac{\Phi}{W}\right)\frac{a_1^2 + b_1^2}{4W^2},\end{aligned} \tag{5.83}$$

$$\hat{\psi}^{\sin}_{\mathrm{c},1} = \frac{3Ha_1}{2W}\left[1 - \left(1 - \frac{\Phi}{W}\right)^2 - \frac{a_1^2 + b_1^2}{4W^2}\right], \tag{5.84}$$

$$\hat{\psi}^{\cos}_{\mathrm{c},1} = \frac{3Hb_1}{2W}\left[1 - \left(1 - \frac{\Phi}{W}\right)^2 - \frac{a_1^2 + b_1^2}{4W^2}\right]. \tag{5.85}$$

Now, substituting (5.70) and (5.83)-(5.85) into (5.71), (5.72), (5.77), and (5.78), we obtain

$$\left(m+\frac{1}{a}\right)\frac{\mathrm{d}a_1}{\mathrm{d}\xi}-\frac{1}{2a}b_1=\frac{3Ha_1}{2W}\left[1-\left(1-\frac{\Phi}{W}\right)^2-\frac{a_1^2+b_1^2}{4W^2}\right], \tag{5.86}$$

$$\left(m+\frac{1}{a}\right)\frac{\mathrm{d}b_1}{\mathrm{d}\xi}+\frac{1}{2a}a_1=\frac{3Hb_1}{2W}\left[1-\left(1-\frac{\Phi}{W}\right)^2-\frac{a_1^2+b_1^2}{4W^2}\right], \tag{5.87}$$

$$\frac{\mathrm{d}\Phi}{\mathrm{d}\xi}=\frac{1}{l_{\mathrm{c}}}\left[\psi_{\mathrm{c}}(\Phi)-\Psi+3H\left(1-\frac{\Phi}{W}\right)\frac{a_1^2+b_1^2}{4W^2}\right], \tag{5.88}$$

$$\frac{\mathrm{d}\Psi}{\mathrm{d}\xi}=\frac{1}{4B^2l_{\mathrm{c}}}[\Phi(\xi)-\gamma_{\mathrm{th}}\sqrt{\Psi}]. \tag{5.89}$$

Introducing the new state variables

$$J\triangleq\frac{a_1^2+b_1^2}{4W^2},\qquad r_1\triangleq\tan^{-1}\frac{a_1}{b_1}, \tag{5.90}$$

(5.86) and (5.87) collapse to

$$\frac{\mathrm{d}J}{\mathrm{d}\xi}=J\left[1-\left(1-\frac{\Phi}{W}\right)^2-J\right]\frac{3aH}{W(1+ma)}, \tag{5.91}$$

$$\frac{\mathrm{d}r_1}{\mathrm{d}\xi}=\frac{1}{2(1+ma)}. \tag{5.92}$$

Equations (5.88), (5.89), and (5.91) form the standard three-state Moore-Greitzer model [104].

5.4 Finite Element Multi-Mode State Space Model

Since the state space model given in Section 5.3 requires the computation of $\hat{\psi}_{\mathrm{c},0}$, $\hat{\psi}^{\mathrm{sin}}_{\mathrm{c},k}$, and $\hat{\psi}^{\mathrm{cos}}_{\mathrm{c},k}$, $k=1,\ldots,n_{\mathrm{m}}$, which involve integrals of transcendental functions, in this section we give an alternative state space basis which eliminates this complexity. Our state transformation only involves the state variables Φ, a_k, and b_k, $k=1,\ldots,n_{\mathrm{m}}$, so that we need only consider the truncated state vector

$$x_{\mathrm{t}}\triangleq\begin{bmatrix}\hat{x}\\ \Phi\end{bmatrix}. \tag{5.93}$$

Specifically, we consider $n_{\mathrm{f}}\triangleq 2n_{\mathrm{m}}+1$ flow state variables given by

$$\phi_i\triangleq\Phi+\sum_{k=1}^{n_{\mathrm{m}}}[a_k(\xi)\sin(k\theta_i)+b_k(\xi)\cos(k\theta_i)], \tag{5.94}$$

where

$$\theta_i \triangleq \frac{2\pi i}{n_{\mathrm{f}}}, \qquad i = 1, \ldots, n_{\mathrm{f}}. \tag{5.95}$$

Hence, define the flow state vector $\hat{\phi}$ by

$$\hat{\phi} \triangleq \begin{bmatrix} \phi_1 \\ \phi_2 \\ \vdots \\ \phi_{n_{\mathrm{f}}} \end{bmatrix}, \tag{5.96}$$

so that $\hat{\phi} = Sx_{\mathrm{t}}$, where

$$S = \begin{bmatrix} \sin(\theta_1) & \cos(\theta_1) & \ldots & \sin(n_{\mathrm{m}}\theta_1) & \cos(n_{\mathrm{m}}\theta_1) & 1 \\ \sin(\theta_2) & \cos(\theta_2) & \ldots & \sin(n_{\mathrm{m}}\theta_2) & \cos(n_{\mathrm{m}}\theta_2) & 1 \\ \vdots & \vdots & \ddots & \vdots & \vdots & \vdots \\ \sin(\theta_{n_{\mathrm{f}}}) & \cos(\theta_{n_{\mathrm{f}}}) & \ldots & \sin(n_{\mathrm{m}}\theta_{n_{\mathrm{f}}}) & \cos(n_{\mathrm{m}}\theta_{n_{\mathrm{f}}}) & 1 \end{bmatrix}. \tag{5.97}$$

Now, applying this change of variables, the new state space description for (5.76) and (5.77) is

$$\frac{\mathrm{d}\hat{\phi}}{\mathrm{d}\xi} = A_s\hat{\phi} + D_s^{-1}\left(\begin{bmatrix} \hat{\psi}_{\mathrm{c}}(\theta_1) \\ \vdots \\ \hat{\psi}_{\mathrm{c}}(\theta_{n_{\mathrm{f}}}) \end{bmatrix} - \begin{bmatrix} \Psi \\ \vdots \\ \Psi \end{bmatrix}\right), \tag{5.98}$$

where

$$A_s \triangleq S\begin{bmatrix} \hat{A} & 0_{2n_{\mathrm{m}}\times 1} \\ 0_{1\times 2n_{\mathrm{m}}} & 0 \end{bmatrix}S^{-1}, \qquad D_s \triangleq S\begin{bmatrix} D_\alpha & 0_{2n_{\mathrm{m}}\times 1} \\ 0_{1\times 2n_{\mathrm{m}}} & l_{\mathrm{c}} \end{bmatrix}S^{-1}, \tag{5.99}$$

and

$$\hat{\psi}_{\mathrm{c}}(\theta_i) = \hat{\psi}_{\mathrm{c},0} + \sum_{k=1}^{n_{\mathrm{m}}}[\hat{\psi}_{\mathrm{c},k}^{\mathrm{sin}}\sin(k\theta_i) + \hat{\psi}_{\mathrm{c},k}^{\mathrm{cos}}\cos(k\theta_i)]. \tag{5.100}$$

It is interesting to note that since (5.100) is a truncated Fourier expansion of $\psi_{\mathrm{c}}(\phi_i)$ we can introduce an approximation in (5.98) by replacing $\hat{\psi}_{\mathrm{c}}(\theta_i)$ with $\psi_{\mathrm{c}}(\phi_i)$. Hence, including the equation for the state variable Ψ, the new state space model becomes

$$\frac{\mathrm{d}\hat{\phi}}{\mathrm{d}\xi} = A_s\hat{\phi} + D_s^{-1}\psi_{\mathrm{c}}(\hat{\phi}) - e\Psi, \tag{5.101}$$

$$\frac{\mathrm{d}\Psi}{\mathrm{d}\xi} = \frac{1}{4B^2 l_{\mathrm{c}}}\left(\frac{e^{\mathrm{T}}\hat{\phi}}{n_{\mathrm{f}}} - \gamma_{\mathrm{th}}\sqrt{\Psi}\right), \tag{5.102}$$

where

$$e \triangleq \begin{bmatrix} 1 \\ 1 \\ \vdots \\ 1 \end{bmatrix} \in \mathbb{R}^{n_\gamma \times 1}, \qquad \psi_C(\hat{\phi}) \triangleq \begin{bmatrix} \psi_C(\phi_1) \\ \vdots \\ \psi_C(\phi_{n_\gamma}) \end{bmatrix}. \tag{5.103}$$

Note that $\frac{e^T\hat{\phi}}{n_\gamma} = \Phi$, A_s is skew symmetric, and D_s is non-singular with positive eigenvalues. Furthermore, e is an eigenvector of A_s and D_s associated with the eigenvalues 0 and l_C, respectively. This state space representation is similar to that obtained in [98] using discrete Fourier transforms.

5.5 Control for Single-Mode versus Multi-Mode Model

As noted in Section 5.1, a fundamental shortcoming of the low-order three-state Moore-Greitzer model and, as a consequence, the control design methodologies based on the model, is the fact that only a one-mode expansion of the disturbance velocity potential in the compression system is considered. In this section we use the multi-mode state space model obtained in Section 5.3 to show that the second and higher-order disturbance velocity potential harmonics strongly interact with the first harmonic during compressor stall inception and hence must be accounted for in the control design processes. Specifically, using the *globally* stabilizing recursive backstepping controller predicated on the one-mode Moore-Greitzer model given in [77], we show that in the two-mode case the same controller drives the system to a stalled condition.

The backstepping controller obtained in [77] is given by, in our notation,

$$\begin{aligned} u = c_2 &\left[\frac{\Psi - \psi_{C_0}}{H} - 2 - c_1\left(\frac{\Phi}{W} - 2\right) + \tfrac{3}{2}\left(\frac{\Phi}{W} - 2\right)^2 + 3\frac{a_1^2 + b_1^2}{4W^2}\right] \\ &- \left(\frac{\Phi}{W} - 2\right) - \left[c_1 - 3\left(\frac{\Phi}{W} - 2\right)\right]\left[-\frac{\Psi - \psi_{C_0}}{H} + 2\right. \\ &\left. -\tfrac{3}{2}\left(\frac{\Phi}{W} - 2\right)^2 - \tfrac{1}{2}\left(\frac{\Phi}{W} - 2\right)^3 - 3\left(\frac{\Phi}{W} - 1\right)\frac{a_1^2 + b_1^2}{4W^2}\right] \\ &- \frac{9al_C}{1 + ma}\frac{a_1^2 + b_1^2}{4W^2}\left[2\left(\frac{\Phi}{W} - 2\right) + \left(\frac{\Phi}{W} - 2\right)^2 + \frac{a_1^2 + b_1^2}{4W^2}\right], \end{aligned} \tag{5.104}$$

where $c_1 \geq 0$ and $c_2 > 0$ are arbitrary constants. This control law guarantees global asymptotic stability of the equilibrium $a_1^e = 0$, $b_1^e = 0$ ($J^e = 0$), $\Phi^e = 2W$, and $\Psi^e = \psi_C(\Phi^e) = \psi_{C_0} + 2H$.

Using the parameter values $c_1 = c_2 = 1$, $a = 1/3$, $l_c = 6$, $m = 2$, $H = 0.32$, $W = 0.18$, $\psi_{c_0} = 0.23$, and $B = 0.1$ with initial conditions $a_{10} = b_{10} = 0.1$, $\Phi_0 = 0.36$, and $\Psi_0 = 0.87$ corresponding to perturbation in the first mode disturbance velocity potential, Figure 5.3 shows the controlled responses for the squared stall cell amplitude (J), the compressor flow (Φ), and the pressure rise (Ψ). Note that the backstepping controller (5.104) drives the system to the desired equilibrium point $(J^e, \Phi^e, \Psi^e) = (0, 0.36, 0.87)$.

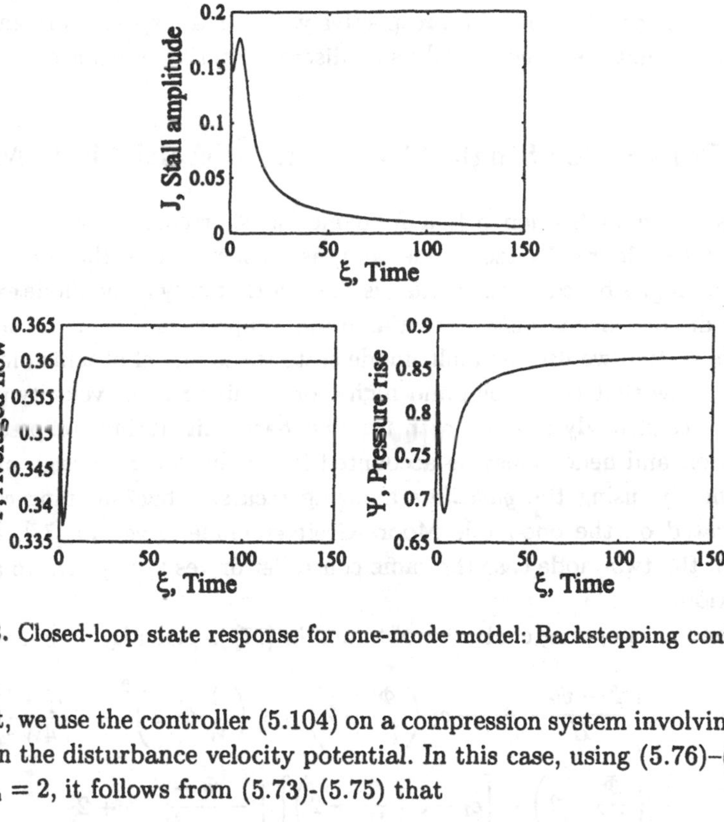

Fig. 5.3. Closed-loop state response for one-mode model: Backstepping controller

Next, we use the controller (5.104) on a compression system involving two modes in the disturbance velocity potential. In this case, using (5.76)–(5.78) with $n_{\mathrm{m}} = 2$, it follows from (5.73)-(5.75) that

$$\hat{\psi}_{c,0} = \psi_c(\Phi) + 3H\left(1 - \frac{\Phi}{W}\right)(J_1 + J_2) + \frac{3H}{4W^3}\left(b_2\frac{a_1^2 - b_1^2}{2} - a_1 a_2 b_1\right),$$

$$\hat{\psi}_{c,1}^{\sin} = a_1 Y(a_1, b_1, a_2, b_2, \Phi) + \frac{3H}{2}\left[\frac{a_2 b_1 - a_1 b_2}{W^2}\left(1 - \frac{\Phi}{W}\right) - \frac{a_1}{W}J_2\right],$$

$$\hat{\psi}_{c,1}^{\cos} = b_1 Y(a_1, b_1, a_2, b_2, \Phi) + \frac{3H}{2}\left[\frac{a_1 a_2 + b_1 b_2}{W^2}\left(1 - \frac{\Phi}{W}\right) - \frac{b_1}{W}J_2\right],$$

$$\hat{\psi}_{c,2}^{\sin} = a_2 Y(a_1, b_1, a_2, b_2, \Phi) + \frac{3H}{2}\left[\frac{a_1 b_1}{W^2}\left(1 - \frac{\Phi}{W}\right) - \frac{a_2}{W}J_1\right],$$

$$\hat{\psi}_{c,2}^{\cos} = b_2 Y(a_1, b_1, a_2, b_2, \Phi) + \frac{3H}{2}\left[\frac{b_1^2 - a_1^2}{2W^2}\left(1 - \frac{\Phi}{W}\right) - \frac{b_2}{W}J_1\right],$$

where

$$Y(a_1, b_1, a_2, b_2, \Phi) \triangleq \frac{3H}{2W}\left[1 - \left(1 - \frac{\Phi}{W}\right)^2 - J_1 - J_2\right].$$

$$J_1 \triangleq \frac{a_1^2 + b_1^2}{4W^2}, \qquad J_2 \triangleq \frac{a_2^2 + b_2^2}{4W^2}.$$

Using the initial conditions $a_{10} = b_{10} = 0.1$, $a_{20} = b_{20} = 0$, $\Phi_0 = 0.36$, and $\Psi_0 = 0.87$, corresponding to a perturbation in the first-mode disturbance velocity potential the controller (5.104) drives the system to the stalled state $(J_1^e, J_2^e, \Phi^e, \Psi^e) = (0.357, 0.105, 0.319, 0.481)$ (see Figure 5.4). This clearly shows that a multi-mode model that accounts for the higher mode interactions with the first mode is necessary for achieving control objectives during stall inception.

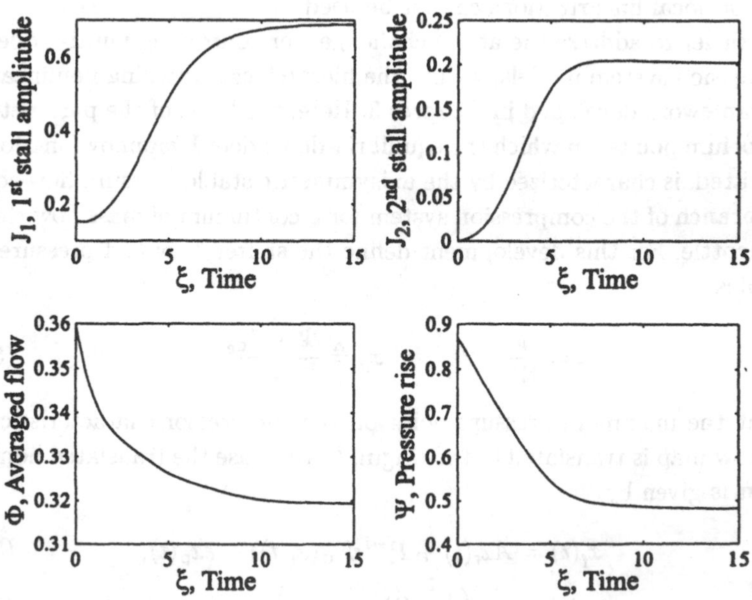

Fig. 5.4. Closed-loop state response for two-mode model: Backstepping controller

5.6 Stabilization of Multi-Mode Axial Flow Compression System Models

In this section we apply the hierarchical switching nonlinear control framework developed in Chapter 3 to the control of rotating stall and surge in jet engine compression systems using the finite element multi-mode state space model developed in Section 5.4. It is important to note that unlike the one-mode Moore-Greitzer model [77], the multi-mode model given by (5.101) and (5.102) cannot be represented in strict-feedback form and hence backstepping and inverse optimal designs are not applicable. Furthermore, (5.101) and (5.102) possess unstable zero dynamics and hence are not feedback linearizable. In addition, since by definition multi-mode models are higher dimensional models, Hamilton-Jacobi-Bellman solutions are intractable and hence optimal controllers cannot be easily designed. Finally, linearizing the multi-mode compressor model about any operating point on the pressure-flow equilibrium branch (including the maximum pressure point) results in an unstabilizable linear system and hence linear-quadratic stabilization schemes based on local linearizations cannot be used.

In order to address the above challenges for controlling multi-mode axial compression system models, we use the hierarchical switching nonlinear control framework developed in Chapter 3. Here, the locus of the parameterized equilibrium points, on which the equilibria-dependent Lyapunov functions are predicated, is characterized by the axisymmetric stable pressure-flow equilibrium branch of the compression system for a continuum of mass flow through the throttle. For this development define the shifted flow and pressure state variables

$$x_{\rm f} \triangleq \frac{\hat{\phi}}{W} - 2e, \qquad x_{\rm p} \triangleq \frac{\Psi - \psi_{\rm C_0}}{H} - 2, \tag{5.105}$$

so that the maximum pressure point on the compressor characteristic pressure-flow map is translated to the origin. In this case the translated nonlinear system is given by

$$\dot{x}_{\rm f}(t) = Ax_{\rm f}(t) + P^{-1}\psi_{\rm sc}(x_{\rm f}(t)) - ex_{\rm p}(t), \tag{5.106}$$

$$\dot{x}_{\rm p}(t) = \frac{1}{\beta^2}\left(\frac{e^{\rm T}x_{\rm f}(t)}{n_{\rm f}} - u(t)\right), \tag{5.107}$$

where

$$A \triangleq \frac{Wl_{\rm c}}{H}A_{\rm s}, \qquad P \triangleq \frac{1}{l_{\rm c}}D_{\rm s}, \qquad \beta \triangleq \frac{2BH}{W}, \qquad u \triangleq \frac{\gamma_{\rm th}\sqrt{\Psi}}{W} - 2, \tag{5.108}$$

$$\psi_{\rm sc}(x_{\rm f}) \triangleq \left[\psi_{\rm sc}(x_{\rm f1}) \cdots \psi_{\rm sc}(x_{{\rm f}n_{\rm f}})\right]^{\rm T}, \qquad \psi_{\rm sc}(x_{{\rm f}i}) \triangleq -\tfrac{3}{2}x_{{\rm f}i}^2 - \tfrac{1}{2}x_{{\rm f}i}^3, \tag{5.109}$$

and $(\dot{\ })$ represents differentiation with respect to nondimensional scaled time $t \triangleq \frac{H}{Wl_C}\xi$. To carry out Step 1 of Algorithm 3.1, let $q = m = 1$ and $\varphi(x_f, x_p, \lambda) = \lambda$ so that the system equilibria are parameterized by the constant control $u(t) = \lambda$. In this case, (5.106), (5.107) have an equilibrium point at $(x_{f_\lambda}, x_{p_\lambda})$, where

$$x_{f_\lambda} \triangleq \lambda e, \qquad x_{p_\lambda} \triangleq \psi_{sc}(\lambda) = -\tfrac{3}{2}\lambda^2 - \tfrac{1}{2}\lambda^3. \tag{5.110}$$

Next, we carry out Step 2 of Algorithm 3.1. Specifically, we show that for $\lambda > 0$ there exists a control law such that the equilibrium point $(x_{f_\lambda}, x_{p_\lambda})$ of (5.106), (5.107) is locally asymptotically stable with an estimate of the domain of attraction given by $\mathcal{D}_\lambda$. To show this, consider the equilibrium-dependent Lyapunov function candidate

$$V_\lambda(x_f, x_p) = \tfrac{1}{2n_f}(x_f - x_{f_\lambda})^{\mathrm{T}} P (x_f - x_{f_\lambda}) + \tfrac{1}{2}\beta^2[x_p - x_{p_\lambda}]^2, \tag{5.111}$$

with Lyapunov derivative

$$\begin{aligned}\dot{V}_\lambda(x_f, x_p) &= \tfrac{1}{n_f}(x_f - \lambda e)^{\mathrm{T}} P (A x_f + P^{-1}\psi_{sc}(x_f) - e x_p) \\ &\quad + [x_p - \psi_{sc}(\lambda)]\left(\frac{e^{\mathrm{T}} x_f}{n_f} - u\right) \\ &= -\tfrac{1}{n_f}\sum_{i=1}^{n_f}(x_{f_i} - \lambda)[\psi_{sc}(\lambda) - \psi_{sc}(x_{f_i})] - (u - \lambda)[x_p - \psi_{sc}(\lambda)].\end{aligned} \tag{5.112}$$

Substituting (5.109) into (5.112) and choosing the *feedback* control law $u = \phi_\lambda(x_f, x_p) \triangleq \lambda + h_\lambda(x_f, x_p)$, where $h_\lambda : \mathbb{R}^{n_f} \times \mathbb{R} \to \mathbb{R}$ is such that $h_\lambda(x_{f_\lambda}, x_{p_\lambda}) = 0$, we obtain

$$\begin{aligned}\dot{V}_\lambda(x_f, x_p) &= -\tfrac{1}{2n_f}\sum_{i=1}^{n_f}(x_{f_i} - \lambda)^2[x_{f_i}^2 + (\lambda + 3)x_{f_i} + \lambda(\lambda + 3)] \\ &\quad - h_\lambda(x_f, x_p)[x_p - \psi_{sc}(\lambda)].\end{aligned} \tag{5.113}$$

Now, a sufficient condition guaranteeing that $\dot{V}_\lambda(x_f, x_p) < 0$ is given by

$$x_{f_i}^2 + (\lambda + 3)x_{f_i} + \lambda(\lambda + 3) > 0, \qquad i = 1, \ldots, n_f, \tag{5.114}$$

$$h_\lambda(x_f, x_p)[x_p - \psi_{sc}(\lambda)] > 0, \qquad (x_f, x_p) \neq (x_{f_\lambda}, x_{p_\lambda}). \tag{5.115}$$

Note that (5.114) holds for all x_{f_i}, $i = 1, \ldots, n_f$, when $\lambda > 1$. If $0 < \lambda \leq 1$ then (5.114) holds for

$$x_{f_i} - \lambda > -d_\lambda, \quad d_\lambda \triangleq \frac{3(\lambda + 1) - \sqrt{3(\lambda + 3)(1 - \lambda)}}{2}, \quad i = 1, \ldots, n_f, \tag{5.116}$$

while a particular choice of $h_\lambda(\cdot,\cdot)$ satisfying (5.115) is given by

$$h_\lambda(x_{\rm f}, x_{\rm p}) \triangleq x_{\rm p} - \psi_{\rm sc}(\lambda). \tag{5.117}$$

In this case $\dot{V}_\lambda(x_{\rm f}, x_{\rm p}) < 0$, $(x_{\rm f}, x_{\rm p}) \in \mathbb{R}^{n_{\rm f}+1} \setminus (x_{{\rm f}\lambda}, x_{{\rm p}\lambda})$, and hence the equilibrium point $(x_{{\rm f}\lambda}, x_{{\rm p}\lambda})$ of (5.106), (5.107) with $u = \phi_\lambda(x_{\rm f}, x_{\rm p})$ is locally asymptotically stable for $\lambda \in (0,1]$ and globally asymptotically stable for $\lambda > 1$. An estimate of the domain of attraction for (5.106), (5.107) with $u = \phi_\lambda(x_{\rm f}, x_{\rm p})$ is given by

$$\mathcal{D}_\lambda = \begin{cases} \{(x_{\rm f}, x_{\rm p}) : \quad V_\lambda(x_{\rm f}, x_{\rm p}) \le c_\lambda\}, & 0 \le \lambda \le 1, \\ \mathbb{R}^{n_{\rm f}} \times \mathbb{R}, & \lambda > 1, \end{cases} \tag{5.118}$$

where $c_\lambda = \frac{\mu}{2n_{\rm f}} d_\lambda^2$ and $\mu \triangleq (\max_i\{P_{ii}^{-1}\})^{-1}$. The contour level surfaces $V_\lambda(x_{\rm f}, x_{\rm p}) = c_\lambda$ are defined such that the intersection of the boundary of $\mathcal{D}_\lambda$ with the plane $x_{\rm p} = x_{{\rm p}\lambda}$ is a closed surface contained in the region $\{x_{\rm f} : -d_\lambda < x_{{\rm f}_i} - \lambda < d_\lambda,\ i = 1, \ldots, n_{\rm f}\}$ so that $\dot{V}_\lambda(x_{\rm f}, x_{\rm p}) < 0$ for all $(x_{\rm f}, x_{\rm p}) \in \mathcal{D}_\lambda \setminus (x_{{\rm f}\lambda}, x_{{\rm p}\lambda})$. Note that $c_\lambda > 0$ for $\lambda > 0$.

To carry out Step 3 of Algorithm 3.1, we consider two topologies for the switching set $\mathcal{S}$; namely an isolated point topology and a hybrid topology. For $\mathcal{S}$ consisting of countably finite isolated points let $\mathcal{S} = \{\lambda_0, \ldots, \lambda_q\}$ be such that $0 < \lambda_q < \cdots < \lambda_1 \le 1$, $\lambda_0 > 1$, and $(x_{{\rm f}\lambda_{i+1}}, x_{{\rm p}\lambda_{i+1}}) \in \mathcal{D}_{\lambda_i}$, $i \in \{0, \ldots, q-1\}$, and let $p(\lambda) = \lambda$, $\lambda \in \mathcal{S}$. To guarantee that $p(\cdot)$ satisfies Assumption 3.1 construct λ_k, $k = 0, 1, \ldots, q$, online by considering the smallest solution to the equation $V_{\lambda_k}(x(t_k)) = c_{\lambda_k}$, $t_k \triangleq k\Delta T$, where $\Delta T > 0$ and $k = 0, 1, \ldots, q$, and define $\mathcal{S} \triangleq \{\lambda_k\}_{k=0}^q$. Now, with the feedback switching control law $u = \phi_{\lambda_\mathcal{S}(x_{\rm f}, x_{\rm p})}(x_{\rm f}, x_{\rm p})$, where $\lambda_\mathcal{S}(x_{\rm f}, x_{\rm p})$ is obtained as described in Step 4 of Algorithm 3.1, it follows from Theorem 3.5 that the equilibrium point $(x_{{\rm f}\lambda_q}, x_{{\rm p}\lambda_q})$ is globally asymptotically stable. In particular, choosing $\lambda_q > 0$ arbitrarily small, global asymptotic stability of an equilibrium point on the compressor characteristic pressure-flow map arbitrarily close to the maximum pressure point is established. Furthermore, note that $\lambda_\mathcal{S}(x(t))$, $t \ge 0$, is piecewise constant and hence the feedback switching control law $u = \phi_{\lambda_\mathcal{S}(x_{\rm f}, x_{\rm p})}(x_{\rm f}, x_{\rm p})$ is piecewise continuous.

For $\mathcal{S}$ consisting of a hybrid topology let $\mathcal{S} = [0,1] \cup \{\bar{\lambda}\}$, where $\bar{\lambda} > 1$ is such that $(x_{{\rm f}\hat{\lambda}}, x_{{\rm p}\hat{\lambda}}) \in \overset{\circ}{\mathcal{D}}_{\bar{\lambda}}$, for at least one $\hat{\lambda} \in [0,1]$, and let $p(\lambda) = \lambda$, $\lambda \in \mathcal{S}$. Since $p(\cdot)$ does not have a local minimum in $\mathcal{S}$ (other than the origin) and every $\lambda \in [0,1]$ is an accumulation point for $\mathcal{S}$, we are guaranteed, by Step 3b of Algorithm 3.1, that Assumption 3.1 is satisfied. Now global attraction of the origin of the nonlinear dynamical system (5.106), (5.107) is guaranteed by Theorem 3.5 with the feedback control law $u = \phi_{\lambda_\mathcal{S}(x_{\rm f}, x_{\rm p})}(x_{\rm f}, x_{\rm p})$, where $\lambda_\mathcal{S}(x_{\rm f}, x_{\rm p})$ is obtained as described in Step 4 of Algorithm 3.1. In particular, if

$(x_f(0), x_p(0)) \in \hat{\mathcal{D}} \triangleq \cup_{\lambda \in [0,1]} \mathcal{D}_\lambda$ then $\lambda_S(x(t))$, $t \geq 0$, is a continuous function. Alternatively, if $(x_f(0), x_p(0)) \notin \hat{\mathcal{D}}$ then $\lambda_S(x(t)) = \bar{\lambda}$, $t \in [0, \hat{t})$, where $\hat{t} > 0$ is such that $(x_f(\hat{t}), x_p(\hat{t})) \in \partial\hat{\mathcal{D}}$. In this case, $\lambda_S(x(t))$, $t \geq 0$, is continuous modulo one discontinuity at $t = \hat{t}$. Note that since $c_0 = 0$ and the origin is on the boundary of $\hat{\mathcal{D}}$, the origin is a global attractor but not Lyapunov stable.

Next, if $(x_f(0), x_p(0)) \in \hat{\mathcal{D}}$, the on-line fixed-order dynamic compensation procedure given in Section 3.5 can be employed to compute $\lambda_S(x(t))$, $t \geq 0$, using the update law (3.21). Specifically, in this case $s = \lambda$, $w_\lambda = 1$, and $Q_\lambda(x) = \frac{1}{v_\lambda(x)} V'_\lambda(x)$. Now, using (3.17), (3.21), (5.111), and (5.116), we obtain

$$\dot{\lambda} = \frac{\dot{V}_\lambda(x_f, x_p)}{\frac{\mu}{n_f} d_\lambda \frac{dd_\lambda}{d\lambda} - \lambda + \frac{e^T x_f}{n_f} - 3\lambda\beta^2(x_p + \frac{3}{2}\lambda^2 + \frac{1}{2}\lambda^3)(1 + \frac{1}{2}\lambda)}, \tag{5.119}$$

where $\dot{V}_\lambda(x_f, x_p)$ is given by (5.113) and $\lambda(0) = \lambda_S(x_f(0), x_p(0))$. Note that the compensator dynamics given by (5.119) characterize the admissible rate of the compensator state $\lambda(t)$ such that the switching nonlinear controller guarantees that $(x_f(t), x_p(t)) \in \partial\mathcal{D}_{\lambda(t)}$, $t \geq 0$.

It is important to note that the proposed switching nonlinear controller framework can be incorporated to address practical actuator limitations such as control amplitude and rate saturation constraints. Specifically, since the dynamic compensator state $\lambda(t)$ is proportional to the throttle opening (actuator) and since the dynamics given by (5.119) indirectly characterize the fastest admissible rate at which the control throttle can open while maintaining stability of the controlled system, it follows that by constraining the rate at which the dynamics of $\lambda(t)$ can evolve on the equilibrium branch effectively places a rate constraint on the throttle opening. This corresponds to the case where the switching rate of the nonlinear controller is decreased so that the trajectory $(x_f(t), x_p(t))$, $t \geq 0$, is allowed to enter $\mathcal{D}_{\lambda(t)}$. Additionally, amplitude saturation constraints and state constraints can also be enforced by simply choosing $\lambda_{\max} < 1$ such that $\mathcal{D}_{\max} \triangleq \cup_{0 \leq \lambda \leq \lambda_{\max}} \mathcal{D}_\lambda$ is contained in the region where the system is constrained to operate. In this case, the switching nonlinear controller guarantees attraction of the origin with an estimate of the domain of attraction given by $\mathcal{D}_{\max}$. Of course, in practice it is sufficient to implement controllers with adequate domains of attraction and *a priori* saturation constraint guarantees rather than implementing global controllers without realistic actuator limitations.

To show the efficacy of the proposed control approach, we consider a two-mode compressor model so that the state-space model given by (5.101) and (5.102), or, equivalently, (5.106) and (5.107) is of sixth order. Using the same parameter values and the initial condition introduced in Section 5.5, the dynamic compensation controller discussed above was used to mitigate the

aerodynamic instabilities of rotating stall and surge in the two-mode axial compressor model. Figure 5.5 shows the squared stall cell first and second mode amplitudes versus time.

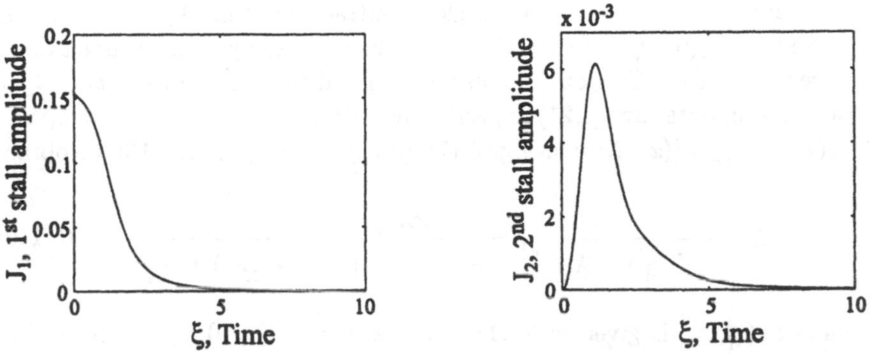

Fig. 5.5. Closed-loop state response for two-mode model: Switching nonlinear controller

Finally, to reflect a more realistic design we impose a rate saturation constraint on the system actuator throttle opening. In particular, we assume that the system actuator throttle opening has a rate constraint of $|\dot{\gamma}_{th}| < 1$. To illustrate the behavior of the closed-loop system with the switching dynamic controller designed to guarantee stability in the face of actuator rate saturation consider the initial condition $\hat{\phi}_0 = S\,[0.15\ 0.15\ 0\ 0\ 0.5]$, $\Psi_0 = 0.3$. Figures 5.6 and 5.7 show the controlled stall cell first and second mode amplitudes versus time and the control throttle opening amplitude and rate versus time. These figures show that the proposed rate saturation switching dynamic controller guarantees stability with *no* degradation in system performance.

5.7 Robust Stabilization of Axial Flow Compressors with Uncertain Pressure-Flow Maps

In this section we address the problem of nonlinear robust control for rotating stall and surge in axial flow compressors with uncertain performance characteristic pressure-flow maps. As shown in [57], feedback controllers that do not account for the presence of uncertainty in the compressor-flow map can have adverse effects on compression system performance by driving the compression system to a stalled equilibrium or a surge limit cycle. Hence, it is of paramount importance that modeling pressure-flow map system uncertainty be accounted for in the control-system design process. System modeling errors such as uncertainty in the compressor performance pressure-flow

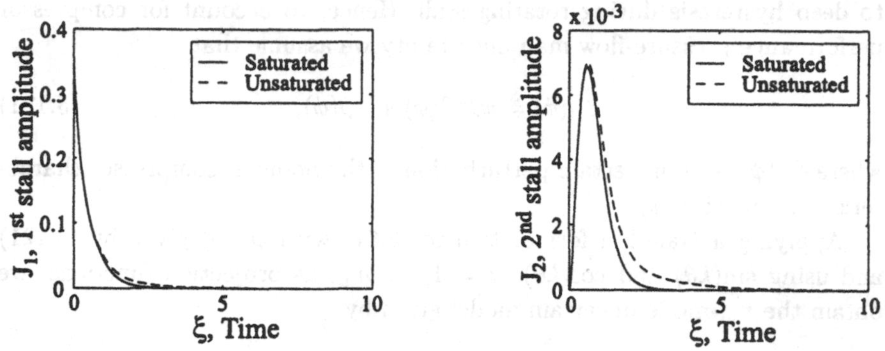

Fig. 5.6. Closed-loop state response for two-mode model: Rate saturated versus rate unsaturated control

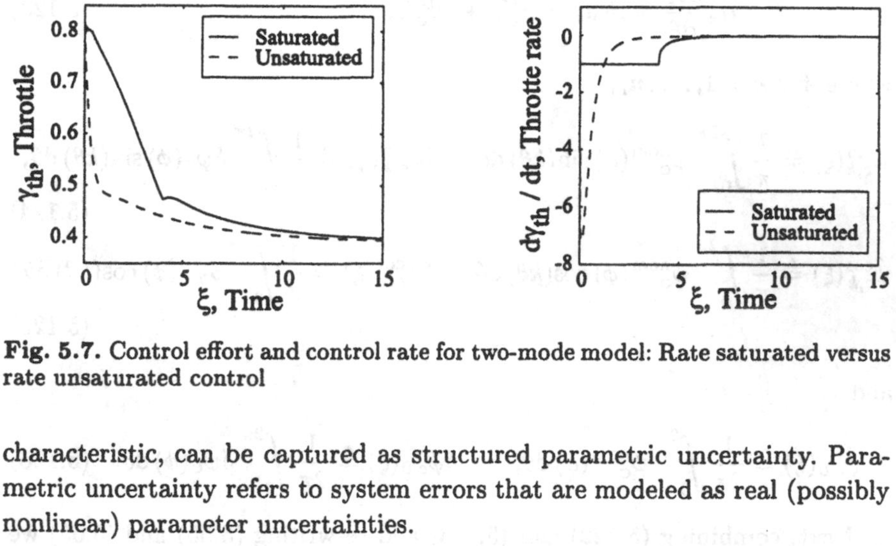

Fig. 5.7. Control effort and control rate for two-mode model: Rate saturated versus rate unsaturated control

characteristic, can be captured as structured parametric uncertainty. Parametric uncertainty refers to system errors that are modeled as real (possibly nonlinear) parameter uncertainties.

5.7.1 Uncertain Finite Element Multi-Mode State Space Model

In this subsection we extend the multi-mode model for rotating stall and surge developed in Sections 5.3 and 5.4 to include uncertainty in the compressor characteristic pressure-flow map. As discussed in Section 5.3, the standard nominal model considered in the literature [104] for the compressor pressure-flow characteristic map $\psi_{\rm C}^{\rm nom}(\phi)$ is a cubic function given by (5.81)

$$\psi_{\rm C}^{\rm nom}(\phi) = \psi_{{\rm C}_0} + H\left[1 + \tfrac{3}{2}\left(\frac{\phi}{W} - 1\right) - \tfrac{1}{2}\left(\frac{\phi}{W} - 1\right)^3\right]. \qquad (5.120)$$

In actual compressor data [98, 20] however, the compressor characteristic map exhibits a non-cubic morphology that can drive the compression system

to deep hysteresis during rotating stall. Hence, to account for compressor performance pressure-flow map uncertainty we assume that

$$\psi_{\mathrm{C}}(\phi) \triangleq \psi_{\mathrm{C}}^{\mathrm{nom}}(\phi) + \delta\psi(\phi), \tag{5.121}$$

where $\delta\psi(\phi)$ is an uncertain perturbation of the nominal compressor characteristic map $\psi_{\mathrm{C}}^{\mathrm{nom}}(\phi)$.

Applying a Galerkin formulation to (5.67) with $\psi_{\mathrm{C}}(\phi)$ given by (5.121) and using $\sin(k\theta)$ and $\cos(k\theta)$, $k = 1, \ldots, n_{\mathrm{m}}$, as projection functions, we obtain the n_{m}-mode uncertain model given by

$$\alpha_k \frac{\mathrm{d}a_k}{\mathrm{d}\xi} - \beta_k b_k = \hat{\psi}_{\mathrm{c},k}^{\sin} + \delta\psi_{\mathrm{c},k}^{\sin}, \qquad k = 1, \ldots, n_{\mathrm{m}}, \tag{5.122}$$

$$\alpha_k \frac{\mathrm{d}b_k}{\mathrm{d}\xi} + \beta_k a_k = \hat{\psi}_{\mathrm{c},k}^{\cos} + \delta\psi_{\mathrm{c},k}^{\cos}, \tag{5.123}$$

where, for $k = 1, \ldots, n_{\mathrm{m}}$,

$$\hat{\psi}_{\mathrm{c},k}^{\sin}(\xi) \triangleq \frac{1}{\pi}\int_0^{2\pi} \psi_{\mathrm{C}}^{\mathrm{nom}}(\phi)\sin(k\theta)\,\mathrm{d}\theta, \quad \delta\psi_{\mathrm{c},k}^{\sin}(\xi) \triangleq \frac{1}{\pi}\int_0^{2\pi} \delta\psi_{\mathrm{C}}(\phi)\sin(k\theta)\,\mathrm{d}\theta, \tag{5.124}$$

$$\hat{\psi}_{\mathrm{c},k}^{\cos}(\xi) \triangleq \frac{1}{\pi}\int_0^{2\pi} \psi_{\mathrm{C}}^{\mathrm{nom}}(\phi)\cos(k\theta)\,\mathrm{d}\theta, \quad \delta\psi_{\mathrm{c},k}^{\cos}(\xi) \triangleq \frac{1}{\pi}\int_0^{2\pi} \delta\psi_{\mathrm{C}}(\phi)\cos(k\theta)\,\mathrm{d}\theta, \tag{5.125}$$

and

$$\hat{\psi}_{\mathrm{c},0}(\xi) \triangleq \frac{1}{2\pi}\int_0^{2\pi} \psi_{\mathrm{C}}^{\mathrm{nom}}(\phi)\,\mathrm{d}\theta, \qquad \delta\psi_{\mathrm{c},0}(\xi) \triangleq \frac{1}{2\pi}\int_0^{2\pi} \delta\psi_{\mathrm{C}}(\phi)\,\mathrm{d}\theta. \tag{5.126}$$

Next, combining (5.122) and (5.123) and re-writing (5.55) and (5.65) we obtain

$$\frac{\mathrm{d}\hat{x}}{\mathrm{d}\xi} = \hat{A}\hat{x} + \hat{f}(\hat{x}, \Phi) + \Delta\hat{f}(\hat{x}, \Phi), \tag{5.127}$$

$$\frac{\mathrm{d}\Phi}{\mathrm{d}\xi} = \frac{1}{l_{\mathrm{C}}}\left(\hat{\psi}_{\mathrm{c},0} + \delta\psi_{\mathrm{c},0} - \Psi\right), \tag{5.128}$$

$$\frac{\mathrm{d}\Psi}{\mathrm{d}\xi} = \frac{1}{4B^2 l_{\mathrm{C}}}[\Phi - \gamma_{\mathrm{th}}\sqrt{\Psi}], \tag{5.129}$$

where

$$\hat{f}(\Phi, \hat{x}) \triangleq D_\alpha^{-1}\left[\hat{\psi}_{\mathrm{c},1}^{\sin}\ \hat{\psi}_{\mathrm{c},1}^{\cos} \cdots \hat{\psi}_{\mathrm{c},n_{\mathrm{m}}}^{\sin}\ \hat{\psi}_{\mathrm{c},n_{\mathrm{m}}}^{\cos}\right]^{\mathrm{T}}, \tag{5.130}$$

$$\Delta\hat{f}(\Phi, \hat{x}) \triangleq D_\alpha^{-1}\left[\delta\psi_{\mathrm{c},1}^{\sin}\ \delta\psi_{\mathrm{c},1}^{\cos} \cdots \delta\psi_{\mathrm{c},n_{\mathrm{m}}}^{\sin}\ \delta\psi_{\mathrm{c},n_{\mathrm{m}}}^{\cos}\right]^{\mathrm{T}}. \tag{5.131}$$

Equations (5.127)–(5.129) give a $2n_{\mathrm{m}}+2$ order state space model for the compression problem with state variables $\hat{x}$, Φ, and Ψ and control variable γ_{th}. In the case where $n_{\mathrm{m}} = 1$ and $\psi_{\mathrm{C}}(\phi) = \psi_{\mathrm{C}}^{\mathrm{nom}}(\phi)$ (i.e., $\delta\psi_{\mathrm{C}}(\phi) = 0$), (5.127)–(5.129) collapse to the standard three-state Moore-Greitzer model [104].

Now, applying the change of variables given in Section 5.4 to (5.127) and (5.128), we obtain the finite element multi-mode state space model

$$\frac{\mathrm{d}\hat{\phi}}{\mathrm{d}\xi} = A_s\hat{\phi} + D_s^{-1}\left(\begin{bmatrix} \hat{\psi}_{\mathrm{C}}^{\mathrm{nom}}(\theta_1) \\ \vdots \\ \hat{\psi}_{\mathrm{C}}^{\mathrm{nom}}(\theta_{n_{\mathrm{f}}}) \end{bmatrix} + \begin{bmatrix} \delta\hat{\psi}_{\mathrm{C}}(\theta_1) \\ \vdots \\ \delta\hat{\psi}_{\mathrm{C}}(\theta_{n_{\mathrm{f}}}) \end{bmatrix} - \begin{bmatrix} \Psi \\ \vdots \\ \Psi \end{bmatrix}\right), \tag{5.132}$$

where

$$\hat{\psi}_{\mathrm{C}}^{\mathrm{nom}}(\theta_i) = \hat{\psi}_{\mathrm{c},0} + \sum_{k=1}^{n_{\mathrm{m}}}[\hat{\psi}_{\mathrm{c},k}^{\sin}\sin(k\theta_i) + \hat{\psi}_{\mathrm{c},k}^{\cos}\cos(k\theta_i)], \tag{5.133}$$

$$\delta\hat{\psi}_{\mathrm{C}}(\theta_i) = \delta\psi_{\mathrm{c},0} + \sum_{k=1}^{n_{\mathrm{m}}}[\delta\psi_{\mathrm{c},k}^{\sin}\sin(k\theta_i) + \delta\psi_{\mathrm{c},k}^{\cos}\cos(k\theta_i)]. \tag{5.134}$$

It is interesting to note that (5.133) and (5.134) are truncated Fourier expansions of $\psi_{\mathrm{C}}^{\mathrm{nom}}(\phi_i)$ and $\delta\psi_{\mathrm{C}}(\phi_i)$, respectively, therefore we can introduce an approximation in (5.132) by replacing $\hat{\psi}_{\mathrm{C}}^{\mathrm{nom}}(\theta_i)$ with $\psi_{\mathrm{C}}^{\mathrm{nom}}(\phi_i)$ and $\delta\hat{\psi}_{\mathrm{C}}(\theta_i)$ with $\delta\psi_{\mathrm{C}}(\phi_i)$. Hence, including the equation for the state variable Ψ, the new nonlinear state space model becomes

$$\frac{\mathrm{d}\hat{\phi}}{\mathrm{d}\xi} = A_s\hat{\phi} + D_s^{-1}\left[\psi_{\mathrm{C}}^{\mathrm{nom}}(\hat{\phi}) + \Delta\psi_{\mathrm{C}}(\hat{\phi})\right] - \mathrm{e}\Psi, \tag{5.135}$$

$$\frac{\mathrm{d}\Psi}{\mathrm{d}\xi} = \frac{1}{4B^2l_{\mathrm{C}}}\left(\frac{\mathrm{e}^{\mathrm{T}}\hat{\phi}}{n_{\mathrm{f}}} - \gamma_{\mathrm{th}}\sqrt{\Psi}\right), \tag{5.136}$$

where

$$\psi_{\mathrm{C}}^{\mathrm{nom}}(\hat{\phi}) \triangleq \begin{bmatrix} \psi_{\mathrm{C}}^{\mathrm{nom}}(\phi_1) \\ \vdots \\ \psi_{\mathrm{C}}^{\mathrm{nom}}(\phi_{n_{\mathrm{f}}}) \end{bmatrix}, \quad \Delta\psi_{\mathrm{C}}(\hat{\phi}) \triangleq \begin{bmatrix} \delta\psi_{\mathrm{C}}(\phi_1) \\ \vdots \\ \delta\psi_{\mathrm{C}}(\phi_{n_{\mathrm{f}}}) \end{bmatrix}.$$

In the nominal case, i.e., $\Delta\psi_{\mathrm{C}}(\hat{\phi}) \equiv 0$, this state space representation collapses to the one obtained in Section 5.6.

5.7.2 Hierarchical Robust Control for Propulsion Systems

In order to address the challenges for controlling uncertain multi-mode axial compression system models, we use the hierarchical robust switching

nonlinear control framework developed in Chapter 4. Here, the locus of the parameterized equilibrium points, on which the equilibria-dependent Lyapunov functions are predicated, is characterized by the axisymmetric stable pressure-flow equilibrium branch of the *nominal* system for a continuum of mass flow through the throttle. For this development we use the shifted variables defined in (5.105) so that the maximum pressure point on the nominal compressor characteristic map is translated to the origin. In this case the translated nonlinear uncertain system is given by

$$\dot{x}_{\mathrm{f}}(t) = Ax_{\mathrm{f}}(t) + P^{-1}\left[\psi_{\mathrm{sc}}^{\mathrm{nom}}(x_{\mathrm{f}}(t)) + \Delta\psi_{\mathrm{s}}(x_{\mathrm{f}}(t))\right] - ex_{\mathrm{p}}(t), \quad (5.137)$$

$$\dot{x}_{\mathrm{p}}(t) = \frac{1}{\beta^2}\left(\frac{e^{\mathrm{T}}x_{\mathrm{f}}(t)}{n_{\mathrm{f}}} - u(t)\right), \quad (5.138)$$

where A, P, β, and u are defined as in (5.108) and

$$\psi_{\mathrm{sc}}^{\mathrm{nom}}(x_{\mathrm{f}}) \triangleq \begin{bmatrix} \psi_{\mathrm{sc}}^{\mathrm{nom}}(x_{\mathrm{f}1}) \\ \vdots \\ \psi_{\mathrm{sc}}^{\mathrm{nom}}(x_{\mathrm{f}n_{\mathrm{f}}}) \end{bmatrix}, \qquad \Delta\psi_{\mathrm{s}}(x_{\mathrm{f}}) \triangleq \begin{bmatrix} \delta\psi_{\mathrm{s}}(x_{\mathrm{f}1}) \\ \vdots \\ \delta\psi_{\mathrm{s}}(x_{\mathrm{f}n_{\mathrm{f}}}) \end{bmatrix}.$$

Next, it follows from (5.121) that the actual compressor characteristic $\psi_{\mathrm{sc}}(x_{\mathrm{f}_i})$, $i = 1,\ldots,n_{\mathrm{f}}$, is given by

$$\psi_{\mathrm{sc}}(x_{\mathrm{f}_i}) = \psi_{\mathrm{sc}}^{\mathrm{nom}}(x_{\mathrm{f}_i}) + \delta\psi_{\mathrm{s}}(x_{\mathrm{f}_i}), \qquad i = 1,\ldots,n_{\mathrm{f}}, \quad (5.139)$$

where $\psi_{\mathrm{sc}}^{\mathrm{nom}}(x_{\mathrm{f}_i}) = -\frac{3}{2}x_{\mathrm{f}i}^{2} - \frac{1}{2}x_{\mathrm{f}i}^{3}$ is the nominal compressor characteristic and $\delta\psi_{\mathrm{s}}(x_{\mathrm{f}_i})$, $i = 1,\ldots,n_{\mathrm{f}}$, is an uncertain perturbation of the nominal characteristic $\psi_{\mathrm{sc}}^{\mathrm{nom}}(x_{\mathrm{f}_i})$, $i = 1,\ldots,n_{\mathrm{f}}$. Here, we assume

$$\delta\psi_{\mathrm{s}}(\cdot) \in \Delta \triangleq \{\delta\psi_{\mathrm{s}} : \mathbb{R} \to \mathbb{R} : [\delta\psi_{\mathrm{s}}(y) - m_1(y)][\delta\psi_{\mathrm{s}}(y) - m_2(y)] \le 0,\ y \in \mathbb{R}\}, \quad (5.140)$$

where $m_1, m_2 : \mathbb{R} \to \mathbb{R}$ are given arbitrary bounding functions.

To carry out Step 1 of Algorithm 4.1, let $q = m = 1$ and $\varphi(x_{\mathrm{f}}, x_{\mathrm{p}}, \lambda) = \lambda$ so that the nominal system equilibria are parameterized by the constant control $u(t) = \lambda$. In this case, (5.137) and (5.138) with $\Delta\psi_{\mathrm{s}}(x_{\mathrm{f}}(t)) \equiv 0$ have a nominal equilibrium point at $(x_{\mathrm{f}\lambda}, x_{\mathrm{p}\lambda})$, where

$$x_{\mathrm{f}\lambda} \triangleq \lambda e, \qquad x_{\mathrm{p}\lambda} \triangleq \psi_{\mathrm{sc}}^{\mathrm{nom}}(\lambda) = -\tfrac{3}{2}\lambda^2 - \tfrac{1}{2}\lambda^3. \quad (5.141)$$

Next, we carry out Step 2 of Algorithm 4.1. Specifically, for the uncertain compression system (5.137) and (5.138), we show that there exists $\lambda > 0$ and a robust control law such that a neighborhood $\mathcal{N}_\lambda$ of the nominal equilibrium point $(x_{\mathrm{f}\lambda}, x_{\mathrm{p}\lambda})$ is locally robustly asymptotically stable with domain

of attraction $\mathcal{D}_\lambda$. Specifically, consider the equilibrium-dependent Lyapunov function candidate predicated on the nominal pressure-flow axisymmetric stable equilibria given by

$$V_\lambda(x_{\rm f}, x_{\rm p}) = \frac{1}{2n_{\rm f}}(x_{\rm f} - x_{{\rm f}\lambda})^{\rm T} P(x_{\rm f} - x_{{\rm f}\lambda}) + \frac{1}{2}\beta^2[x_{\rm p} - x_{{\rm p}\lambda}]^2, \quad (5.142)$$

with Lyapunov derivative

$$\begin{aligned}\dot V_\lambda(x_{\rm f}, x_{\rm p}) &= \frac{1}{n_{\rm f}}(x_{\rm f} - \lambda e)^{\rm T} P\left[Ax_{\rm f} + P^{-1}\left(\psi_{\rm SC}^{\rm nom}(x_{\rm f}) + \Delta\psi_{\rm s}(x_{\rm f})\right) - e x_{\rm p}\right] \\ &\quad +[x_{\rm p} - \psi_{\rm SC}^{\rm nom}(\lambda)]\left(\frac{e^{\rm T} x_{\rm f}}{n_{\rm f}} - u\right) \\ &= -\frac{1}{n_{\rm f}}(x_{\rm f} - \lambda e)^{\rm T}[\psi_{\rm SC}^{\rm nom}(\lambda)e - \psi_{\rm SC}^{\rm nom}(x_{\rm f}) - \Delta\psi_{\rm s}(x_{\rm f})] \\ &\quad -h_\lambda(x_{\rm f}, x_{\rm p})[x_{\rm p} - \psi_{\rm SC}^{\rm nom}(\lambda)], \end{aligned} \quad (5.143)$$

where $u(x_{\rm f}, x_{\rm p}) = u_\lambda(x_{\rm f}, x_{\rm p}) \triangleq \lambda + h_\lambda(x_{\rm f}, x_{\rm p})$ and $h_\lambda : \mathbb{R}^{n_{\rm f}} \times \mathbb{R} \to \mathbb{R}$ is such that $h_\lambda(x_{{\rm f}\lambda}, x_{{\rm p}\lambda}) = 0$. Now, it follows from Theorem 4.1 that requiring $\dot V_\lambda(x_{\rm f}, x_{\rm p}) < 0$, $(x_{\rm f}, x_{\rm p}) \in \mathcal{D}_\lambda \setminus \mathcal{N}_\lambda$, guarantees local robust stability of the compact positively invariant set $\mathcal{N}_\lambda$ for all $\delta\psi_{\rm s}(\cdot) \in \boldsymbol{\Delta}$. However, (5.143) is dependent on the system uncertainty and needs to be checked for all $\delta\psi_{\rm s}(x_{{\rm f}_i}) \in \boldsymbol{\Delta}$, $i = 1, \ldots, n_{\rm f}$, and hence is unverifiable. To obtain verifiable conditions for robust stability we utilize Conditions (4.11) and (4.12) and introduce an equilibrium-dependent bounding function $\Gamma_\lambda(\cdot)$ for the uncertainty set $\boldsymbol{\Delta}$ such that $\Gamma_\lambda(\cdot)$ bounds $\boldsymbol{\Delta}$. Specifically, define $\Gamma_\lambda : \mathbb{R}^{n_{\rm f}} \to \mathbb{R}$ by

$$\begin{aligned}\Gamma_\lambda(x_{\rm f}) &\triangleq \tfrac{1}{4}[m_2(x_{\rm f}) - m_1(x_{\rm f})]^{\rm T}[m_2(x_{\rm f}) - m_1(x_{\rm f})] + \tfrac{1}{4}(x_{\rm f} - \lambda e)^{\rm T}(x_{\rm f} - \lambda e) \\ &\quad +\tfrac{1}{2}(x_{\rm f} - \lambda e)^{\rm T}[m_1(x_{\rm f}) + m_2(x_{\rm f})], \end{aligned} \quad (5.144)$$

where

$$m_1(x_{\rm f}) \triangleq \begin{bmatrix} m_1(x_{{\rm f}1}) \\ \vdots \\ m_1(x_{{\rm f}n_{\rm f}}) \end{bmatrix}, \qquad m_2(x_{\rm f}) \triangleq \begin{bmatrix} m_2(x_{{\rm f}1}) \\ \vdots \\ m_2(x_{{\rm f}n_{\rm f}}) \end{bmatrix}.$$

Now, note that if $\delta\psi(\cdot) \in \boldsymbol{\Delta}$ then

$$\begin{aligned}0 &\le \tfrac{1}{4}\left[m_1(x_{\rm f}) + m_2(x_{\rm f}) + x_{\rm f} - \lambda e - 2\Delta\psi_{\rm s}(x_{\rm f})\right]^{\rm T}\left[m_1(x_{\rm f}) + m_2(x_{\rm f}) + x_{\rm f}\right. \\ &\quad \left. -\lambda e - 2\Delta\psi_{\rm s}(x_{\rm f})\right] - [\Delta\psi_{\rm s}(x_{\rm f}) - m_1(x_{\rm f})]^{\rm T}[\Delta\psi_{\rm s}(x_{\rm f}) - m_2(x_{\rm f})] \\ &= \tfrac{1}{4}\left[m_2(x_{\rm f}) - m_1(x_{\rm f})\right]^{\rm T}\left[m_2(x_{\rm f}) - m_1(x_{\rm f})\right] + \tfrac{1}{4}(x_{\rm f} - \lambda e)^{\rm T}(x_{\rm f} - \lambda e) \\ &\quad +\tfrac{1}{2}(x_{\rm f} - \lambda e)^{\rm T}\left[m_1(x_{\rm f}) + m_2(x_{\rm f})\right] - (x_{\rm f} - \lambda e)^{\rm T}\Delta\psi_{\rm s}(x_{\rm f}),\end{aligned}$$

and hence $(x_{\rm f} - \lambda e)^{\rm T}\Delta\psi(x_{\rm f}) \le \Gamma_\lambda(x_{\rm f})$, $\delta\psi(x_{{\rm f}_i}) \in \boldsymbol{\Delta}$, $i = 1, \ldots, n_{\rm f}$. Now, requiring

$$-\frac{1}{n_\mathrm{f}}(x_\mathrm{f}-\lambda e)^\mathrm{T}[\psi_\mathrm{sc}^\mathrm{nom}(\lambda)e-\psi_\mathrm{sc}^\mathrm{nom}(x_\mathrm{f})]$$
$$-h_\lambda(x_\mathrm{f},x_\mathrm{p})[x_\mathrm{p}-\psi_\mathrm{sc}^\mathrm{nom}(\lambda)]+\frac{1}{n_\mathrm{f}}\Gamma_\lambda(x_\mathrm{f})<0,\quad (x_\mathrm{f},x_\mathrm{p})\in\mathcal{D}_\lambda\setminus\mathcal{N}_\lambda,\quad (5.145)$$

it follows from (5.143) that $\dot{V}_\lambda(x_\mathrm{f},x_\mathrm{p})<0$, $(x_\mathrm{f},x_\mathrm{p})\in\mathcal{D}_\lambda\setminus\mathcal{N}_\lambda$, so that all assumptions of Theorem 4.1 are satisfied.

Next, for simplicity of exposition we set $m_1(\cdot)=-m_2(\cdot)=m(\cdot)$, where $m:\mathbb{R}\to\mathbb{R}$ is a given arbitrary function. In this case, it follows from (5.141) and (5.145) that

$$\dot{V}_\lambda(x_\mathrm{f},x_\mathrm{p})\le-\frac{1}{2n_\mathrm{f}}\sum_{i=1}^{n_\mathrm{f}}\{(x_{\mathrm{f}_i}-\lambda)^2\left[(x_{\mathrm{f}_i})^2+(\lambda+3)x_{\mathrm{f}_i}+\lambda(\lambda+3)-\tfrac{1}{2}\right]-2m^2(x_{\mathrm{f}_i})\}$$
$$-h_\lambda(x_\mathrm{f},x_\mathrm{p})[x_\mathrm{p}-\psi_\mathrm{sc}^\mathrm{nom}(\lambda)]<0,\qquad (x_\mathrm{f},x_\mathrm{p})\in\mathcal{D}_\lambda\setminus\mathcal{N}_\lambda.\quad (5.146)$$

Now, a sufficient condition guaranteeing $\dot{V}_\lambda(x_\mathrm{f},x_\mathrm{p})<0$, $(x_\mathrm{f},x_\mathrm{p})\in\mathcal{D}_\lambda\setminus\mathcal{N}_\lambda$, is given by

$$\tfrac{1}{2n_\mathrm{f}}\textstyle\sum_{i=1}^{n_\mathrm{f}}(x_{\mathrm{f}_i}-\lambda)^2p_{1\lambda}(x_{\mathrm{f}_i})>0,\qquad (x_\mathrm{f},x_\mathrm{p})\in\mathcal{D}_\lambda,\quad (5.147)$$
$$\tfrac{1}{2n_\mathrm{f}}\textstyle\sum_{i=1}^{n_\mathrm{f}}(x_{\mathrm{f}_i}-\lambda)^2p_{2\lambda}(x_{\mathrm{f}_i})+h_\lambda(x_\mathrm{f},x_\mathrm{p})[x_\mathrm{p}-\psi_\mathrm{sc}^\mathrm{nom}(\lambda)]>\tfrac{1}{n_\mathrm{f}}\sum_{i=1}^{n_\mathrm{f}}m^2(x_{\mathrm{f}_i}),$$
$$(x_\mathrm{f},x_\mathrm{p})\notin\mathcal{N}_\lambda,\quad (5.148)$$

where $p_{1\lambda}(x_{\mathrm{f}_i})\triangleq a_{1\lambda}(x_{\mathrm{f}_i})^2+b_{1\lambda}x_{\mathrm{f}_i}+c_{1\lambda}$ and $p_{2\lambda}(x_{\mathrm{f}_i})\triangleq a_{2\lambda}(x_{\mathrm{f}_i})^2+b_{2\lambda}x_{\mathrm{f}_i}+c_{2\lambda}$ are such that

$$p_{1\lambda}(x_{\mathrm{f}_i})+p_{2\lambda}(x_{\mathrm{f}_i})=(x_{\mathrm{f}_i})^2+(\lambda+3)x_{\mathrm{f}_i}+\lambda(\lambda+3)-\tfrac{1}{2}.\quad (5.149)$$

Note that (5.147) is satisfied in a domain $\mathcal{D}_\lambda\neq\emptyset$ only if there exists $d_\lambda>0$ such that $p_{1\lambda}(x_{\mathrm{f}_i})>0$, $-d_\lambda<x_{\mathrm{f}_i}-\lambda<d_\lambda$, $i=1,\ldots,n_\mathrm{f}$, and, in order to satisfy (5.148), we require that $p_{2\lambda}(x_{\mathrm{f}_i})>0$, $i=1,\ldots,n_\mathrm{f}$. Hence, we require that $p_{1\lambda}(\lambda)>0$ and $p_{2\lambda}(\lambda)>0$. A particular choice of $h_\lambda(\cdot,\cdot)$ satisfying (5.148) is given by

$$h_\lambda(x_\mathrm{f},x_\mathrm{p})\triangleq w[x_\mathrm{p}-\psi_\mathrm{sc}^\mathrm{nom}(\lambda)]p(x_\mathrm{f}-\lambda e),\quad (5.150)$$

where $w:\mathbb{R}\to\mathbb{R}$ is such that $xw(x)>0, x\neq 0$, and $p:\mathbb{R}^{n_\mathrm{f}}\to\mathbb{R}$ is positive definite. However, note that for $x_\mathrm{f}=\lambda e$ it is not possible to satisfy (5.148) and hence by continuity there exists a neighborhood of this point where (5.146) *cannot* be satisfied. Thus, we construct a robust control law such that a neighborhood $\mathcal{N}_\lambda$ of the equilibrium point $(x_\mathrm{f},x_\mathrm{p})$ is robustly stabilized with a given domain of attraction.

Next, note that it follows from (5.149) that for all $0<\lambda\le\sqrt{\frac{7}{6}}-1$, $p_{1\lambda}(\lambda)+p_{2\lambda}(\lambda)\le 0$ and hence the necessary conditions $p_{1\lambda}(\lambda)>0$ and

$p_{2\lambda}(\lambda) > 0$ for satisfying (5.147) and (5.148) are violated. Furthermore, if $\lambda > \sqrt{\frac{14}{3}} - 1$, then $p_{1\lambda}(x_{f_i}) + p_{2\lambda}(x_{f_i}) > 0$, $i = 1, \ldots, n_f$, which implies that it is always possible to choose $p_{1\lambda}(\cdot)$ such that $p_{1\lambda}(x_{f_i}) > 0$, $i = 1, \ldots, n_f$. More generally, there exists $\lambda_0 \geq \sqrt{\frac{7}{6}} - 1$ and $\lambda_{\text{global}} > \sqrt{\frac{14}{3}} - 1$ such that $\mathcal{D}_{\lambda_0}$ collapses to the equilibrium point and $\mathcal{D}_{\lambda_{\text{global}}}$ coincides with the whole state space. Note that λ_0 and λ_{global} are dependent on the particular choice of the coefficients $a_{1\lambda}, b_{1\lambda}, c_{1\lambda}, a_{2\lambda}, b_{2\lambda}$, and $c_{2\lambda}$.

Next, with $u(x_f, x_p) = u_\lambda(x_f, x_p)$, we provide an estimate of the domain of attraction for (5.137), (5.138). In particular, define

$$\mathcal{D}_\lambda \triangleq \begin{cases} \{(x_f, x_p) : \quad V_\lambda(x_f, x_p) \leq k_{1\lambda}\}, & \lambda_0 \leq \lambda \leq \lambda_{\text{global}}, \\ \mathbb{R}^{n_f} \times \mathbb{R}, & \lambda > \lambda_{\text{global}}, \end{cases} \tag{5.151}$$

$$\mathcal{N}_\lambda \triangleq \{(x_f, x_p) : \quad V_\lambda(x_f, x_p) \leq k_{2\lambda}\}, \qquad \lambda \geq \lambda_0, \tag{5.152}$$

where

$$k_{1\lambda} \triangleq \frac{\mu}{2n_f} d_\lambda^2, \qquad \mu \triangleq \left(\max_i \{P_{ii}^{-1}\}\right)^{-1}, \tag{5.153}$$

and

$$k_{2\lambda} \triangleq \max_{(x_f, x_p) \in \mathcal{D}_\lambda} \frac{1}{2n_f}(x_f - x_{f\lambda})^{\mathrm{T}} P (x_f - x_{f\lambda}) + \tfrac{1}{2}\beta^2 [x_p - x_{p\lambda}]^2, \tag{5.154}$$

subject to

$$\frac{1}{2n_f}\sum_{i=1}^{n_f}(x_{f_i} - \lambda)^2 p_{2\lambda}(x_{f_i}) + h_\lambda(x_f, x_p)[x_p - \psi_{\text{sc}}^{\text{nom}}(\lambda)] = \frac{1}{n_f}\sum_{i=1}^{n_f} m^2(x_{f_i}). \tag{5.155}$$

The Lyapunov level surfaces $V_\lambda(x_f, x_p) = k_{1\lambda}$ and $V_\lambda(x_f, x_p) = k_{2\lambda}$ are constructed such that the intersection of the boundary of $\mathcal{D}_\lambda$ with the plane $x_p = x_{p\lambda}$ is a closed surface contained in the region $\{x_f : -d_\lambda < x_{f_i} - \lambda < d_\lambda, i = 1, \ldots, n_f\}$ and $\mathcal{N}_\lambda$ contains the region where (5.148) is *not* satisfied, so that $\dot{V}_\lambda(x_f, x_p) < 0$ for all $(x_f, x_p) \in \mathcal{D}_\lambda \setminus \mathcal{N}_\lambda$. Note that for $\lambda_0 \leq \lambda \leq \lambda_{\text{global}}$, $k_{1\lambda} \geq 0$ and $k_{2\lambda} \geq 0$. Furthermore, since $V_\lambda(x_f, x_p)$ is continuous and radially unbounded $\mathcal{N}_\lambda$ and $\mathcal{D}_\lambda$ are compact sets for $\lambda \in [\lambda_0, \lambda_{\text{global}}]$, and hence positively invariant. Thus, if the state space trajectories of (5.137), (5.138) enter $\mathcal{D}_\lambda$, then $\mathcal{N}_\lambda$ serves as an attractor. Now, to ensure that $\mathcal{N}_\lambda \subset \mathcal{D}_\lambda$ we require that $k_{1\lambda} > k_{2\lambda}$. A typical plot for the level set values $k_{1\lambda}$ and $k_{2\lambda}$ as functions of λ is shown in Figure 5.8. Note that there exists $\lambda_{\min}$ such that $k_{1\lambda_{\min}} = k_{2\lambda_{\min}}$ and hence $\mathcal{D}_{\lambda_{\min}} = \mathcal{N}_{\lambda_{\min}}$. Hence, requiring $\lambda > \lambda_{\min}$ assures the necessary condition that $\mathcal{N}_\lambda \subset \mathcal{D}_\lambda$.

The coefficients of the two parabolas $p_{1\lambda}(\cdot)$ and $p_{2\lambda}(\cdot)$ must be such that (5.149) is satisfied along with the above stated necessary conditions. This

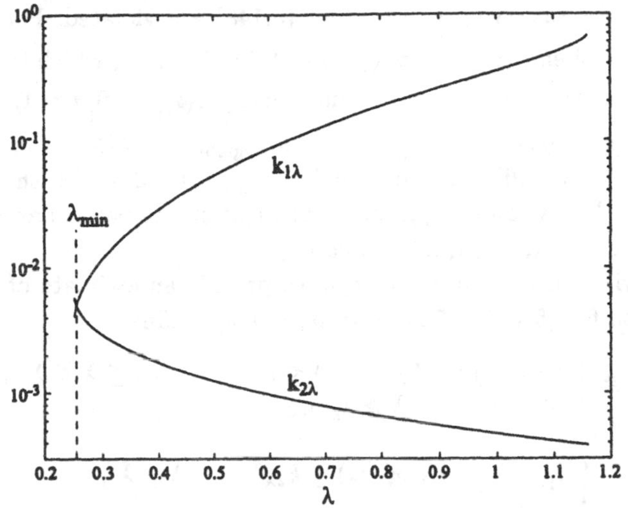

Fig. 5.8. Level set values $k_{1\lambda}$ and $k_{2\lambda}$ as functions of λ

leaves some degree of freedom in the choice of the coefficients $a_{1\lambda}$, $b_{1\lambda}$, $c_{1\lambda}$, $a_{2\lambda}$, $b_{2\lambda}$, and $c_{2\lambda}$, which can be used to maximize the domain of attraction $\mathcal{D}_\lambda$ and minimize the attractor $\mathcal{N}_\lambda$. This leads to the following optimization problem for each λ:

$$\max_{a_{1\lambda},b_{1\lambda},c_{1\lambda},a_{2\lambda},b_{2\lambda},c_{2\lambda}} \left(\lambda^2 - \frac{c_{1\lambda}}{a_{1\lambda}}\right), \tag{5.156}$$

subject to

$$a_{1\lambda} + a_{2\lambda} = 1, \tag{5.157}$$
$$b_{1\lambda} + b_{2\lambda} = \lambda + 3, \tag{5.158}$$
$$c_{1\lambda} + c_{2\lambda} = \lambda(\lambda + 3) - \tfrac{1}{2}, \tag{5.159}$$
$$(q_\lambda - \lambda)^2(a_{2\lambda}q_\lambda^2 + b_{2\lambda}q_\lambda + c_{2\lambda}) = 2n_f\kappa^2, \tag{5.160}$$
$$2a_{1\lambda}\lambda + b_{1\lambda} = 0, \tag{5.161}$$
$$a_{1\lambda} < 0, \tag{5.162}$$
$$b_{1\lambda}^2 - 4a_{1\lambda}c_{1\lambda} > 0, b_{2\lambda}^2 - 4a_{2\lambda}c_{2\lambda} < 0, \tag{5.163}$$

where $q_\lambda \triangleq \frac{2a_{2\lambda}\lambda - 3b_{2\lambda} - \sqrt{(2a_{2\lambda}\lambda - 3b_{2\lambda})^2 - 16a_{2\lambda}(2c_{2\lambda} - b_{2\lambda}\lambda)}}{8a_{2\lambda}}$ and $m(x_{f_i})$ is chosen to be a constant value $\kappa \in \mathbb{R}$, $i = 1, \ldots, n_f$. Note that, under the assumption that $p_{1\lambda}(\cdot)$ achieves a maximum at λ, the objective function given by (5.156) corresponds to maximizing d_λ^2. Furthermore, conditions (5.157)–(5.159) are obtained by equating the coefficients of equal powers in (5.149). Condition (5.160) guarantees that $(x_{f_i} - \lambda)^2 p_{2\lambda}(x_{f_i})$, $i = 1, \ldots, n_f$, is a convex function

for all x_{f_i} so that $\mathcal{N}_\lambda$ is minimized, while conditions (5.161)–(5.163) guarantee that $p_{1\lambda}(\cdot)$ achieves a maximum at λ and $p_{1\lambda}(\lambda) > 0$. Finally, (5.163) guarantees that $p_{2\lambda}(\cdot) > 0$.

To carry out Step 3 of Algorithm 4.1, we consider two topologies for the switching set $\mathcal{S}$; namely an isolated point topology and a hybrid topology. For $\mathcal{S}$ consisting of countably finite isolated points let $\mathcal{S} = \{\lambda_0, \ldots, \lambda_q\}$ be such that $\lambda_{\min} < \lambda_q < \cdots < \lambda_1 \leq \lambda_{\text{global}}$, $\lambda_0 > \lambda_{\text{global}}$, and $\mathcal{N}_{\lambda_{i+1}} \subset \mathcal{D}_{\lambda_i}$, $i \in \{0, \ldots, q-1\}$, and let $p(\lambda) = \lambda$, $\lambda \in \mathcal{S}$. To guarantee that $p(\cdot)$ satisfies Assumption 4.1 construct λ_k, $k = 0, 1, \ldots, q$, online by considering the smallest solution to the equation $V_{\lambda_k}(x(t_k)) = c_{\lambda_k}$, $t_k \triangleq k\Delta T$, where $\Delta T > 0$ and $k = 0, 1, \ldots, q$, and define $\mathcal{S} \triangleq \{\lambda_k\}_{k=0}^q$. Now, with the robust feedback switching control law $u = \phi_{\lambda_{\mathcal{S}}(x_f, x_p)}(x_f, x_p)$, where $\lambda_{\mathcal{S}}(x_f, x_p)$ is obtained as described in Step 4 of Algorithm 4.1, it follows from Theorem 4.4 that the compact positively invariant set $\mathcal{N}_{\lambda_q}$ is globally asymptotically stable for all $\delta\psi_s(\cdot) \in \boldsymbol{\Delta}$. Furthermore, note that $\lambda_{\mathcal{S}}(x(t))$, $t \geq 0$, is piecewise constant and hence the robust feedback switching control law $u = \phi_{\lambda_{\mathcal{S}}(x_f, x_p)}(x_f, x_p)$ is piecewise continuous.

For $\mathcal{S}$ consisting of a hybrid topology let $\mathcal{S} = [\lambda_{\min}, \lambda_{\text{global}}] \cup \{\bar{\lambda}\}$, where $\bar{\lambda} > \lambda_{\text{global}}$ is such that $\mathcal{N}_{\bar{\lambda}} \in \mathcal{D}_{\hat{\lambda}}$, for at least one $\hat{\lambda} \in [\lambda_{\min}, \lambda_{\text{global}}]$, and let $p(\lambda) = \lambda$, $\lambda \in \mathcal{S}$. Since $p(\cdot)$ does not have a local minimum in $\mathcal{S}$ (other than the origin) and every $\lambda \in [\lambda_{\min}, \lambda_{\text{global}}]$ is an accumulation point for $\mathcal{S}$, we are guaranteed, by Step 3b of Algorithm 4.1, that Assumption 4.1 is satisfied. Now global robust asymptotic stability of $\mathcal{N}_{\lambda_{\min}}$ for all $\delta\psi_s(\cdot) \in \boldsymbol{\Delta}$ is guaranteed by Theorem 4.4 with the feedback control law $u = \phi_{\lambda_{\mathcal{S}}(x_f, x_p)}(x_f, x_p)$, where $\lambda_{\mathcal{S}}(x_f, x_p)$ is obtained as described in Step 4 of Algorithm 4.1. In particular, if $(x_f(0), x_p(0)) \in \hat{\mathcal{D}} \triangleq \cup_{\lambda \in [\lambda_{\min}, \lambda_{\text{global}}]} \mathcal{D}_\lambda$ then $\lambda_{\mathcal{S}}(x(t))$, $t \geq 0$, is a continuous function. Alternatively, if $(x_f(0), x_p(0)) \notin \hat{\mathcal{D}}$ then $\lambda_{\mathcal{S}}(x(t)) = \bar{\lambda}$, $t \in [0, \hat{t})$, where $\hat{t} > 0$ is such that $(x_f(\hat{t}), x_p(\hat{t})) \in \partial\hat{\mathcal{D}}$. In this case, $\lambda_{\mathcal{S}}(x(t))$, $t \geq 0$, is continuous modulo one discontinuity at $t = \hat{t}$. Note that since $\mathcal{N}_{\lambda_{\min}} \equiv \mathcal{D}_{\lambda_{\min}}$, $\mathcal{N}_{\lambda_{\min}}$ is a global attractor but not Lyapunov stable.

As in the nominal case, the proposed robust switching nonlinear controller framework can be incorporated to address practical actuator limitations such as control amplitude and rate saturation constraints. Specifically, since $\lambda_t \triangleq \lambda_{\mathcal{S}}(x_f(t), x_p(t))$ is proportional to the throttle opening (actuator) and since the dynamics of λ_t indirectly characterize the fastest admissible rate at which the control throttle can open while maintaining stability of the controlled system, it follows that by constraining how fast λ_t can change on the nominal equilibrium branch effectively places a rate constraint on the throttle opening. This corresponds to the case where the switching rate of the nonlinear controller is decreased so that the trajectory $(x_f(t), x_p(t))$, $t \geq 0$, is allowed to enter $\mathcal{D}_{\lambda_t}$. Additionally, amplitude saturation constraints and

state constraints can also be enforced by simply choosing $\lambda_{\max} < \lambda_{\text{global}}$ such that $\mathcal{D}_{\max} \triangleq \cup_{\lambda_{\min}\leq\lambda\leq\lambda_{\max}} \mathcal{D}_\lambda$ is contained in the region where the system is constrained to operate. In this case, the hierarchical robust switching nonlinear controller guarantees local robust asymptotic stability of $\mathcal{N}_{\lambda_{\min}}$ with an estimate of the domain of attraction given by $\mathcal{D}_{\max}$.

To show the efficacy of the proposed robust control approach, we consider a two-mode compressor model so that the state space model given by (5.101) and (5.102), or, equivalently, (5.137) and (5.138) is of sixth order. Using the same parameter values and the initial conditions given in Section 5.5, the non-robust controller developed in Section 5.6 and the proposed robust globally stabilizing controller were used to compare the closed-loop system response. Here we model the uncertain perturbation to the nominal pressure-flow compressor characteristic map by

$$\delta\psi(x_{f_i}) = 0.1\cos[10(x_{f_i} - 1)], \qquad i = 1,\ldots,5. \tag{5.164}$$

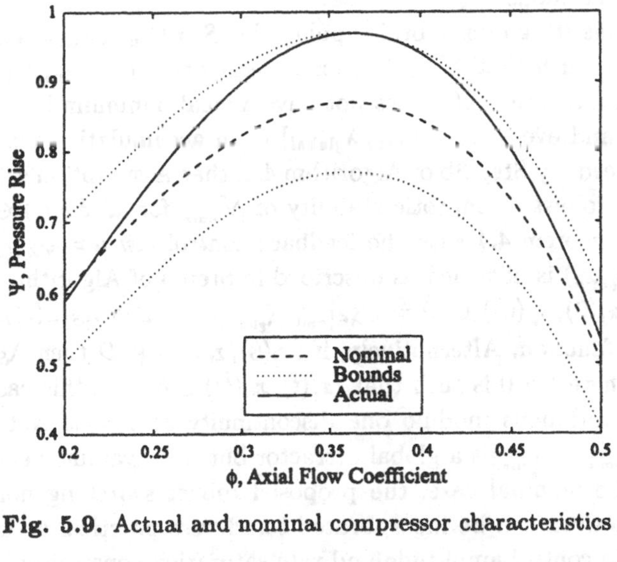

Fig. 5.9. Actual and nominal compressor characteristics

Figure 5.9 shows the nominal ($\psi_{\text{C}}^{\text{nom}}(\phi)$) and actual ($\psi_{\text{C}}(\phi)$) pressure-flow compressor characteristic maps for $\kappa = 0.1$. For this value of κ the optimization problem outlined above for maximizing the domain of attraction $\mathcal{D}_\lambda$ and minimizing the attractor $\mathcal{N}_\lambda$ yields

$$\lambda_{\min} = 0.2547, \quad \lambda_{\text{global}} = 1.1604, \quad d_{\lambda_{\min}} = 0.2236, \quad k_{\lambda_{\min}} = 0.0050.$$

Finally, we use $u(x_{\text{f}}(t), x_{\text{p}}(t)) = \lambda + h_\lambda(x_{\text{f}}(t), x_{\text{p}}(t))$, where $h_\lambda(x_{\text{f}}(t), x_{\text{p}}(t)) = x_{\text{p}}(t) - \psi_{\text{SC}}^{\text{nom}}(\lambda)$.

Figure 5.10 shows the controlled responses for the squared stall cell amplitudes J_1 and J_2, the compressor flow Φ, and the pressure rise Ψ for both designs. This comparison illustrates that the robust controller globally stabilizes the axisymmetric operating point corresponding to $(J_1, J_2, \Phi, \Psi) = (0, 0, 0.4133, 0.8471)$. Alternatively, the non-robust controller drives the system to a limit-cycle instability induced by the control action. Finally, Figure 5.11 shows the throttle opening versus time of the proposed robust controller.

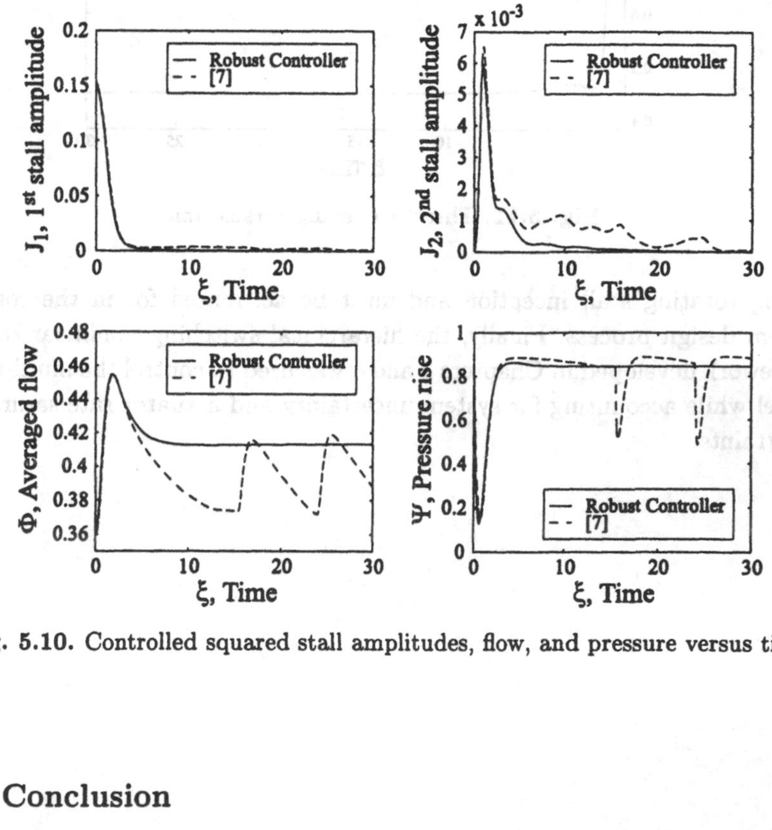

Fig. 5.10. Controlled squared stall amplitudes, flow, and pressure versus time

5.8 Conclusion

A multi-mode state space model for rotating stall and surge in axial flow compression systems that lends itself to the application of nonlinear control design was developed. In particular, to account for the interaction between higher-order disturbance velocity potential harmonics and the first harmonic during rotating stall inception an n_m-mode expansion of the disturbance potential in the flow field was considered. The multi-mode model was used to show that the second and higher-order disturbance velocity potential harmonics in the governing flow equations strongly interact with the first harmonic

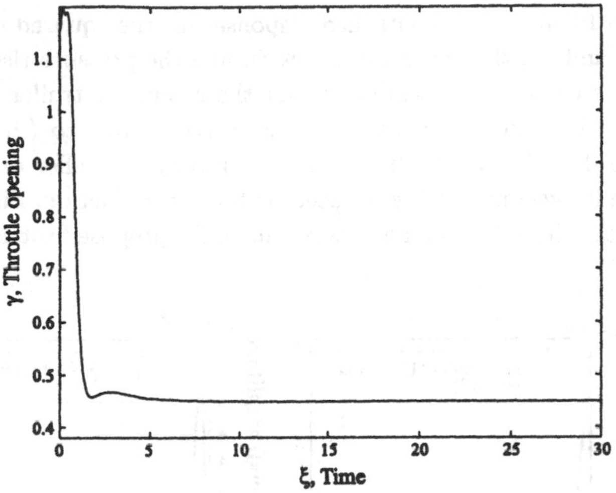

Fig. 5.11. Throttle opening versus time

during rotating stall inception and must be accounted for in the control-system design process. Finally, the hierarchical switching nonlinear control framework developed in Chapters 3 and 4 was used to control the multi-mode model while accounting for system uncertainty and actuator rate saturation constraints.

6. Hierarchical Switching Control for Centrifugal Flow Compressor Models

6.1 Introduction

While the literature on modeling and control of compression systems predominantly focuses on axial flow compression systems, the research literature on centrifugal flow compression systems is rather limited in comparison. Notable exceptions include [17, 42, 61, 112, 45, 69, 48] which address modeling and control of centrifugal compressors. In contrast to axial flow compression systems involving the aerodynamic instabilities of rotating stall and surge, a common feature of [17, 42, 61, 112, 45, 69, 48] is the realization that surge and deep surge is the predominant aerodynamic instability arising in centrifugal compression systems. Surge within centrifugal compressors is a one-dimensional axisymmetric global compression system oscillation which involves radial flow oscillations and in some case even radial flow reversal (deep surge) which can damage engine components.

In this chapter we address the problem of nonlinear stabilization for centrifugal compression systems. First, however, we obtain a three-state lumped parameter model for surge in centrifugal flow compression systems that is accessible to control-system designers requiring state space models for modern nonlinear control. The low-order centrifugal compression system model presented here closely parallels the model developed in [48] and hence only salient portions of the model are presented which are relevant for the proposed control design framework. Specifically, the authors in [48] develop a centrifugal compression system model involving pressure and mass flow compression system dynamics using principles of conservation of mass and momentum. Furthermore, in order to account for the influence of speed transients on the compression surge dynamics, turbocharger spool dynamics are also considered.

Next, using the hierarchical nonlinear control framework developed in Chapter 3, we develop globally stabilizing control laws for the lumped parameter centrifugal compressor surge model. The locus of equilibrium points, on which the nonlinear switching controller is predicated, is characterized by the axisymmetric pressure-flow equilibria of the compression system as well

as by the family of compressor characteristic maps for different spool speeds. As in the axial compressor case, to reflect a more realistic design we account for a rate saturation constraint on the system actuator throttle opening.

Finally, even though for simplicity of exposition we do not address system parametric uncertainty in the centrifugal compressor model, the proposed centrifugal compressor controller can be extended as described in Chapter 4 to provide robust stability guarantees in the face of system modeling uncertainty.

6.2 Governing Fluid Dynamic Equations for Centrifugal Compression Systems

In this section we develop a low-order, three-state surge model for centrifugal compressors. Specifically, we consider the basic centrifugal compression system shown in Figure 6.1, consisting of a short inlet duct, a compressor, an outlet duct, a plenum, an exit duct, and a control throttle. We assume that the plenum dimensions are large as compared to the compressor-duct dimensions so that the fluid velocity and acceleration in the plenum are negligible. In this case the pressure in the plenum is spatially uniform. Furthermore, we assume that the flow is controlled by a throttle at the plenum exit and a driving torque that affects the spool dynamics. In addition, we assume a low speed compression system with oscillation frequencies much lower than the acoustic resonance frequencies so that the flow can be considered incompressible. However, we do assume that the gas in the plenum is compressible and acts as a gas spring. Finally, we assume isentropic process dynamics in the plenum and negligible gas angular momentum in the compressor passages as compared to the impeller angular momentum.

6.2.1 Conservation of Mass in the Plenum

Using continuity it follows that mass conservation in the plenum is given by

$$m_{\rm c} - m_{\rm t} = \frac{{\rm d}(\rho_{\rm p} V_{\rm p})}{{\rm d}t}, \tag{6.1}$$

where $m_{\rm c}$ is the mass flow rate at the plenum entrance, $m_{\rm t}$ is the mass flow rate through the throttle, $V_{\rm p}$ is the plenum volume, and $\rho_{\rm p}$ is the gas density in the plenum. Assuming that the plenum is a rigid volume, it follows from (6.1) that

$$m_{\rm c} - m_{\rm t} = V_{\rm p} \frac{{\rm d}\rho_{\rm p}}{{\rm d}t}. \tag{6.2}$$

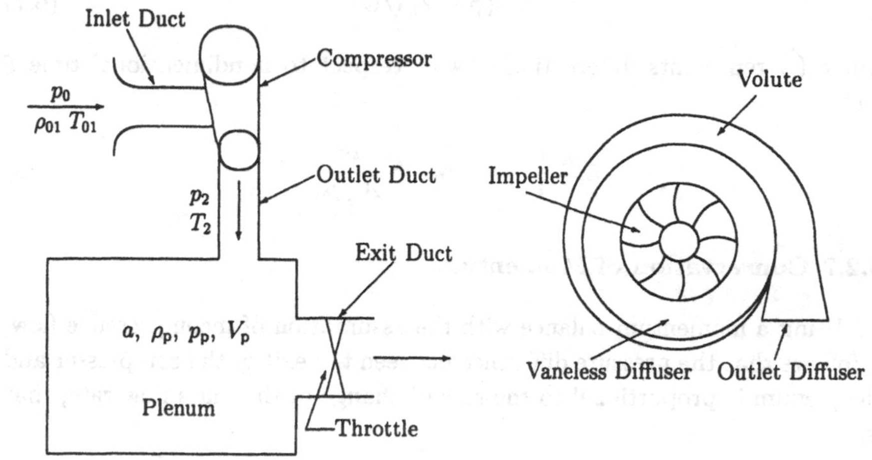

Fig. 6.1. Centrifugal compressor system geometry

Since the plenum flow dynamics are assumed to be isentropic, it follows that ([63])

$$\frac{d\rho_p}{dt} = \frac{1}{a^2}\frac{dp_p}{dt}, \tag{6.3}$$

where p_p is the flow pressure inside the plenum and a is the ambient sonic velocity. Substituting (6.3) into (6.2), we obtain

$$\frac{dp_p}{dt} = \frac{a^2}{V_p}(m_c - m_t). \tag{6.4}$$

Next, assuming that the throttle discharges to an infinite reservoir with pressure p_0 it follows that the pressure difference $p_p - p_0$ must balance both the throttle pressure loss and the net difference in pressure due to the flow acceleration through the throttle duct. Here we model the flow through the throttle by ([104])

$$m_t = k_t\sqrt{p_p - p_0}, \tag{6.5}$$

where the parameter k_t is proportional to the throttle opening and p_0 is the downstream pressure. If the plenum exit duct is short, then p_0 can be regarded as the ambient pressure. Now substituting (6.5) into (6.4) and defining the nondimensional pressure, mass flow, and time, respectively, by

$$\psi \triangleq \frac{p_p - p_0}{p_0}, \qquad \phi \triangleq \frac{a}{Ap_0}m_c, \qquad \xi \triangleq \frac{Aa}{L^3}t, \tag{6.6}$$

where A is the cross sectional area of compressor exit duct and L is the length of the compressor duct, it follows that

$$\dot{\psi} = \bar{a}(\phi - \gamma_{\rm th}\sqrt{\psi}), \tag{6.7}$$

where $\dot{( )}$ represents differentiation with respect to nondimensional time ξ and

$$\bar{a} \triangleq \frac{L^3}{V_p}, \qquad \gamma_{\rm th} \triangleq \frac{ak_{\rm t}}{A\sqrt{p_0}}. \tag{6.8}$$

6.2.2 Conservation of Momentum

Using a momentum balance with the assumption of incompressible flow, it follows that the pressure difference between the exit of the compressor and the plenum is proportional to the rate of change of the mass flow rate, that is,

$$A(p_2 - p_{\rm p}) = L\frac{{\rm d}m_{\rm c}}{{\rm d}t}, \tag{6.9}$$

where p_2 is the pressure rise at the exit of the compressor. Next, assuming isentropic process dynamics with a constant specific heat $c_{\rm p}$, it follows that ([63])

$$\frac{p_2}{p_0} = \left(\frac{T_2}{T_{01}}\right)^{\frac{\gamma}{\gamma-1}}, \tag{6.10}$$

where T_{01} is the compressor inlet temperature, T_2 is the fluid temperature at compressor rotor exit, and γ is the specific heat ratio. Now, using $\Delta h_{\rm ideal} = c_{\rm p}(T_2 - T_{01})$, where $\Delta h_{\rm ideal}$ is the ideal change in fluid specific enthalpy which, for a conservative system, is equal to the work done by the compressor rotor, it follows from (6.10) that

$$\frac{p_2}{p_0} = \left(1 + \frac{\Delta h_{\rm ideal}}{c_{\rm p}T_{01}}\right)^{\frac{\gamma}{\gamma-1}}. \tag{6.11}$$

Now, using the fact that the change in angular momentum of fluid is equal to the compressor torque $\tau_{\rm c}$ it follows that ([40, 99])

$$\tau_{\rm c} = m_{\rm c}(r_2 c_{\theta_2} - r_1 c_{\theta_1}), \tag{6.12}$$

where $r_1 \triangleq \frac{1}{4}\sqrt{D_{\rm t1}^2 + D_{\rm h1}^2}$, r_2 is the radius of the rotor tip, $D_{\rm t1}$ is the inducer tip diameter, $D_{\rm h1}$ is the inducer hub diameter, and c_{θ_1} and c_{θ_2} are the absolute tangential velocity of fluid at the rotor inlet and rotor outlet, respectively. Next, we assume that there is no pre-whirl at the rotor inlet so that $c_{\theta_1} = 0$ and define the slip factor which is the ratio between the tangential velocity of the fluid at the rotor outlet and the rotor tip velocity by ([40])

$$\sigma \triangleq \frac{c_{\theta 2}}{r_2 \omega}, \tag{6.13}$$

where ω is the angular velocity of the compressor spool. Here we assume that σ is constant. Substituting (6.13) into (6.12), we obtain

$$\tau_c(m_c, \omega) = \sigma r_2^2 m_c \omega, \tag{6.14}$$

so that the work done to the fluid by the compressor, which is equal to the change of enthalpy of the fluid, is given by

$$\Delta h(\omega) = \frac{\tau_c \omega}{m_c} = \sigma r_2^2 \omega^2. \tag{6.15}$$

Now, we consider incidence losses at the inducer and the diffuser which, respectively, can be expressed as ([48])

$$\Delta h_{ii}(m_c, \omega) = \tfrac{1}{2}\left(r_1 \omega - \frac{\cot \beta_{1b} m_c}{\rho A}\right)^2,$$

$$\Delta h_{di}(m_c, \omega) = \tfrac{1}{2}\left(\sigma r_2 \omega - \frac{\cot \alpha_{2b} m_c}{\rho A}\right)^2,$$

where ρ is the gas density in the compressor stage, α_{2b} is the inducer inlet angle, and β_{1b} is the rotor blade angle. In addition, we consider friction losses at the inducer and the diffuser, denoted as $\Delta h_{if}(m_c)$ and $\Delta h_{df}(m_c)$, respectively, assumed to be quadratic functions of the mass flow rate at the plenum entrance given by ([43])

$$\Delta h_{if}(m_c) = k_{if} m_c^2, \qquad \Delta h_{df}(m_c) = k_{df} m_c^2, \tag{6.16}$$

where k_{if} and k_{df} are the friction coefficients. Combining (6.15)–(6.16) we obtain the energy delivered to the fluid by the compressor as

$$\Delta h(m_c, \omega) = \Delta h_{ideal}(\omega) + \Delta h_{loss}(m_c, \omega), \tag{6.17}$$

where

$$\Delta h_{loss}(m_c, \omega) \triangleq \Delta h_{ii}(m_c, \omega) + \Delta h_{di}(m_c, \omega) + \Delta h_{if}(m_c) + \Delta h_{df}(m_c). \tag{6.18}$$

In order to capture compressor efficiency, define the isentropic efficiency as ([48])

$$\eta_c(m_c, \omega) \triangleq \frac{\Delta h_{ideal}(\omega)}{\Delta h_{loss}(m_c, \omega) + \Delta h_{ideal}(\omega)}. \tag{6.19}$$

Substituting (6.11) and (6.19) into (6.9) yields

$$\frac{\mathrm{d}m_{\mathrm{c}}}{\mathrm{d}t} = \frac{A}{L}\left[\left(1+\eta_{\mathrm{c}}(m_{\mathrm{c}},\omega)\frac{\Delta h(\omega)}{c_{\mathrm{p}}T_{01}}\right)^{\frac{\gamma}{\gamma-1}} p_0 - p_{\mathrm{p}}\right]. \tag{6.20}$$

Next, defining the nondimensional angular velocity of compressor spool by $\tilde{\omega} \triangleq \frac{r_2\omega}{a}$ and using (6.6), it follows that

$$\dot{\phi} = \bar{b}(\psi_{\mathrm{c}}(\phi,\tilde{\omega}) - \psi), \tag{6.21}$$

where $\bar{b} \triangleq \frac{L^2}{A}$ and $\psi_{\mathrm{c}}(\phi,\tilde{\omega})$ is the compressor characteristic pressure-flow/angular velocity map given by

$$\psi_{\mathrm{c}}(\phi,\tilde{\omega}) \triangleq \left(1+\eta_{\mathrm{c}}(\phi,\tilde{\omega})\sigma\bar{d}\tilde{\omega}^2\right)^{\frac{\gamma}{\gamma-1}} - 1, \tag{6.22}$$

where

$$\eta_{\mathrm{c}}(\phi,\tilde{\omega}) = \frac{\sigma\tilde{\omega}^2}{\sigma\tilde{\omega}^2 + \frac{1}{2}(f_1\tilde{\omega} - f_2\phi)^2 + \frac{1}{2}(\sigma\tilde{\omega} - f_3\phi)^2 + f_4\phi^2 + f_5\phi^2}, \tag{6.23}$$

and

$$\bar{d} \triangleq \frac{a^2}{c_{\mathrm{p}}T_{01}}, \qquad f_1 \triangleq \frac{r_1}{r_2}, \qquad f_2 \triangleq \frac{p_0}{a^2\rho\tan\beta_{1\mathrm{b}}}, \tag{6.24}$$

$$f_3 \triangleq \frac{p_0}{a^2\rho\tan\alpha_{2\mathrm{b}}}, \qquad f_4 \triangleq k_{\mathrm{fi}}\left(\frac{Ap_0}{a^2}\right)^2, \qquad f_5 \triangleq k_{\mathrm{fd}}\left(\frac{Ap_0}{a^2}\right)^2. \tag{6.25}$$

It is important to note that the compressor characteristic map given by (6.22) holds for the case where the flow through the compressor is positive. In the case of deep surge involving negative mass flow, it is assumed that the pressure rise in the compressor is proportional to the square of the mass flow so that [61, 97]

$$\psi_{\mathrm{c}}(\phi,\tilde{\omega}) = \mu\phi^2 + \psi_{\mathrm{c}_0}(\tilde{\omega}), \qquad \phi < 0, \tag{6.26}$$

where μ is a constant and

$$\psi_{\mathrm{c}_0}(\tilde{\omega}) \triangleq \psi_{\mathrm{c}}(\phi,\tilde{\omega})\Big|_{\phi=0} = \left(1+\sigma\eta_{\mathrm{c}_0}\bar{d}\tilde{\omega}^2\right)^{\frac{\gamma}{\gamma-1}} - 1, \tag{6.27}$$

where

$$\eta_{\mathrm{c}_0} \triangleq \eta_{\mathrm{c}}(\phi,\tilde{\omega})\Big|_{\phi=0} = \frac{2\sigma}{\sigma^2 + 2\sigma + f_1^2}. \tag{6.28}$$

Now, for a fixed $\tilde{\omega}$, taking the gradient of $\psi_{\mathrm{c}}(\phi,\tilde{\omega})$ with respect to the nondimensional flow ϕ it follows that the flow corresponding to the maximum

pressure point of the compressor characteristic map is directly proportional to the nondimensional angular velocity of the compressor spool and is given by $\phi_{\max} = k_f\tilde{\omega}$, where

$$k_f \triangleq \frac{f_1 f_2 + \sigma f_3}{f_2^2 + f_3^2 + 2f_4 + 2f_5}. \tag{6.29}$$

Similarly, for a fixed $\tilde{\omega}$ taking the gradient of $\eta_c(\phi, \tilde{\omega})$ with respect to the nondimensional flow ϕ we obtain that the maximum value for the isentropic efficiency is given by

$$\eta_{c_{\max}} = \frac{2\sigma(f_2^2 + f_3^2 + 2(f_4 + f_5))}{\sigma(2+\sigma)f_2^2 - 2\sigma f_1 f_2 f_3 + (2\sigma + f_1^2)f_3^2 + 2(2\sigma + \sigma^2 + f_1^2)(f_4 + f_5)}. \tag{6.30}$$

Note that $\eta_{c_{\max}}$ is constant for all spool speeds. This indicates that the compressor achieves the same maximum isentropic efficiency at each maximum pressure point for all spool speeds. However, since these points are critically stable, the need for active control is severe to guarantee stable compression system operation for peak compressor performance. Figure 6.2 shows a typical family of compressor characteristic maps for different spool speeds along with the corresponding constant isentropic efficiency lines. The stone wall depicted in Figure 6.2 corresponds to choked flow at a given cross-section of the compression system. Specifically, assuming that area choking occurs at the impeller and the process is isentropic, it follows [40, p. 211] that

$$\frac{T_{\mathrm{ch}}}{T_{01}} = \frac{2}{\gamma - 1}\left(1 + \frac{r_1^2\omega^2}{2c_p T_{01}}\right), \tag{6.31}$$

$$\frac{\rho_{\mathrm{ch}}}{\rho_{01}} = \left(\frac{T_{\mathrm{ch}}}{T_{01}}\right)^{\frac{1}{\gamma-1}}, \tag{6.32}$$

where T_{ch} is the stagnation temperature, ρ_{ch} is the stagnation density, and ρ_{01} is the flow density at the compressor inlet. Using a mass balance (6.31) and (6.32) yield

$$m_{\mathrm{ch}} = A_e \rho_{01} a_{01} \left[\frac{2a_{01}^2 + (\gamma - 1)r_1^2\omega^2}{a_{01}^2(\gamma + 1)}\right]^{\frac{\gamma+1}{2(\gamma-1)}}, \tag{6.33}$$

where m_{ch} is the choked mass flow, A_e is the area of the impeller eye, and a_{01} is the speed of sound at the inlet. Using $\tilde{\omega} = \frac{r_2\omega}{a}$, the nondimensional form of (6.33) is given by

$$\phi_{\mathrm{ch}} = \bar{e}\left[\frac{2 + (\gamma - 1)f_1^2\tilde{\omega}^2}{\gamma + 1}\right]^{\frac{\gamma+1}{2(\gamma-1)}}, \tag{6.34}$$

where $\phi_{\mathrm{ch}} \triangleq \frac{a}{A\rho_0}m_{\mathrm{ch}}$ and $\bar{e} \triangleq \frac{\gamma A_e}{A}$.

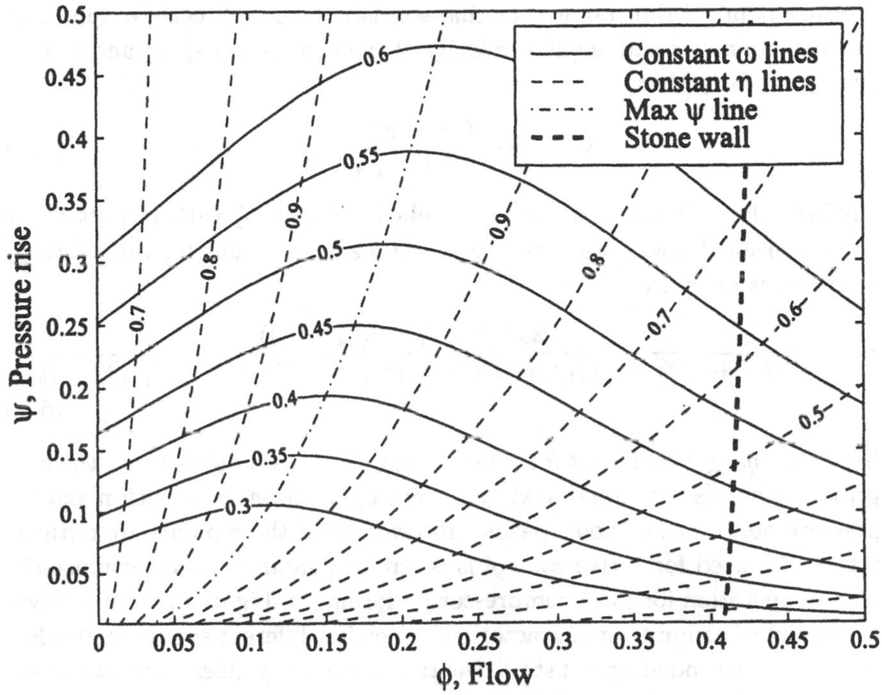

Fig. 6.2. Compressor characteristic maps and efficiency lines for different spool speeds

6.2.3 Turbocharger Spool Dynamics

Using conservation of angular momentum in the turbocharger spool it follows that the spool dynamics are given by

$$I_s \frac{d\omega}{dt} = \tau_d - \tau_c(m_c, \omega), \tag{6.35}$$

where I_s is the spool mass moment of inertia, τ_d is the driving torque, and $\tau_c(m_c, \omega)$ is the compressor torque. Substituting (6.14) into (6.35) we obtain

$$I_s \frac{d\omega}{dt} = \tau_d - \sigma r_2^2 m_c \omega, \tag{6.36}$$

which, using (6.6) and $\tilde{\omega} = \frac{r_2 \omega}{a}$, can be written in nondimensional form as

$$\dot{\tilde{\omega}} = \bar{c}(\tau - \sigma|\phi|\tilde{\omega}), \tag{6.37}$$

where

$$\bar{c} \triangleq \frac{L^3 r_2^2 p_0}{I_s a^2}, \qquad \tau \triangleq \frac{\tau_d}{A p_0 r_2}. \tag{6.38}$$

Note that in (6.37) the fact that the compressor may enter deep surge has been taken into account. In this case reverse flow can occur, during which the centrifugal compressor can be viewed as a throttling device, and hence can be approximated as a turbine [45].

6.3 Parameterized System Equilibria and Local Set Point Designs

In this section we develop Lyapunov-based subcontroller designs for local set points parameterized by the flow through the throttle and the nondimensional driving torque. It is important to note that even though a Lyapunov-based framework [87] can be used to stabilize the compression system, the resulting controller may generate unnecessarily large control amplitude and rate signals that can amplitude and rate saturate the control actuators resulting in system performance degradation and even instability (see [1] and the references therein). To proceed with the local set point designs, first note that with control inputs $u_1 \triangleq \gamma_{\rm th}\sqrt{\psi}$ and $u_2 \triangleq \tau$ it follows from (6.7), (6.21), and (6.37), that a state space model for the centrifugal compressor is given by

$$\dot{\psi} = \bar{a}(\phi - u_1), \tag{6.39}$$

$$\dot{\phi} = \bar{b}(\psi_{\rm c}(\phi, \tilde{\omega}) - \psi), \tag{6.40}$$

$$\dot{\tilde{\omega}} = \bar{c}(u_2 - \sigma\phi\tilde{\omega}). \tag{6.41}$$

To carry out Step 1 of Algorithm 3.1, let $q = m = 2$ and $\varphi(\psi, \phi, \tilde{\omega}, \lambda) = \frac{1}{\alpha}\lambda$, where α is a scaling factor and $\lambda \triangleq [\lambda_1 \, \lambda_2]^{\rm T} \in \mathbb{R}^2$, so that the system equilibria are parameterized by the constant control $u(t) = \frac{1}{\alpha}\lambda$, where $u(t) \triangleq [u_1(t) \, u_2(t)]^{\rm T}$. In this case, (6.39)–(6.41) have an equilibrium point at $(\psi_\lambda, \phi_\lambda, \tilde{\omega}_\lambda)$, where

$$(\psi_\lambda, \ \phi_\lambda, \ \tilde{\omega}_\lambda) = \left(\psi_{\rm c}(\tfrac{\lambda_1}{\alpha}, \tfrac{\lambda_2}{\sigma\lambda_1}), \ \tfrac{\lambda_1}{\alpha}, \ \tfrac{\lambda_2}{\sigma\lambda_1}\right). \tag{6.42}$$

Next, we carry out Step 2 of Algorithm 3.1. Specifically we show that for $\lambda_1 > 0$ and $\lambda_2 > 0$ there exists a control law such that the equilibrium point $(\psi_\lambda, \phi_\lambda, \tilde{\omega}_\lambda)$ of (6.39)–(6.41) is *globally* asymptotically stable. To show this, define the shifted variables $x_{\lambda 1} \triangleq \psi - \psi_\lambda$, $x_{\lambda 2} \triangleq \phi - \phi_\lambda$, and $x_{\lambda 3} \triangleq \tilde{\omega} - \tilde{\omega}_\lambda$, so that the given equilibrium point is translated to the origin. Furthermore, with the shifted controls $\tilde{u}_1 \triangleq u_1 - \frac{\lambda_1}{\alpha}$ and $\tilde{u}_2 \triangleq u_2 - \frac{\lambda_2}{\alpha}$, it follows that the parameterized translated nonlinear system is given by

$$\dot{x}_{\lambda 1} = \bar{a}(x_{\lambda 2} - \tilde{u}_1), \tag{6.43}$$

$$\dot{x}_{\lambda 2} = \bar{b}(\psi_{{\rm c}\lambda}(x_{\lambda 2}, x_{\lambda 3}) - x_{\lambda 1}), \tag{6.44}$$

$$\dot{x}_{\lambda 3} = \bar{c}(\tilde{u}_2 - f(x_{\lambda 2}, x_{\lambda 3})), \tag{6.45}$$

where

$$\psi_{c\lambda}(x_{\lambda 2}, x_{\lambda 3}) \triangleq \psi_c(\phi_\lambda + x_{\lambda 2}, \tilde{\omega}_\lambda + x_{\lambda 3}) - \psi_c(\phi_\lambda, \tilde{\omega}_\lambda), \tag{6.46}$$

$$f(x_{\lambda 2}, x_{\lambda 3}) \triangleq \sigma|\phi_\lambda + x_{\lambda 2}|(\tilde{\omega}_\lambda + x_{\lambda 3}) - \sigma\phi_\lambda\tilde{\omega}_\lambda. \tag{6.47}$$

Now, setting

$$\hat{u}_\lambda = x_{\lambda 2} - \tilde{u}_1, \tag{6.48}$$

$$\tilde{u}_2 = -k_3 x_{\lambda 3} + f(x_{\lambda 2}, x_{\lambda 3}), \tag{6.49}$$

where $k_3 > 0$, and substituting (6.48) and (6.49) into (6.43)–(6.45) yields

$$\dot{x}_{\lambda 1} = \bar{a}\hat{u}_\lambda, \tag{6.50}$$

$$\dot{x}_{\lambda 2} = \bar{b}(\psi_{c\lambda}(x_{\lambda 2}, x_{\lambda 3}) - x_{\lambda 1}), \tag{6.51}$$

$$\dot{x}_{\lambda 3} = -k_3 \bar{c} x_{\lambda 3}. \tag{6.52}$$

Next, consider the equilibria-dependent Lyapunov function candidate

$$\tilde{V}_\lambda(x_{\lambda 1}, x_{\lambda 2}, x_{\lambda 3}) = \frac{\alpha_1}{2}(x_{\lambda 1} - \psi_{c\lambda}(x_{\lambda 2}, x_{\lambda 3}) - k_2 x_{\lambda 2})^2 + \frac{\alpha_2}{2}x_{\lambda 2}^2 + \frac{\alpha_3}{2}x_{\lambda 3}^2, \tag{6.53}$$

where $k_2 > 0$, $\alpha_i > 0$, $i = 1, 2, 3$. The corresponding Lyapunov derivative is given by

$$\begin{aligned}\dot{\tilde{V}}_\lambda(x_{\lambda 1}, x_{\lambda 2}, x_{\lambda 3}) &= \alpha_1(x_{\lambda 1} - \psi_{c\lambda}(x_{\lambda 2}, x_{\lambda 3}) - k_2 x_{\lambda 2})(\dot{x}_{\lambda 1} - \dot{\psi}_{c\lambda}(x_{\lambda 2}, x_{\lambda 3}) \\ &\quad - k_2\dot{x}_{\lambda 2}) + \alpha_2 x_{\lambda 2}\dot{x}_{\lambda 2} + \alpha_3 x_{\lambda 3}\dot{x}_{\lambda 3} \\ &= -(x_{\lambda 1} - \psi_{c\lambda}(x_{\lambda 2}, x_{\lambda 3}) - k_2 x_{\lambda 2})(-\alpha_1\bar{a}\hat{u}_\lambda \\ &\quad + \alpha_1\psi_{c\lambda, x_{\lambda 2}}(x_{\lambda 2}, x_{\lambda 3})\dot{x}_{\lambda 2} + \alpha_1\psi_{c\lambda, x_{\lambda 3}}(x_{\lambda 2}, x_{\lambda 3})\dot{x}_{\lambda 3} \\ &\quad + \alpha_1 k_2\dot{x}_{\lambda 2} + \alpha_2\bar{b}x_{\lambda 2}) - \alpha_2 k_2\bar{b}x_{\lambda 2}^2 - \alpha_3 k_3\bar{c}x_{\lambda 3}^2,\end{aligned}$$

where

$$\begin{aligned}\psi_{c\lambda, x_{\lambda 2}}(x_{\lambda 2}, x_{\lambda 3}) &\triangleq \frac{\partial\psi_{c\lambda}(x_{\lambda 2}, x_{\lambda 3})}{\partial x_{\lambda 2}} = \left.\frac{\partial\psi_c(\phi, \tilde{\omega})}{\partial\phi}\right|_{(\phi,\tilde{\omega}) = (\phi_\lambda + x_{\lambda 2},\, \tilde{\omega}_\lambda + x_{\lambda 3})} \\ &= \frac{\gamma\sigma\bar{d}\tilde{\omega}^2}{\gamma - 1}\left[\psi_c(\phi_\lambda + x_{\lambda 2}, \tilde{\omega}_\lambda + x_{\lambda 3}) + 1\right]^{\frac{1}{\gamma}} f_1(x_{\lambda 2}, x_{\lambda 3}), \\ \psi_{c\lambda, x_{\lambda 3}}(x_{\lambda 2}, x_{\lambda 3}) &\triangleq \frac{\partial\psi_{c\lambda}(x_{\lambda 2}, x_{\lambda 3})}{\partial x_{\lambda 3}} = \left.\frac{\partial\psi_c(\phi, \tilde{\omega})}{\partial\tilde{\omega}}\right|_{(\phi,\tilde{\omega}) = (\phi_\lambda + x_{\lambda 2},\, \tilde{\omega}_\lambda + x_{\lambda 3})} \\ &= \frac{\gamma\sigma\bar{d}}{\gamma - 1}\left[\psi_c(\phi_\lambda + x_{\lambda 2}, \tilde{\omega}_\lambda + x_{\lambda 3}) + 1\right]^{\frac{1}{\gamma}} f_2(x_{\lambda 2}, x_{\lambda 3}),\end{aligned}$$

with

$$f_1(x_{\lambda 2}, x_{\lambda 3}) \triangleq \left.\frac{\partial \eta_c(\phi,\tilde{\omega})}{\partial \phi}\right|_{(\phi,\tilde{\omega})=(\phi_\lambda + x_{\lambda 2},\, \tilde{\omega}_\lambda + x_{\lambda 3})}$$

$$\frac{\partial \eta_c(\phi,\tilde{\omega})}{\partial \phi} = \frac{\sigma\tilde{\omega}^2(f_2(\tilde{\omega}f_1 - \phi f_2) + f_3(\sigma\tilde{\omega} - \phi f_3) - 2\phi(f_4 + f_5))}{(\sigma\tilde{\omega}^2 + \frac{1}{2}(\tilde{\omega}f_1 - \phi f_2)^2 + \frac{1}{2}(\sigma\tilde{\omega} - \phi f_3)^2 + \phi^2(f_4 + f_5))^2},$$

and

$$f_2(x_{\lambda 2}, x_{\lambda 3}) \triangleq \left.\frac{\partial \left[\tilde{\omega}^2\eta_c(\phi,\tilde{\omega})\right]}{\partial \tilde{\omega}}\right|_{(\phi,\tilde{\omega})=(\phi_\lambda + x_{\lambda 2},\, \tilde{\omega}_\lambda + x_{\lambda 3})}$$

$$\frac{\partial \left[\tilde{\omega}^2\eta_c(\phi,\tilde{\omega})\right]}{\partial \tilde{\omega}} = \frac{\sigma\tilde{\omega}^3(2\sigma\tilde{\omega}^2 + \tilde{\omega}f_1(\tilde{\omega}f_1 - \phi f_2) + \sigma\tilde{\omega}(\sigma\tilde{\omega} - \phi f_3))}{(\sigma\tilde{\omega}^2 + \frac{1}{2}(\tilde{\omega}f_1 - \phi f_2)^2 + \frac{1}{2}(\sigma\tilde{\omega} - \phi f_3)^2 + \phi^2(f_4 + f_5))^2} - 2\tilde{\omega}\phi\frac{\partial \eta_c(\phi,\tilde{\omega})}{\partial \phi}.$$

Now, choosing the nonlinear control law

$$\hat{u}_\lambda = -k_1(x_{\lambda 1} - \psi_{c\lambda}(x_{\lambda 2}, x_{\lambda 3}) - k_2 x_{\lambda 2}) + \frac{1}{\bar{a}}\Big[\bar{b}(k_2 + \psi_{c\lambda, x_{\lambda 2}}(x_{\lambda 2}, x_{\lambda 3})) \cdot(\psi_{c\lambda}(x_{\lambda 2}, x_{\lambda 3}) - x_{\lambda 1}) - k_3\bar{c}\psi_{c\lambda, x_{\lambda 3}}(x_{\lambda 2}, x_{\lambda 3})x_{\lambda 3} + \frac{\alpha_2}{\alpha_1}\bar{b}x_{\lambda 2}\Big],$$

where $k_1 > 0$, it follows that

$$\dot{V}_\lambda(x_{\lambda 1}, x_{\lambda 2}, x_{\lambda 3}) = -\alpha_1 k_1 \bar{a}(x_{\lambda 1} - \psi_{c\lambda}(x_{\lambda 2}, x_{\lambda 3}) - k_2 x_{\lambda 2})^2 - \alpha_2 k_2 \bar{b} x_{\lambda 2}^2 - \alpha_3 k_3 \bar{c} x_{\lambda 3}^2 < 0,$$

for $(x_{\lambda 1}, x_{\lambda 2}, x_{\lambda 3}) \neq (0,0,0)$, which guarantees that all parameterized system equilibria given by (6.42) of the nonlinear system (6.39)–(6.41) are globally asymptotically stable when using the equilibria-dependent nonlinear feedback controller

$$\begin{bmatrix} \tilde{u}_1 \\ \tilde{u}_2 \end{bmatrix} = \tilde{\phi}_\lambda(x_{\lambda 1}, x_{\lambda 2}, x_{\lambda 3}) \triangleq \begin{bmatrix} x_{\lambda 2} - \hat{u}_\lambda \\ -k_3 x_{\lambda 3} + f(x_{\lambda 2}, x_{\lambda 3}) \end{bmatrix}. \quad (6.54)$$

As mentioned above, however, the nonlinear Lyapunov-based controller (6.54) may generate unnecessarily large control amplitude and rate signals leading to actuator saturation. In the next two sections, we develop globally stabilizing switching control strategies that directly address actuator amplitude and rate saturation constraints, as well as inverse optimality notions.

6.4 Hierarchical Nonlinear Switching Control for Centrifugal Compression Systems

In this section we use the hierarchical nonlinear control framework developed in Chapter 3 to design globally stabilizing controllers for controlling

the centrifugal compressor model (6.39)–(6.41). Define the shifted variables $\psi_s \triangleq \psi - \psi_m$, $\phi_s \triangleq \phi - \phi_m$, $\tilde{\omega}_s \triangleq \tilde{\omega} - \tilde{\omega}_m$, where $(\psi_m, \phi_m, \tilde{\omega}_m)$ are the coordinates of the desired equilibrium point. Now, rewriting the control law (6.54) and the Lyapunov function (6.53) in terms of the shifted variables $(\psi_s, \phi_s, \tilde{\omega}_s)$, we obtain the shifted control law

$$\phi_\lambda(\psi_s, \phi_s, \tilde{\omega}_s) \triangleq \tilde{\phi}_\lambda(\psi_s + \psi_m - \psi_\lambda, \phi_s + \phi_m - \phi_\lambda, \tilde{\omega}_s + \tilde{\omega}_m - \tilde{\omega}_\lambda), \quad (6.55)$$

which globally stabilizes the equilibrium point $(\psi_\lambda, \phi_\lambda, \tilde{\omega}_\lambda)$ with an associated Lyapunov function

$$V_\lambda(\psi_s, \phi_s, \tilde{\omega}_s) \triangleq \tilde{V}_\lambda(\psi_s + \psi_m - \psi_\lambda, \phi_s + \phi_m - \phi_\lambda, \tilde{\omega}_s + \tilde{\omega}_m - \tilde{\omega}_\lambda), \quad (6.56)$$

and an estimate of the domain of attraction given by $\mathcal{D}_\lambda \triangleq \{(\psi_s, \phi_s, \tilde{\omega}_s) : V_\lambda(\psi_s, \phi_s, \tilde{\omega}_s) \leq c_\lambda\}$, where $c_\lambda > 0$ is the finite value of $V_\lambda(\cdot)$ on the boundary of $\mathcal{D}_\lambda \neq \emptyset$.

Next, let the diffeomorphism $\sigma : [0, a] \to \mathbb{R}^2$, $a > 0$, be such that $(\psi_{\sigma(s)}, \phi_{\sigma(s)}, \tilde{\omega}_{\sigma(s)})$, $s \in [0, a]$, is an equilibrium point of (6.43)–(6.45) in the shifted variables $(\psi_s, \phi_s, \tilde{\omega}_s)$. In particular, $s = 0$ corresponds to the origin in the shifted variables and hence corresponds to the desired maximum pressure point in the original variables. Furthermore, assume that $c_{\sigma(s)}$ is a C^1 function of $s \in [0, a]$ and note that since $c_{\sigma(s)} > 0$ and $V_{\sigma(s)}(\cdot)$ is continuous and radially unbounded, $\mathcal{D}_{\sigma(s)}$ is a compact set for $s \in [0, a]$, which further implies that $\mathcal{D}_{\sigma(s)}$ is a positively invariant set of (6.39)–(6.41) with feedback control law $\phi_{\sigma(s)}(\psi_s, \phi_s, \tilde{\omega}_s)$, $s \in [0, a]$. Finally, since $\mathcal{D}_{\sigma(s)}$ is not empty for all $s \in [0, a]$, there exists $s_1, s_2 > 0$ such that $s_2 < s_1$ and $(\psi_{\sigma(s_1)}, \phi_{\sigma(s_1)}, \tilde{\omega}_{\sigma(s_1)}) \in \mathcal{D}_{\sigma(s_2)}$.

To carry out Step 3 of Algorithm 3.1, we consider two topologies for the switching set $\mathcal{S}$; namely an isolated point topology and a continuous topology. For $\mathcal{S}$ consisting of countably finite isolated points, let $\mathcal{S} = \{\lambda_0, \ldots, \lambda_q\}$ where $\lambda_k \triangleq \sigma(s_k)$, $k = 0, 1, \ldots, q$, and $0 = s_q < \cdots < s_0 \leq a$, $a > 0$, such that $(\psi_{\lambda_i}, \phi_{\lambda_i}, \tilde{\omega}_{\lambda_i}) \in \mathcal{D}_{\lambda_{i+1}}$, $i \in \{0, \ldots, q-1\}$, and let $p(\lambda) = \sigma^{-1}(\lambda) = s$, $\lambda \in \mathcal{S}$. To guarantee that $p(\cdot)$ satisfies Assumption 3.1 construct s_k, $k = 0, 1, \ldots, q$, online by considering the smallest solution $s_k \geq 0$ to the equation $V_{\sigma(s_k)}(x(t_k)) = c_{\sigma(s_k)}$, $t_k \triangleq k\Delta T$, where $\Delta T > 0$ and $k = 0, 1, \ldots, q$, and define $\mathcal{S} \triangleq \{\lambda_k\}_{k=0}^{q}$, where $\lambda_k = \sigma(s_k)$. Now, with the feedback switching control law $u = \phi_{\lambda_S(\psi_s, \phi_s, \tilde{\omega}_s)}(\psi_s, \phi_s, \tilde{\omega}_s)$, where $\lambda_S(\psi_s, \phi_s, \tilde{\omega}_s)$ is obtained as described in Step 4 of Algorithm 3.1, it follows from Theorem 3.5 that the equilibrium point $(\psi_m, \phi_m, \tilde{\omega}_m)$ is globally asymptotically stable. Furthermore, note that $\lambda_S(\psi_s(t), \phi_s(t), \tilde{\omega}_s(t))$, $t \geq 0$, is piecewise constant and hence the feedback switching control law $u = \phi_{\lambda_S(\psi_s, \phi_s, \tilde{\omega}_s)}(\psi_s, \phi_s, \tilde{\omega}_s)$ is piecewise continuous.

For $\mathcal{S}$ consisting of a continuous topology let $\mathcal{S} \triangleq \sigma([0,a])$ and let $p(\lambda) = \sigma^{-1}(\lambda) = s$, $\lambda \in \mathcal{S}$. By requiring that $p(\cdot)$ does not have a local minimum in $\mathcal{S}$ (other than the origin) and since every $\lambda \in \mathcal{S}$ is an accumulation point for $\mathcal{S}$, we are guaranteed, by Step 3b of Algorithm 3.1, that Assumption 3.1 is satisfied. Now global attraction of $(\psi_{\rm m}, \phi_{\rm m}, \tilde{\omega}_{\rm m})$ is guaranteed by Theorem 3.5 with the feedback control law $u = \phi_{\lambda_{\mathcal{S}}(\psi_{\rm s},\phi_{\rm s},\tilde{\omega}_{\rm s})}(\psi_{\rm s}, \phi_{\rm s}, \tilde{\omega}_{\rm s})$, where $\lambda_{\mathcal{S}}(\psi_{\rm s}, \phi_{\rm s}, \tilde{\omega}_{\rm s})$ is obtained as described in Step 4 of Algorithm 3.1. In particular, if $(\psi_{\rm s}(0), \phi_{\rm s}(0), \tilde{\omega}_{\rm s}(0)) \in \mathcal{D} \triangleq \cup_{s\in[0,a]} \mathcal{D}_{\sigma(s)}$ then $\lambda_{\mathcal{S}}(\psi_{\rm s}(t), \phi_{\rm s}(t), \tilde{\omega}_{\rm s}(t))$, $t \geq 0$, is a continuous function. Alternatively, if $(\psi_{\rm s}(0), \phi_{\rm s}(0), \tilde{\omega}_{\rm s}(0)) \notin \mathcal{D}$ then $\lambda_{\mathcal{S}}(\psi_{\rm s}(t), \phi_{\rm s}(t), \tilde{\omega}_{\rm s}(t)) = \sigma(a)$, $t \in [0, \hat{t})$, where $\hat{t} > 0$ is such that $(\psi_{\rm s}(\hat{t}), \phi_{\rm s}(\hat{t}), \tilde{\omega}_{\rm s}(\hat{t})) \in \partial\mathcal{D}$. In this case, $\lambda_{\mathcal{S}}(\psi_{\rm s}(t), \phi_{\rm s}(t), \tilde{\omega}_{\rm s}(t))$, $t \geq 0$, is continuous modulo one discontinuity at $t = \hat{t}$.

Next, if $(\psi_{\rm s}(0), \phi_{\rm s}(0), \tilde{\omega}_{\rm s}(0)) \in \mathcal{D}$, the online fixed-order dynamic compensation procedure given in Section 3.5 can be employed to compute $\lambda_{\mathcal{S}}(\psi_{\rm s}(t), \phi_{\rm s}(t), \tilde{\omega}_{\rm s}(t))$, $t \in [0, T_{x_0}]$, using the update law (3.21). Note that the compensator dynamics given by (3.21) characterize the admissible rate of the compensator state $\lambda(t)$ such that the switching nonlinear controller guarantees that $(\psi_{\rm s}(t), \phi_{\rm s}(t), \tilde{\omega}_{\rm s}(t)) \in \partial\mathcal{D}_{\lambda(t)}$, $t \in [0, T_{x_0}]$.

Once again, it is important to note that the proposed switching nonlinear controller framework can be incorporated to address practical actuator limitations such as control amplitude and rate saturation constraints. Specifically, since the dynamic compensator state $\lambda(t)$ is proportional to the throttle opening and the nondimensional driving torque and since the dynamics given by (3.21) indirectly characterize the fastest admissible rate at which the control variables can change while maintaining stability of the controlled system, it follows that by constraining the rate at which the dynamics of $\lambda(t)$ can evolve on the equilibrium branch effectively places a rate constraint on the throttle opening and the nondimensional driving torque. This corresponds to the case where the switching rate of the nonlinear controller is decreased so that the trajectory $(\psi_{\rm s}(t), \phi_{\rm s}(t), \tilde{\omega}_{\rm s}(t))$, $t \geq 0$, is allowed to enter $\mathcal{D}_{\lambda(t)}$. Additionally, amplitude saturation constraints and state constraints can also be enforced by simply choosing $s_{\max} > 0$ such that $\mathcal{D}_{\max} \triangleq \cup_{s\in[0,s_{\max}]} \mathcal{D}_{\sigma(s)}$ is contained in the region where the system is constrained to operate. In this case, the switching nonlinear controller guarantees attraction of $(\psi_{\rm m}, \phi_{\rm m}, \tilde{\omega}_{\rm m})$ with an estimate of the domain of attraction given by $\mathcal{D}_{\max}$.

Next, we apply the hierarchical nonlinear switching control framework developed in Chapter 3 to the control of surge in centrifugal compression systems. Specifically, we use the three-state centrifugal compressor model derived in this chapter with $(\bar{a}, \bar{b}, \bar{c}, \bar{d}) = (9.37, 310.81, 23.70, 0.38)$, $(f_1, f_2, f_3, f_4, f_5) = (0.44, 1.07, 2.18, 0.17, 0.12)$, $\gamma = 1.4$, $\mu = 5$, and $\sigma = 0.9$.

We compare the open-loop response when the compression system is taken from an operating speed of 20,000 rpm to 25,000 rpm, corresponding to the initial conditions $(\psi_0, \phi_0, \tilde{\omega}_0) = (0.305, 0.177, 0.493)$, with the closed-loop responses obtained using the design parameters $(\alpha_1, \alpha_2, \alpha_3) = (1, 0.1, 1)$ and $(k_1, k_2, k_3) = (1, 3, 1)$ and the scaling factor $\alpha = 10$.

For the standard nonlinear switching control framework, we use the diffeomorphism $\sigma(s) = (\alpha s, 0)$, $s \in [0, 0.5]$, and $c_s = 0.01 + 2s$. Furthermore, we consider the closed-loop responses obtained with and without a rate saturation constraint on the throttle opening ($|\dot{\gamma}_{th}| \leq 5$). Figure 6.3 shows the ψ–ϕ phase portrait of the state trajectories. The pressure rise, mass flow, and spool speed variations for the open-loop and controlled system are shown in Figures 6.4, 6.5, and 6.6, respectively. Figures 6.7 and 6.8 show the control effort versus time. This comparison illustrates that open-loop control drives the compression system into deep surge while the proposed nonlinear switching controller drives the system to the desired maximum pressure-flow equilibrium point $(\psi_\lambda, \phi_\lambda, \tilde{\omega}_\lambda) = (0.656, 0.248, 0.690)$. Note that the switching controller with a rate saturation constraint guarantees stability with minimal degradation in system performance.

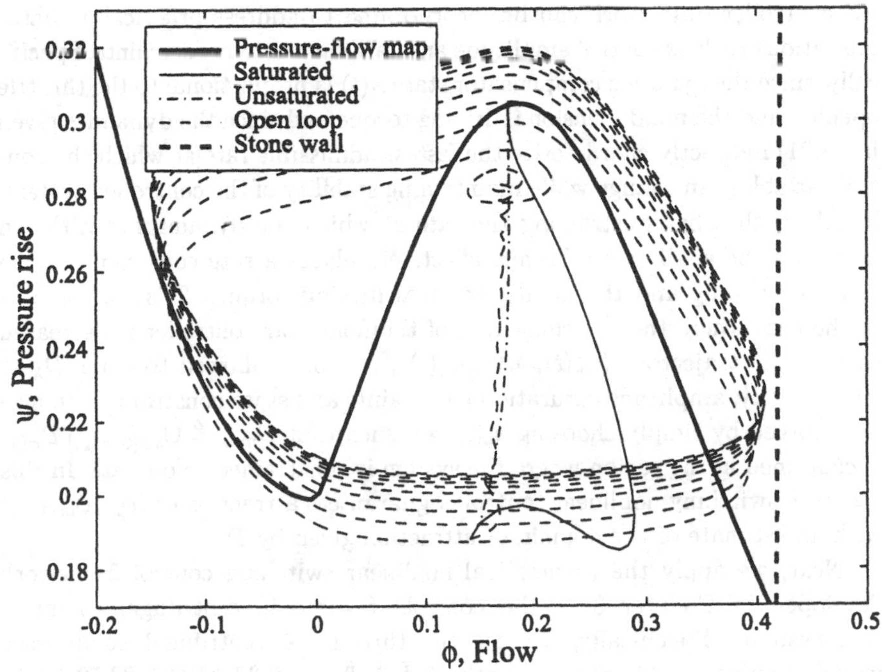

Fig. 6.3. Phase portrait of pressure-flow map

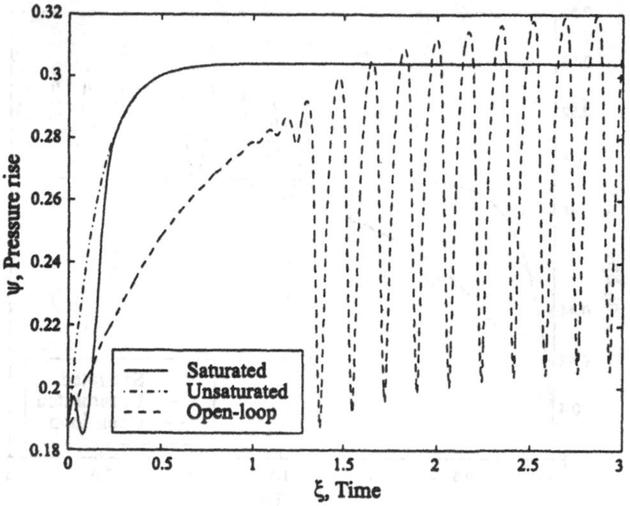

Fig. 6.4. Pressure rise versus time

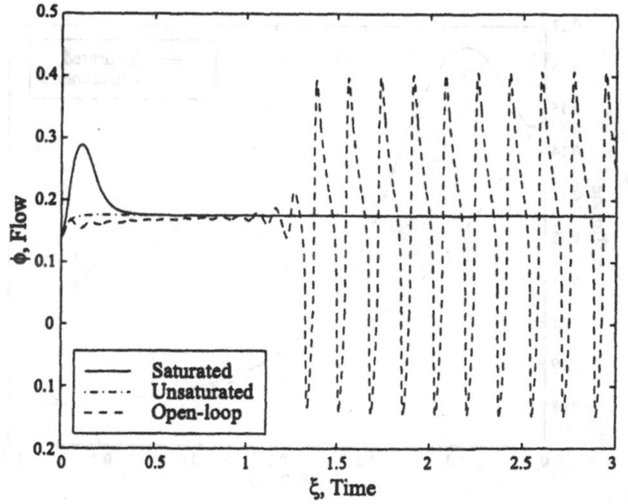

Fig. 6.5. Mass flow versus time

For the inverse optimal nonlinear switching control framework developed in Section 3.6, we construct a positive-definite potential function $p(\cdot)$ and solve the Extended Optimal Switching Control Problem by implementing the dynamic controller (3.38)–(3.40). This yields a switching function $\lambda_S(\cdot)$ such that the feedback control law $u \triangleq \phi_{\lambda_S(\psi_s,\phi_s,\omega_s)}(\psi_s, \phi_s, \omega_s)$ globally asymptotically stabilizes the operating condition $(\psi_s, \phi_s, \omega_s) = (0, 0, 0)$. Furthermore, since the dynamic compensator state $\lambda(t)$ is related to the throttle opening

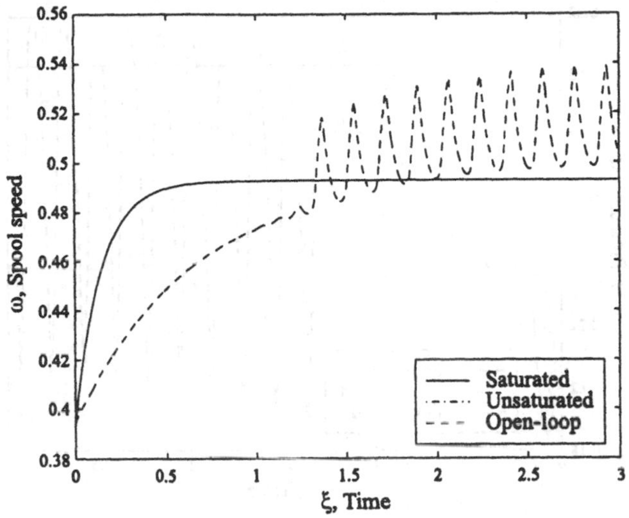

Fig. 6.6. Compressor spool speed versus time

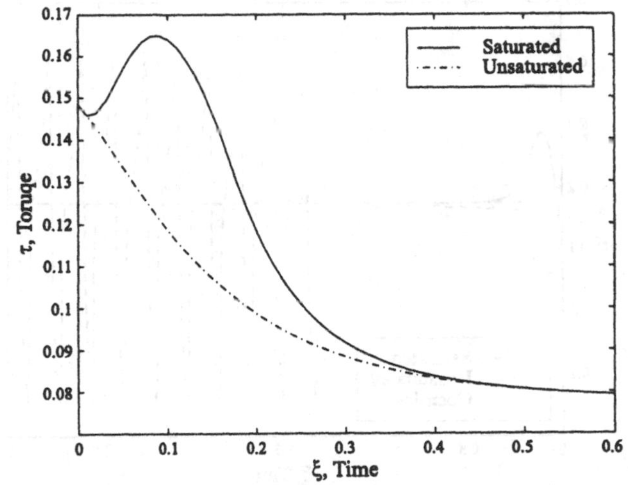

Fig. 6.7. Control effort versus time: Torque

and to the nondimensional driving torque, it follows that by constraining the rate at which the dynamics of $\lambda(t)$ can evolve effectively places a rate constraint on the control variables. Amplitude saturation constraints can be enforced by assigning a higher potential value to the parameters corresponding to the control values that are magnitude limited. Hence, by appropriately choosing the potential function, the proposed inverse optimal switching non-

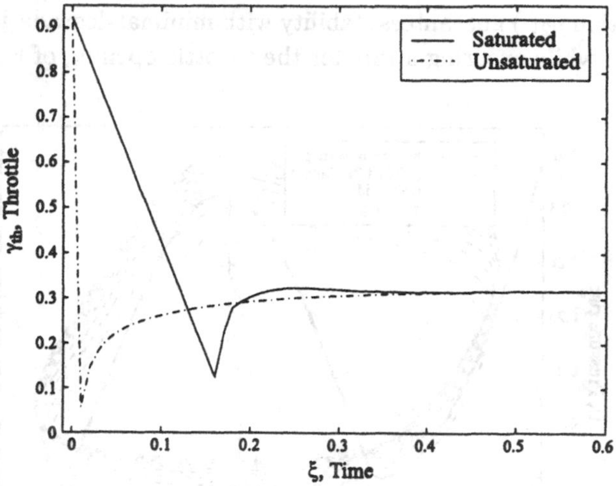

Fig. 6.8. Control effort versus time: Throttle opening

linear controller can address the practical limitations of control amplitude and rate saturation constraints.

To show the efficacy of the proposed inverse optimal switching control approach, we choose $c_\lambda = 0.4\,V_\lambda(\psi_m, \phi_m, \omega_m)$ and consider the potential functions

$$p_1(\lambda_1, \lambda_2) = \lambda_1^2 + \lambda_2^2,$$

and

$$p_2(\lambda_1, \lambda_2) = \begin{cases} \lambda_1^2 + \lambda_2^2, & d(\lambda_1, \lambda_2) > r, \\ \lambda_1^2 + \lambda_2^2 + h\left[1 - (\frac{\lambda_1 - \lambda_{1c}}{r}) - (\frac{\lambda_2 - \lambda_{2c}}{r})^2\right]^3, & d(\lambda_1, \lambda_2) \le r, \end{cases}$$

where $d(\lambda_1, \lambda_2) \triangleq \sqrt{(\lambda_1 - \lambda_{1c})^2 + (\lambda_2 - \lambda_{2c})^2}$. The potential function $p_1(\cdot)$ is a classical paraboloid-shaped function and the potential function $p_2(\cdot)$ is a modified paraboloid-shaped function obtained by adding to $p_1(\cdot)$ a "bump" whose geometry is determined by the parameters $h = 10$, $r = 1$, and $(\lambda_{1c}, \lambda_{2c}) = (-0.1, -1.0)$.

Figure 6.9 shows the ψ–ϕ phase portrait of the state trajectories. The pressure rise, mass flow, and spool speed variations for the open-loop and controlled system are shown in Figures 6.10, 6.11, and 6.12, respectively. Figures 6.13 and 6.14 show the control effort versus time. This comparison illustrates that open-loop control drives the compression system into deep surge while the proposed inverse optimal switching controller drives the system to the desired maximum pressure-flow equilibrium point $(\psi_\lambda, \phi_\lambda, \omega_\lambda) = (0.656, 0.248, 0.690)$. Note that the switching controller obtained using the po-

tential function $p_2(\cdot)$ guarantees stability with minimal degradation in system performance while ensuring a rate for the throttle opening of $|\dot{\gamma}_{th}| \leq 0.5$.

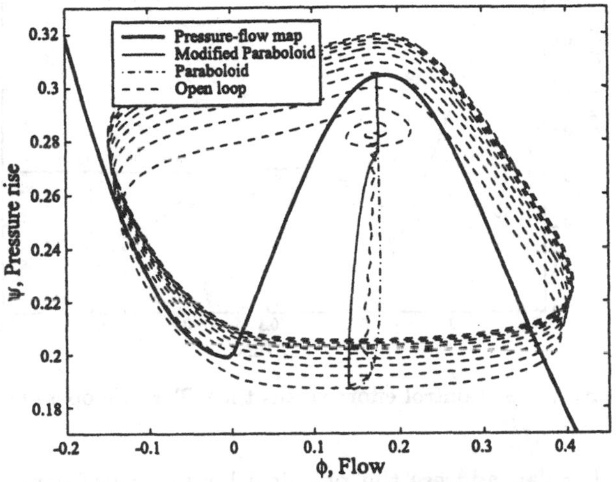

Fig. 6.9. Phase portrait of pressure-flow map

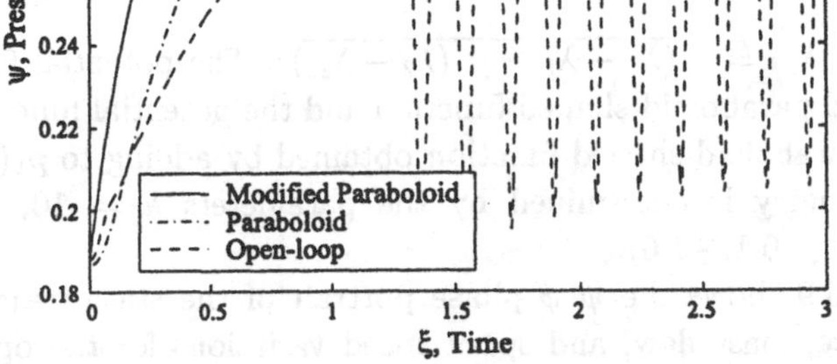

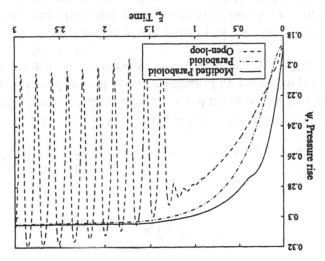

Fig. 6.10. Pressure rise versus time

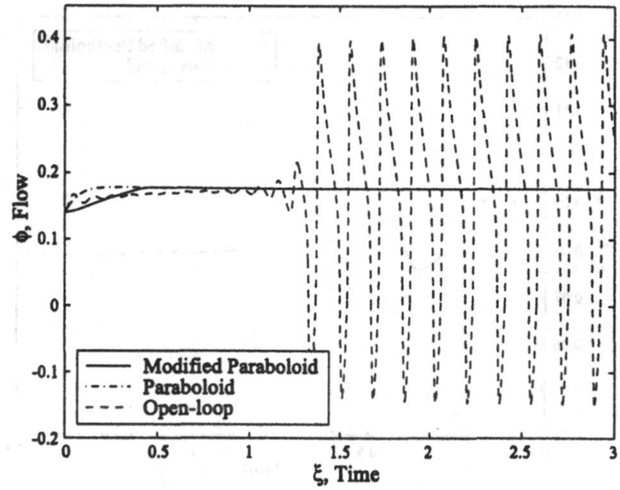

Fig. 6.11. Mass flow versus time

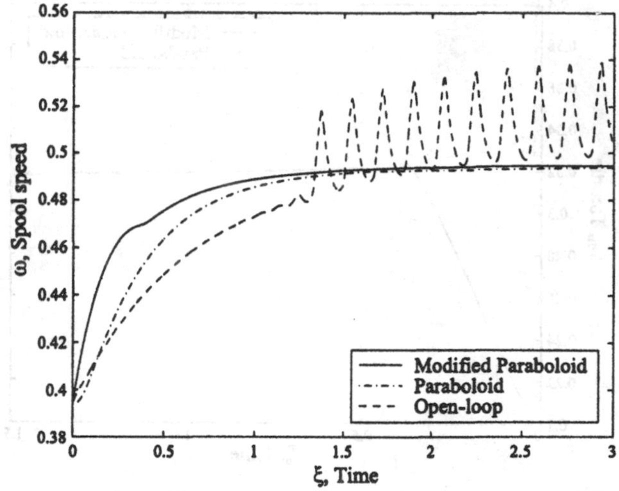

Fig. 6.12. Compressor spool speed versus time

6.5 Conclusion

A three-state centrifugal compressor surge model involving pressure and mass flow compression system dynamics as well as spool dynamics to account for the influence of speed transients on the compression surge dynamics was developed. Using the control framework developed in Chapter 3, a nonlinear globally stabilizing switching control law based on equilibria-dependent Lyapunov functions was developed. The proposed nonlinear switching control

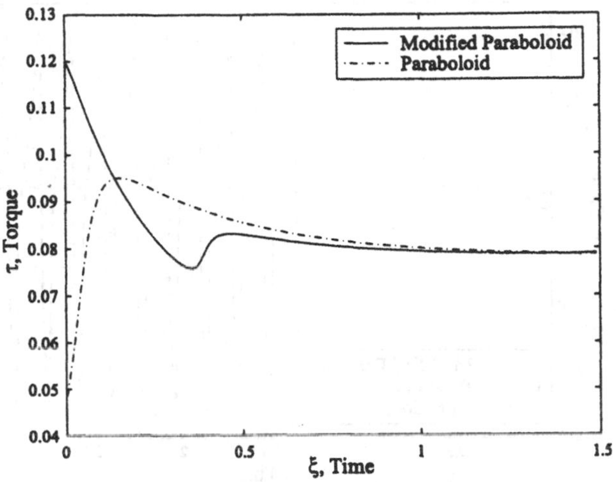

Fig. 6.13. Control effort versus time: Torque

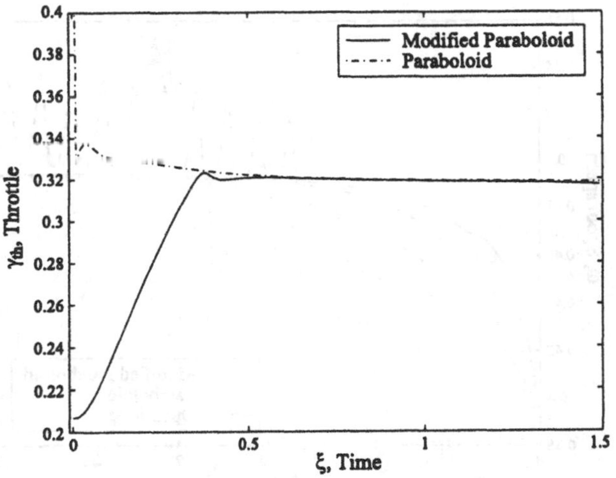

Fig. 6.14. Control effort versus time: Throttle opening

framework was shown to be directly applicable to centrifugal compression systems with actuator amplitude and rate saturation constraints.

7. Conclusions

The focus of this monograph was the synthesis of a hierarchical nonlinear switching control design framework for general nonlinear dynamical systems. The proposed approach was shown to account for actuator amplitude and rate saturation constraints, inverse optimality notions, and robustness to system modeling uncertainty. The effectiveness of this control framework was shown by addressing the control of the compressor aerodynamic instabilities of rotating stall and surge in jet engine propulsion systems.

To develop the hierarchical nonlinear switching control framework, we first developed generalized Lyapunov and invariant set stability theorems for nonlinear dynamical systems. In particular, local and global stability theorems were presented using generalized lower semicontinuous Lyapunov functions, providing a transparent generalization of standard Lyapunov and invariant set theorems. Using the generalized Lyapunov and invariant set theorems, a nonlinear control-system design framework predicated on a hierarchical switching controller architecture parameterized over a set of system equilibria was developed. Specifically, a hierarchical switching nonlinear control strategy was constructed to stabilize a given nonlinear system by stabilizing a collection of nonlinear controlled subsystems. The switching nonlinear controller architecture was designed based on a generalized Lyapunov function obtained by minimizing a potential function over a given switching set induced by the parameterized system equilibria. An online procedure for computing the switching scheme was proposed by constructing an initial value problem having a fixed-order dynamic compensator structure. Furthermore, an inverse optimal control strategy was obtained by constructing a hierarchical controller parameterized with respect to a given system equilibrium manifold wherein an inverse optimal morphing strategy was developed to coordinate the hierarchical switching. Finally, the proposed control framework was extended to account for nonlinear system parametric uncertainty.

Next, the proposed hierarchical nonlinear switching control framework was applied to propulsion systems. Specifically, a multi-mode state space model for rotating stall and surge in axial flow compression systems that

lends itself to the application of nonlinear control design was developed. In particular, to account for the interaction between higher-order disturbance velocity potential harmonics and the first harmonic during rotating stall inception, an n_{m}-mode expansion of the disturbance potential in the flow field was considered. The multi-mode model was used to show that the second and higher-order disturbance velocity potential harmonics in the governing flow equations strongly interact with the first harmonic during rotating stall inception and must be accounted for in the control-system design process. Finally, the hierarchical switching nonlinear control framework was used to control the multi-mode model while accounting for system uncertainty and actuator rate saturation constraints.

Then, the hierarchical nonlinear control framework was applied to a three-state centrifugal compressor surge model involving pressure and mass flow compression system dynamics as well as spool dynamics to account for the influence of speed transients on the compression surge dynamics. Using the hierarchical nonlinear switching control framework, a nonlinear globally stabilizing switching control law based on equilibria-dependent Lyapunov functions was developed. As in the axial compressor case, actuator amplitude and rate saturation constraints were accounted for.

The hierarchical nonlinear switching control framework developed in Chapter 3 provides a rigorous alternative to gain scheduling control for general nonlinear systems. A key extension of this work is to address disturbance rejection guarantees via switching controllers. Furthermore, optimality considerations of the proposed hierarchical control approach involving minimum time and minimum control energy can also be explored. In addition, inverse optimality notions, wherein lower semicontinuous solutions to the Hamilton-Jacobi-Bellman equation [46] can serve as generalized Lyapunov functions for the nonlinear controlled system, can prove to be fruitful directions of research. For example, constructing a lower semicontinuous solution to the Hamilton-Jacobi-Isaacs equation can lead to hierarchical nonlinear controllers possessing generalized Lyapunov functions that are dissipative with respect to a weighted input-output energy supply rate. This of course would guarantee a nonexpansivity (bounded gain) constraint on the nonlinear closed-loop input-output map providing disturbance rejection guarantees. Finally, the inverse optimality result given in Section 3.6 clearly demonstrates that the *derived* cost functional is intimately related to the potential function and hence a *given* cost functional can be minimized provided that a framework can be developed that allows one to directly relate performance specifications to the potential function.

The proposed hierarchical nonlinear switching control framework relies on the design of subcontrollers for a collection of controlled subsystems. In

particular, these subcontrollers correspond to local set point designs and can be obtained using any appropriate standard linear or nonlinear stabilization scheme. For example, appropriate nonlinear stabilization techniques such as feedback linearization [66], nonlinear H_∞ control [145], constructive nonlinear control [119], optimal nonlinear control [24], and nonlinear regulation via state-dependent Riccati techniques [34], as well as linear-quadratic stabilization schemes based on locally approximated linearizations, can be used to design the subcontrollers. In each case, an estimate of the domain of attraction is required for each set point design. In this monograph, we computed estimates of the domain of attraction for local set point designs by using closed Lyapunov sublevel sets, which may be conservative. To reduce conservatism in estimating a subset of the domain of attraction, several alternative methods can be used. For example, maximal Lyapunov functions [136], Zubov's method [142, 59], ellipsoidal estimate mappings [37], Carlemann linearizations [94], computer generated Lyapunov functions [102], iterative Lyapunov function constructions [32], trajectory-reversing methods [73], and open Lyapunov sublevel sets [56], can be used to construct less conservative estimates of the domain of attraction. These extensions can serve to hone the proposed hierarchical switching control framework.

A different formulation for addressing the problem of system parametric uncertainty is to combine the robust hierarchical switching control approach with adaptive control ideas to develop a robust adaptive switching controller framework. In particular, the construction of generalized Lyapunov functions that are predicated on local *adaptive* controllers to construct ***hierarchical adaptive control*** schemes can also be investigated. Finally, in order to address system parameter convergence issues, the proposed scheme should be considered for nonlinear systems with bounded amplitude persistent L_∞ disturbances.

As discussed in the Introduction, gain scheduled controllers have been extensively used by control practitioners for general nonlinear system stabilization and in particular, aerospace applications. A fruitful area of research is to unify the hierarchical nonlinear switching control framework developed in this monograph with gain scheduling control. Specifically, the parameterization in the proposed framework is with respect to the system operating conditions (system equilibria) rather than physical system parameters as in traditional gain scheduling control. Even though the proposed approach gives a rigorous *alternative* to gain scheduling control for general nonlinear systems, a fruitful area of research is to extend the ideas in this monograph to develop a hierarchical switching controller framework wherein the parameterization is with respect to the physical system parameters. This would rigorize the

classical *ad hoc* gain scheduling control framework that has been the creed of practicing control engineers.

The Moore-Greitzer model for axial flow compression systems developed in [104], as well as the multi-mode model extension developed in this monograph, is valid for a low-speed compressor with inviscid and irrotational flow. Such assumptions are not realistic for modern high-performance axial compressors. More realistic models for high-speed axial flow compression systems should include compressibility, viscosity, and vorticity effects. Developments in modeling of compressible 3-D flow phenomena and the modeling of unsteady blade-row behavior would allow for more realistic models [50, 91].

As seen in this monograph, compressor surge can be stabilized by using a single throttle actuator, often called 1-D actuation. For practical applications, it is worth investigating what type of sensor-actuator topology is most suitable for surge control. In [125], the effect of several actuators and sensors is studied, and advantages are recognized in using a closed-coupled valve actuator in combination with a mass flow sensor. Preliminary studies regarding the effect of position, number, and type of sensors and actuators can be found in [69, 106], but, as seen from the quantitative analysis in [144], this issue seems worth of further investigation. Control of rotating stall requires information about the nonuniformity of the flow. This of course requires an array of sensors placed around the circumference of the compressor (2-D sensing). Common realizations of the 2-D sensing architecture are circular arrays of pressure transducers [139, 13, 20] and hot wire anemometers [109, 62, 38]. To stabilize rotating stall, movable inlet guide vanes [109, 62] and an array of air injectors has been investigated in [38, 20, 52]. Important drawbacks of the use of 2-D actuation and sensing are, the complexity and cost due to the large number of required sensors and actuators, the relatively high bandwidth requirements, and the reduced reliability. In this monograph we use 2-D sensing with a single low-bandwidth 1-D actuator for rotating stall control. A problem that needs further investigation is the possibility of minimizing sensor complexity for controlling rotating stall with a 1-D sensing scheme. Recent progress in this direction has been reported in [54] wherein a nonlinear control framework based on absolute stability theory is developed.

Finally, since actual variable-cycle gas turbine engines involve compressor, combustor, and turbine component coupling, it is of paramount importance to develop global engine models that take into account the intrinsic coupling between these engine components. In recent research the authors in [128] present a simplified model of a jet engine based on the basic principles of each engine component. The hierarchical nonlinear switching control framework presented herein can be effectively used to design controllers for such global engine models. In particular, the proposed control framework can be directly

used to stabilize any operating condition close to the engine operating limit by simply switching between other operating conditions within the global operating range.

Bibliography

1. Special Issue on Saturating Actuators. *Int. J. Robust and Nonlinear Control*, vol. 5, 1995.
2. Special Issue on Hybrid Control Systems. *IEEE Trans. Autom. Contr.*, vol. 43, no. 4, 1998.
3. Special Issue on Hybrid Control Systems. *Automatica*, vol. 35, no. 3, 1999.
4. R. A. Adomaitis and E. H. Abed, "Bifurcation analysis of nonuniform flow patterns in axial-flow gas compressors," in *Proc. 1st World Congress of Nonlinear Analysts*, (Tampa, FL), pp. 1597–1608, 1992.
5. R. A. Adomaitis and E. H. Abed, "Local nonlinear control of stall inception in axial flow compressors," in *29th Joint Propulsion Conference and Exhibit – Paper No. AIAA 93-2230*, (Monterey, CA), 1993.
6. H. Amann, *Ordinary Differential Equations: An Introduction to Nonlinear Analysis*, vol. 13 of *De Gruyter Studies in Mathematics*. De Gruyter, Berlin-New York: Addison-Wesley, 1990.
7. J.-P. Aubin, "Smallest Lyapunov functions of differential inclusions," *Differ. Integral Equ.*, vol. 2, pp. 333–343, 1989.
8. J.-P. Aubin, *Viability Theory*. Boston, MA: Birkhäuser, 1991.
9. J.-P. Aubin and A. Cellina, *Differential Inclusions: Set-Valued Maps and Viability Theory*. Springer-Verlag, 1984.
10. A. Bacciotti, *Local Stabilizability of Nonlinear Control Systems*, vol. 8 of *Series on Advances in Math. for Applied Sciences*. Singapore: World Scientific, 1992.
11. A. Back, J. Guckenheimer, and M. Myers, "A dynamical simulation facility for hybrid systems," in *Hybrid Systems* (R. Grossman, A. Nerode, A. Ravn, and H. Rischel, eds.), pp. 255–267, New York: Springer-Verlag, 1993.
12. O. O. Badmus, S. Chowdhury, K. M. Eveker, and C. N. Nett, "Control-oriented high-frequency turbomachinery modeling: Single-stage compression system 1D model," *ASME J. of Turbomachinery*, vol. 117, no. 1, pp. 47–61, 1995.
13. O. O. Badmus, S. Chowdhury, K. M. Eveker, C. N. Nett, and C. J. Rivera, "A simplified approach for control of rotating stall - Part 1 and 2," *Journal of Propulsion and Power*, vol. 11, no. 6, pp. 1195–1223, 1995.
14. O. O. Badmus, S. Chowdhury, and C. N. Nett, "Nonlinear control of surge in axial compression systems," *Automatica*, vol. 32, no. 1, pp. 59–70, 1996.
15. O. O. Badmus, K. M. Eveker, and C. N. Nett, "Control-oriented high-frequency turbomachinery modeling, Part 1: Theoretical foundations," in *28th Joint Propulsion Conference and Exhibit – Paper No. AIAA 92-3314*, (Nashville, TN), 1992.
16. J. Baillieul, S. Dahlgren, and B. Lehman, "Nonlinear control designs for systems with bifurcations with applications to stabilization and control of compressors," in *Proc. IEEE Conf. Dec. Contr.*, (New Orleans, LA), pp. 3062–3067, 1995.

17. O. E. Balje, "A contribution to the problem of designing radial turbomachines," *Transactions of the ASME*, vol. 74, pp. 451–472, 1952.
18. E. A. Barbashin and N. N. Krasovskii, "On the stability of motion in the large," *Dokl. Akad. Nauk.*, vol. 86, pp. 453–456, 1952.
19. M. Bardi and I. C. Dolcetta, *Optimal Control and Viscosity Solutions of Hamilton-Jacobi-Bellman Equations.* Boston, MA: Birkhäuser, 1997.
20. R. L. Behnken, R. D'Andrea, and R. M. Murray, "Control of rotating stall in a low-speed axial flow compressor using pulsed air," in *Proc. IEEE Conf. Dec. Contr.*, (New Orleans, LA), pp. 3056–3061, 1995.
21. R. Bellman, *Dynamic Programming.* Princeton, NJ: Princeton Univ. Press, 1957.
22. D. S. Bernstein, "Nonquadratic cost and nonlinear feedback control," *Int. J. Robust and Nonlinear Control*, vol. 3, pp. 211–229, 1993.
23. N. P. Bhatia and G. P. Szegö, *Stability Theory of Dynamical Systems.* New York-Berlin: Springer-Verlag, 1970.
24. A. E. Bryson, Jr. and Y.-C. Ho, *Applied Optimal Control: Optimization, Estimation and Control.* New York, NY: Hemisphere, 1975.
25. C. I. Byrnes and A. Isidori, "Local stabilization of minimum-phase nonlinear systems," *Sys. Contr. Lett.*, vol. 11, pp. 9–17, 1988.
26. C. I. Byrnes and A. Isidori, "New results and examples in nonlinear feedback stabilization," *Sys. Contr. Lett.*, vol. 12, pp. 437–442, 1989.
27. C. I. Byrnes and A. Isidori, "Asymptotic stabilization of minimum phase nonlinear systems," *IEEE Trans. Autom. Contr.*, vol. 36, no. 10, pp. 1122–1137, 1991.
28. C. I. Byrnes, A. Isidori, and J. C. Willems, "Passivity, feedback equivalence, and the global stabilization of minimum phase nonlinear systems," *IEEE Trans. Autom. Contr.*, vol. 36, no. 11, pp. 1228–1240, 1991.
29. C. I. Byrnes and C. F. Martin, "An integral-invariance principle for nonlinear systems," *IEEE Trans. Autom. Contr.*, vol. 40, pp. 983–994, 1995.
30. P. E. Caines and Y.-J. Wei, "Hierarchical hybrid control systems: A lattice theoretic formulation," *IEEE Trans. Autom. Contr.*, vol. 43, no. 4, pp. 501–508, 1998.
31. C.-T. Chen, *Linear System Theory and Design.* New York: Holt, Rinehart, and Winston, 1984.
32. H.-D. Chiang and J. S. Thorp, "Stability regions of nonlinear dynamical systems: A constructive methodology," *IEEE Trans. Autom. Contr.*, vol. 34, no. 12, pp. 1229–1241, 1989.
33. F. H. Clarke, Y. S. Ledyaev, E. D. Sontag, and A. I. Subbotin, "Asymptotic controllability implies feedback stabilization," *IEEE Trans. Autom. Contr.*, vol. 42, no. 10, pp. 1394–1407, 1997.
34. J. R. Cloutier, C. N. D'Souza, and C. P. Mracek, "Nonlinear regulation and nonlinear h_∞ control via the state-dependent riccati equation technique; Part 1: Theory; Part 2: Examples," in *Proc. Int. Conf. on Nonlinear Problems in Aviation and Aerospace*, (Daytona Beach, FL), pp. 117–142, 1996.
35. E. A. Coddington and N. Levinson, *Theory of Ordinary Differential Equations.* McGraw-Hill, 1955.
36. J. E. R. Cury, B. H. Krogh, and T. Niinomi, "Synthesis of supervisory controllers for hybrid systems based on approximating automata," *IEEE Trans. Autom. Contr.*, vol. 43, no. 4, pp. 564–568, 1998.
37. E. J. Davison and E. M. Kurak, "A computational method for determining quadratic Lyapunov functions for nonlinear systems," *Automatica*, vol. 7, pp. 627–636, 1971.

38. I. Day, "Active suppression of rotating stall and surge in axial compressors," *ASME J. of Turbomachinery*, vol. 115, no. 1, pp. 40–47, 1993.
39. R. A. DeCarlo, S. H. Zak, and G. P. Mathews, "Variable structure control of nonlinear multivariable systems: A tutorial," *Proc. IEEE*, vol. 76, no. 3, pp. 212–232, 1988.
40. S. L. Dixon, *Fluid Mechanics, Thermodynamics of Turbomachinery*. Oxford: Pergamon Press, 1975.
41. I. C. Dolcetta and L. C. Evans, "Optimal switching for ordinary differential equations," *SIAM J. Control Optim.*, vol. 22, pp. 1133–1148, 1984.
42. J. L. Dussourd, G. W. Pfannebecker, and S. K. Singhania, "An experimental investigation of the control of surge in radial compressors using close coupled resistances," *Journal of Fluid Engineering*, vol. 99, pp. 64–76, 1977.
43. T. B. Ferguson, *The Centrifugal Compressor Stage*. London: Butterworth, 1963.
44. A. F. Filippov, *Differential Equations with Discontinuous Right-Hand Sides*. Mathematics and its Applications (Soviet series), Dordrecht, The Netherlands: Kluwer Academic Publishers, 1988.
45. D. A. Fink, N. A. Cumptsy, and E. M. Greitzer, "Surge dynamics in a free-spool centrifugal compressor system," *ASME J. of Turbomachinery*, vol. 114, pp. 321–332, 1992.
46. H. Frankowska, "Lower semicontinuous solutions of Hamilton-Jacobi-Bellman equations," *SIAM J. Control Optim.*, vol. 31, no. 1, pp. 257–272, 1993.
47. R. A. Freeman and P. V. Kokotović, "Inverse optimality in robust stabilization," *SIAM J. Control Optim.*, vol. 34, no. 4, pp. 1365–1391, 1996.
48. J. T. Gravdahl and O. Egeland, "Centrifugal compressor surge and speed control," *IEEE Trans. Contr. Syst. Tech.*, vol. 7, no. 5, pp. 567–579, 1999.
49. E. M. Greitzer, "Surge and rotating stall in axial flow compressors: Part 1 and 2," *Journal of Engineering for Power*, vol. 98, pp. 190–217, 1976.
50. G. Gu, S. Banda, and A. Sparks, "An overview of rotating stall and surge control for axial flow compressors," in *Proc. IEEE Conf. Dec. Contr.*, (Kobe, Japan), pp. 2786–2791, 1996.
51. J. Guckenheimer, "A robust hybrid stabilization strategy for equilibria," *IEEE Trans. Autom. Contr.*, vol. 40, no. 2, pp. 321–326, 1995.
52. D. L. Gysling and E. M. Greitzer, "Dynamic control of rotating stall in axial flow compressors using aeromechanical feedback," *ASME J. of Turbomachinery*, vol. 117, no. 3, pp. 307–319, 1995.
53. W. M. Haddad, V. Chellaboina, J. L. Fausz, and A. Leonessa, "Optimal nonlinear robust control for nonlinear uncertain systems," *Int. J. of Contr.*, vol. 73, pp. 329–342, 2000.
54. W. M. Haddad, J. R. Corrado, and A. Leonessa, "Pressure feedback reduced-order dynamic compensation for axial flow compression systems," in *Proc. IEEE Conf. on Contr. Appl.*, (Kohala Coast, HI), pp. 371–376, 1999.
55. W. M. Haddad, J. L. Fausz, V. Chellaboina, and C. T. Abdallah, "A unification between nonlinear-nonquadratic optimal control and integrator backstepping," *Int. J. Robust and Nonlinear Control*, pp. 879–906, 1998.
56. W. M. Haddad, V. Kapila, and V. Chellaboina, "Guaranteed domains of attraction for multivariable Luré systems via open Lyapunov surfaces," *Int. J. Robust and Nonlinear Control*, vol. 7, pp. 935–949, 1997.
57. W. M. Haddad, A. Leonessa, V. Chellaboina, and J. L. Fausz, "Nonlinear robust disturbance rejection controllers for rotating stall and surge in axial flow compressors," *IEEE Trans. Contr. Syst. Tech.*, vol. 7, pp. 391–398, 1999.

58. T. Hagiwara and M. Araki, "Design of a stable feedback controller based on the multirate sampling of the plant output," *IEEE Trans. Autom. Contr.*, vol. 33, pp. 812–819, 1988.
59. W. Hahn, *Stability of Motion.* Berlin, DE: Springer-Verlag, 1967.
60. J. K. Hale, *Ordinary Differential Equations.* Malabar, FL: Krieger Publishing Company, 1980.
61. K. E. Hansen, P. Jorgensen, and P. S. Larsen, "Experimental and theoretical study of surge in a small centrifugal compressor," *Journal of Fluid Engineering*, vol. 103, pp. 391–394, 1981.
62. J. M. Haynes, G. J. Hendricks, and A. H. Epstein, "Active stabilization of rotating stall in a three-stage axial compressor," *ASME J. of Turbomachinery*, vol. 116, pp. 226–239, 1994.
63. P. G. Hill and C. R. Peterson, *Mechanics and Thermodynamics of Propulsion.* Reading, MA: Addison-Wesley, 1992.
64. M. Ikeda, Y. Ohta, and D. D. Siljak, "Parametric stability," in *New Trends in Systems Theory*, G. Conte, A. M. Perdon, and B. Wyman, eds, Birkhauser, Boston, MA, pp. 1–20, 1991.
65. M. Ikeda and D. D. Siljak, "Hierarchical Lyapunov functions," *J. Math. Anal. Appl.*, vol. 112, no. 1, pp. 110–128, 1985.
66. A. Isidori, *Nonlinear Control Systems.* Berlin: Springer-Verlag, 2nd ed., 1989.
67. D. H. Jacobson, *Extensions of Linear-Quadratic Control Optimization and Matrix Theory.* New York, NY: Academic Press, 1977.
68. M. Jamshidi, *Large-Scale Systems: Modeling and Control*, vol. 9 of *North-Holland Series in System Science and Engineering.* New York: North-Holland, 1983.
69. W. M. Jungovski, M. H. Weiss, and G. R. Price, "Pressure oscillations occurring in a centrifugal compressor system with and without passive and active surge control," *ASME J. of Turbomachinery*, vol. 118, no. 1, pp. 29–40, 1996.
70. V. Jurdjevic and J. P. Quinn, "Controllability and stability," *J. of Diff. Eqs.*, vol. 28, pp. 381–389, 1978.
71. R. E. Kalman and J. E. Bertram, "Control system analysis and design via the second method of Lyapunov, Part I: Continuous-time systems," *J. Basic Engr. Trans. ASME*, vol. 80, pp. 371–393, 1960.
72. I. Kanellakopoulos, P. V. Kokotović, and A. S. Morse, "Systematic design of adaptive controllers for feedback linearizable systems," *IEEE Trans. Autom. Contr.*, vol. 36, no. 11, pp. 1241–1253, 1991.
73. H. K. Khalil, *Nonlinear Systems.* Upper Saddle River, NJ: Prentice Hall, 1996.
74. D. E. Kirk, *Optimal Control Theory: An Introduction.* Englewood Cliffs, NJ: Prentice-Hall, 1970.
75. N. N. Krasovskii, *Problems of the Theory of Stability of Motion.* Stanford, CA: Stanford Univ. Press, 1959. English translation in 1963.
76. M. Krstić, I. Kanellakopoulos, and P. V. Kokotović, "Adaptive nonlinear control without overparametrization," *Sys. Contr. Lett.*, vol. 19, pp. 177–185, 1992.
77. M. Krstić, I. Kanellakopoulos, and P. V. Kokotović, *Nonlinear and Adaptive Control Design.* New York, NY: John Wiley and Sons, 1995.
78. M. Krstić and P. V. Kokotović, "Lean backstepping design for a jet engine compressor model," in *Proc. IEEE Conf. on Contr. Appl.*, (Piscataway, NJ), pp. 1047–1052, 1995.
79. M. Krstić, J. M. Protz, J. D. Paduano, and P. V. Kokotović, "Backstepping designs for jet engine stall and surge control," in *Proc. IEEE Conf. Dec. Contr.*, (New Orleans, LA), pp. 3049–3055, 1995.

80. T. Kuang-Hsuan and J. S. Shamma, "Nonlinear gain-scheduled control design using set-valued methods," in *Proc. Amer. Contr. Conf.*, (Philadelphia, PA), pp. 1195–1199, 1998.
81. H. G. Kwatny and B.-C. Chang, "Constructing linear families from parameter-dependent nonlinear dynamics," *IEEE Trans. Autom. Contr.*, vol. 43, no. 8, pp. 1143–1147, 1998.
82. J. P. LaSalle, "Some extensions of Liapunov's second method," *IRE Trans. Circ. Thy.*, vol. 7, pp. 520–527, 1960.
83. J. P. LaSalle and S. Lefschetz, *Stability by Lyapunov's Direct Method.* New York, NY: Academic Press, 1961.
84. D. A. Lawrence and W. J. Rugh, "Gain scheduling dynamic linear controllers for a nonlinear plants," *Automatica*, vol. 31, pp. 381–390, 1995.
85. A. Leonessa, V. Chellaboina, and W. M. Haddad, "Globally stabilizing controllers for multi-mode axial flow compressors via equilibria-dependent Lyapunov functions," in *Proc. Amer. Contr. Conf.*, (Albuquerque, NM), pp. 993–997, 1997.
86. A. Leonessa, W. M. Haddad, and V. Chellaboina, "Nonlinear system stabilization via stability-based switching," in *Proc. IEEE Conf. Dec. Contr.*, (Tampa, FL), pp. 2983–2996, 1998.
87. H. Li, A. Leonessa, and W. M. Haddad, "Globally stabilizing controllers for a centrifugal compressor model with spool dynamics," in *Proc. American Control Conf.*, (Philadelphia, PA), pp. 2160–2164, 1998.
88. D. C. Liaw and E. H. Abed, "Stability analysis and control of rotating stall," *IFAC Nonlinear Control Systems*, pp. 295–300, 1992.
89. V. Lin, E. Sontag, and Y. Wang, "Recent results on Lyapunov-theoretic techniques for nonlinear stability," in *Proc. Amer. Contr. Conf.*, (Baltimore, MD), pp. 1771–1775, 1994.
90. X. Liu, "Stability results for impulsive differential systems with applications to population growth models," *Dyn. Stab. Sys.*, vol. 9, pp. 163–174, 1994.
91. J. Longley, "A review of nonsteady flow models for compressor stability," *ASME J. of Turbomachinery*, vol. 116, no. 2, pp. 202–215, 1994.
92. K. A. Loparo, J. T. Aslanis, and O. Hajek, "Analysis of switched linear systems in the plane, Part 1: Local behavior of trajectories and local cycle geometry," *J. of Opt. Theory and Appl.*, vol. 52, no. 3, pp. 365–394, 1987.
93. K. A. Loparo, J. T. Aslanis, and O. Hajek, "Analysis of switched linear systems in the plane, Part 2: Global behavior of trajectories, controllability and attainability," *J. of Opt. Theory and Appl.*, vol. 52, no. 3, pp. 395–427, 1987.
94. K. A. Loparo and G. L. Blankenship, "Estimating the domain of attraction of nonlinear feedback systems," *IEEE Trans. Autom. Contr.*, vol. 23, no. 4, pp. 602–608, 1978.
95. A. M. Lyapunov, *The General Problem of Stability of Motion.* Washington, DC: Taylor and Francis, 1892. translated and edited by A. T. Fuller in 1992.
96. J. Malmborg, B. Bernhardsson, and K. J. Astrom, "A stabilizing switching scheme for multi controller systems," in *Proc. 13th IFAC World Congress*, (San Francisco, CA), pp. 229–234, 1996.
97. C. A. Mansoux, D. L. Gysling, and J. D. Paduano, "Distributed nonlinear modeling and stability analysis of axial compressor stall and surge," *Proc. American Control Conf.*, pp. 2305–2316, 1994.
98. C. A. Mansoux, D. L. Gysling, J. D. Setiawan, and J. D. Paduano, "Distributed nonlinear modeling and stability analysis of axial compressor stall and surge," in *Proc. American Control Conf.*, (Baltimore, MD), pp. 2305–2316, 1994.
99. J. D. Mattingly, *Elements of Gas Turbine Propulsion.* New York: McGraw-Hill, 1996.

100. F. E. McCaughan, "Bifurcation analysis of axial flow compressor stability," *SIAM J. Applied Mathematics*, vol. 20, pp. 1232–1253, 1990.
101. M. W. McConley, B. D. Appleby, M. A. Dahleh, and E. Feron, "A control Lyapunov function approach to robust stabilization of nonlinear systems," in *Proc. Amer. Contr. Conf.*, (Albuquerque, NM), pp. 329–333, 1997.
102. A. N. Michel, R. K. Miller, and B. H. Nam, "Stability analysis of interconnected systems using computer generated Lyapunov functions," *IEEE Trans. Circuits Syst.*, vol. 29, no. 7, pp. 431–440, 1982.
103. F. K. Moore, "A theory of rotating stall of multistage axial compressors: Part 1, 2, and 3," *Journal of Engineering for Gas Turbines and Power*, vol. 106, pp. 313–336, 1984.
104. F. K. Moore and E. M. Greitzer, "A theory of post-stall transients in axial compression systems: Part 1 and 2," *Journal of Engineering for Gas Turbines and Power*, vol. 108, pp. 68–76, 231–239, 1986.
105. P. J. Moylan and B. D. O. Anderson, "Nonlinear regulator theory and an inverse optimal control problem," *IEEE Trans. Autom. Contr.*, vol. 18, pp. 460–465, 1973.
106. K. Nakagawa, M. Fujiware, T. Nishiota, S. Tanaka, and Y. Kashiwara, "Experimental and numerical analysis of active suppression of centrifugal compressor surge by sction-side valve control," *JSME Int. J. Series B*, vol. 37, no. 4, pp. 878–885, 1994.
107. Y. Ohta and D. D. Siljak, "Parametric quadratic stabilizability of uncertain nonlinear systems," *Sys. Contr. Lett.*, pp. 437–444, 1994.
108. A. Packard and M. Kantner, "Gain scheduling the LPV way," in *Proc. IEEE Conf. Dec. Contr.*, (Kobe, Japan), pp. 3938–3941, 1996.
109. J. D. Paduano, L. Valavani, A. H. Epstein, E. M. Greitzer, and G. R. Guenette, "Modeling for control of rotating stall," *Automatica*, vol. 30, no. 9, pp. 1357–1373, 1994.
110. K. M. Passino, A. N. Michel, and P. J. Antsaklis, "Lyapunov stability of a class of discrete event systems," *IEEE Trans. Autom. Contr.*, vol. 39, pp. 269–279, 1994.
111. P. Peleties and R. A. DeCarlo, "Asymptotic stability of m-switched systems using Lyapunov-like functions," in *Proc. Amer. Contr. Conf.*, (Boston, MA), pp. 1679–1684, 1991.
112. J. E. Pinsley, G. R. Guenette, and E. M. Greitzer, "Active stabilization of centrifugal compressor surge," *ASME J. of Turbomachinery*, vol. 113, pp. 723–732, 1991.
113. K. M. Przyluski, "Controllability does not imply stabilizability," *Sys. Contr. Lett.*, vol. 10, pp. 119–121, 1988.
114. H. L. Royden, *Real Analysis*. New York, NY: Macmillan Publishing Company, 1988.
115. W. J. Rugh and J. S. Shamma, "A survey of research on gain scheduling," *Automatica*, to appear.
116. E. P. Ryan, "An integral invariance principle for differential inclusions with applications in adaptive control," *SIAM J. Control Optim.*, vol. 36, pp. 960–980, 1998.
117. A. Saberi, P. Kokotović, and H. Sussman, "Global stabilization of partially linear composite systems," *SIAM J. Control Optim.*, vol. 28, pp. 1491–1503, 1990.
118. C. Samson, "Robust nonlinear control of robotic manipulators," in *Proc. IEEE Conf. Dec. Contr.*, pp. 1211–1216, 1983.
119. R. Sepulchre, M. Jankovic, and P. V. Kokotović, *Constructive Nonlinear Control*. London, GB: Springer-Verlag, 1997.

120. J. Serrin, *Mathematical Principles of Classical Fluid Mechanics*, vol. VIII/1 of *Encyclopedia of Physics*, pp. 125–263. Springer Verlag, 1959.
121. S. Shahruz and S. Behtash, "Design of controllers for linear parameter-varying systems by the gain scheduling technique," *J. Math. Anal. and Appl.*, vol. 168, pp. 195–217, 1992.
122. J. S. Shamma and M. Athans, "Analysis of gain scheduled control for nonlinear plants," *IEEE Trans. Autom. Contr.*, vol. 35, no. 8, pp. 898–907, 1990.
123. J. S. Shamma and M. Athans, "Guaranteed properties of gain scheduled control for linear parameter-varying plants," *Automatica*, vol. 27, no. 3, pp. 559–564, 1991.
124. J. S. Shamma and M. Athans, "Gain scheduling: Potential hazards and possible remedies," *IEEE Contr. Syst. Mag.*, vol. 12, pp. 101–107, 1992.
125. J. Simon, L. Valavani, A. Epstein, and E. Greitzer, "Evaluation of approaches to active compressor surge stabilization," *ASME J. of Turbomachinery*, vol. 115, no. 1, pp. 57–67, 1993.
126. J. J. Slotine and S. S. Sastry, "Tracking control of nonlinear systems using sliding surfaces with applications to robot manipulators," *Int. J. of Contr.*, vol. 2, pp. 465–492, 1983.
127. E. D. Sontag, *Mathematical Control Theory: Deterministic Finite-Dimensional Systems.* New York, NY: Springer-Verlag, 1990.
128. H. A. Spang III and H. Brown, "Control of jet engines," *Control Engineering Practice*, vol. 7, pp. 1043–1059, 1999.
129. M. W. Spong, "Robust stabilization for a class of nonlinear systems," in *Theory and Application of Nonlinear Control Systems* (C. I. Byrnes and A. Lindquist, eds.), Amsterdam: North-Holland, 1986.
130. M. W. Spong, J. S. Thorp, and J. M. Kleinwaks, "The control of robotic manipulators with bounded control. Part 2: Robustness and disturbance rejection," in *Proc. IEEE Conf. Dec. Contr.*, (Las Vegas, NW), pp. 1047–1052, 1984.
131. M. W. Spong and M. Vidyasagar, "Robust nonlinear control of robot manipulators," in *Proc. IEEE Conf. Dec. Contr.*, pp. 1767–1772, 1985.
132. G. Stein, "Adaptive flight control - a pragmatic view," in *Application of Adaptive Control*, K. S. Narendra and R. V. Monopoli, Eds., New York: Academic Press, 1980.
133. D. Trim, *Applied Partial Differential Equations.* Boston, MA: PWS-Kent, 1990.
134. J. Tsinias, "Existence of control Lyapunov functions and applications to state feedback stabilizability of nonlinear systems," *SIAM J. Control Optim.*, vol. 29, pp. 457–473, 1991.
135. V. Utkin, "Variable structure systems with sliding modes," *IEEE Trans. Autom. Contr.*, vol. 22, no. 2, pp. 212–222, 1977.
136. A. Vannelli and M. Vidyasagar, "Maximal Lyapunov functions and domains of attraction for autonomous nonlinear systems," *Automatica*, vol. 21, no. 1, pp. 69–80, 1985.
137. M. Vidyasagar, *Nonlinear Systems Analysis.* Englewood Cliffs, NJ: Prentice-Hall, 1993.
138. H. O. Wang, R. A. Adomaitis, and E. H. Abed, "Nonlinear analysis and control of rotating stall in axial flow compressors," in *Proc. American Control Conf.*, (Baltimore, MD), pp. 2317–2321, 1994.
139. J. F. Williams, M. Harper, and D. Alwright, "Active stabilization of compressor instability and surge in a working engine," *ASME J. of Turbomachinery*, vol. 115, no. 1, pp. 68–75, 1993.

140. H. Ye, A. N. Michel, and L. Hou, "Stability theory for hybrid dynamical systems," *IEEE Trans. Autom. Contr.*, vol. 43, pp. 461–474, 1998.
141. T. Yoshizawa, *Stability Theory by Liapunov's Second Method.* The Mathematical Society of Japan, 1966.
142. V. I. Zubov, *Methods of A. M. Lyapunov and Their Application.* Groningen, NL: P. Noordhoff, 1964.
143. B. de Jager, "Rotating stall and surge control: A survey," in *Proc. IEEE Conf. Dec. Contr.*, (New Orleans, LA), pp. 1857–1862, 1995.
144. M. van de Wal, F. Willems, and B. de Jager, "Selection of actuators and sensors for active surge control," in *Proc. IEEE Conf. on Contr. Appl.*, (Hartford, CT), pp. 121–126, 1997.
145. A. J. van der Schaft, L_2-*Gain and Passivity Techniques in Nonlinear Control.* London, UK: Springer-Verlag, 1996.

Index

Lecture Notes in Control and Information Sciences

Edited by M. Thoma

1997–2000 Published Titles:

Vol. 222: Morse, A.S.
Control Using Logic-Based Switching
288 pp. 1997 [3-540-76097-0]

Vol. 223: Khatib, O.; Salisbury, J.K.
Experimental Robotics IV: The 4th International Symposium, Stanford, California, June 30 - July 2, 1995
596 pp. 1997 [3-540-76133-0]

Vol. 224: Magni, J.-F.; Bennani, S.; Terlouw, J. (Eds)
Robust Flight Control: A Design Challenge
664 pp. 1997 [3-540-76151-9]

Vol. 225: Poznyak, A.S.; Najim, K.
Learning Automata and Stochastic Optimization
219 pp. 1997 [3-540-76154-3]

Vol. 226: Cooperman, G.; Michler, G.; Vinck, H. (Eds)
Workshop on High Performance Computing and Gigabit Local Area Networks
248 pp. 1997 [3-540-76169-1]

Vol. 227: Tarbouriech, S.; Garcia, G. (Eds)
Control of Uncertain Systems with Bounded Inputs
203 pp. 1997 [3-540-76183-7]

Vol. 228: Dugard, L.; Verriest, E.I. (Eds)
Stability and Control of Time-delay Systems
344 pp. 1998 [3-540-76193-4]

Vol. 229: Laumond, J.-P. (Ed.)
Robot Motion Planning and Control
360 pp. 1998 [3-540-76219-1]

Vol. 230: Siciliano, B.; Valavanis, K.P. (Eds)
Control Problems in Robotics and Automation
328 pp. 1998 [3-540-76220-5]

Vol. 231: Emel'yanov, S.V.; Burovoi, I.A.; Levada, F.Yu.
Control of Indefinite Nonlinear Dynamic Systems
196 pp. 1998 [3-540-76245-0]

Vol. 232: Casals, A.; de Almeida, A.T. (Eds)
Experimental Robotics V: The Fifth International Symposium Barcelona, Catalonia, June 15-18, 1997
190 pp. 1998 [3-540-76218-3]

Vol. 233: Chiacchio, P.; Chiaverini, S. (Eds)
Complex Robotic Systems
189 pp. 1998 [3-540-76265-5]

Vol. 234: Arena, P.; Fortuna, L.; Muscato, G.; Xibilia, M.G.
Neural Networks in Multidimensional Domains: Fundamentals and New Trends in Modelling and Control
179 pp. 1998 [1-85233-006-6]

Vol. 235: Chen, B.M.
H_∞ Control and Its Applications
361 pp. 1998 [1-85233-026-0]

Vol. 236: de Almeida, A.T.; Khatib, O. (Eds)
Autonomous Robotic Systems
283 pp. 1998 [1-85233-036-8]

Vol. 237: Kreigman, D.J.; Hagar, G.D.; Morse, A.S. (Eds)
The Confluence of Vision and Control
304 pp. 1998 [1-85233-025-2]

Vol. 238: Elia, N. ; Dahleh, M.A.
Computational Methods for Controller Design
200 pp. 1998 [1-85233-075-9]

Vol. 239: Wang, Q.G.; Lee, T.H.; Tan, K.K.
Finite Spectrum Assignment for Time-Delay Systems
200 pp. 1998 [1-85233-065-1]

Vol. 240: Lin, Z.
Low Gain Feedback
376 pp. 1999 [1-85233-081-3]

Vol. 241: Yamamoto, Y.; Hara S.
Learning, Control and Hybrid Systems
472 pp. 1999 [1-85233-076-7]

Vol. 242: Conte, G.; Moog, C.H.; Perdon A.M.
Nonlinear Control Systems
192 pp. 1999 [1-85233-151-8]

Vol. 243: Tzafestas, S.G.; Schmidt, G. (Eds)
Progress in Systems and Robot Analysis and Control Design
624 pp. 1999 [1-85233-123-2]

Vol. 244: Nijmeijer, H.; Fossen, T.I. (Eds)
New Directions in Nonlinear Observer Design
552pp: 1999 [1-85233-134-8]

Vol. 245: Garulli, A.; Tesi, A.; Vicino, A. (Eds)
Robustness in Identification and Control
448pp: 1999 [1-85233-179-8]

Vol. 246: Aeyels, D.; Lamnabhi-Lagarrigue, F.; van der Schaft, A. (Eds)
Stability and Stabilization of Nonlinear Systems
408pp: 1999 [1-85233-638-2]

Vol. 247: Young, K.D.; Özgüner, Ü. (Eds)
Variable Structure Systems, Sliding Mode and Nonlinear Control
400pp: 1999 [1-85233-197-6]

Vol. 248: Chen, Y.; Wen C.
Iterative Learning Control
216pp: 1999 [1-85233-190-9]

Vol. 249: Cooperman, G.; Jessen, E.; Michler, G. (Eds)
Workshop on Wide Area Networks and High Performance Computing
352pp: 1999 [1-85233-642-0]

Vol. 250: Corke, P. ; Trevelyan, J. (Eds)
Experimental Robotics VI
552pp: 2000 [1-85233-210-7]

Vol. 251: van der Schaft, A. ; Schumacher, J.
An Introduction to Hybrid Dynamical Systems
192pp: 2000 [1-85233-233-6]

Vol. 252: Salapaka, M.V.; Dahleh, M.
Multiple Objective Control Synthesis
192pp. 2000 [1-85233-256-5]

Vol. 253: Elzer, P.F.; Kluwe, R.H.; Boussoffara, B.
Human Error and System Design and Management
240pp. 2000 [1-85233-234-4]

Vol. 254: Hammer, B.
Learning with Recurrent Neural Networks
160pp. 2000 [1-85233-343-X]

9 781852 333355

Contents

tions for axial and centrifugal flow compression systems that is accessible to control-system designers requiring state space models for modern nonlinear control is developed. The hierarchical switching control framework is then applied to control rotating stall and surge in jet engine compression systems. To reflect a more realistic design we account for uncertainty in the pressure-flow compressor performance characteristic map as well as impose a rate saturation constraint on the system actuator throttle opening.

The appropriate background for this monograph is a first course in state space methods along with a first course on nonlinear systems at the level of Khalil [73]. Chapters 2–4 are suitable for an advanced course in nonlinear switching feedback control design while Chapters 5 and 6 are suitable for students and researchers having an interest in propulsion control.

The results reported in this monograph were obtained at the School of Aerospace Engineering, Georgia Institute of Technology, Atlanta, between September 1995 and December 1999. Financial support for this work was in part provided by the National Science Foundation under Grant ECS-9496249, the Air Force Office of Scientific Research under Grant F49620-96-1-0125, and the Army Research Office under Grant DAAH04-96-1-0008.

Boca Raton, Florida, USA, April 2000 — *Alexander Leonessa*
Atlanta, Georgia, USA, April 2000 — *Wassim M. Haddad*
Columbia, Missouri, USA, April 2000 — *VijaySekhar Chellaboina*

oped in Chapters 2–4 while Chapters 5 and 6 present applications of the proposed hierarchical control framework to axial and centrifugal compression systems, respectively. Specifically, in Chapter 2 we develop generalized Lyapunov and invariant set theorems for nonlinear dynamical systems wherein all regularity assumptions on the Lyapunov function and the system dynamics are removed. In particular, local and global stability theorems are given using lower semicontinuous Lyapunov functions. Furthermore, generalized invariant set theorems are derived wherein system trajectories converge to a union of largest invariant sets contained in intersections over finite intervals of the closure of generalized Lyapunov level surfaces. The proposed results provide transparent generalizations to standard Lyapunov and invariant set theorems.

Using the generalized Lyapunov and invariant set theorems developed in Chapter 2, in Chapter 3 a nonlinear control-system design framework predicated on a hierarchical switching controller architecture parameterized over a set of moving system equilibria is developed. Specifically, using equilibria-dependent Lyapunov functions, a hierarchical nonlinear control strategy is developed that stabilizes a given nonlinear system by stabilizing a collection of nonlinear controlled subsystems. The switching nonlinear controller architecture is designed based on a generalized lower semicontinuous Lyapunov function obtained by minimizing a potential function over a given switching set induced by the parameterized system equilibria. The proposed framework provides a rigorous alternative to ***designing*** gain scheduled feedback controllers and guarantees local and global closed-loop system stability for general nonlinear systems. Furthermore, the hierarchical switching control framework is extended to include inverse optimality notions. Specifically, the hierarchical controller is parameterized with respect to a given system equilibrium manifold wherein an inverse optimal morphing strategy is constructed to coordinate the hierarchical switching. The overall approach is quite different from the quasivariational inequality methods for optimal switching systems developed in the literature in that our results provide hierarchical homotopic feedback controllers guaranteeing closed-loop stability via an underlying Lyapunov function. Finally, in Chapter 4 the proposed control framework is extended to account for system parametric uncertainty wherein the hierarchical switching architecture is parameterized over a set of moving ***nominal*** system equilibria.

In Chapters 5 and 6 we apply the proposed hierarchical nonlinear control framework to propulsion systems. First, however, we develop models for rotating stall and surge in axial and centrifugal flow compression systems that lend themselves to the application of nonlinear control design. Specifically, a self-contained first principles derivation of the governing fluid dynamic equa-

Preface

General nonlinear system stabilization is notoriously difficult and remains an open problem. If the operating range of the control system is small and if the system nonlinearities are smooth, then the control system can be locally approximated by a linearized system about a given operating condition and well established linear multivariable control methods can be used to maintain local stability and performance. However, in modern high performance engineering applications such as advanced tactical fighter aircraft, large flexible space structures, and variable-cycle gas turbine engines, to cite but a few examples, the locally approximated linearized system does not cover the operating range of the nonlinear system. In this case, gain scheduled controllers can be designed over several fixed operating points covering the system's operating range and controller gains interpolated over this range. However, due to approximation linearization errors and neglected operating point transitions, the resulting gain scheduled system does not have any guarantees of performance or stability. The main objective of this monograph is to develop a general nonlinear control design methodology for nonlinear uncertain dynamical systems. Specifically, a hierarchical nonlinear switching control framework is developed that provides a rigorous alternative to gain scheduling control for general nonlinear uncertain systems. The main tool for establishing a hierarchical nonlinear switching control framework is a generalized Lyapunov and invariant set framework that address stability of switched feedback systems.

The desire for developing an integrated control system-design methodology for advanced propulsion systems has led to significant activity in modeling and control of flow compression systems in recent years. In this monograph we apply the hierarchical switching control framework to axial and centrifugal flow compression systems to address the compressor aerodynamic instabilities of rotating stall and surge. The proposed control framework accounts for the coupling between higher-order modes while explicitly addressing actuator rate saturation constraints and system modeling uncertainty.

After the introductory chapter, the presentation is organized in two major parts. The basic hierarchical nonlinear switching control framework is devel-

To my fiancée; Keri N. Swaby

A. L.

To my family and the memory of my father; Mikhael S. Haddad

W. M. H.

To my family and the memory of my mother; Andhra Jayashree

V. C.

Author

Alexander Leonessa, Assistant Professor
Department of Ocean Engineering, Florida Atlantic University, USA

Wassim M. Haddad, Professor
School of Aerospace Engineering, Georgia Instituute of Technology, Atlanta, GA 30332-0150, USA

VijaySekhar Chellaboina, Assistant Professor
Department of Mechanical and Aerospace Engineering, University of Missouri, Columbia

ISBN 1-85233-335-9 Springer-Verlag London Berlin Heidelberg

British Library Cataloguing in Publication Data
Leonessa, A.
Hierarchical nonlinear switching control design with applications to propulsion systems. - (Lecture notes in control and information sciences ; 255)
1.Nonlinear systems 2.Nonlinear control theory 3.Switching theory
I.Title II.Haddad, W.M. III.Chellaboina, V.
629.8'36
ISBN 1852333359

Library of Congress Cataloging-in-Publication Data
A catalog record for this book is available from the Library of Congress

Typesetting: Camera ready by authors
Printed and bound at the Athenæum Press Ltd., Gateshead, Tyne & Wear
69/3830-543210 Printed on acid-free paper SPIN 10770348

Alexander Leonessa, Wassim M. Haddad
and VijaySekhar Chellaboina

Hierarchical Nonlinear Switching Control Design with Applications to Propulsion Systems

With 28 Figures

Springer

Springer
London
Berlin
Heidelberg
New York
Barcelona
Hong Kong
Milan
Paris
Singapore
Tokyo

Lecture Notes in Control and Information Sciences 255

Editor: M. Thoma

Contents

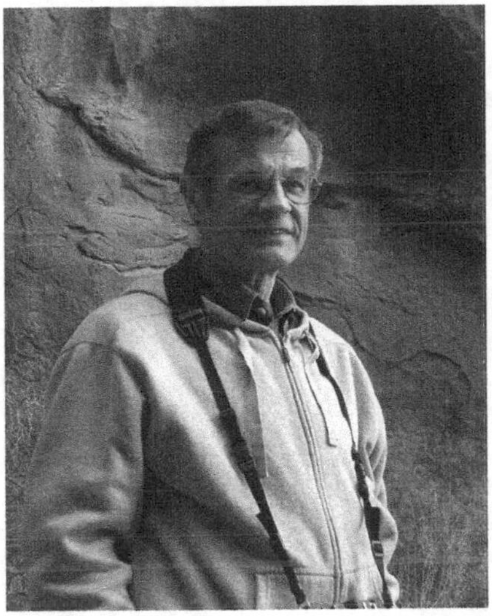

Professor Sergei Kuznetsov

Markov Processes and Their Applications to Partial Differential Equations: Kuznetsov's Contributions

E.B. Dynkin

Abstract We describe some directions of research in probability theory and related problems of analysis to which S. E. Kuznetsov has made fundamental contributions.

A Markov process (understood as a random path $X_t, 0 \leq t < \infty$ such that past before t and future after t are independent given X_t) is determined by a probability measure P on a path space. This measure can be constructed starting from a transition function and probability distribution of X_0 . For a number of applications, it is also important to consider a path in both, forward and backward directions which leads to a concept of dual processes. In 1973, Kuznetsov constructed, as a substitute for such a pair of processes, a single random process $(X_t, \mathbb{P})$ determined on a random time interval (α, β). The corresponding forward and backward transition functions define a dual pair of processes. A σ-finite measure $\mathbb{P}$ became, under the name "Kuznetsov measure," an important tool for research on Markov processes and their applications.

In 1980, Kuznetsov proved that every Markov process in a Borel state space has a transition function (a problem that was open for many years). In 1992, he used this result to obtain simple necessary and sufficient conditions for existence of a unique decomposition of excessive functions into extreme elements—a significant extension of a classical result on positive superharmonic functions.

Intimate relations between the Brownian motion and differential equations involving the Laplacian Δ were known for a long time. Applications of probabilistic tools to classical potential theory and to study of linear PDEs are more recent. Even more recent is application of such tools to nonlinear PDEs. In a series of publications, starting from 1994, Dynkin and Kuznetsov investigated a class of semilinear elliptic equations by using super-Brownian motion and more general measure-valued Markov processes called superdiffusions. The main directions

E.B. Dynkin (✉)
Professor Emeritus Department of Mathematics Malott Hall Cornell University, Ithaca, NY 14853, 4201
e-mail: ebd1@cornell.edu

J. Englander and B. Rider (eds.), *Advances in Superprocesses and Nonlinear PDEs*, Springer Proceedings in Mathematics & Statistics 38,
DOI 10.1007/978-1-4614-6240-8_1,

of this work were (a) description of removable singularities of solutions and (b) characterization of all positive solutions. One of the principal tools for solving the second problem was the fine trace of a solution on the boundary invented by Kuznetsov.

The same class of semilinear equations was the subject of research by Le Gall who applied a path-valued process Brownian snake instead of the super-Brownian motion. A slightly more general class of equations was studied by analysts including H. Brezis, M. Marcus and L. Veron. In the opinion of Brezis: "it is amazing how useful for PDEs are the new ideas coming from probability. This is an area where the interaction of probability and PDEs is most fruitful and exiting".

Keywords Markov processes • Transition function • Excessive functions • Superdiffusions • Semilinear PDE • Fine trace • Kuznetsov measures

1 Dual Markov Processes and Kuznetsov Measures

The idea of duality plays an important role in application of stochastic analysis to potential theory and partial differential equations.

Markov processes (X_t, P) and $(\hat{X}_t, \hat{P})$ with stationary transition functions $p, \hat{p}$ are in duality relative to a given σ-finite measure m if

$$m(dx)p_t(x,dy) = m(dy)\hat{p}_t(y,dx) \tag{1}$$

The processes can be defined on different spaces Ω and $\hat{\Omega}$ and there exists no relation between their paths. However the real source of duality is the fact that each Markov process can be considered in two time directions (this is the way the Kolmogorov forward and backward differential equations are deduced).

An alternative approach was suggested in [Dyn72] (see also [Dyn75] and [Dyn76]): functions p and $\hat{p}$ are interpreted as a forward and a backward transition functions of a single Markov process with random birth time α and death time β. This approach was applied also to nonstationary transition functions $p(s,x;t,dy), \hat{p}(s,x;t,dy)$ and measures m depending on the time interval (s,t). The process $X_t, t \in (\alpha, \beta)$ could be described by its two-dimensional distributions $m_{st}(dx,dy) = P\{\alpha < s, X_s \in dx, X_t \in dy, t < \beta\}$. The family m_{st} (called determining function) satisfies certain conditions including a normalization condition that guarantees that P is a probability measure. In the stationary case, this condition holds if m is a probability measure invariant for both dual processes. Since the definition of duality requires only σ-finiteness of m, this is a too restrictive assumption. It was removed by Kuznetsov in [Kuz73]. Of course, the measure $\mathbb{P}$ corresponding to a σ-finite m is only σ-finite. However considering a process $(X_t, \mathbb{P})$ with random birth and death times instead of a pair of dual Markov processes proved to be very useful in the theory of Markov processes and its applications. The name "Kuznetsov measure" for $\mathbb{P}$ is commonly used in the literature. In [Mey00],[1] P.-A. Meyer called its construction "the most remarkable result...of the Dynkin school which was in advance of its time."

2 Markov Property and the Existence of a Transition Function

A stochastic process (X_t, P) is Markovian if, for every t, the past $X_s, s < t$ and future $X_u, u > t$ are independent given X_t. This property is preserved under the time reversal. However, interactions between Markov processes and mathematical analysis are based on a possibility to start a process from a given point x of the state space E at a given time t which requires the existence of a transition function, that is of a family of (sub)-probability measures

$$p(s,x;t,\cdot) \quad s < t, x \in E, \tag{2}$$

subject to the Chapman–Kolmogorov equation, such that

$$P\{X_t \in \Gamma | X_s\} = p(s,x;t,\Gamma) \qquad P-a.s. \tag{3}$$

Markov property implies the existence of a quasi-transition function, subject to a weaker condition: the Chapman–Kolmogorov equation holds only off an exceptional set depending on three times $s < t < u$. A transition function exists for such important processes as diffusions and Markov chains. (In fact, these processes are defined in terms of certain transition functions.) A natural question—does it exist for any Markov process—was asked by Kolmogorov in 1930s. In [Wal72] Walsh proved that transition function exists for time reversed strong Markov right continuous processes.

In [Kuz80], Kuznetsov proved that Markov property implies the existence of transition function if the state space E is Borel (i.e., a measurable space isomorphic to a Borel subset in a separable complete metric space). In [Kuz86] he established the existence of a stationary transition function for every stochastically continuous Markov process in E with a stationary quasi-transition function and with one-dimensional distributions absolutely continuous with respect to a reference measure ν. In [Kuz92], he applied this result to investigate an existence of a stationary transition function $\hat{p}_t(x,dy)$ dual to a given stationary transition function $p_t(x,dy)$ relative to a given measure m that is related to p by (1).

[1] Based on his talk given in 1997 at Université Pierre et Marie Curie (Paris VI).

The measure ν must be excessive.[2] Under this condition on ν, it is sufficient that p is normal[3] and separates points.[4]

A substitute for a transition function for a random field $X_t, t \in T$ is its specification defined as a family of compatible conditional distributions for a value of the field given its restriction to subsets of the parameter set T. The existence of a specification for a wide class of random fields was proved by Kuznetsov in [Kuz84].

3 Existence and Uniqueness of Decomposition of Excessive Functions into Extremes

A class of positive functions f similar to the class of classical positive superharmonic functions is associated with every stationary Markov transition function p. This class, called excessive functions, is defined by the conditions

$$\int_E p_t(x,dy)f(y) \leq f(x) \quad \text{for every } x \in E\,, \int_E p_t(x,dy)f(y) \uparrow f(x) \quad \text{as } t \downarrow 0. \quad (4)$$

Every positive superharmonic function f in a smooth domain $D \subset R^d$ has a unique representation

$$f(x) = \int_D g(x,y)\mu(dy) + \int_{\partial D} k(x,y)\nu(dy) \quad (5)$$

where $g(x,y)$ is the Green function, $k(x,y)$ is the Poisson kernel and μ, ν are finite measures. To every $x \in D$ there corresponds an extreme function $g_x(\cdot) = g(x,\cdot)$ and to every $x \in \partial D$ there corresponds an extreme function $k_x(\cdot) = k(x,\cdot)$. In the early 1940s, Martin [Mar41] deduced a similar formula for non-smooth domains by replacing ∂D and $k(x,y)$ by what we call now the Martin boundary and the Martin kernel. In the 1950s Doob [Doo59] interpreted Martin's results in terms of the final behavior of the Brownian motion in D and he obtained analogous results for Markov chains in a countable state space. In the 1960s Hunt [Hun68] deduced the Martin representation of excessive functions by investigating the initial behavior of approximate Markov chains—a concept introduced by him for this purpose. In 1969 Dynkin [Dyn69a, Dyn69b] suggested to use instead the space of paths X_t defined for all positive and negative integers t and a σ-finite measure $\mathbb{P}$ on this space that can be considered as a precursor of the Kuznetsov measure. Martin boundary theory was extended to some classes of Markov processes with continuous time parameter by Kunita and T. Watanabe [KW65].

---

[2]Which means for every Γ, t, $\int_E \nu(dx)p_t(x,\Gamma) \leq \nu(\Gamma)$ and for every Γ, $\int_E \nu(dx)p_t(x,\Gamma) \uparrow \nu(\Gamma)$ as $t \downarrow 0$.

[3]That is $p_t(x,E) \uparrow 1$ as $t \downarrow 0$ for every $x \in E$.

[4]That is, if $p(\cdot,x,\cdot) = p(\cdot,y,\cdot)$, then $x = y$.

The problem—to describe the most general conditions on a Markov process under which the related excessive functions have a unique decomposition into extreme elements—was discussed in several papers of Dynkin. In his talk at the International Congress of Mathematicians (Nice, 1970) he suggested to start from the description of time-dependent excessive measures (also called entrance laws). He proved in [Dyn72] that such measures can be uniquely decomposed into extremes and he deduced from here an analogous property of time-dependent excessive functions assuming the existence of a transition density. In 1974, Kuznetsov [Kuz74] established: if this condition is violated, then there exists a time-dependent excessive function not decomposable into extremes. In [Kuz92b], Kuznetsov combined these ideas with the existence of a dual transition function proved in [Kuz92] to get the following final result. Every γ-integrable excessive function has a unique decomposition into extremes if and only if, for every $x \in E$, the measure

$$U_x(\cdot) = \int_0^\infty e^{-t} p_t(x,\cdot)\,dt, \quad x \in E$$

is absolutely continuous with respect to $U_\gamma(\cdot) = \int_E \gamma(dx) U_x(\cdot)$.

4 Superdiffusions and Semilinear PDEs

A diffusion is a model of a random motion of a single particle. It is characterized by a second-order elliptic differential operator L. A special case is the Brownian motion corresponding to the Laplacian Δ. A superdiffusion describes a random evolution of a cloud of particles. It is closely related to equations involving an operator $Lu - \psi(u)$. Here ψ belongs to a class of functions which contains, in particular $\psi(u) = u^\alpha$ with $\alpha > 1$. Fundamental contributions to the analytic theory of equations

$$Lu = \psi(u) \tag{6}$$

and the corresponding parabolic equations were made by Keller [Kel57], Osserman [Oss57], Brezis and Strauss [BS], Loewner and Nirenberg [LN74], Brezis and Véron [BV], Baras and Pierre [BP84,BP84b], and Marcus and Véron [MV,MV98b, MV07]. A relation between the Eq. (6) and superdiffusions was established, first, by S. Watanabe [Wat68]. Dawson [Daw75] and Perkins [Per89, Per91] obtained deep results on the path behavior of the super-Brownian motion.

These equations were investigated in a series of papers of Dynkin and Kuznetsov by a combination of probabilistic and analytic tools. A systematic presentation of the results can be found in monographs [Dyn02] and [Dyn05]. In our exposition we concentrate on a special case of equations

$$Lu = u^\alpha, \qquad 1 < \alpha \le 2 \tag{7}$$

in a bounded smooth domain $D \subset R^d$.

4.1 Removable Boundary Singularities

A subset Γ of the boundary ∂D is called a removable boundary singularity if every positive solution of (7) in D that is equal to zero on the boundary off Γ is equal to 0 in D. In [GV91], Gmira and Véron proved that single points are removable if and only if the dimension $d \geq \frac{\alpha+1}{\alpha-1}$. Later in [Le94], Le Gall gave a complete characterization of the class of removable sets in the case $\alpha = 2$ and $L = \Delta$. Namely, a set is removable if and only if it has certain capacity zero. Probabilistically, it is removable if and only if it is polar for the corresponding superdiffusion X.[5]

In [DK96], Dynkin and Kuznetsov established similar result for an arbitrary $\alpha \in (1,2]$. The proof consists of two parts. The first is probabilistic. For a set Γ with nonzero capacity we construct a required solution as a log-potential of a certain linear additive functional of X. The more difficult is the second analytic part: to show that, if the capacity of Γ is 0, then every solution $u \geq 0$ such that $u = 0$ on ∂D off Γ vanishes in D. Here analytic tools used by Le Gall are insufficient[6] and need to be supplemented by a new construction suggested by Kuznetsov. It involves a sequence of truncating functions h_n vanishing near Γ , with a Sobolev-type norm tending to zero. By using an estimate of $||uh_n||_\alpha$ we get that $u = 0$.

4.2 Positive Solutions and Their Boundary Traces

Every positive solution of a linear equation $Lu = 0$ in a bounded smooth domain D has a unique representation

$$u(x) = \int_{\partial D} k(x,y)\nu(dy) \tag{8}$$

where k is the Poisson kernel and ν is a finite measure on ∂D. We call ν the boundary trace of u.[7] Formula (8) establishes a 1–1 correspondence between the set of all positive solutions and the set of all finite measures on ∂D.

The situation is much more complicated for Eq. (7). A positive solution can explode on a part of the boundary and even if this part is empty, not every finite measure can serve as the boundary trace. A description of all positive solutions in the case $L = \Delta, \alpha = 2, d = 2$ was given by Le Gall in [Le97]. He defined a trace of u as a pair (Γ, ν) where Γ is a closed subset of ∂D on which u blows up rapidly, and ν is a σ-finite measure on $\partial D \setminus \Gamma$. He also described the class of pairs (Γ, ν) that are in 1–1 correspondence with solutions.

---

[5] Intuitively, it is not hit by the random cloud modeled by X.

[6] Conditions $\alpha = 2, L = \Delta$ simplify the situation drastically.

[7] If u is continuous in $D \cup \partial D$, then its boundary value is the density of ν with respect to the surface area.

The definition of trace in [Le97] is applicable for all dimensions d however, if $\alpha = 2$ and $d > 2$, then there exist infinite many solutions with the same trace. A principal difference between these cases is that there exist no polar sets except the empty set in the first case and such sets exist in the second case. For Eq. (7), the first case (called subcritical) takes place if and only if $\alpha < (d+1)/(d-1)$. In this case positive solutions can be characterized by a boundary trace introduced in [DK98a] and in a slightly different form in [MV]. However, this is not true for $\alpha \geq (d+1)/(d-1)$. A breakthrough in investigating the latter case was an idea of a fine trace suggested by Kuznetsov in [Kuz98]. There he defined a special topology on the boundary associated with the equation (the so-called fine topology), and he constructed a "fine trace" (Γ, ν) where the set Γ is finely closed and ν is still a σ-finite measure that does not charge polar sets. He then described the class of possible fine traces and showed that a wide class of solutions, called σ-moderate solutions, can be uniquely characterized by their fine traces. In [DK98] these results were extended to Eq. (6) under weak conditions on a monotone increasing convex nonlinear term $\psi(u)$. The crucial problem: "are all solutions σ-moderate?" (stated in the Epilog in [Dyn02]) was solved positively by Mselati in [Ms04] for $L = \Delta, \alpha = 2$ by using the Brownian snake. In a series of articles by Dynkin and Kuznetsov, Mselati's result was extended, by using superdiffusions instead of the Brownian snake to the equation $\Delta u = u^\alpha$ with $1 < \alpha \leq 2$. A self-contained presentation of the proofs can be found in [Dyn05].

References

BP84. P. Baras and M. Pierre, *Problems paraboliques semi-lineáres avec donnees measures*, Applicable Analysis **18** (1984), 111–149.

BP84b. ——, *Singularités éliminable pour des équations semi-linéares*, Ann. Inst. Fourier Grenoble **34** (1984), 185–206.

BS. H. Brezis and V. Strauss, *Semilinear second-order Elliptic equations in L^1*, J. Math. Soc. Japan **25** (1973), 565–590.

BV. H. Brezis and L. Véron, *Removable singularities of some removable equations*, Arch. Rat. Mech. Anal. **75** (1980), 1–6.

Daw75. D. A. Dawson, *Stochastic evolution equations and related measure processes*, J. Multivariate Anal. **3** (1975), 1–52.

Doo59. J. L. Doob, *Discrete Potential Theory and Boundaries*, J. Math. Mech **8** (1959), 433–458.

Dyn69a. E. B. Dynkin, *The boundary theory for Markov processes (discrete case)*, Russian Math. Surveys **24, 2** (1969), 89–157.

Dyn69b. ——, *Exit space of a Markov process*, Russian Math. Surveys **24, 4** (1969), 89–152.

Dyn72. ——, *Integral representation of excessive measures and excessive functions*, Russian Math. Surveys **27,1** (1972), 43–84.

Dyn75. ——, *Markov representation of stochastic systems*, Russian Math. Surveys **30,1** (1975), 64–104.

Dyn76. ——, *On a new approach to Markov processes*, Proceedings of the Third Japan-USSR Symposium on Probability Theory (1976), 42–62. Lecture Notes in Mathematics, vol. 550.

Dyn02. ——, *Diffusions, Superdiffusions and Partial Differential Equations*, 2002. American Math Society, Colloquium Publications, Vol. 50.

Dyn05. ——, *Superdiffusions and Positive Solutions of Nonlinear Partial Differential Equations*, 2004. American Math Society, University Lecture Series, Vol. 34.

DK96. E. B. Dynkin and S.E. Kuznetsov, *Superdiffusions and removable singularities for quasilinear partial differential equations*, Comm. Pure Appl.Math. **49** (1996), 125–176.

DK98a. ——, *Trace on the boundary for solutions of nonlinear partial differential equations*, Trans. Amer. Math. Soc. **350** (1998), 4499–4519.

DK98. ——, *Fine topology and fine trace on the boundary associated with a class of quasilinear partial differential equations*, Comm. Pure Appl.Math. **51** (1998), 897–936.

GV91. A. Gmira and L. Véron, *Boundary singularities of solutions of some nonlinear elliptic equations*, Duke Math. J. **64** (1991), 271–324.

Hun68. G. A. Hunt, *Markov Chains and Martin Boundaries*, Illinois J. Math. **4** (1968), 233–278, 365–410.

Kel57. J. B. Keller, *On the solutions of $\Delta u = f(u)$*, Comm. Pure Appl. Math. **10** (1957), 503–510.

KW65. H. Kunita and T. Watanabe, *Markov processes and Martin boundaries*, Illinois J. Math. **9** (1965), 386–391.

Kuz73. S. E. Kuznetsov, *Construction of Markov processes with random birth and death times*, Teoriya Veroyatn. i ee Primen. **18** (1973), 596–601. English Transl. in Theor. Prob. Appl., Vol 18 (1974).

Kuz74. ——, *On decomposition of excessive functions*, Dokl. AN SSSR **214** (1974), 276–278. English Transl. in Soviet Math. Dokl., Vol 15 (1974).

Kuz80. ——, *Any Markov process in a Borel space has a transition function*, Teoriya Veroyatn. i ee Primen. **25** (1980), 389–393. English Transl. in Theor. Prob. Appl., Vol 25 (1980).

Kuz84. ——, *Specifications and stopping theorem for random fields*, Teoriya Veroyatn. i ee Primen. **29** (1984), 65–77. English Transl. in Theor. Prob. Appl., Vol 29 (1985).

Kuz86. ——, *On the existence of a homogeneous transition function*, Teoriya Veroyatn. i ee Primen. **31** (1986), 290–300. English Transl. in Theor. Prob. Appl., Vol 31 (1987).

Kuz92. ——, *More on existence and uniqueness of decomposition of excessive functions and measures to extremes*, Seminaire de Probabilites XXVI. Lecture Notes in Mathematics **1526** (1992), 445–472.

Kuz92b. ——, *On existence of a dual semigroup*, Seminaire de Probabilites XXVI. Lecture Notes in Mathematics **1526** (1992), 473–484.

Kuz98. ——, *σ-moderate solutions of $Lu = u^\alpha$ and fine trace on the boundary*, C.R. Acad. Sci. Paris, Série I **326** (1998), 1189–1194.

Oss57. R. Osserman, *On the inequality $\Delta u \geq f(u)$, Pacific J. Math.* **7** (1957), 1641–1647.

Le94. J.-F. Le Gall, *The Brownian snake and solutions of $\Delta u = u2$ in a domain*, Probab. Theory Rel. Fields **102** (1995), 393–402.

Le97. ——, *A probabilistic Poisson representation for positive solutions of $\Delta u = u2$ in a planar domain*, Comm. Pure Appl. Math. **50** (1997), 69–103.

LN74. C. Loevner and L. Nirenberg, *Partial differential equations invariant under conformal or projective transformations*, (1974), 245–272. Contributions to Analysis, Academic Press, Orlando, FL.

Mar41. R. S. Martin, *Minimal positive harmonic functions*, Trans. Amer. Math. Soc. **49** (1941), 137–172.

Mey00. P.-A. Meyer, *Dynkin and theory of Markov processes*, Selected Papers of E. B. Dynkin with Commentary (2000), 763–772. Amer. Math. Soc., Providence, RI.

Ms04. B. Mselaty, *Classification and probabilistic representation of the positive solutions of a semilinear elliptic equation*, Mem. Amer. Math. Soc. **168** (2004).

MV. M. Marcus and L. Véron, *The boundary trace of positive solutions of semilinear elliptic equations, I. The subcritical case*, Arch. Rat. Mech. Anal. **144** (1998), 201–231.

MV98b. ——, *The boundary trace of positive solutions of semilinear elliptic equations: The supercritical case*, J. Math. Pures Appl. **77** (1998), 481–524.

MV07. ——, *The precise boundary trace of positive solutions of $\Delta u = u^q$ in the supercritical case*, **446** (2007), 345–383. Perspectives in nonlinear partial differential equations, Contemp. Math., Amer. Math. Soc., Providence, RI.

Per89. E. A. Perkins, *The Hausdorff measure of the closed support of super-Brownian motion*, Ann. Inst. H. Poincaré **25** (1989), 205–224.

Per91. ——, *On the continuity of measure-valued processes*, (1991), 261–268. Seminar on Stochastic Processes, Boston/Basel/Berlin.

Wal72. J. B. Walsh, *Transition functions of Markov processes*, Sminaire de Probabilits **VI** (1972), 215–232. Lecture Notes in Math., Vol. 258, Springer, Berlin, 1972.

Wat68. S. Watanabe, *A limit theorem on branching processes and continuous state branching processes*, J. Math. Kyoto Univ. **8** (1968), 141–167.

Stochastic Equations on Projective Systems of Groups

Steven N. Evans and Tatyana Gordeeva

Abstract We consider stochastic equations of the form $X_k = \phi_k(X_{k+1})Z_k$, $k \in \mathbb{N}$, where X_k and Z_k are random variables taking values in a compact group G_k, $\phi_k : G_{k+1} \to G_k$ is a continuous homomorphism, and the noise $(Z_k)_{k\in\mathbb{N}}$ is a sequence of independent random variables. We take the sequence of homomorphisms and the sequence of noise distributions as given and investigate what conditions on these objects result in a unique distribution for the "solution" sequence $(X_k)_{k\in\mathbb{N}}$ and what conditions permit the existence of a solution sequence that is a function of the noise alone (i.e., the solution does not incorporate extra input randomness "at infinity"). Our results extend previous work on stochastic equations on a single group that was originally motivated by Tsirelson's example of a stochastic differential equation that has a unique solution in law but no strong solutions.

Keywords Group representation • Uniqueness in law • Strong solution • Extreme point • Lucas theorem • Toral automorphism

Mathematics Subject Classification (2010): 60B15, 60H25.

SNE supported in part by NSF grant DMS-0907630. TG supported in part by a VIGRE grant awarded to the Department of Statistics, University of California at Berkeley.

S.N. Evans (✉) • T. Gordeeva
Department of Statistics #3860, University of California at Berkeley,
367 Evans Hall, Berkeley, CA 94720-3860, U.S.A
e-mail: evans@stat.Berkeley.EDU; gordeeva@stat.Berkeley.EDU

J. Englander and B. Rider (eds.), *Advances in Superprocesses and Nonlinear PDEs*,
Springer Proceedings in Mathematics & Statistics 38,
DOI 10.1007/978-1-4614-6240-8_2,

1 Introduction

The following stochastic process was considered by Yor in [Yor92] in order to clarify the structure underpinning Tsirelson's celebrated example [Cir75] of a stochastic differential equation that does not have a strong solution even though all solutions have the same law.

Let $\mathbb{T}$ be the usual circle group; that is, $\mathbb{T}$ can be thought of as the interval $[0,1)$ equipped with addition modulo 1. Suppose for each $k \in \mathbb{N}$ that μ_k is a Borel probability measure on $\mathbb{T}$. Write $\mu = (\mu_k)_{k\in\mathbb{N}}$. We say that sequence of $\mathbb{T}$-valued random variables $(X_k)_{k\in\mathbb{N}}$ defined on some probability space $(\Omega, \mathcal{F}, \mathbb{P})$ *solves the stochastic equation associated with* μ if

$$\mathbb{P}[f(X_k) \mid (X_j)_{j>k}] = \int_{\mathbb{T}} f(X_{k+1} + z)\, \mu_k(dz)$$

for all bounded Borel function $f : \mathbb{T} \to \mathbb{R}$, where we use the notation $\mathbb{P}[\cdot \mid \cdot]$ for condition expectations with respect to $\mathbb{P}$. In other words, if for each $k \in \mathbb{N}$ we define a $\mathbb{T}$-valued random variable Z_k by requiring

$$X_k = X_{k+1} + Z_k, \tag{1}$$

then $(X_k)_{k\in\mathbb{N}}$ solves the stochastic equation associated with μ if and only if for all $k \in \mathbb{N}$ the distribution of Z_k is μ_k and Z_k is independent of $(X_j)_{j>k}$.

Yor addressed the existence of solutions $(X_k)_{k\in\mathbb{N}}$ that are *strong* in the sense that the random variable X_k is measurable with respect to $\sigma((Z_j)_{j\geq k})$ for each $k \in \mathbb{N}$; that is, speaking somewhat informally, a solution is strong if it can be reconstructed from the "noise" $(Z_j)_{j\in\mathbb{N}}$ without introducing additional randomness "at infinity." It turns out that strong solutions exist if and only if

$$\lim_{m\to\infty} \lim_{n\to\infty} \prod_{\ell=m}^{n} \left| \int_{\mathbb{T}} \exp(2\pi i h x)\, \mu_\ell(dx) \right| > 0$$

for all $h \in \mathbb{Z}$ or, equivalently,

$$\sum_{k=1}^{\infty} \left[1 - \left| \int_{\mathbb{T}} \exp(2\pi i h x)\, \mu_k(dx) \right| \right] < \infty.$$

Yor's investigation was extended in [AUY08], where the group $\mathbb{T}$ is replaced by an arbitrary, possibly non-abelian, compact Hausdorff group. As one would expect, the role of the complex exponentials $\exp(2\pi i h \cdot)$, $h \in \mathbb{Z}$, in this more general setting is played by group representations. Interesting new phenomena appear when the group is non-abelian due to the fact that there are irreducible representations which are no longer one-dimensional. Several of the results in [AUY08] are framed in terms of properties of the set of extremal solutions (i.e., solutions that cannot be written as mixtures of others), and the structure of such solutions was elucidated further in [HY10].

We further extend the work in [Yor92, AUY08] by considering the following more general setup.

Fix a sequence $(G_k)_{k\in\mathbb{N}}$ of compact Hausdorff groups with countable bases. Suppose for each $k \in \mathbb{N}$ that there is a continuous homomorphism $\phi_k : G_{k+1} \to G_k$. Define a compact subgroup $H \subseteq G := \prod_{k\in\mathbb{N}} G_k$ by

$$H := \{g = (g_k)_{k\in\mathbb{N}} \in G : g_k = \phi_k(g_{k+1}) \text{ for all } k \in \mathbb{N}\}, \tag{2}$$

For example, if we take $G_k = \mathbb{T}$ for all $k \in \mathbb{N}$, then the homomorphism ϕ_k is necessarily of the form $\phi_k(x) = N_k x$ for some $N_k \in \mathbb{Z}$ and

$$H = \{g = (g_k)_{k\in\mathbb{N}} \in G : g_k = N_k g_{k+1} \text{ for all } k \in \mathbb{N}\}.$$

For a more interesting example, fix a compact group abelian group Γ, put $G_k := G_{1,k} \times G_{2,k-1} \cdots \times G_{k,1}$, where each group $G_{i,j}$ is a copy of Γ, and define the homomorphism ϕ_k by

$$\phi_k(g_{1,k+1}, g_{2,k}, \ldots, g_{k+1,1}) := (g_{1,k+1} + g_{2,k}, g_{2,k} + g_{3,k-1}, \ldots, g_{k,2} + g_{k+1,1})$$

(where we write the group operation in Γ additively). Note that in this case H is isomorphic to the infinite product $\Gamma^{\mathbb{N}}$, because an element $h = (h_{i,j})_{(i,j)\in\mathbb{N}\times\mathbb{N}}$ is uniquely specified by the values $(h_{i,1})_{i\in\mathbb{N}}$ and there are no constraints on these elements. The following picture shows a piece of an element of H when Γ is the group $\{0,1\}$ equipped with addition modulo 2.

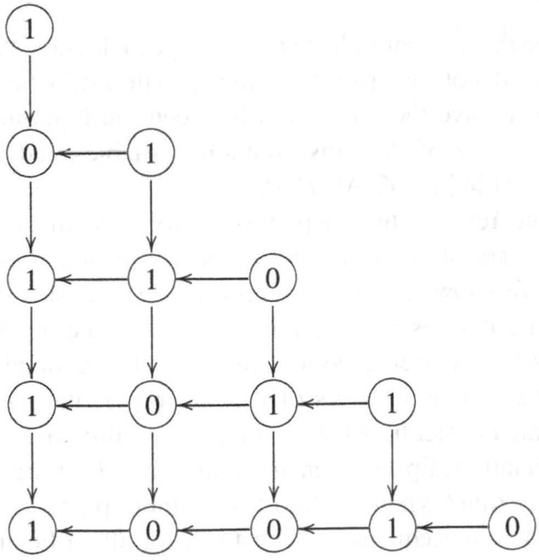

Assume for each $k \in \mathbb{N}$ that μ_k is a Borel probability measure G_k and write $\mu = (\mu_k)_{k\in\mathbb{N}}$. We say that sequence of random variables $(X_k)_{k\in\mathbb{N}}$ defined on some probability space $(\Omega, \mathcal{F}, \mathbb{P})$, where X_k takes values in G_k, *solves the stochastic equation associated with* μ if

$$\mathbb{P}[f(X_k) \mid (X_j)_{j>k}] = \int_{G_k} f(\phi_k(X_{k+1})z)\, \mu_k(dz)$$

for all bounded Borel function $f : G_k \to \mathbb{R}$. In other words, if for each $k \in \mathbb{N}$ we define a G_k-valued random variable Z_k by requiring

$$X_k = \phi_k(X_{k+1})Z_k, \tag{3}$$

then $(X_k)_{k\in\mathbb{N}}$ solves the stochastic equation if and only if for all $k \in \mathbb{N}$ the distribution of Z_k is μ_k and Z_k is independent of $(X_j)_{j>k}$. In particular, if $(X_k)_{k\in\mathbb{N}}$ solves the stochastic equation, then the sequence of random variables $(Z_k)_{k\in\mathbb{N}}$ is independent.

Certain special cases of this setup when $G_k = \Gamma$, $k \in \mathbb{N}$, for some fixed group Γ and $\phi_k = \psi$, $k \in \mathbb{N}$ for a fixed automorphism $\psi : \Gamma \to \Gamma$ were considered in [Tak09, Raj11].

Note that whether or not a sequence $(X_k)_{k\in\mathbb{N}}$ solves the stochastic equation associated with μ is solely a feature of the distribution of the sequence, and so we say that a probability measure on the product group $\prod_{k\in\mathbb{N}} G_k$ is a solution of the stochastic equation if it is the distribution of a sequence that solves the equation and write $\mathcal{P}_\mu$ for the set of such measures.

In keeping with the terminology above, we say that a solution $(X_k)_{k\in\mathbb{N}}$ is *strong* if X_k is measurable with respect to $\sigma((Z_j)_{j\geq k})$ for each $k \in \mathbb{N}$. Note that whether or not a solution is strong also depends only on its distribution, and so we define strong elements of $\mathcal{P}_\mu$ in the obvious manner and denote the set of such probability measures by $\mathcal{P}_\mu^{\text{strong}}$.

Because applying the homomorphism ϕ_k to X_{k+1} can degrade the "signal" present in X_{k+1} (e.g., ϕ_k need not be invertible), the question of whether or not strong solutions exist will involve the interaction between the homomorphisms $(\phi_k)_{k\in\mathbb{N}}$ and distributions $(\mu_k)_{k\in\mathbb{N}}$ of the noise random variables and it introduces new phenomena not present in [Yor92, AUY08].

An outline of the rest of this chapter is as follows. In Sect. 2 we examine the compact, convex set of solutions and show that strong solutions are extreme points of this set. We show that the subgroup H acts transitively on the extreme points of the set of solutions and we relate the existence of strong solutions to properties of the set of extreme points. In Sect. 3, we obtain criteria for the existence of strong solutions in terms of the the representations of the group G_k and the corresponding Fourier transforms of the probability measures μ_k. In Sect. 3, we determine the relationship between the existence of strong solutions and the phenomenon of "freezing" wherein almost all sample paths of the random noise sequence agree with some sequence of constants for all sufficiently large indices. Finally, in Sects. 5 and 6, respectively, we investigate the example considered above of random variables indexed by the nonnegative quadrant of the two-dimensional integer lattice and another example where each group G_k is the two-dimensional torus and each homomorphisms ϕ_k is a fixed ergodic toral automorphism.

2 Extreme Points of $\mathcal{P}_\mu$ and Strong Solutions

It is natural to first inquire whether $\mathcal{P}_\mu$ is nonempty and, if so, whether it consists of a single point that is, whether there exist probability measures that solve the stochastic equation associated with μ and, if so, whether there is a single such measure. The question of existence is easily disposed of by Proposition 2.1 below. Note that because the group $G = \prod_{k\in\mathbb{N}} G_k$ is compact and metrizable, the set of probability measures on G equipped with the topology of weak convergence is also compact and metrizable.

Proposition 2.1. *For any sequence μ, the set $\mathcal{P}_\mu$ is non-empty.*

Proof. Construct on some probability space a sequence $(Z_k)_{k\in\mathbb{N}}$ of independent random variables such that Z_k has distribution μ_k. For each $N \in \mathbb{N}$, define random variables $X_1^{(N)},\ldots,X_{N+1}^{(N)}$ recursively by

$$X_{N+1}^{(N)} := e_{N+1} := \text{identity in } G_{N+1}$$

and

$$X_k^{(N)} = \phi_k(X_{k+1}^{(N)})Z_k, \quad 1 \le k \le N,$$

so that for $1 \le k \le N$ the random variable $\phi_k(X_{k+1}^{(N)})^{-1}X_k^{(N)}$ has distribution μ_k and is independent of $X_{k+1}^{(N)}, X_{k+2}^{(N)}, \ldots, X_N^{(N)}$.

Write $\mathbb{P}_N$ for the distribution of the sequence $(X_1^{(N)},\ldots,X_N^{(N)}, e_{N+1}, e_{N+2},\ldots)$. Because the space of probability measures on the group $\prod_{k\in\mathbb{N}} G_k$ equipped with the weak topology is compact and metrizable, there exists a subsequence $(N_n)_{n\in\mathbb{N}}$ and a probability measure $\mathbb{P}_\infty$ such that $\mathbb{P}_{N_n} \to \mathbb{P}_\infty$ weakly as $n \to \infty$. It is clear that $\mathbb{P}_\infty \in \mathcal{P}_\mu$. □

The question of uniqueness (i.e., whether or not $\#\mathcal{P}_\mu = 1$) is more demanding and will occupy much of our attention in the remainder of this chapter.

As a first indication of what is involved, consider the case where each measure μ_k is simply the unit point mass at the identity e_k of G_k. In this case $(X_k)_{k\in\mathbb{N}}$ solves the stochastic equation if $X_k = \phi_k(X_{k+1})$ for all $k \in \mathbb{N}$. Recall the definition of the compact subgroup $H \subseteq G := \prod_{k\in\mathbb{N}} G_k$ from (2). It is clear that $\mathcal{P}_\mu$ coincides with the set of probability measures that are supported on H, and hence $\#\mathcal{P}_\mu = 1$ if and only if H consists of just the single identity element. Note that if $\#H > 1$ and $(X_k)_{k\in\mathbb{N}}$ is a solution with distribution $\mathbb{P} \in \mathcal{P}_\mu$ that is not a point mass, then X_k is certainly not a function of $(Z_j)_{j\ge k} = (e_j)_{j\ge k}$ and the solution $(X_k)_{k\in\mathbb{N}}$ is not strong. Moreover, the probability measures $\mathbb{P} \in \mathcal{P}_\mu$ that are distributions of strong solutions $(X_k)_{k\in\mathbb{N}}$ are the point masses at elements of H and $\mathcal{P}_\mu$ is the closed convex hull of this set of measures.

An elaboration of the argument we have just given establishes the following result.

Proposition 2.2. *If H is nontrivial (i.e, contains elements other than the identity), then $\mathcal{P}_\mu \setminus \mathcal{P}_\mu^{\text{strong}} \neq \emptyset$. In particular, if H is nontrivial and $\#\mathcal{P}_\mu = 1$, then $\mathcal{P}_\mu^{\text{strong}} = \emptyset$.*

Proof. Suppose that all solutions are strong. Let $(X_k)_{k\in\mathbb{N}}$ be a strong solution.

By extending the underlying probability space if necessary, construct an H-valued random variable $(U_k)_{k\in\mathbb{N}}$ that is independent of $(X_k)_{k\in\mathbb{N}}$ and is not almost surely constant. Note that $(U_k)_{k\in\mathbb{N}}$ is not $\sigma((X_k)_{k\in\mathbb{N}})$-measurable and hence, *a fortiori*, $(U_k)_{k\in\mathbb{N}}$ is not $\sigma((Z_k)_{k\in\mathbb{N}})$-measurable.

Observe that

$$\phi_k(U_{k+1}X_{k+1})Z_k = \phi_k(U_{k+1})\,\phi_k(X_{k+1})Z_k = U_kX_k,$$

because $\phi_k(U_{k+1}) = U_k$ for all $k \in \mathbb{N}$ by definition of H. Hence, $(U_kX_k)_{k\in\mathbb{N}}$ is also a solution. Thus, $(U_kX_k)_{k\in\mathbb{N}}$ is a strong solution by our assumption that all solutions are strong. In particular, U_kX_k is $\sigma((Z_j)j \geq k)$-measurable for all $k \in \mathbb{N}$. However, $U_k = (U_kX_k)X_k^{-1}$ is $\sigma((Z_j)_{j\geq k})$-measurable, and we arrive at a contradiction. □

Remark 2.3. Consider the particular setting of [AUY08], where $G_k = \Gamma$, $k \in \mathbb{N}$, for some fixed group Γ, each homomorphism ϕ_k is the identity, and $H = \{(g,g,\ldots) : g \in \Gamma\}$. In this case, one can choose the sequence $(U_k)_{k\in\mathbb{N}}$ in the proof of Proposition 2.2 to be $(U,U,\cdots)$, where U is distributed according to Haar measure on Γ; that is, $(U_k)_{k\in\mathbb{N}}$ is distributed according to Haar measure on H. Each marginal distribution of the solution $(X_k)_{k\in\mathbb{N}}$ is then Haar measure on $G_k = \Gamma$. In our more general setting it will not generally be the case that if $(U_k)_{k\in\mathbb{N}}$ is distributed according to Haar measure on H, then X_k will be distributed according to Haar measure on G_k for each $k \in \mathbb{N}$. For example, fix a compact group Γ, put $G_k = \Gamma^{\mathbb{N}}$ for all $k \in \mathbb{N}$ and define $\phi_k : G_{k+1} \to G_k$ by $\phi_k(g_1,g_2,g_3,\ldots) = (g_1,g_1,g_2,g_2,g_3,g_3,\ldots)$ for all $k \in \mathbb{N}$. It is clear that $H = \{((g,g,\ldots),(g,g,\ldots),\ldots) : g \in \Gamma\}$, so that $\{x_k : (x_1,x_2,\ldots) \in H\} \subseteq G_k$ is just the diagonal subgroup $\{(g,g,\ldots) : g \in \Gamma\}$ of the group G_k. Hence, for example, if μ_k is the point mass at the identity of G_k for each $k \in \mathbb{N}$, the possible solutions $(X_k)_{k\in\mathbb{N}}$ are just arbitrary random elements of H, and it is certainly not possible to construct a solution such that the marginal distribution of X_k is Haar measure on G_k for some $k \in \mathbb{N}$.

From now on, we let $X_k : G \to G_k$, $k \in \mathbb{N}$, denote the random variable defined by $X_k((x_j)_{j\in\mathbb{N}}) := x_k$ and define $Z_k : G \to G_k$, $k \in \mathbb{N}$, by $Z_k := \phi_n(X_{k+1})^{-1}X_k$.

Notation 2.4. Given a sequence of random variables $S = (S_1,S_2,\ldots)$ and $k \in \mathbb{N}$, set $\mathcal{F}_k^S := \sigma((S_j)_{j\geq k})$. Similarly, set $\mathcal{F}^S := \mathcal{F}_1^S$ and $\mathcal{F}_\infty^S := \bigcap_{k\in\mathbb{N}} \mathcal{F}_k^S$.

Notation 2.5. For any sequence $\mu = (\mu_k)_{k\in\mathbb{N}}$, the set of solutions $\mathcal{P}_\mu$ is clearly a compact convex subset. Let $\mathcal{P}_\mu^{\text{ex}}$ denote the extreme points of $\mathcal{P}_\mu$.

Lemma 2.6. *A probability measure $\mathbb{P} \in \mathcal{P}_\mu$ belongs to $\mathcal{P}_\mu^{\text{ex}}$ if and only if the remote future $\mathcal{F}_\infty^X$ is trivial under $\mathbb{P}$.*

Proof. Our proof follows that of an analogous result in [AUY08].

Suppose that $\mathbb{P} \in \mathcal{P}_\mu$ and the σ-field $\mathcal{F}_\infty^X$ is not trivial under $\mathbb{P}$.

Fix a set $A \in \mathcal{F}_\infty^X$ with $0 < \mathbb{P}(A) < 1$. Then,

$$\mathbb{P}(\cdot) = \mathbb{P}(A)\mathbb{P}(\cdot\,|\,A) + (1-\mathbb{P}(A))\mathbb{P}(\cdot\,|\,A^c).$$

Observe that $\mathbb{P}(\cdot\,|\,A) \neq \mathbb{P}(\cdot\,|\,A^c)$, since $\mathbb{P}(A\,|\,A) = 1 \neq \mathbb{P}(A\,|\,A^c) = 0$.

Note for each $k \in \mathbb{N}$ and $B \subseteq G_k$ that

$$\begin{aligned}\mathbb{P}\{X_k\phi_k(X_{k+1})^{-1} \in B\,|\,A\} &= \frac{\mathbb{P}(\{X_k\phi_k(X_{k+1})^{-1} \in B\} \cap A)}{\mathbb{P}(A)} \\ &= \frac{\mu_k(B)\mathbb{P}(A)}{\mathbb{P}(A)} = \mu_k(B)\end{aligned}$$

because $\mathbb{P} \in \mathcal{P}_\mu$ and hence $X_k\phi_k(X_{k+1})^{-1}$ is independent of $\mathcal{F}_\infty^X$ under $\mathbb{P}$. Similarly, if $C \in \mathcal{F}_{k+1}^X$,

$$\begin{aligned}\mathbb{P}(\{X_k\phi_k(X_{k+1})^{-1} \in B\} \cap C\,|\,A) &= \frac{\mu_k(B)\mathbb{P}(C\cap A)}{\mathbb{P}(A)} \\ &= \mathbb{P}\{X_k\phi_k(X_{k+1})^{-1} \in B\,|\,A\}\mathbb{P}(C\,|\,A)\end{aligned}$$

Thus, $\mathbb{P}(\cdot\,|\,A) \in \mathcal{P}_\mu$. The analogous argument establishes $\mathbb{P}(\cdot\,|\,A^c) \in \mathcal{P}_\mu$. Since $\mathbb{P}(\cdot\,|\,A) \neq \mathbb{P}(\cdot\,|\,A^c)$, the probability measure $\mathbb{P}$ cannot belong to $\mathcal{P}_\mu^{\text{ex}}$.

Now assume that $\mathbb{P} \in \mathcal{P}_\mu$ and $\mathcal{F}_\infty^X$ is trivial under $\mathbb{P}$. To show $\mathbb{P}$ is an extreme point, it suffices to show that if $\mathbb{P}' \in \mathcal{P}_\mu$ is absolutely continuous with respect to $\mathbb{P}$, then $\mathbb{P} = \mathbb{P}'$.

Note that a solution X is a time-inhomogeneous Markov chain (indexed in backwards time with index set starting at infinity) with the following transition probability:

$$\mathbb{P}\{X_k \in A\,|\,X_{k+1}\} = \mu_k\{g \in G_k : \phi_k(X_{k+1})g \in A\}.$$

Since $\mathbb{P}$ and $\mathbb{P}'$ are the distributions of Markov chains with common transition probabilities and $\mathbb{P}'$ is absolutely continuous with respect to $\mathbb{P}$, it follows that for any measurable set A the random variables $\mathbb{P}(A\,|\,\mathcal{F}_\infty^X)$ and $\mathbb{P}'(A\,|\,\mathcal{F}_\infty^X)$ are equal $\mathbb{P}$-a.s. Because $\mathcal{F}_\infty^X$ is trivial under both $\mathbb{P}$ and $\mathbb{P}'$, it must be the case that $\mathbb{P}(A) = \mathbb{P}'(A)$. □

Corollary 2.7. *All strong solutions $\mathbb{P} \in \mathcal{P}_\mu$ are extreme; that is, $\mathcal{P}_\mu^{\text{strong}} \subseteq \mathcal{P}_\mu^{\text{ex}}$.*

Proof. By definition, if $\mathbb{P} \in \mathcal{P}_\mu$ is strong, then $X_k \in \mathcal{F}_k^Z$ for all $k \in \mathbb{N}$. Thus, $\mathcal{F}_k^X = \mathcal{F}_k^Z$ for all $k \in \mathbb{N}$ and hence $\mathcal{F}_\infty^X = \mathcal{F}_\infty^Z$. The last σ-field is trivial by the Kolmogorov zero-one law. □

Remark 2.8. There can be extreme solutions that are not strong. For example, suppose that the $G_k = \Gamma$, $k \in \mathbb{N}$, for some nontrivial group Γ, each ϕ_k is the identity map, and each μ_k is the Haar measure on Γ. It is clear that $\mathcal{P}_\mu$ consists of just the measure $\bigotimes_{k\in\mathbb{N}} \mu_k$ (i.e., Haar measure on G), and so this solution is extreme. However, it follows from Proposition 2.2 that this solution is not strong.

It is clear that if $\mathbb{P} \in \mathcal{P}_\mu$ and $h = (h_k)_{k\in\mathbb{N}} \in H$, then the distribution of the sequence $(h_k X_k)_{k\in\mathbb{N}}$ also belongs to $\mathbb{P} \in \mathcal{P}_\mu$. Moreover, if $\mathbb{P} \in \mathcal{P}_\mu^{\mathrm{ex}}$, then it follows from Lemma 2.6 that the distribution of the sequence $(h_k X_k)_{k\in\mathbb{N}}$ also belongs to $\mathcal{P}_\mu^{\mathrm{ex}}$. Similarly, if $\mathbb{P} \in \mathcal{P}_\mu^{\mathrm{strong}}$, then the distribution of the sequence $(h_k X_k)_{k\in\mathbb{N}}$ also belongs to $\mathcal{P}_\mu^{\mathrm{strong}}$. We record these observations for future reference.

Lemma 2.9. *The collection of maps $T_h : \mathcal{P}_\mu \to \mathcal{P}_\mu$, $h \in H$, defined by $T_h(\mathbb{P})(\cdot) = \mathbb{P}\{(h_k X_k)_{k\in\mathbb{N}} \in \cdot\}$ constitute a a group action of H on $\mathcal{P}_\mu$. The set $\mathcal{P}_\mu^{\mathrm{ex}}$ of extreme solutions and the set $\mathcal{P}_\mu^{\mathrm{strong}}$ of strong solutions are both invariant for this action.*

It follows from the next result that either $\mathcal{P}_\mu^{\mathrm{strong}} = \emptyset$ or $\mathcal{P}_\mu^{\mathrm{strong}} = \mathcal{P}_\mu^{\mathrm{ex}}$. For the purposes of the proof and later it is convenient to introduce the following notation.

Notation 2.10. For $k, \ell \in \mathbb{N}$ with $k < \ell$, define $\phi_k^\ell : G_\ell \to G_k$ by

$$\phi_k^\ell = \phi_k \circ \phi_{k+1} \circ \cdots \circ \phi_{\ell-1},$$

and adopt the convention that ϕ_k^k is the identity map from G_k to itself.

Theorem 2.11. *The group action $(T_h)_{h\in H}$ is transitive on $\mathcal{P}_\mu^{\mathrm{ex}}$.*

Proof. For $k \in \mathbb{N}$, define $X_k' : \prod_{k\in\mathbb{N}}(G_k \times G_k \times G_k) \to G_k$ (resp. $X_k'' : \prod_{k\in\mathbb{N}}(G_k \times G_k \times G_k) \to G_k$) and $Y_k : \prod_{k\in\mathbb{N}}(G_k \times G_k \times G_k) \to G_k$) by $X_k'((x_j', x_j'', y_j)_{j\in\mathbb{N}}) = x_k'$ (resp. $X_k''((x_j', x_j'', y_j)_{j\in\mathbb{N}}) = x_k''$ and $Y_k((x_j', x_j'', y_j)_{j\in\mathbb{N}}) = y_k$).

Suppose that $\mathbb{P}', \mathbb{P}'' \in \mathcal{P}_\mu$. Write $\mathbb{P}_z'(\cdot)$ (resp. $\mathbb{P}_z''(\cdot)$) for the regular conditional probability of $\mathbb{P}'\{X \in \cdot \mid Z = z\}$ (resp. $\mathbb{P}''\{X \in \cdot \mid Z = z\}$).

Define a probability measure $\mathbb{Q}$ on $\prod_{k\in\mathbb{N}}(G_k \times G_k \times G_k)$ by

$$\mathbb{Q}\{(X', X'', Y) \in A' \times A'' \times B\} = \int_G \mathbb{P}_z'(A')\mathbb{P}_z''(A'')1_B(z)(\bigotimes_{k\in\mathbb{N}} \mu_k)(dz).$$

By construction, $\phi_k(X_{k+1}')^{-1} X_k' = \phi_k(X_{k+1}'')^{-1} X_k'' = Y_k$ for all $k \in \mathbb{N}$, $\mathbb{Q}$-a.s., the distribution of the pair (X', Y) under $\mathbb{Q}$ is the same as that of the pair (X, Z) under $\mathbb{P}'$, and the distribution of the pair (X'', Y) under $\mathbb{Q}$ is the same as that of the pair (X, Z) under $\mathbb{P}''$. In particular, the distributions of X' and X'' under $\mathbb{Q}$ are, respectively, $\mathbb{P}'$ and $\mathbb{P}''$.

Suppose for some $k \in \mathbb{N}$ that $\Phi' : G \to \mathbb{R}$ and $\Phi'' : G \to \mathbb{R}$ are both bounded $\mathcal{F}^X_{k+1}$-measurable functions and $\Psi : G_k \to \mathbb{R}$ is a bounded Borel function. Then, $\Phi' \circ X' : \prod_{j\in\mathbb{N}}(G_j \times G_j \times G_j) \to \mathbb{R}$ is $\mathcal{F}^{X'}_{k+1}$-measurable and $\Phi'' \circ X'' : \prod_{j\in\mathbb{N}}(G_j \times G_j \times G_j) \to \mathbb{R}$ is $\mathcal{F}^{X''}_{k+1}$-measurable, and hence, by the construction of $\mathbb{Q}$ (using the notations $\nu[\cdot]$ and $\nu[\cdot\,|\,\cdot]$ for expectation and conditional expectation with respect to a probability measure ν),

$$\begin{aligned}\mathbb{Q}[\Phi' \circ X' \Phi'' \circ X'' \,|\, \mathcal{F}^Y] &= \mathbb{Q}[\Phi' \circ X' \,|\, \mathcal{F}^Y]\, \mathbb{Q}[\Phi'' \circ X'' \,|\, \mathcal{F}^Y] \\ &= \mathbb{P}'_Y[\Phi' \circ X]\, \mathbb{P}''_Y[\Phi'' \circ X]\end{aligned}$$

is $\mathcal{F}^Y_{k+1}$-measurable. Thus, by the construction of $\mathbb{Q}$ and the independence of the elements of the sequence $(Y_j)_{j\in\mathbb{N}}$ under $\mathbb{Q}$,

$$\begin{aligned}\mathbb{Q}[\Phi' \circ X' \Phi'' \circ X'' \Psi \circ Y_k] &= \mathbb{Q}[\mathbb{Q}[\Phi' \circ X' \Phi'' \circ X'' \Psi \circ Y_k \,|\, \mathcal{F}^Y]] \\ &= \mathbb{Q}[\mathbb{Q}[\Phi' \circ X' \Phi'' \circ X'' \,|\, \mathcal{F}^Y] \Psi \circ Y_k] \\ &= \mathbb{Q}[\mathbb{P}'_Y[\Phi' \circ X]\, \mathbb{P}''_Y[\Phi'' \circ X]]\, \mathbb{Q}[\Psi \circ Y_k] \\ &= \mathbb{Q}[\Phi' \circ X' \Phi'' \circ X'']\, \mathbb{Q}[\Psi \circ Y_k].\end{aligned}$$

Therefore, by a standard monotone class argument, Y_k is independent of $\mathcal{F}^{(X',X'')}_{k+1}$. Consequently, the sub-$\sigma$-fields $\mathcal{F}_Y$ and $\mathcal{F}^{(X',X'')}_\infty$ are independent.

Suppose now that $\mathbb{P}', \mathbb{P}'' \in \mathcal{P}^{\mathrm{ex}}_\mu$. Observe for $k < n$ that

$$\begin{aligned}X'_k(X''_k)^{-1} &= \left[\phi^n_k(X'_n) \prod_{m=k}^{n-1} \phi^m_k(Y_m) Y_k\right] \left[\phi^n_k(X''_n) \prod_{m=k}^{n-1} \phi^m_k(Y_m) Y_k\right]^{-1} \quad \mathbb{Q}-\text{a.s.} \\ &= \phi^n_k(X'_n)\, \phi^n_k(X''_n)^{-1},\end{aligned} \tag{4}$$

and so there exists a G-valued random variable $W \in \mathcal{F}^{X',X''}_\infty$ such that $W_k = X'_k(X''_k)^{-1}$, $\mathbb{Q}$-a.s. From the above, W is independent of the sub-σ-field $\mathcal{F}_Y$. By construction, W takes values in the subgroup H.

Let $\mathbb{Q}(\cdot\,|\,W = h)$ be the regular conditional probability for $\mathbb{Q}$ given $W = h \in H$, so that

$$\mathbb{Q}(\cdot) = \int_H \mathbb{Q}(\cdot\,|\,W = h)\, \mathbb{Q}\{W \in dh\}. \tag{5}$$

It follows that

$$\mathbb{Q}\{X'_k = \phi_k(X'_{k+1}) Y_k, \forall k \in \mathbb{N} \,|\, W = h\} = 1$$

for $\mathbb{Q}\{W \in dh\}$-almost every $h \in H$. Moreover, because W is independent of $\mathcal{F}_Y$ it follows that $\mathbb{Q}\{Y \in \cdot\} = \mathbb{Q}\{Y \in \cdot\,|\,W = h\} = \bigotimes_{k\in\mathbb{N}} \mu_k$ for $\mathbb{Q}\{W \in dh\}$-almost every $h \in H$. Thus, $\mathbb{Q}\{X' \in \cdot\,|\,W = h\} \in \mathcal{P}_\mu$ for $\mathbb{Q}\{\varepsilon \in dh\}$-almost every $h \in H$ and, by (5),

$$\mathbb{P}'(\cdot) = \mathbb{Q}\{X' \in \cdot\} = \int_H \mathbb{Q}\{X' \in \cdot\,|\,W = h\}\, \mathbb{Q}\{W \in dh\}.$$

This would contradict the extremality of $\mathbb{P}'$ unless

$$\mathbb{P}'(\cdot) = \mathbb{Q}\{X' \in \cdot \,|\, W = h\}, \text{ for } \mathbb{Q}\{W \in dh\}\text{-almost every } h \in H.$$

Similarly,

$$\mathbb{P}''(\cdot) = \mathbb{Q}\{X'' \in \cdot \,|\, W = h\}, \text{ for } \mathbb{Q}\{W \in dh\}\text{-almost every } h \in H.$$

By (4),

$$\mathbb{Q}\{X'_k = h_k X''_k \,\forall k \in \mathbb{N} \,|\, W = h\} = 1, \text{ for } \mathbb{Q}\{W \in dh\}\text{-almost every } h \in H.$$

Therefore,

$$\mathbb{P}' = T_h(\mathbb{P}''), \text{ for } \mathbb{Q}\{W \in dh\}\text{-almost every } h \in H. \qquad \square$$

Notation 2.12. Given $\mathbb{P}^0 \in \mathcal{P}_\mu^{\text{ex}}$, let $H_\mu^{\text{stab}}(\mathbb{P}^0) := \{h \in H : T_h(\mathbb{P}^0) = \mathbb{P}^0\}$ be the stabilizer subgroup of the point $\mathbb{P}^0$ under the group action $(T_h)_{h \in H}$.

Remark 2.13. It follows from the transitivity of H on $\mathcal{P}_\mu^{\text{ex}}$ that for any two probability measures $\mathbb{P}', \mathbb{P}'' \in \mathcal{P}_\mu^{\text{ex}}$ the subgroups $H_\mu^{\text{stab}}(\mathbb{P}')$ and $H_\mu^{\text{stab}}(\mathbb{P}'')$ are conjugate.

Corollary 2.14. *A necessary and sufficient condition for $\#\mathcal{P}_\mu = 1$ is that $H_\mu^{\text{stab}}(\mathbb{P}^0) = H$ for some, and hence all, $\mathbb{P}^0 \in \mathcal{P}_\mu^{\text{ex}}$.*

Proof. This is immediate from Theorem 2.11 and the observation that $\#\mathcal{P}_\mu = 1$ if and only if $\#\mathcal{P}_\mu^{\text{ex}} = 1$. $\square$

Corollary 2.15. *If $H_\mu^{\text{stab}}(\mathbb{P}^0)$ is nontrivial for some, and hence all, $\mathbb{P}^0 \in \mathcal{P}_\mu^{\text{ex}}$, then $\mathcal{P}_\mu^{\text{strong}} = \emptyset$.*

Proof. As we observed prior to the statement of Theorem 2.11, it is a consequence of that result that either $\mathcal{P}_\mu^{\text{strong}} = \emptyset$ or $\mathcal{P}_\mu^{\text{strong}} = \mathcal{P}_\mu^{\text{ex}}$.

Suppose that $\mathbb{P}^0 \in \mathcal{P}_\mu^{\text{strong}}$ is such that $H_\mu^{\text{stab}}(\mathbb{P}^0)$ is nontrivial. By working on an extended probability space, we may assume that there is an $H_\mu^{\text{stab}}(\mathbb{P}^0)$-valued random variable $(U_k)_{k \in \mathbb{N}}$ that is independent of $(X_k)_{k \in \mathbb{N}}$ and is not almost surely constant. The distribution of the solution $(U_k X_k)_{k \in \mathbb{N}}$ is also $\mathbb{P}^0$ and, in particular, this solution is strong. However, this implies that

$$\begin{aligned}
\sigma(U_k X_k) &\subseteq \sigma((\phi_j(U_{j+1} X_{j+1})^{-1} U_j X_j)_{j \geq k}) \\
&= \sigma((\phi_j(X_{j+1})^{-1} X_j)_{j \geq k}) \\
&= \mathcal{F}_k^Z
\end{aligned}$$

for all $k \in \mathbb{N}$, and hence U_k is $\mathcal{F}_k^Z$-measurable for all $k \in \mathbb{N}$, because X_k is $\mathcal{F}_k^Z$-measurable by the assumption that $\mathbb{P}^0 \in \mathcal{P}_\mu^{\text{strong}}$. However, because the sequence $(U_k)_{k\in\mathbb{N}}$ is independent of the sequence of $(X_k)_{k\in\mathbb{N}}$ and not almost surely constant, it follows that $(U_k)_{k\in\mathbb{N}}$ is not $\sigma((X_k)_{k\in\mathbb{N}})$-measurable and hence *a fortiori* $(U_k)_{k\in\mathbb{N}}$ is not $\sigma((Z_k)_{k\in\mathbb{N}})$-measurable. We thus arrive at a contradiction. □

3 Representation Theory and the Existence of Strong Solutions

Notation 3.1. Let $\mathcal{G}$ be the set of all unitary, finite-dimensional representations of the compact group $G = \prod_{k\in\mathbb{N}} G_k$.

Any irreducible representations of G are equivalent to a tensor product representation of the form

$$(g_k)_{k\in\mathbb{N}} \mapsto \rho^{(k_1)}(g_{k_1}) \otimes \cdots \otimes \rho^{(k_n)}(g_{k_n}),$$

where $\{k_1, \ldots, k_n\}$ is a finite subset of $\mathbb{N}$ and $\rho^{(k_j)}$ is an (necessarily finite-dimensional) irreducible representation of G_{k_j} for $1 \le j \le n$. Furthermore, an arbitrary element of $\mathcal{G}$ is equivalent to a (finite) direct sum of irreducible representations.

Notation 3.2. For $k \in \mathbb{N}$ write $\iota_k : G_k \mapsto G$ for the map that sends $h \in G_k$ to $(e_1, \ldots, e_{k-1}, h, e_{k+1}, \ldots)$, where, as above, e_j is the identity element of G_j for $j \in \mathbb{N}$.

Consider an arbitrary representation $\rho \in \mathcal{G}$. It is clear from the above that if $\mathbb{P} \in \mathcal{P}_\mu^{\text{strong}}$, then $\rho \circ \iota_k(X_k)$ is $\mathcal{F}_k^Z$-measurable for all $k \in \mathbb{N}$. Note that $\rho \circ \iota_k$ is a representation of G_k and all representations of G_k arise this way. On the other hand, because, by the Peter–Weyl theorem, the closure in the uniform norm of the (complex) linear span of matrix entries of the irreducible representations of G_k is the vector space of continuous complex-valued functions on G_k, it follows that if $\rho \circ \iota_k(X_k)$ is $\mathcal{F}_k^Z$-measurable for all $k \in \mathbb{N}$ for an arbitrary representation $\rho \in \mathcal{G}$, then $\mathbb{P} \in \mathcal{P}_\mu^{\text{strong}}$. This observation leads to the following definition and theorem.

Notation 3.3. Set

$$\mathcal{H}_\mu^{\text{strong}} := \{\rho \in \mathcal{G} : \exists \mathbb{P} \in \mathcal{P}_\mu^{\text{ex}} \text{ such that } \rho \circ \iota_k(X_k) \text{ is } \mathcal{F}_k^Z\text{-measurable } \mathbb{P}\text{-a.s. } \forall k \in \mathbb{N}\}.$$

Theorem 3.4. *The set $\mathcal{P}_\mu^{\text{strong}}$ of strong solutions is non-empty (and hence equal to $\mathcal{P}_\mu^{\text{ex}}$) if and only if $\mathcal{H}_\mu^{\text{strong}} = \mathcal{G}$.*

Proof. The result is immediate from the discussion preceding the statement of the theorem once we note that if $\mathbb{P}'$ and $\mathbb{P}''$ both belong to $\mathcal{P}_\mu^{\text{ex}}$ then, by Theorem 2.11, there exists $h \in H$ such that $\mathbb{P}''$ is the distribution of $hX = (h_k X_k)_{k\in\mathbb{N}}$ under $\mathbb{P}'$ and so $\rho \circ \iota_k(X_k)$ is $\mathcal{F}_k^Z$-measurable $\mathbb{P}''$-a.s. if and only if $\rho \circ \iota_k(h_k X_k)$ is $\mathcal{F}_k^Z$-measurable $\mathbb{P}'$-a.s. (recall that $Z_k = \phi(X_{k+1})^{-1} X_k = \phi(h_k X_{k+1})^{-1} h_k X_k$ when $h \in H$); therefore, $\rho \circ \iota_k(X_k)$ is $\mathcal{F}_k^Z$-measurable $\mathbb{P}''$-a.s. if and only if $[\rho \circ \iota_k(h_k)]\,[\rho \circ \iota_k(X_k)]$ is $\mathcal{F}_k^Z$-measurable $\mathbb{P}'$-a.s., which is in turn equivalent to $\rho \circ \iota_k(X_k)$ being $\mathcal{F}_k^Z$-measurable $\mathbb{P}'$-a.s. by the invertibility of the matrix $\rho \circ \iota_k(h_k)$. Thus,

$$\mathcal{H}_\mu^{\text{strong}} = \{\rho \in \mathcal{G} : \rho \circ \iota_k(X_k) \text{ is } \mathcal{F}_k^Z\text{-measurable } \mathbb{P}\text{-a.s. } \forall k \in \mathbb{N}\}$$

for any $\mathbb{P} \in \mathcal{P}_\mu^{\text{ex}}$. □

Theorem 3.4 is still somewhat unsatisfactory as a criterion for the existence of strong solutions because it requires a knowledge of the set $\mathcal{P}_\mu^{\text{ex}}$ of extreme solutions. We would prefer a criterion that was directly in terms of the sequence $(\mu_k)_{k\in\mathbb{N}}$. In order to (partly) remedy this situation, we introduce the following objects.

Notation 3.5. Fix $\rho \in \mathcal{G}$. For $k, \ell \in \mathbb{N}$ with $k \le \ell$, set

$$R_k^\ell := \int_{G_\ell} \rho \circ \iota_k \circ \phi_k^\ell(z)\, \mu_\ell(dz).$$

Let

$$\mathcal{H}_\mu^{\text{det}} := \{\rho \in \mathcal{G} : \lim_{m\to\infty}\lim_{n\to\infty} \left|\det(R_k^n R_k^{n-1} \cdots R_k^m)\right| > 0\ \forall k \in \mathbb{N}\}$$

and

$$\mathcal{H}_\mu^{\text{norm}} := \{\rho \in \mathcal{G} : \lim_{m\to\infty}\lim_{n\to\infty} \|R_k^n R_k^{n-1} \cdots R_k^m\| > 0\ \forall k \in \mathbb{N}\},$$

where $\|\cdot\|$ is the ℓ^2 operator norm on the appropriate space of matrices.

Proposition 3.6. *Fix* $\mathbb{P} \in \mathcal{P}_\mu$.

(i) If $\rho \in \mathcal{H}_\mu^{\text{det}}$*, then*

$$\mathbb{P}[\rho \circ \iota_k(X_k) \,|\, \mathcal{F}_\infty^X \vee \mathcal{F}_k^Z] = \rho \circ \iota_k(X_k)$$

for all $k \in \mathbb{N}$*. In particular, if* $\mathbb{P} \in \mathcal{P}_\mu^{\text{ex}}$*, then* $\rho \circ \iota_k(X_k)$ *is* $\mathcal{F}_k^Z$*-measurable for all* $k \in \mathbb{N}$*.*

(ii) If $\rho \notin \mathcal{H}_\mu^{\text{norm}}$*, then*

$$\mathbb{P}[\rho \circ \iota_k(X_k) \,|\, \mathcal{F}_\infty^X \vee \mathcal{F}_k^Z] = 0$$

for some $k \in \mathbb{N}$*. In particular, if* $\mathbb{P} \in \mathcal{P}_\mu^{\text{ex}}$*, then* $\rho \circ \iota_k(X_k)$ *is not* $\mathcal{F}_k^Z$*-measurable for some* $k \in \mathbb{N}$*.*

Proof. The proof follows that of an analogous result in [AUY08] with modifications required by the greater generality in which we are working.

Consider claim (i). Fix $\rho \in \mathcal{H}_\mu^{\text{det}}$ and $k \in \mathbb{N}$. For $\ell > k$ we have

$$\rho \circ \iota_k(X_k) = \rho \circ \iota_k \circ \phi_k^\ell(X_\ell)\, \rho \circ \iota_k \circ \phi_k^{\ell-1}(Z_{\ell-1}) \cdots \rho \circ \iota_k \circ \phi_k^k(Z_k). \tag{6}$$

For $k \leq m \leq n$ put

$$\Xi_n^m := \rho \circ \iota_k \circ \phi_k^n(Z_m) \cdots \rho \circ \iota_k \circ \phi_k^m(Z_m).$$

Note that

$$\mathbb{P}[\Xi_n^m] = R_k^n \cdots R_k^m.$$

For any $p \geq k$, the matrix $\rho \circ \iota_k \circ \phi_k^p$ is unitary, and so $\|\rho \circ \iota_k \circ \phi_k^p(h)\| = 1$ for all $h \in G_p$. By Jensen's inequality, $\|R_k^p\| \leq 1$. In particular, $|\det(R_k^p)| \leq 1$. Hence,

$$\lim_{m\to\infty} \lim_{n\to\infty} |\det(\mathbb{P}[\Xi_n^m])|$$

exists and is given by

$$\sup_m \inf_{n\geq m} |\det(R_k^n)| \cdots |\det(R_k^m)|.$$

Moreover, there are constants $\varepsilon > 0$ and $M \in \mathbb{N}$ such that $|\det(\mathbb{P}[\Xi_n^m])| \geq \varepsilon$ whenever $n \geq m \geq M$. It follows from Cramer's rule that the matrices $\mathbb{P}[\Xi_n^m]$ are invertible with uniformly bounded entries for $n \geq m \geq M$.

Set $\Phi_n^m := \mathbb{P}[\Xi_n^m]^{-1}\Xi_n^m$ for $n \geq m \geq M$. The matrices Φ_n^m have uniformly bounded entries and

$$\mathbb{P}\left[\Phi_{n+1}^m \,|\, \sigma((Z_p)_{p=m}^n)\right] = \Phi_n^m,$$

so that $(\Phi_n)_{n\geq m}$ is a bounded matrix-valued martingale with respect to the filtration $(\sigma((Z_p)_{p=m}^n))_{n\geq m}$. Thus, $\lim_{n\to\infty} \Phi_n^m =: \Phi_\infty^m$ exists and is $\mathcal{F}_m^Z$-measurable $\mathbb{P}$-a.s. for each $m \geq M$. Consequently, $\lim_{n\to\infty} \Xi_n^m =: \Xi_\infty^m$ also exists and is $\mathcal{F}_m^Z$-measurable $\mathbb{P}$-a.s. for each $m \geq M$. Part (i) is now clear from (6).

Now consider part (ii). Fix $\rho \notin \mathcal{H}_\mu^{\text{norm}}$ and $k \in \mathbb{N}$ such that

$$\lim_{m\to\infty} \lim_{n\to\infty} \|R_k^n R_k^{n-1} \cdots R_k^m\| = 0.$$

It follows from (6) that for $n \geq m \geq k$

$$\mathbb{P}\left[\rho \circ \iota_k(X_k) \,|\, \mathcal{F}_n^X \vee \sigma((Z_j)_{j=k}^m)\right] = \rho \circ \iota_k \circ \phi_k^n(X_n) R_k^{n-1} \cdots R_k^{m+1}$$
$$\rho \circ \iota_k \circ \phi_m^k(Z_m) \cdots \rho \circ \iota_k \circ \phi_k^k(Z_k).$$

Since $\rho(g)$ is a unitary matrix for all $g \in G$, the norm of the right-hand side is at most $\|R_k^{n-1} \cdots R_k^{m+1}\|$, which, by assumption, converges to 0 as $n \to \infty$ followed by $m \to \infty$. Thus, by the reverse martingale convergence theorem and the martingale convergence theorem,

$$\mathbb{P}\left[\rho \circ \iota_k(X_k) \,|\, \mathcal{F}_\infty^X \vee \mathcal{F}_k^Z\right] = \lim_{m\to\infty} \lim_{n\to\infty} \mathbb{P}\left[\rho \circ \iota_k(X_k) \,|\, \mathcal{F}_n^X \vee \sigma((Z_j)_{j=k}^m)\right] = 0. \qquad \square$$

The following result is immediate from Theorem 3.4 and Proposition 3.6.

Theorem 3.7. *The following containments hold*

$$\mathcal{H}_\mu^{\mathrm{norm}} \supseteq \mathcal{H}_\mu^{\mathrm{strong}} \supseteq \mathcal{H}_\mu^{\mathrm{det}}.$$

Thus, $\mathcal{H}_\mu^{\mathrm{det}} = \mathcal{G}$ *implies that* $\mathcal{P}_\mu^{\mathrm{strong}} \neq \emptyset$ *and* $\mathcal{H}_\mu^{\mathrm{norm}} \neq \mathcal{G}$ *implies that* $\mathcal{P}_\mu^{\mathrm{strong}} = \emptyset$.

The following is a straightforward equivalent of Theorem 3.7 and we omit the proof.

Corollary 3.8. *If*

$$\lim_{m\to\infty}\lim_{n\to\infty}\left|\det\left(\prod_{\ell=m}^{n}\int_{G_\ell}\rho\circ\phi_k^\ell(z)\,\mu_\ell(dz)\right)\right| > 0$$

for all irreducible representations ρ *of* G_k *for all* $k \in \mathbb{N}$, *then* $\mathcal{P}_\mu^{\mathrm{strong}} \neq \emptyset$. *If*

$$\lim_{m\to\infty}\lim_{n\to\infty}\left\|\prod_{\ell=m}^{n}\int_{G_\ell}\rho\circ\phi_k^\ell(z)\,\mu_\ell(dz))\right\| = 0$$

for some irreducible representation ρ *of* G_k *for some* $k \in \mathbb{N}$, *then* $\mathcal{P}_\mu^{\mathrm{strong}} = \emptyset$.

Under a further assumption, we get a representation theoretic necessary and sufficient condition for the existence of strong solutions.

Definition 3.9. A Borel probability measure ν on a compact Hausdorff group Γ is *conjugation invariant* if

$$\int_\Gamma f(g^{-1}xg)\,\nu(dx) = \int_\Gamma f(x)\,\nu(dx)$$

for all $g \in \Gamma$ and bounded Borel functions $f : \Gamma \to \mathbb{R}$.

Remark 3.10. Note that if Γ is abelian, then any Borel probability measure ν on Γ is conjugation invariant.

Corollary 3.11. *Suppose that each probability measure* μ_k, $k \in \mathbb{N}$, *is conjugation invariant. Then,*

$$\mathcal{H}_\mu^{\mathrm{norm}} = \mathcal{H}_\mu^{\mathrm{strong}} = \mathcal{H}_\mu^{\mathrm{det}}$$

and $\mathcal{P}_\mu^{\mathrm{strong}} \neq \emptyset$ *if and only if each of these sets is* $\mathcal{G}$ *or, equivalently,*

$$\lim_{m\to\infty}\lim_{n\to\infty}\left|\prod_{\ell=m}^{n}\int_{G_\ell}\chi\circ\phi_k^\ell(z)\,\mu_\ell(dz)\right| > 0$$

for each character χ *of an irreducible representation of* G_k *for all* $k \in \mathbb{N}$.

Proof. The result is immediate from Corollary 3.8 and Lemma 3.12 below. □

The following lemma is well known, but we include a proof for the sake of completeness.

Lemma 3.12. *If ν is a conjugation invariant Borel probability measure on a compact Hausdorff group Γ and ρ is an irreducible representation of Γ with character χ, then*

$$\int_\Gamma \rho(x)\,\nu(dx) = \int_\Gamma \chi(x)\,\nu(dx) \times I,$$

where I is the identity matrix.

Proof. Let λ be the normalized Haar measure on Γ. By assumption,

$$\int_\Gamma \rho(x)\,\nu(dx) = \int_\Gamma \int_\Gamma \rho(g^{-1}xg)\,\lambda(dg)\,\nu(dx).$$

Now, for $x, y \in \Gamma$ we have

$$\begin{aligned}\int_\Gamma \rho(g^{-1}xg)\,\lambda(dg)\,\rho(y) &= \int_\Gamma \rho(g^{-1}xgy)\,\lambda(dg)\\ &= \int_\Gamma \rho(yh^{-1}xh)\,\lambda(dh)\\ &= \rho(y)\int_\Gamma \rho(h^{-1}xh)\,\lambda(dh),\end{aligned}$$

and so the matrix $\int_\Gamma \rho(g^{-1}xg)\,\lambda(dg)$ commutes with the matrix $\rho(y)$ for all $y \in \Gamma$. It follows from Schur's lemma that $\int_\Gamma \rho(g^{-1}xg)\,\lambda(dg) = cI$ for some constant c, and taking traces of both sides gives $c = \chi(x)$. □

4 Freezing

Recall that the Hilbert–Schmidt norm of a matrix A is given by $\|A\|_{HS} := \mathrm{tr}(A^*A)^{\frac{1}{2}}$, where A^* is the adjoint of A (this norm is also called the Frobenius norm and the Schur norm). Write $d(\rho)$ for the dimension of a unitary representation $\rho \in \mathcal{G}$, and note that $\|\rho(x)\|_{HS}^2 = \mathrm{tr}(I) = d(\rho)$. If ν is a probability measure on G, then $\|\int_G \rho(x)\,\nu(dx)\|_{HS}^2 \le d(\rho)$ by Jensen's inequality.

Notation 4.1. Set

$$\mathcal{H}_\mu^{\mathrm{freeze}} := \left\{\rho \in \mathcal{G} : \sum_{m=k}^{\infty}\left[d(\rho) - \left\|\int_{G_k} \rho \circ \iota_k \circ \phi_k^m(z)\,\mu_m(dz)\right\|_{HS}^2\right] < \infty\ \forall k \in \mathbb{N}\right\}.$$

Proposition 4.2. *The sets $\mathcal{H}_\mu^{\text{freeze}}$ and $\mathcal{H}_\mu^{\text{det}}$ are equal, and so $\mathcal{H}_\mu^{\text{freeze}} = \mathcal{H}_\mu^{\text{det}} = \mathcal{G}$ implies that $\mathcal{P}_\mu^{\text{strong}} \neq \emptyset$. Moreover, if each probability measure μ_k, $k \in \mathbb{N}$, is conjugation invariant, then*

$$\mathcal{H}_\mu^{\text{norm}} = \mathcal{H}_\mu^{\text{strong}} = \mathcal{H}_\mu^{\text{det}} = \mathcal{H}_\mu^{\text{freeze}}$$

and $\mathcal{P}_\mu^{\text{strong}} \neq \emptyset$ if and only if each of these sets is $\mathcal{G}$ or, equivalently,

$$\lim_{m\to\infty}\lim_{n\to\infty}\left|\prod_{\ell=m}^{n}\int_{G_\ell}\chi\circ\phi_k^\ell(z)\,\mu_\ell(dz)\right| > 0$$

for each character χ of an irreducible representation of G_k for all $k \in \mathbb{N}$.

Proof. It suffices to show that $\mathcal{H}_\mu^{\text{freeze}} = \mathcal{H}_\mu^{\text{det}}$, because the remainder of the result will then follow from Theorem 3.7 and Corollary 3.11.

Fix $\rho \in \mathcal{G}$. Write $0 \le \lambda_k^\ell(1) \le \cdots \le \lambda_k^\ell(d(\rho))$ for the eigenvalues of the matrix

$$\left(\int_{G_k}\rho(z)\,\mu_k^\ell(dz)\right)^*\left(\int_{G_k}\rho(z)\,\mu_k^\ell(dz)\right).$$

Observe that

$$\lim_{m\to\infty}\lim_{n\to\infty}\prod_{\ell=m}^{n}\left|\det\int_{G_k}\rho(z)\,\mu_k^\ell(dz)\right| > 0$$

$$\iff$$

$$\lim_{m\to\infty}\lim_{n\to\infty}\prod_{\ell=m}^{n}\left|\det\int_{G_k}\rho(z)\,\mu_k^\ell(dz)\right|^2 > 0$$

$$\iff$$

$$\lim_{m\to\infty}\lim_{n\to\infty}\prod_{\ell=m}^{n}\lambda_k^\ell(1)\cdots\lambda_k^\ell(d(\rho)) > 0$$

$$\iff$$

$$\sum_{m=k}^{\infty}[(1-\lambda_k^m(1))+\cdots+(1-\lambda_k^m(d(\rho)))] < \infty$$

$$\iff$$

$$\sum_{m=k}^{\infty}[d(\rho)-(\lambda_k^m(1))+\cdots+\lambda_k^m(d(\rho)))] < \infty$$

$$\iff$$

$$\sum_{m=k}^{\infty}\left[d(\rho)-\left\|\int_{G_k}\rho\circ\iota_k\circ\phi_k^m(z)\,\mu_m(dz)\right\|_{HS}^2\right] < \infty,$$

as required. □

Given Proposition 4.2, the reader may wonder why we introduced the set $\mathcal{H}_\mu^{\text{freeze}}$. The equivalence established in Proposition 4.2 makes the proof of the following result considerably more transparent.

Proposition 4.3. *Suppose that each group G_k, $k \in \mathbb{N}$, is finite. Then, $\mathcal{H}_\mu^{\text{det}} = \mathcal{H}_\mu^{\text{freeze}} = \mathcal{G}$ if and only if for some (equivalently, all) $\mathbb{P} \in \mathcal{P}_\mu$ there are constants $c_{k,m} \in G_k$, $k,m \in \mathbb{N}$, $k \le m$, such that*

$$\mathbb{P}\{\phi_k^m(Z_m) \neq c_{k,m}\,\text{i.o.}\} = 0$$

for all $k \in \mathbb{N}$.

Proof. Write μ_k^m for the probability measure on G_k that is the push-forward of the probability measure μ_m on G_m by the map $\phi_k^m : G_m \to G_k$. For simplicity, we write $\mu_k^m(g)$ instead of $\mu_k^m(\{g\})$ for $g \in G_k$. It is clear that $\mathbb{P}\{\phi_k^m(Z_m) \neq c_{k,m}\,\text{i.o.}\} = 0$ $k \le m$ for all $k \in \mathbb{N}$ for some family of constants $c_{k,m} \in G_k$, $k,m \in \mathbb{N}$, if and only if $\mathbb{P}\{\phi_k^m(Z_m) \neq c_{k,m}^*\,\text{i.o.}\} = 0$ where $c_{k,m}^*$ is any family with the property

$$\mu(c_{k,m}^*) = \max\{\mu_k^m(g) : g \in G_k\}$$

and, by the Borel–Cantelli lemma, this in turn occurs if and only if

$$\sum_{m=k}^{\infty} \mu(G_k \backslash \{c_{k,m}^*\}) < \infty$$

for all $k \in \mathbb{N}$.

Now,

$$\left(\sum_{g \in G_k} \mu_k^m(g)^2\right)^{1/2} \ge \max_{g \in G_k} \mu_k^m(g) = \mu_k^m(c_{k,m}) = \mu_k^m(c_{k,m}) \sum_{g \in G_k} \mu_k^m(g) \ge \sum_{g \in G_k} \mu_k^m(g)^2.$$

By Parseval's equality,

$$\sum_{g \in G_k} \mu_k^m(g)^2 = \frac{1}{\#G_k} \sum_{\rho \in \hat{G}_k} d(\rho) \left\| \sum_{g \in G_k} \rho(g)\mu_k^m(g) \right\|_{HS}^2,$$

and hence

$$\begin{aligned} 1 - &\left(\frac{1}{\#G_k} \sum_{\rho \in \hat{G}_k} d(\rho) \left\| \sum_{g \in G_k} \rho(g)\mu_k^m(g) \right\|_{HS}^2\right) \\ &\ge \mu_k^m(G_k \backslash \{c_{k,m}\}) \\ &\ge 1 - \left(\frac{1}{\#G_k} \sum_{\rho \in \hat{G}_k} d(\rho) \left\| \sum_{g \in G_k} \rho(g)\mu_k^m(g) \right\|_{HS}^2\right)^{1/2}. \end{aligned}$$

Note for a sequence of constant $(a_n)_{n\in\mathbb{N}} \subset [0,1]$ that $\sum_{n\in\mathbb{N}}(1-a_n) < \infty$ if and only if $\sum_{n\in\mathbb{N}}(1-a_n^2) < \infty$. Note also that

$$1 = \frac{1}{\#G_k} \sum_{\rho\in\hat{G}_k} d(\rho)^2.$$

Thus,

$$\sum_{m=k}^{\infty} \mu(G_k \backslash \{c_{k,m}^*\}) < \infty$$

for all $k \in \mathbb{N}$ if and only if

$$\sum_{m=k}^{\infty} \frac{1}{\#G_k} \sum_{\rho\in\hat{G}_k} d(\rho) \left[d(\rho) - \left\| \sum_{g\in G_k} \rho(g)\mu_k^m(g) \right\|_{HS}^2 \right] < \infty$$

for all $k \in \mathbb{N}$, which is in turn equivalent to

$$\sum_{m=k}^{\infty} \sum_{\rho\in\hat{G}_k} \left[d(\rho) - \left\| \sum_{g\in G_k} \rho(g)\mu_k^m(g) \right\|_{HS}^2 \right] < \infty$$

for all $\rho \in \hat{G}_k$ for all $k \in \mathbb{N}$.

A decomposition of the representation $\rho \circ \iota_k$ of G_k for some $\rho \in \mathcal{G}$ into irreducibles shows that the last condition is equivalent to the one in the statement. □

Remark 4.4. It follows from Proposition 4.2 and Proposition 4.3 that if each group G_k, $k \in \mathbb{N}$, is finite and for some (equivalently, all) $\mathbb{P} \in \mathcal{P}_\mu$ there are constants $c_{k,m} \in G_k$, $k,m \in \mathbb{N}$, $k \le m$, such that

$$\mathbb{P}\{\phi_k^m(Z_m) \neq c_{k,m} \, \text{i.o.}\} = 0$$

for all $k \in \mathbb{N}$, then $\mathcal{P}_\mu^{\text{strong}} \neq \emptyset$. Moreover, these two conditions are equivalent when each probability measure μ_k, $k \in \mathbb{N}$, is conjugation invariant. Also, for the special case when $G_k = \Gamma$, $k \in \mathbb{N}$, for some fixed finite group Γ and each homomorphism $\phi_k : \Gamma \to \Gamma$ is the identity, it follows from Corollary 2.6 of [HY10] that the two conditions are equivalent. It would be interesting to know the status of the reverse implication in general.

5 Groups Indexed by the Lattice

Recall from the Introduction the example of our general setup where $G_k := G_{1,k} \times G_{2,k-1} \cdots \times G_{k,1}$ with each group $G_{i,j}$ a copy of some fixed compact abelian group Γ and the homomorphism ϕ_k is given by

$$\phi_k(g_{1,k+1}, g_{2,k}, \ldots, g_{k+1,1}) := (g_{1,k+1} + g_{2,k}, g_{2,k} + g_{3,k-1}, \ldots, g_{k,2} + g_{k+1,1}).$$

We will consider the particular case where Γ is $\mathbb{Z}_p$, the group of integers modulo some prime number p.

Because $\mathbb{Z}_p$ is abelian, all its irreducible representations of G are one-dimensional. The irreducible representations are the trivial one and those of the form $\rho(g) = \prod_{n=1}^{m} \exp\left(\frac{2\pi i z_n}{p} g_{i_n,j_n}\right)$ for some m, pairs $(i_1, j_1), \ldots, (i_m, j_m) \in \mathbb{N}^2$, and $1 \le z_n \le p-1$.

The homomorphism ϕ_k^ℓ maps $(g_{1,\ell}, \ldots, g_{\ell,1}) \in G_\ell$ to $(h_{1,k}, \ldots, h_{k,1}) \in G_k$ where

$$h_{i,k+1-i} = \sum_{j=0}^{\ell-k} \binom{\ell-k}{j} g_{i+j,\ell+1-i-j} \in \mathbb{Z}_p.$$

Set $f(m,n) := \binom{m}{n} \mod p$. When we restrict to G_k, the representation $\rho \circ \iota_k$ is of the form $\prod_{i=1}^{k} \exp\left(\frac{2\pi z_i}{p} g_{i,k+1-i}\right)$ with $0 \le z_i \le p-1$. We therefore need to evaluate

$$R_k^\ell = \int_{G_\ell} \prod_{i=1}^{k} \prod_{j=0}^{\ell-k} \exp\left(\frac{2\pi z_i}{p} f(\ell-k, j) g_{i+j,\ell+1-i-j}\right) \mu_\ell(dg_\ell)$$

to determine whether or not $\mathcal{P}_\mu^{\text{strong}} = \emptyset$. The following theorem of Lucas (see [Gra97]) gives the value of f.

Theorem 5.1. *Let m, n be nonnegative integers and p a prime number. Suppose*

$$m = m_k p^k + \ldots + m_1 p + m_0$$

and

$$n = n_k p^k + \ldots + n_1 p + n_0.$$

Then,

$$\binom{m}{n} = \prod_{i=0}^{k} \binom{m_i}{n_i} \mod p.$$

Equivalently, if m_0 and n_0 are the least nonnegative residues of m and n mod p, then $\binom{m}{n} = \binom{\lfloor m/p \rfloor}{\lfloor n/p \rfloor}\binom{m_0}{n_0}$.

Rather than use Theorem 5.1 directly to construct interesting examples, we consider a consequence of it for the case $p = 2$. Suppose that $\mu_k = \mu_{1,k} \otimes \cdots \otimes \mu_{k,1}$ where $\mu_{i,k+1-i}\{1\} = \pi_k = 1 - \mu_{i,k+1-i}\{0\}$ for some $0 \le \pi_k \le 1$.

Define $x = (x_{m,\ell+1-m})_{m=1}^{\ell} \in G_\ell = G_{1,\ell} \times \cdots \times G_{\ell,1} \cong \mathbb{Z}_2^\ell$ by

$$x := \sum_{i=1}^{k} \sum_{j=0}^{\ell-k} z_i f(\ell-k, j) e^{(i+j,\ell+1-i-j)},$$

where the arithmetic is performed modulo 2 and $e^{(m,\ell+1-m)} \in G_\ell$ is the vector with $e^{(m,\ell+1-m)}_{m,\ell+1-m} = 1$ and $e^{(m,\ell+1-m)}_{n,\ell+1-n} = 0$ for $n \neq m$. Then,

$$\int_{G_\ell} \prod_{i=1}^{k} \prod_{j=0}^{\ell-k} \exp\left(\frac{2\pi z_i}{p} f(\ell-k,j) g_{i+j,\ell+1-i-j}\right) \mu_\ell(dg_\ell) = (1-2\pi_\ell)^{M(k,\ell,z)},$$

where

$$M(k,\ell,z) := \#\{1 \leq m \leq \ell : x_{m,\ell+1-m} = 1\}.$$

Observe that if $x_{m,\ell+1-m} = 1$, then

$$\sum_{j=0}^{\ell-k} f(\ell-k,j) e^{(i+j,\ell+1-i-j)}_{m,\ell+1-m} = 1$$

for some $1 \leq i \leq k$ with $z_i = 1$. Now

$$\begin{aligned}
&\#\{1 \leq m \leq \ell : \sum_{j=0}^{\ell-k} f(\ell-k,j) e^{(i+j,\ell+1-i-j)}_{m,\ell+1-m} = 1\} \\
&= \#\{1 \leq m \leq \ell : f(\ell-k,m-i) = 1, i \leq m \leq i+\ell-k\} \\
&= \#\{i \leq m \leq i+\ell-k : f(\ell-k,m-i) = 1\} \\
&= \#\{0 \leq m \leq \ell-k : f(\ell-k,m) = 1\}.
\end{aligned}$$

As remarked in [Gra97], a consequence of the following theorem of Kummer from 1852 that the number of the binomial coefficients $\binom{m}{n}$, $0 \leq n \leq m$, which are odd is $2^{N(m)}$, where $N(m)$ is the number of times that the digit 1 appears in the base 2 representation of m.

Theorem 5.2. *Let m,n be nonnegative integers and p a prime number. The greatest power of p that divides $\binom{m}{n}$ is given by the number of "carries" that are necessary when we add m and $n-m$ in base p.*

Thus,

$$M(k,\ell,z) \leq k2^{N(\ell-k)}$$

and $M(k,\ell,z) = 2^{N(\ell-k)}$ when $\#\{1 \leq i \leq k : z_i = 1\} = 1$.

Therefore, if we assume $\pi_n \to 0$ as $n \to \infty$, then we are interested in whether

$$\lim_{\ell\to\infty} \prod_{r=1}^{\ell} (1-2\pi_{h+r})^{2^{N(r)}} \neq 0$$

for all $h \in \mathbb{N}$ or, equivalently, whether

$$\sum_{r=1}^{\infty} 2^{N(r)} \pi_{h+r} < \infty$$

for all $h \in \mathbb{N}$.

For example, fix a positive integer a and an increasing function $b : \mathbb{N} \to \mathbb{N}$ such that $a \le b(m) < m$ and $\lim_{m\to\infty} b(m) = \infty$. Suppose that $\pi_n = 0$ unless $2^m + 2^{b(m)} - 2^a \le n \le 2^m + 2^{b(m)}$ for some $m \in \mathbb{N}$. Note for any $h \in \mathbb{N}$ that

$$\sum_{r=1}^{\infty} 2^{N(r)} \pi_{h+r} = \sum_{s=k+1}^{\infty} 2^{N(s-h)} \pi_s$$

and this sum is finite if and only if

$$\sum_{n=1}^{\infty} 2^{b(\log_2 n)} \pi_n$$

is finite.

Thus, $\mathcal{P}_\mu^{\mathrm{strong}} \neq \emptyset$ if and only if $\sum_{n=1}^{\infty} 2^{b(\log_2 n)} \pi_n < \infty$ in this case. On the other hand, $\mathbb{P}\{Z_k \neq 0 \,\mathrm{i.o.}\} > 0$ (equivalently, $\mathbb{P}\{Z_k \neq 0 \,\mathrm{i.o.}\} = 1$) if and only if $\sum_{n=1}^{\infty} n\pi_n < \infty$. Therefore, when $\lim_{m\to\infty} m - b(m) = \infty$ it is possible to construct $(\pi_n)_{n\in\mathbb{N}}$ such that almost surely infinitely many "bits" are "corrupted" and yet strong solutions still exist.

6 Automorphisms of the Torus

Consider the torus group $\mathbb{T}^2 = \mathbb{R}^2/\mathbb{Z}^2$. We write an element $x \in \mathbb{T}^2$ as a column vector $x = (x_1, x_2)^\top \in [0,1)^2$, where $\top$ denotes the transpose of a vector.

Any 2×2 $\mathbb{Z}$-valued matrix S defines a homomorphism $x \mapsto Sx$ from $\mathbb{T}^2$ to itself if we do ordinary matrix multiplication modulo $\mathbb{Z}^2$. If the matrix S has determinant 1, then this homomorphism is invertible. Such a transformation is called a *linear toral automorphism.*

Note that if

$$S = \begin{pmatrix} a & b \\ c & d \end{pmatrix},$$

then the eigenvalues of S are

$$\frac{1}{2}(a+d \pm \sqrt{a^2+4bc-2ad+d^2}) = \frac{1}{2}(a+d \pm \sqrt{(a+d)^2-4}),$$

Thus, the eigenvalues are real and distinct unless $a+d$ is 0, ± 1 or ± 2, in which case the pairs of eigenvalues are, respectively $\{\pm i\}$, $\{\frac{1}{2}(1 \pm i\sqrt{3})\}$, $\{\frac{1}{2}(-1 \pm i\sqrt{3})\}$, $\{1,1\}$, and $\{-1,-1\}$. Note that in each of the latter cases the eigenvalues lie on the unit circle.

Definition 6.1. A *ergodic toral automorphism* is a linear toral automorphism given by a matrix S with no eigenvalues on the unit circle.

For some of the more probabilistic properties of ergodic toral automorphisms, see [Kat71]. Such mappings are the prototypical examples of Anosov systems that have been the subject of intensive study dynamical systems world (see [Fra69]).

A hyperbolic linear toral automorphism has two real eigenvalues $\lambda_1 > 1 > \lambda_1^{-1} = \lambda_2$. These eigenvalues are irrational and the corresponding (right) eigenvectors v^1 and v^2 have irrational slope (see, e.g., Sect. 5.6 of [LT93]).

Theorem 6.2. *Suppose for every $i \in \mathbb{N}$ that the group G_i is a copy of $\mathbb{T}^2$ and that the homomorphism ϕ_i is a fixed ergodic toral automorphism given by a matrix S. Suppose the noise distribution μ_k is a fixed measure μ^* that satisfies $\mu^*(A) \geq \varepsilon\lambda(A \cap B)$ for every Borel set A, where $\varepsilon > 0$, λ is normalized Haar measure, and B is a fixed Borel set B with $\lambda(B) > 0$. Then, $\mathcal{P}_\mu^{strong} = \emptyset$.*

Proof. We need to evaluate $R_k^\ell = \int_{\mathbb{T}^2} \rho \cdot \iota_k \cdot \phi_k^\ell(z)\mu_\ell(dz)$. Let ν be the measure defined by $\nu(A) = \varepsilon\lambda(A \cap B)$ a Borel set A, where ε, λ and B are as in the statement. Observe that

$$|R_k^\ell| \leq \int_{\mathbb{T}^2 G_\ell} |\rho \cdot \iota_k \cdot \phi_k^\ell(z)|\,(\mu_\ell - \nu)(dz) + \int_{\mathbb{T}^2} |\rho \cdot \iota_k \cdot \phi_k^\ell(z)|\,\nu(dz)|$$

$$\leq \int_{\mathbb{T}^2} (\mu_\ell - \nu)(dz) + \left| \int_{\mathbb{T}^2} \rho \cdot \iota_k \cdot \phi_k^\ell(z)\,\nu(dz) \right|,$$

and note that the last term on the right-hand side is $\left|\int_{\mathbb{T}^2} \rho \cdot \iota_k(z)\,(\nu \cdot \phi_k^\ell)^{-1})(dz)\right|$.

As noted in Sect. 5.6 of [LT93], any ergodic toral automorphism S exhibits *topological mixing*: for any Borel sets $A, B \subseteq \mathbb{R}^2$, $\lim_{n\to\infty} \frac{\lambda(S^n B) \cap A}{\lambda(B)} = \lambda(A)$. Because ϕ_k^ℓ is an ergodic toral automorphism, so is $(\phi_k^\ell)^{-1}$. Therefore, $\lim_{\ell\to\infty} \left|\int_{\mathbb{T}^2} \rho \cdot \iota_k(z)(\nu \cdot \phi_k^\ell)^{-1}(dz)\right| = |\int_{\mathbb{T}^2} \rho \cdot \iota_k(z)\varepsilon\lambda(dz)| = 0$. Consequently, $|R_k^\ell| \leq \int_{\mathbb{T}^2} (\mu_\ell - \nu)(dz) = 1 - \varepsilon\lambda(B)$ for every nontrivial representation ρ, and hence

$$\lim_{m\to\infty} \lim_{n\to\infty} |R_k^n R_k^{n-1} \cdots R_k^m| = 0 \;\forall k \in \mathbb{N},$$

showing that $\mathcal{P}_\mu^{strong} = \emptyset$. □

Every finite-dimensional unitary representation of G_i is of the form

$$x \mapsto e^{2\pi i (z \cdot x)},$$

where z is a vector $(z_1, z_2) \in \mathbb{Z}^2$ and $z \cdot x$ is the usual inner product. Hence, if we lift this representation to a representation of G we have

$$R_k^\ell = \int_{\mathbb{T}^2} e^{2\pi i (z \cdot S^{\ell-k} x)}\,\mu_\ell(dx).$$

Suppose that the probability measure μ_ℓ is concentrated on the set of multiples of the eigenvector v^2 associated with the eigenvalue $\lambda_2 \in (0,1)$. Then,

$$R_k^\ell = \int_{\mathbb{R}} e^{2\pi i (t\lambda_2^{\ell-k} z\cdot v^2)} \, \nu_\ell(dt)$$

for some probability measure ν_ℓ on $\mathbb{R}$. It is clear that under appropriate hypotheses

$$\lim_{m\to\infty} \lim_{n\to\infty} |R_k^n R_k^{n-1} \cdots R_k^m| > 0 \, \forall k \in \mathbb{N}$$

and hence, by Corollary 3.8, $\mathcal{P}_\mu^{\text{strong}} \neq \emptyset$. For example, if $\nu_\ell = \nu$ for all $\ell \in \mathbb{N}$ for some fixed probability measure ν on $\mathbb{R}$, then it suffices that $\int_{\mathbb{R}} |t| \, \nu(dt) < \infty$. In particular, it is possible to construct examples where $\mu_1 = \mu_2 = \ldots$ is a measure that has all of $\mathbb{T}^2$ as its closed support and yet $\mathcal{P}_\mu^{\text{strong}} \neq \emptyset$.

References

AUY08. Jirô Akahori, Chihiro Uenishi, and Kouji Yano, *Stochastic equations on compact groups in discrete negative time*, Probab. Theory Related Fields **140** (2008), no. 3–4, 569–593. MR 2365485 (2009d:60173)

Cir75. B. S. Cirel′son, *An example of a stochastic differential equation that has no strong solution*, Teor. Verojatnost. i Primenen. **20** (1975), no. 2, 427–430. MR 0375461 (51 #11654)

Fra69. John Franks, *Anosov diffeomorphisms on tori*, Trans. Amer. Math. Soc. **145** (1969), 117–124. MR 0253352 (40 #6567)

Gra97. Andrew Granville, *Arithmetic properties of binomial coefficients. I. Binomial coefficients modulo prime powers*, Organic mathematics (Burnaby, BC, 1995), CMS Conf. Proc., vol. 20, Amer. Math. Soc., Providence, RI, 1997, pp. 253–276. MR 1483922 (99h:11016)

HY10. Takao Hirayama and Kouji Yano, *Extremal solutions for stochastic equations indexed by negative integers and taking values in compact groups*, Stochastic Process. Appl. **120** (2010), no. 8, 1404–1423. MR 2653259 (2011j:60222)

Kat71. Yitzhak Katznelson, *Ergodic automorphisms of T^n are Bernoulli shifts*, Israel J. Math. **10** (1971), 186–195. MR 0294602 (45 #3672)

LT93. Ding Jun Luo and Li Bang Teng, *Qualitative theory of dynamical systems*, Advanced Series in Dynamical Systems, vol. 12, World Scientific Publishing Co. Inc., River Edge, NJ, 1993. MR 1249274 (94k:58043)

Raj11. Chandiraraj Robinson Edward Raja, *A stochastic difference equation with stationary noise on groups*, 2011, *Canad. J. Math.*, to appear. http://dx.doi.org/10.4153/CJM-2011-094-6.

Tak09. Yoichiro Takahashi, *Time evolution with and without remote past*, Advances in discrete dynamical systems, Adv. Stud. Pure Math., vol. 53, Math. Soc. Japan, Tokyo, 2009, pp. 347–361. MR 2582432 (2011b:60019)

Yor92. Marc Yor, *Tsirel′ son's equation in discrete time*, Probab. Theory Related Fields **91** (1992), no. 2, 135–152. MR 1147613 (93d:60104)

Modeling Competition Between Two Influenza Strains

Rinaldo B. Schinazi

Abstract We use spatial and nonspatial models to argue that competition alone may explain why two influenza strains do not usually coexist during a given flu season. The more virulent strain is likely to crowd out the less virulent one. This can be seen as a consequence of the Exclusion Principle of Ecology. We exhibit, however, a spatial model for which coexistence is possible.

Keywords Competition models • Stochastic process • Influenza • Swine strain • Exclusion principle • Ecology

1 Introduction

The seasonal flu strain was a lot less prevalent during the 2009/2010 influenza season than during the previous years, see Fluview (the weekly CDC influenza report). On the other hand, some time during spring 2009, the new so-called swine strain appeared. There seems to be a relation between these two events. In this chapter we propose to explain this phenomenon using competition models. We will use spatial and nonspatial models to show that in a given flu season coexistence of two strains is unlikely due to competition alone. We will also show that geometry and space may be critical for coexistence. Our models deal with competition over only one flu season. In the real world, because of mutations the fight between two strains may go on for several flu seasons before one strain outcompetes the other. This picture is consistent with the very skinny shape of the phylogenetic tree for influenza; see, for instance Koelle et al. [5] and van Nimwegen [8]. In this chapter, the two competing strains are assumed not to undergo mutations, and therefore the time scale we focus on is one flu season.

R.B. Schinazi (✉)
University of Colorado at Colorado Springs, Colorado Springs CO80933-7150, USA
e-mail: rschinaz@uccs.edu

J. Englander and B. Rider (eds.), *Advances in Superprocesses and Nonlinear PDEs*, Springer Proceedings in Mathematics & Statistics 38,
DOI 10.1007/978-1-4614-6240-8_3,

A competing explanation of the non-coexistence of the two influenza strains is cross immunity. For instance, immunity may explain why older generations have not been as much affected as the younger ones in the swine flu epidemic. It may be due to some previous exposure to a similar strain, see the discussion in Greenbaum et al. [3]. However, using a cross-immunity argument to explain why the swine strain crowds out the seasonal one may be more difficult. The hypothesis would be that the swine strain must confer some immunity against the seasonal flu. But, clearly the seasonal strain does not confer any immunity against the swine strain: after all even young people (the group most severely affected by the swine strain) have usually been exposed to the seasonal strain and do not seem to be protected against the swine strain. Hence, for this argument to work the swine strain must confer some immunity against the seasonal strain, but the seasonal strain cannot confer any immunity against the swine strain. In contrast to this cross immunity hypothesis we argue in this chapter that even in models for which there is no immunity at all (every individual that recovers is immediately susceptible again!), coexistence of two competing strains is rather unlikely.

2 The ODE Model

Our first model is a system of ordinary differential equations. Let $u_1(t)$ and $u_2(t)$ be the density of individuals infected at time t with strains 1 and 2, respectively . We set

$$u_1' = \lambda_1 u_1 u_0 - \delta_1 u_1$$
$$u_2' = \lambda_2 u_2 u_0 - \delta_2 u_2$$

where $u_0(t)$ is the density of susceptible individuals at time t. In words, individuals infected with strain i infect susceptible individuals at rate λ_i and get healthy at rate δ_i, for $i = 1, 2$. We assume that the only possible states are 0, 1 and 2. Hence, at any time $t > 0$ we have $u_0(t) + u_1(t) + u_2(t) = 1$. In particular, as soon as an infected individual gets healthy, it is back in the susceptible pool.

Let 1 be the seasonal and 2 be the swine strains. Some reports indicate that the swine strain may be more virulent than the seasonal strain, see Fraser et al. [2]. Under that assumption,

$$\frac{\lambda_1}{\delta_1} < \frac{\lambda_2}{\delta_2}.$$

Assume also that at some point in time the ODE model is at the equilibrium $(0, 1 - \frac{\delta_2}{\lambda_2})$. That is, there is no seasonal strain and the swine strain is in equilibrium. Now introduce a little bit of seasonal strain (small u_1). Will the seasonal strain be able to grow? Using that u_1 is almost 0 and that u_2 is almost $1 - \frac{\delta_2}{\lambda_2}$ we make the approximation

$$u_0 = 1 - u_1 - u_2 \sim 1 - \left(1 - \frac{\delta_2}{\lambda_2}\right) = \frac{\delta_2}{\lambda_2}.$$

Hence,

$$u_1' \sim \lambda_1 u_1 \frac{\delta_2}{\lambda_2} - \delta_1 u_1 = u_1 \left(\lambda_1 \frac{\delta_2}{\lambda_2} - \delta_1 \right).$$

Since we are assuming that $\frac{\lambda_1}{\delta_1} < \frac{\lambda_2}{\delta_2}$ we get $u_1' < 0$. That is, under these assumptions and according to this model, the seasonal flu will not take hold.

In fact this system of ODE is a particular case of a well-known competition model. For the general version of this model, it is known that one of the strains will vanish; see Exercise 3.3.5 in Hofbauer and Sigmund [4]. The point is that we have two populations (the population of individuals infected with strain 1 and the population of individuals infected with strain 2) that compete for a single resource (the susceptible individuals). It turns out that in such a model, one population will drive the other one out. This is a particular case of the so-called "Exclusion Principle" of Ecology: if the number of populations is larger than the number of resources all the populations cannot subsist in the long run, see 5.4 in Hofbauer and Sigmund [4].

3 The Spatial Stochastic Model

In the preceding model there is no space structure, and all the individuals in the population can be seen as neighbors of each other. In this section, we go to the other extreme where there is a rigid space structure and each individual has a fixed number of neighbors.

We now describe a spatial stochastic model called the multitype contact process, see Neuhauser [7]. Let S be the integer lattice $\mathbf{Z}^d$ (d is the dimension) or the homogeneous tree $\mathbf{T}_d$ for which each site has $d+1$ neighbors. The system is described by a configuration $\xi \in \{0,1,2\}^S$, where $\xi(x) = 0$ means that site x is occupied by a susceptible individual, $\xi(x) = 1$ means that x is occupied by an individual infected by strain 1 and $\xi(x) = 2$ means that x is occupied by an individual infected by strain 2. If S is $\mathbf{Z}^d$, then each site has $2d$ neighbors, if S is $\mathbf{T}_d$, then each site has $d+1$ neighbors. For $x \in S$ and $\xi \in \{0,1,2\}^S$, let $n_1(x,\xi)$ and $n_2(x,\xi)$ denote the number of neighbors of x that are infected by strain 1 and strain 2, respectively.

The multitype contact process ξ_t with birth rates λ_1, λ_2 makes transitions at x when the configuration of the process is ξ

$$1 \to 0 \text{ at rate } 1$$

$$2 \to 0 \text{ at rate } 1$$

$$0 \to 1 \text{ at rate } \lambda_1 n_1(x,\xi),$$

$$0 \to 2 \text{ at rate } \lambda_2 n_2(x,\xi),$$

In words, a susceptible individual gets infected by an infected neighbor at rates λ_1 or λ_2, depending on which strain the neighbor is infected with. An infected individual gets healthy (and is immediately susceptible again) at rate 1. Note that compared to the ODE model, we are assuming in this model that $\delta_1 = \delta_2 = 1$. This is so because most of the mathematical results have been proved under the assumption $\delta_1 = \delta_2$. We take this common value to be 1 to minimize the number of parameters.

The multitype contact process is a generalization of the basic contact process which has only one type. The transitions of the basic contact process are given by

$$1 \to 0 \text{ at rate } 1$$
$$0 \to 1 \text{ at rate } \lambda_1 n_1(x,\xi),$$

For the basic contact process, there exists a critical value λ_c whose exact value is not known and which depends on the graph the model is on. If $\lambda_1 > \lambda_c$, then starting with even a single infected individual, there is a positive probability of having infected individuals at all times somewhere in the graph. On the other hand, if $\lambda_1 \leq \lambda_c$, then starting from any finite number of infected individuals all the infected individuals will disappear after a finite time. See Liggett [6] for more on the basic contact process on the square lattice and on trees.

3.1 *The Space Is the Square Lattice* $\mathbf{Z}^d$

We now go back to the multitype contact process. Assume that $\lambda_2 > \lambda_c$ and $\lambda_2 > \lambda_1$ then there is no coexistence of strains 1 and 2 in the sense that

$$\lim_{t\to\infty} P(\xi_t(x) = 1, \xi_t(y) = 2) = 0$$

for any sites x and y in $\mathbf{Z}^d$ and any initial configuration ξ_0. In fact, strain 2 always drives out strain 1 in the following sense. Let A be the event that strain 2 will not ever disappear. Then,

$$\lim_{t\to\infty} P(\xi_t(x) = 1|A) = 0,$$

for any site x in $\mathbf{Z}^d$ and any initial configuration (see Theorem 2 in Cox and Schinazi [1] and also Neuhauser [7]). Hence, assuming that $\lambda_2 > \lambda_1$ (i.e., strain 2 is more virulent than strain 1) this model too predicts that the seasonal flu will be crowded out by the swine strain. The spatial structure seems to have no influence on the outcome. The next section will show that this is not always so and that a different (more crowded) space structure allows coexistence.

3.2 *The Space Is the Tree* $\mathbf{T}_d$

There is a fundamental difference between the basic contact process on the square lattice and the same model on the tree. There are two (instead of one) critical values for the basic contact process on the tree. The definition of λ_c is as before. We also define another critical value λ_{cc} in the following way. Consider the basic (one type) contact process with birthrate λ_1. Let O be a fixed site on the tree or square lattice. Start the process with a single infected individual at O. The probability that the infection will return to site O infinitely many times is positive if and only if $\lambda_1 > \lambda_{cc}$. It turns out that $\lambda_c < \lambda_{cc}$ on the tree but $\lambda_c = \lambda_{cc}$ on the square lattice.

The fact that the basic contact process has two distinct critical values on the tree has interesting consequences for the multitype contact process on the tree. Let λ_1 and λ_2 be in $(\lambda_c, \lambda_{cc})$, and then strains 1 and 2 may coexist on the tree in the following sense. Under suitable initial configurations we have for any sites x and y

$$\liminf_{t\to\infty} P(\xi_t(x) = 1, \xi_t(y) = 2) > 0.$$

See Theorem 1 in Cox and Schinazi [1]. Note that coexistence occurs even for $\lambda_1 < \lambda_2$ but both parameters need to be in the rather narrow interval $(\lambda_c, \lambda_{cc})$. This result shows that space structure and geometry may be crucial in allowing coexistence.

4 Discussion

Our models show that at least in theory coexistence of two competing strains is unlikely. Coexistence is however possible for the multitype contact process on a tree. The tree can be thought of as a model for high-density populations (in a ball of radius r there are $(d+1)d^{r-1}$ individuals on the tree $\mathbf{T_d}$ but only about r^d on the lattice $\mathbf{Z^d}$). In order to have coexistence, both infection rates cannot be too low or too high but may be unequal. In all other cases, there will be no coexistence on the tree, and there is never coexistence on $\mathbf{Z^d}$ unless λ_1 is exactly equal to λ_2, a rather unlikely possibility, see Neuhauser [7]. Interestingly, the behavior of the mean-field ODE model is the same as the behavior of the model on $\mathbf{Z^d}$ but different from the model on the tree. In general, it is expected that the model on the tree to be closer to the mean-field model than to the model on $\mathbf{Z^d}$. This is not so in this example.

References

1. J.T. Cox and R.B.Schinazi (2009). Survival and coexistence for a multitype contact process. Annals of probability **37**, 853–876
2. C. Fraser et al. (2009).Pandemic Potential of a Strain of Influenza A (H1N1): Early Findings. Science, **324**, 1557–1561
3. J.A. Greenbaum et al. (2009). Pre-existing immunity against swine-origin H1N1 influenza viruses in the general human population. PNAS, **106**, 20365–20370.
4. J. Hofbauer and K. Sigmund (1998). *Evolutionary games and population dynamics.* Cambridge University Press.
5. K.Koelle, S. Cobey, B. Grenfell and M. Pascual (2006) Epochal evolution shapes the phylodynamics of interpandemic influenza A (H3N2) in Humans. Science vol. 314, 1898–1903.

6. T.Liggett (1999). *Stochastic interacting systems: contact, voter and exclusion processes,* Springer-Verlag, Berlin.
7. C. Neuhauser (1992). Ergodic theorems for the multitype contact process. Probab. Theory Related Fields, **91**, 467–506.
8. E. van Nimwegen (2006). Influenza escapes immunity along neutral networks. Science vol. 314, 1884–1886.

Asymptotic Results for Near Critical Bienaymé–Galton–Watson and Catalyst-Reactant Branching Processes

Amarjit Budhiraja and Dominik Reinhold

Abstract Near critical single-type Bienaymé–Galton–Watson (BGW) processes are considered. Results on convergence of Yaglom distributions of suitably scaled BGW processes to that of the corresponding diffusion approximation are given. Convergences of stationary distributions for Q-processes and models with immigration to the corresponding distributions of the associated diffusion approximations are established. Similar results can be obtained in a multitype setting. To illustrate this, a result on convergence of Yaglom distributions of suitably scaled multitype subcritical BGW processes to that of the associated diffusion model is presented.

In the second part, near critical catalyst-reactant branching processes with controlled immigration are considered. The reactant population evolves according to a branching process whose branching rate is proportional to the total mass of the catalyst. The bulk catalyst evolution is that of a classical continuous-time branching process; in addition, there is a specific form of immigration. Immigration takes place exactly when the catalyst population falls below a certain threshold, in which case the population is instantaneously replenished to the threshold. A diffusion limit theorem for the scaled processes is presented, in which the catalyst limit

This research is partially supported by the National Science Foundation (DMS-1004418, DMS-1016441), the Army Research Office (W911NF-10-1-0158), NSF Emerging Frontiers in Research and Innovation (EFRI) (Grant CBE0736007), and the US-Israel Binational Science Foundation (Grant 2008466).

A. Budhiraja
Department of Statistics and Operations Research, University of North Carolina,
Chapel Hill, NC 27599, USA
e-mail: budhiraj@email.unc.edu

D. Reinhold (✉)
Department of Mathematics and Computer Science,
Clark University, Worcester, MA 01610, USA
e-mail: dreinhold@clarku.edu

J. Englander and B. Rider (eds.), *Advances in Superprocesses and Nonlinear PDEs*,
Springer Proceedings in Mathematics & Statistics 38,
DOI 10.1007/978-1-4614-6240-8_4,

is described through a reflected diffusion, while the reactant limit is a diffusion with coefficients that are functions of both the reactant and the catalyst. Stochastic averaging under fast catalyst dynamics is considered next. In the case where the catalyst evolves "much faster" than the reactant, a scaling limit, in which the reactant is described through a one-dimensional SDE with coefficients depending on the invariant distribution of the reflected diffusion, is obtained.

Keywords Branching processes • Catalyst-reactant branching processes • Quasi stationary distributions • Yaglom distributions • Q-processes • Near critical regime • Chemical reaction networks • Diffusion approximations • Stochastic averaging • Multiscale approximations • Reflected diffusions • Constrained martingale problems • Echeverria criterion • Invariant measure convergence

AMS subject classifications (2000): Primary 60J80; secondary 60F05.

1 Introduction

This chapter reviews some recent asymptotic results on near critical Bienaymé–Galton–Watson (BGW) branching processes and on catalyst-reactant branching processes with controlled immigration. Proofs of the former results can be found in [3], whereas those of the latter results are in [2, 18].

It is well known that under suitable assumptions, appropriately scaled near critical BGW processes converge in the large population limit to Feller diffusions (see [5, 8]). We are concerned with relationships between the steady-state behavior of the branching processes and that of their approximating diffusions. One, of course, needs to suitably interpret the term "steady state" since, as is well known, with time each branching process will grow to infinity on the set of non-extinction in the supercritical case and become extinct in the critical and subcritical case (see [1]). A natural approach in the subcritical case is to study probability laws conditioned on the event of non-extinction. In the supercritical case, a common approach is to additionally condition on the event of eventual extinction. Such a conditioning leads to the notion of quasi-stationary distributions. Results establishing convergence of quasi-stationary distributions of the scaled BGW processes to that of the limiting Feller diffusions will be presented. Similar results for the closely related setting of BGW models with immigration will also be given. Analogous properties can be established in a multitype setting, and we will illustrate this through a result for multitype subcritical BGW processes.

Next, we consider near critical catalyst-reactant branching processes with a specific form of controlled immigration. The catalyst population evolves according to a classical continuous-time branching process, while the reactant population evolves according to a branching process whose branching rate is proportional to the total mass of the catalyst. Immigration takes place exactly when the catalyst population falls below a certain threshold, in which case the population is instantaneously replenished to the threshold to ensure a certain level of activity. Our main goal here is to establish diffusion approximations for the catalyst and reactant populations in two settings. In the first setting, both populations evolve on "comparable timescales," while in the second setting, the catalyst evolves "much faster" than the reactant, in a sense made precise in Sect. 3. In the first setting, the limit model is described through a coupled system of stochastic differential equations with reflection in the space $[1,\infty)\times\mathbb{R}$. In the second setting, we establish a stochastic averaging result that says that the limit reactant evolution is given through an autonomous one-dimensional SDE with coefficients described in terms of the invariant distribution of a reflected diffusion in $[1,\infty)$; the reflected diffusion can be interpreted as the limiting dynamics of the catalyst process under a suitable scaling.

This chapter is organized as follows: in Sect. 2 we review results on BGW processes that have appeared in [3, 18], and in Sect. 3 we present results on near critical catalyst-reactant branching processes with controlled immigration, proofs of which can be found in [2, 18].

2 Asymptotic Results on Near Critical Branching Processes

Consider a population consisting of k types of particles whose evolution is described in terms of a discrete time multitype (k type) Bienaymé–Galton–Watson (k-BGW) process—such a process is a Markov chain $\{\mathbf{Z}_p\}_{p\in\mathbb{N}_0}$ on $\mathbb{N}_0^k$, with the vector $\mathbf{Z}_p$ representing the number of particles of each type in generation p. We are interested in the longtime behavior of the scaled process $\frac{1}{p}\mathbf{Z}_{\lfloor pt\rfloor}$, $t\geq 0$, when the k-BGW process is close to criticality. More precisely, we consider a sequence of BGW processes $\{\mathbf{Z}_p^{(n)}, p\in\mathbb{N}_0\}_{n\in\mathbb{N}}$ such that, as n becomes large, the processes approach criticality (in the sense of Condition 2.1). It is well known (see [5, 8]) that, under suitable conditions, the process $\hat{\mathbf{Z}}_t^{(n)}=\frac{1}{n}\mathbf{Z}_{\lfloor nt\rfloor}^{(n)}$, $t\geq 0$, converges weakly to a diffusion $\boldsymbol{\xi}$. Such a result implies convergence of finite time statistics of $\hat{\mathbf{Z}}^{(n)}$ to those of $\boldsymbol{\xi}$, but does not provide any information on relationships between the time asymptotic behaviors of $\hat{\mathbf{Z}}^{(n)}$ and $\boldsymbol{\xi}$. The main goal of this section is to make such relationships mathematically precise. In particular, we show that, under appropriate assumptions, the time asymptotic distribution of $\hat{\mathbf{Z}}_t^{(n)}$ with suitable conditioning converges to that of $\boldsymbol{\xi}_t$ with a similar conditioning, as $n\to\infty$ (see Theorems 2.3 and 2.6). An analogous result for models with immigration (where no conditioning is required) is also established (Theorem 2.8). The results say that the longtime behavior of a BGW process (under suitable conditioning or with immigration) is well approximated by that of the corresponding diffusion limit $\boldsymbol{\xi}$. Most of the results in this section are for single-type BGW processes, namely for the case $k=1$. Similar results can be obtained in multitype settings and we consider one such result in Theorem 2.12.

When $k = 1$, the transition probabilities of a BGW process $\{Z_p\}$ can be written as

$$P(Z_{p+1} = j | Z_p = i) = \begin{cases} \mu^{*i}(j) & \text{if } i \geq 1, \quad j \geq 0, \\ \delta_{0j} & \text{if } i = 0, \quad j \geq 0, \end{cases} \tag{1}$$

where $\{\mu(l)\}_{l \in \mathbb{N}_0}$ is the offspring distribution of a typical particle and $\{\mu^{*i}(l)\}_{l \in \mathbb{N}_0}$ is the i-fold convolution of $\{\mu(l)\}_{l \in \mathbb{N}_0}$. The process starts with Z_0 particles; each of the Z_p particles alive at time p lives for one unit of time and then dies, giving rise to l offspring particles with probability $\mu(l)$, $l \in \mathbb{N}_0$. The particles behave independently of each other and of the past.

Depending on the mean m of the offspring distribution, BGW processes can be divided into three cases: subcritical, critical, and supercritical, according to whether $m < 1$, $m = 1$, or $m > 1$, respectively.

In order to describe near critical BGW processes, we will consider a sequence of processes $Z^{(n)}$ with offspring distributions $\mu^{(n)}$. If $Z_0^{(n)} = 1$, then $Z_1^{(n)}$ has the probability-generating function (pgf)

$$F^{(n)}(s) = \sum_{l=0}^{\infty} \mu^{(n)}(l) s^l, \quad s \in [0,1]. \tag{2}$$

We denote the mean of the offspring distribution by $m^{(n)}$ and the variance by $\kappa^{(n)}$. Denote the pth iterate of $F^{(n)}$ by $F_p^{(n)}$, i.e., for $s \in [0,1]$ and $p \geq 0$

$$F_0^{(n)}(s) = s, \quad F_{p+1}^{(n)}(s) = F^{(n)}(F_p^{(n)}(s)).$$

Let $q^{(n)}$ be the extinction probability of $Z^{(n)}$ starting with a single particle, i.e.

$$q^{(n)} = P(Z_p^{(n)} = 0 \text{ for some } p \in \mathbb{N} | Z_0^{(n)} = 1).$$

Denote by $\mathcal{P}(\mathbb{R}_+)$ the set of probability measures on $\mathbb{R}_+ := [0,\infty)$ with the Borel σ-field.

Condition 2.1. (i) For each n, $\mu^{(n)}(0) > 0$, $\mu^{(n)}(0) + \mu^{(n)}(1) < q^{(n)}$.

(ii) For each n, $m^{(n)} = 1 + \frac{c^{(n)}}{n}, c^{(n)} \in (-n,\infty) \setminus \{0\}$, and as $n \to \infty$, $c^{(n)} \to c \in \mathbb{R} \setminus \{0\}$.

(iii) For each n, $\kappa^{(n)} < \infty$, and as $n \to \infty$, $\kappa^{(n)} \to \kappa \in (0,\infty)$.

(iv) As $n \to \infty$, the distribution of $\frac{Z_0^{(n)}}{n}$ converges to some $\mu_0 \in \mathcal{P}(\mathbb{R}_+)$.

(v) The family of functions $\{F^{(n)''}\}_{n \in \mathbb{N}}$ is equicontinuous at 1.

Condition 2.1 (ii) ensures that, as $n \to \infty$, $m^{(n)} \to 1$, and eventually, the processes approach criticality strictly from above or strictly from below. The case where $c < 0$ will be referred to as the (near critical) subcritical case while $c > 0$ corresponds to the supercritical case. Conditions 2.1 (ii)–(v) are used for the diffusion approximation result in Theorem 2.1. Condition 2.1 (v) will also be used in the study of the supercritical case in Theorem 2.3. Let

$$\hat{Z}_t^{(n)} := \frac{1}{n} Z_{\lfloor nt \rfloor}^{(n)}, \quad t \in \mathbb{R}_+; \tag{3}$$

then $\{\hat{Z}_t^{(n)}\}_{t\in\mathbb{R}_+}$ is an $\mathbb{S}^{(n)} := \{\frac{l}{n} | l \in \mathbb{N}_0\}$-valued (time inhomogeneous) Markov process with sample paths in $D(\mathbb{R}_+ : \mathbb{S}^{(n)})$, the space of RCLL (right continuous, left limit) functions from $\mathbb{R}_+$ to $\mathbb{S}^{(n)}$. Throughout, $\mathbb{S}^{(n)}$ is endowed with the discrete topology and given a metric space S, $D(\mathbb{R}_+ : S)$ is endowed with the usual Skorohod topology. The space of probability measures on a metric space S will be denoted by $\mathcal{P}(S)$.

We now recall a well-known weak convergence result for $\hat{Z}^{(n)}$ (see [6], [14, Theorem 2.1]), which describes the asymptotic behavior of $\hat{Z}^{(n)}$, as $n \to \infty$, over any fixed finite time horizon. A related multidimensional result will be presented later in this section (see also [8]). Denote by $C^l(\mathbb{R}_+)$ the set of l-times differentiable, real-valued functions on $\mathbb{R}_+$. In the following theorem, we do not need part (i) of Condition 2.1, and in part (ii) of the assumption $(-n,\infty) \setminus \{0\}$ and $\mathbb{R} \setminus \{0\}$ can be replaced by $(-n,\infty)$ and $\mathbb{R}$, respectively (see [14, Theorem 2.1]). However, in order to simplify our presentation, we assume Condition 2.1 to hold, even though for some of the results we need only parts of it.

Theorem 2.1. *Assume Condition 2.1. Then $\hat{Z}^{(n)}$ converges weakly in $D(\mathbb{R}_+ : \mathbb{R}_+)$ to the unique (in law) diffusion process ξ with generator*

$$(Lf)(x) = xcf'(x) + \frac{1}{2}x\kappa f''(x), \quad f \in C^2(\mathbb{R}_+), \quad x \in \mathbb{R}_+, \tag{4}$$

and initial distribution (i.e. probability law of ξ_0) equal to μ_0.

We are concerned with the study of relationships between the "steady-state" behavior of $\hat{Z}^{(n)}$ and that of ξ. However, as noted earlier, the term "steady state" needs a careful interpretation. There are two well-studied approaches for formulating time asymptotic questions in the subcritical case. The first is to condition the processes $\hat{Z}^{(n)}$ on non-extinction, where, loosely speaking, the conditioning can either be on non-extinction at the present time or in the distant future. The state process $\hat{Z}^{(n)}$ under these two conditionings has different limiting distributions as $t \to \infty$. The first is called the Yaglom distribution of $\hat{Z}^{(n)}$, while the second is the stationary distribution of the Q-process associated with $\hat{Z}^{(n)}$ (see Sect. I.14 of [1]). The second approach for obtaining a nontrivial time asymptotic behavior is to introduce an immigration component. Namely, in each generation a (random) number of particles that are indistinguishable from the original set of particles are added to the population. The immigration component in particular ensures that the resulting scaled state process, denoted by $\hat{V}^{(n)}$, has a nondegenerate stationary distribution. For the supercritical case, a common approach is to reduce the problem to that of a subcritical setting by conditioning on the event of eventual extinction. The so conditioned state process $\hat{Z}^{(n)}$ has the same law as the state process corresponding to a certain subcritical BGW process. In this section we will show that the time asymptotic distribution of $\hat{Z}_t^{(n)}$ (in both subcritical and supercritical settings), under suitable conditioning, converges to that of ξ_t under a similar conditioning, as $n \to \infty$. For models with immigration, we will prove convergence of stationary distributions.

We begin by describing results for models without immigration. Let $\mathbb{S}$ be a subset of $\mathbb{R}_+^k$, for some $k \in \mathbb{N}$. When $\mathbb{S}$ is endowed with a topology, we will denote by $\mathcal{B}(\mathbb{S})$ the σ-field generated by the open sets of $\mathbb{S}$. Let $\mathbf{Y} \equiv \{\mathbf{Y}_t\}_{t\in\mathbb{R}_+}$ be an $\mathbb{S}$-valued Markov process such that $\mathbf{0} \in \mathbb{S}$ is an absorbing state. If $\mathbf{Y}_0 = \mathbf{y}$, we write $P(\mathbf{Y}_t \in \cdot)$ as $P_{\mathbf{y}}(\mathbf{Y}_t \in \cdot)$. Similarly, when the distribution of $\mathbf{Y}_0$ is μ, we write $P(\mathbf{Y}_t \in \cdot)$ as $P_\mu(\mathbf{Y}_t \in \cdot)$. Similar notations will be used for conditional expectations.

Definition 2.1. (i) A quasi-stationary distribution (qsd) for $\mathbf{Y}$ is a probability distribution μ on $(\mathbb{S}, \mathcal{B}(\mathbb{S}))$ such that $P_\mu(\mathbf{Y}_t \in B | t < T_{\mathbf{Y}} < \infty) = \mu(B)$ for all $B \in \mathcal{B}(\mathbb{S})$ and $t \geq 0$, where $T_{\mathbf{Y}} := \inf\{t | \mathbf{Y}_t = \mathbf{0}\}$.
(ii) If for all $\mathbf{y} \in \mathbb{S} \setminus \{\mathbf{0}\}$, as $t \to \infty$, $P_{\mathbf{y}}(\mathbf{Y}_t \in \cdot | t < T_{\mathbf{Y}} < \infty)$ converges weakly to some probability measure μ on $(\mathbb{S}, \mathcal{B}(\mathbb{S}))$, then μ is called the Yaglom distribution of $\mathbf{Y}$.

The following result is a special case of results in [16] (pp. 77–78); see also [11] and Proposition 2.3.2.1 of [12].

Theorem 2.2. *The Yaglom distribution of ξ exists and is exponential with density*

$$f(x) = \frac{2|c|}{\kappa} \exp\left(-\frac{2|c|}{\kappa} x\right), \quad x \geq 0. \tag{5}$$

Our first result, Theorem 2.3 below, says that the Yaglom distribution of $\hat{Z}^{(n)}$ approaches that of ξ, as $n \to \infty$.

Theorem 2.3. *Assume Condition 2.1. For each n, $\hat{Z}^{(n)}$ has a Yaglom distribution $\nu^{(n)}$. This distribution is also a qsd, and it converges weakly to the Yaglom distribution ν of ξ.*

We now consider the second form of conditioning where one conditions the process on not being extinct in the "distant future." We will see that in this case a somewhat different asymptotic behavior emerges. For this result we restrict ourselves to the subcritical case (i.e., $c < 0$). We begin with the definition of a Q-process (see [1,11]).

Let $\hat{\Omega} = D(\mathbb{R}_+ : \mathbb{R}_+)$ and $\hat{\mathcal{F}}$ be the corresponding Borel σ-field (with the usual Skorohod topology). Denote by $\{\mathcal{F}_t\}_{t\in\mathbb{R}_+}$ the canonical filtration on $(\hat{\Omega}, \hat{\mathcal{F}})$, i.e., $\mathcal{F}_t = \sigma(\pi_s : s \leq t)$, where $\pi_s(x) = x_s$ for $x \in \hat{\Omega}$. We denote by $\hat{P}_\mu^{(n)}$ the measure induced by $\hat{Z}^{(n)}$ on $(\hat{\Omega}, \hat{\mathcal{F}})$ when $Z_0^{(n)}$ has distribution μ (supported on $\mathbb{N}$). Let $T := \inf\{t | \pi_t = 0\}$.

It is easy to check (see [1], p. 58; also [13] and [3]) that there is a probability measure $P_\mu^{(n)\uparrow}$ on $(\hat{\Omega}, \hat{\mathcal{F}})$ such that, as $s \to \infty$, $\hat{P}_\mu^{(n)}(\Theta | T > s) \to P_\mu^{(n)\uparrow}(\Theta)$, for all $\Theta \in \mathcal{F}_t$, $t \in \mathbb{R}_+$. Furthermore, this unique measure on $(\hat{\Omega}, \hat{\mathcal{F}})$ can be characterized as follows. Let $\{Z_k^{(n)\uparrow}\}_{k\in\mathbb{N}_0}$ be a Markov chain with state-space $\mathbb{N}$, transition probabilities

$$P(Z^{(n)}_{l+1} = j | Z^{(n)}_l = i)\frac{j}{im^{(n)}}, \quad i, j \in \mathbb{N}, l \in \mathbb{N}_0,$$

and initial distribution μ, and let $\hat{Z}^{(n)\uparrow}_t := \frac{1}{n} Z^{(n)\uparrow}_{\lfloor nt \rfloor}$, $t \in \mathbb{R}_+$. Then $P^{(n)\uparrow}_{\mu}$ is the law of $\{\hat{Z}^{(n)\uparrow}_t\}_{t\in\mathbb{R}_+}$. The process $Z^{(n)\uparrow}$ [respectively $\hat{Z}^{(n)\uparrow}$] is called the Q-process associated with $Z^{(n)}$ [respectively $\hat{Z}^{(n)}$]. Q-processes associated with branching processes can be interpreted as branching processes conditioned on being never extinct.

Next, we introduce the Q-process associated with the diffusion ξ from Theorem 2.1. Denote by $P_{\xi,x}$ the measure induced by ξ on $(\hat{\Omega}, \hat{\mathcal{F}})$, where $\xi_0 = x > 0$. The following theorem is contained in [11].

Theorem 2.4. *There is a probability measure $P^{\uparrow}_{\xi,x}$ on $(\hat{\Omega}, \hat{\mathcal{F}})$, such that for all $t \in \mathbb{R}_+$ and $\Theta \in \mathcal{F}_t$, $P_{\xi,x}(\Theta | T > s)$ converges to $P^{\uparrow}_{\xi,x}(\Theta)$, as $s \to \infty$. Let $\xi^{\uparrow}$ be the unique weak solution of the SDE*

$$d\xi^{\uparrow}_t = c\xi^{\uparrow}_t dt + \sqrt{\kappa \xi^{\uparrow}_t} dB_t + \kappa dt, \quad \xi^{\uparrow}_0 = x,$$

where B is a standard Brownian motion. Then $P^{\uparrow}_{\xi,x}$ equals the measure induced by $\xi^{\uparrow}$ on $(\hat{\Omega}, \hat{\mathcal{F}})$.

The process $\xi^{\uparrow}$ is referred to as the Q-process associated with ξ. The following result (see [11], Sect. 5.2) says that the process $\xi^{\uparrow}$ has a unique stationary distribution, $\nu^{\uparrow}$, which is given as the convolution of two copies of the exponential distribution ν with density as in (5).

Theorem 2.5. *Assume $c < 0$. As $t \to \infty$, for every initial condition x, $\xi^{\uparrow}_t$ converges in distribution to a random variable $\xi^{\uparrow}_{\infty}$, whose distribution, denoted by $\nu^{\uparrow}$, is the convolution of two copies of the Yaglom distribution ν. In particular, $\nu^{\uparrow}$ has density*

$$f(x) = \left(\frac{2c}{\kappa}\right)^2 x \exp\left(\frac{2c}{\kappa} x\right), \quad x \geq 0. \tag{6}$$

Our next result shows that the time asymptotic behavior of the Q-process associated with $\hat{Z}^{(n)}$ can be well approximated by that of the Q-process associated with the diffusion approximation of $\hat{Z}^{(n)}$.

Theorem 2.6. *Assume Condition 2.1 and that $c_n < 0$ for all $n \in \mathbb{N}$. For each n, $\hat{Z}^{(n)\uparrow}_t$ converges in distribution, as $t \to \infty$, to a random variable $\hat{Z}^{(n)\uparrow}_{\infty}$. The distribution $\nu^{(n)\uparrow}$ of $\hat{Z}^{(n)\uparrow}_{\infty}$ is the unique stationary distribution of the $\mathbb{S}^{(n)}$-valued Markov process $\hat{Z}^{(n)\uparrow}$. As $n \to \infty$, $\nu^{(n)\uparrow}$ converges weakly to $\nu^{\uparrow}$.*

We now describe the results for BGW processes with immigration. Let F and G be pgf's of $\mathbb{N}_0$-valued random variables. A Bienaymé–Galton–Watson branching process with immigration corresponding to (F,G) (referred to as a BPI(F,G) process) is a Markov chain $\{V_n\}$ with state-space $\mathbb{N}_0$ and transition probability function described in terms of the corresponding pgf: Given $V_0 = i \in \mathbb{N}$, the pgf $H(i,\cdot)$ of V_1 is $H(i,s) = \sum_{j=0}^{\infty} P(V_1 = j | V_0 = i) s^j = F(s)^i G(s)$, $s \in [0,1]$.

Let $G^{(n)}$ be a sequence of pgf's of $\mathbb{N}_0$-valued random variables, and let $F^{(n)}$ be as in (2). Let $V^{(n)}$ be a sequence of BPI($F^{(n)}, G^{(n)}$) processes.

Condition 2.2. (i) There is a $\iota_0 \in (0,\infty)$ such that for all $n \in \mathbb{N}$ $G^{(n)\prime}(1) = \iota^{(n)} \geq \iota_0$, and as $n \to \infty$, $\iota^{(n)} \to \iota$.
(ii) There is a $\beta_0 \in (0,\infty)$ such that for all $n \in \mathbb{N}$ $G^{(n)\prime\prime}(1) \leq \beta_0$.
(iii) There is a $\tau_0 \in (0,\infty)$ such that for all $n \in \mathbb{N}$ $F^{(n)\prime\prime\prime}(1) \leq \tau_0$.

Let $\hat{V}_t^{(n)} := \frac{1}{n} V_{\lfloor nt \rfloor}^{(n)}$, $t \in \mathbb{R}_+$. The proof of the following theorem is easy to establish using [11] and [14, Theorem 2.1].

Theorem 2.7. *Assume Conditions 2.1 and 2.2 and that $c < 0$. Suppose that $\hat{V}_0^{(n)}$ converges in distribution to some $\mu \in \mathcal{P}(\mathbb{R}_+)$. Then $\hat{V}^{(n)}$ converges weakly in $D(\mathbb{R}_+ : \mathbb{R}_+)$ to the process ζ which is the unique weak solution of*

$$d\zeta_t = c\zeta_t dt + \sqrt{\kappa \zeta_t} dB_t + \iota dt, \quad t \geq 0,$$

where ζ_0 has distribution μ_0. The Markov process ζ has a unique stationary distribution η, which is a gamma distribution with parameters $2\iota/\kappa$ and $\kappa/(2|c|)$, i.e., η has density g given as

$$g(x) = x^{\frac{2\iota}{\kappa}-1} \frac{\exp\left(-x\frac{2|c|}{\kappa}\right)}{\left(\frac{\kappa}{2|c|}\right)^{\frac{2\iota}{\kappa}} \Gamma\left(\frac{2\iota}{\kappa}\right)}, \quad x > 0.$$

We are interested in the longtime behavior of the scaled processes $\hat{V}^{(n)}$ as they approach criticality. Our main result in the single-type setting is the following.

Theorem 2.8. *Assume Conditions 2.1 and 2.2 and that $c_n < 0$ for all $n \in \mathbb{N}$. For each $n \in \mathbb{N}$, $\hat{V}^{(n)}$ has a unique stationary distribution $\eta^{(n)}$ on $\mathbb{S}^{(n)}$, and as $n \to \infty$, $\eta^{(n)}$ converges weakly to η.*

As noted earlier, results similar to Theorems 2.3, 2.5, and 2.8 can be established for multitype settings as well. We only discuss one case in detail, namely the convergence of the Yaglom distribution in the setting of a subcritical multitype process. We begin with some notation and definitions. Let $\{\mathbf{Z}_j^{(n)}, j \in \mathbb{N}_0\}_{n\in\mathbb{N}}$ be a sequence of k-BGW processes with transition mechanism as described below. Let $C := [0,1]^k$, $\mathbf{e}_\alpha := (\delta_{1\alpha}, \ldots, \delta_{k\alpha})'$ be the αth canonical basis vector and $\mathbf{s}^{\mathbf{i}} := \prod_{\alpha=1}^k s_\alpha^{i_\alpha}$, for $\mathbf{i} = (i_1, \ldots, i_k)' \in \mathbb{N}_0^k$ and $\mathbf{s} = (s_1, \ldots, s_k)' \in \mathbb{R}_+^k$. Similar to the single-type case, the evolution of $\mathbf{Z}_j^{(n)} = (Z_{j,1}^{(n)}, \cdots Z_{j,k}^{(n)})'$ is described as follows. For any

$\alpha = 1,\ldots,k$, each of the $Z_{j,\alpha}^{(n)}$ type α particles alive at time j (if any) lives for one unit of time and then dies, giving rise to a number of offspring particles, represented by $\mathbf{l} = (l_1,\ldots,l_k)$, l_β being the number of type β offspring, with probability $\mu_\alpha^{(n)}(\mathbf{l})$. The particles behave independently of each other and of the past. The probability law of $\mathbf{Z}^{(n)}$ is given in terms of the pgf $\mathbf{F}^{(n)}(\mathbf{s}) := (F_{(1)}^{(n)}(\mathbf{s}),\ldots,F_{(k)}^{(n)}(\mathbf{s}))$, $\mathbf{s} \in C$, where

$$F_{(\alpha)}^{(n)}(\mathbf{s}) := \sum_{\mathbf{j}\in\mathbb{N}_0^k} \mu_\alpha^{(n)}(\mathbf{j})\mathbf{s}^{\mathbf{j}}, \quad 1 \le \alpha \le k.$$

Let $m_{\alpha\beta}^{(n)} = E_{\mathbf{e}_\alpha} Z_{1,\beta}^{(n)}$ be the expected number of type β offspring from a single particle of type α in one generation. Then the $k \times k$ matrix $\mathbf{M}^{(n)} = (m_{\alpha\beta}^{(n)})_{\alpha,\beta=1,\ldots,k}$ is called the *mean matrix* of $\mathbf{Z}^{(n)}$. Note that $m_{\alpha\beta}^{(n)} = \frac{\partial F_{(\alpha)}^{(n)}}{\partial s_\beta}(\mathbf{1})$, where the partial derivative is understood to be the left-hand derivative. The processes $\mathbf{Z}^{(n)}$ will be assumed to have a *uniformly strictly positive* mean matrix $\mathbf{M}^{(n)}$, by which we mean that there exist $U \in \mathbb{N}$ and $a \in (0,\infty)$ such that for every $n \ge 1$ $((\mathbf{M}^{(n)})^U)_{\alpha,\beta} \ge a$ for all $1 \le \alpha,\beta \le k$. From the Perron–Frobenius Theorem, it then follows that $\mathbf{M}^{(n)}$ has a real, positive maximal eigenvalue $\rho^{(n)}$ with associated positive left and right eigenvectors $\mathbf{v}^{(n)}$ and $\mathbf{u}^{(n)}$, respectively, which, without loss of generality, are normalized so that $\mathbf{u}^{(n)\prime}\mathbf{v}^{(n)} = 1$ and $\mathbf{u}^{(n)\prime}\mathbf{1} = 1$ (see [1]). The maximal eigenvalue $\rho^{(n)}$ plays a similar role in the classification of the k-BGW process as the mean played in classifying the (single type) BGW process. The k-BGW process is called subcritical, critical, or supercritical, according to whether $\rho^{(n)} < 1$, $\rho^{(n)} = 1$, or $\rho^{(n)} > 1$, respectively. We will consider the subcritical case, namely for all $n \ge 1$ $\rho^{(n)} \in (0,1)$, and study the behavior of quasi-stationary and Yaglom distributions of the scaled process

$$\hat{\mathbf{Z}}_t^{(n)} = \frac{1}{n}\mathbf{Z}_{\lfloor nt \rfloor}^{(n)}, \quad t \ge 0,$$

as $\rho^{(n)} \to 1$. The existence of the Yaglom distribution of $\hat{\mathbf{Z}}^{(n)}$ is assured by the following result.

Condition 2.3. For each $n \ge 1$, $E_{\mathbf{1}}(||\mathbf{Z}_1^{(n)}||\log||\mathbf{Z}_1^{(n)}||) < \infty$.

Theorem 2.9. *Assume Condition 2.3. For each $n \in \mathbb{N}$, $\hat{\mathbf{Z}}^{(n)}$ has a Yaglom distribution $\nu^{(n)}$. This distribution is also a qsd.*

Condition 2.4. There exist $b,d \in (0,\infty)$ such that for all $n \in \mathbb{N}$

(i) $\sum_{\alpha\beta\gamma} \partial^2 F_{(\alpha)}^{(n)}(\mathbf{1})/\partial s_\beta \partial s_\gamma \ge b$ and

(ii) $\sum_{\alpha,\beta,\gamma,\delta} \partial^3 F_{(\alpha)}^{(n)}(\mathbf{1})/\partial s_\beta \partial s_\gamma \partial s_\delta \le d$,

where $\alpha,\beta,\gamma,\delta$ in the above sums vary over $\{1,\ldots,k\}$.

Part (i) of the condition can be interpreted as a nondegeneracy condition, and part (ii) says that the third moments of the offspring distributions are uniformly bounded in n.

We now introduce a condition, analogous to Condition 2.1 (ii), for the multitype setting.

Condition 2.5. For some strictly positive matrix $\mathbf{M}$ and each $n \in \mathbb{N}$, $\mathbf{M}^{(n)} = \mathbf{M} + \frac{\mathbf{C}^{(n)}}{n}$, and $\lim_{n\to\infty} \mathbf{C}^{(n)} = \mathbf{C}$. The maximal eigenvalues $\rho^{(n)}$ of $\mathbf{M}^{(n)}$ are of the form $\rho^{(n)} = 1 + \frac{c^{(n)}}{n}$, with $c^{(n)} \in (-n, 0)$ and $\lim_{n\to\infty} c^{(n)} = c \in (-\infty, 0)$. Moreover, $\mathbf{M}$ has maximal eigenvalue 1 with corresponding eigenvectors $\mathbf{v} = \lim \mathbf{v}^{(n)}$ and $\mathbf{u} = \lim \mathbf{u}^{(n)}$. Finally, $\mathbf{v}'\mathbf{C}\mathbf{u} = c$.

Example 2.1. Let $\mathbf{C}^{(n)} = c^{(n)}\mathbf{I}$, where $\mathbf{I}$ is the identity matrix and $c^{(n)} \in (-n, 0)$ such that $c^{(n)} \to c \in (-\infty, 0)$. Let $\mathbf{M}$ be a strictly positive matrix with maximal eigenvalue equal to 1. Then $\mathbf{M}^{(n)} = \mathbf{M} + \frac{\mathbf{C}^{(n)}}{n}$ satisfies Condition 2.5.

Let

$$\kappa_{i,j}^{(n)}(l) := \sum_{\mathbf{r}\in\mathbb{N}_0^k} (r_i - m_{li}^{(n)})(r_j - m_{lj}^{(n)})\mu_l^{(n)}(\mathbf{r}).$$

We need the following condition on the variance of the offspring distributions.

Condition 2.6. As $n \to \infty$, $\kappa_{i,j}^{(n)}(l) \to \kappa_{i,j}(l)$ for all $1 \le i, j, l \le k$ and $Q := \frac{1}{2}\sum_{l=1}^k v_l \mathbf{u}'\boldsymbol{\kappa}(l)\mathbf{u} \in (0, \infty)$, where $\boldsymbol{\kappa}(l)$ is the matrix with (i, j)th entry $\kappa_{i,j}(l)$.

The following diffusion approximation result can be established along the lines of Theorem 4.3.1 of [8] and Theorem 9.2.1 of [5]. Let $C_c^\infty(\mathbb{R}_+)$ be the space of infinitely differentiable, real-valued functions with compact support on $\mathbb{R}_+$.

Theorem 2.10. *Assume Conditions 2.4, 2.5, and 2.6. Suppose that the distribution of* $\hat{\mathbf{Z}}_0^{(n)}$ *converges to some* $\mu_0 \in \mathcal{P}(\mathbb{R}_+^k)$. *Let* $\mu_1 \in \mathcal{P}(\mathbb{R}_+)$ *be given as*

$$\mu_1(A) = \mu_0\{\mathbf{x} \in \mathbb{R}_+^k | \mathbf{x}'\mathbf{u} \in A\}, \quad A \in \mathcal{B}(\mathbb{R}_+). \tag{7}$$

Let $\zeta^{(n)} = \hat{\mathbf{Z}}^{(n)\prime}\mathbf{u}^{(n)}$. *Then* $\zeta^{(n)}$ *converges weakly in* $D(\mathbb{R}_+ : \mathbb{R}_+)$ *to the unique (in law) diffusion* ζ *with initial distribution* μ_1 *and generator* $\tilde{L}$ *given as*

$$(\tilde{L}f)(x) = cxf'(x) + Qxf''(x), \quad f \in C_c^\infty(\mathbb{R}_+), \quad x \in \mathbb{R}_+. \tag{8}$$

Furthermore, for any $t_0 \in (0, \infty)$, *the process* $\hat{\mathbf{Z}}^{(n,t_0)}$, *defined by* $\hat{\mathbf{Z}}_t^{(n,t_0)} := \hat{\mathbf{Z}}_{t_0+t}^{(n)}$, $t \ge 0$, *converges weakly to* $\mathbf{Z}^{(t_0)} := \mathbf{v}\zeta^{(t_0)}$, *where* $\zeta_t^{(t_0)} := \zeta_{t_0+t}$, $t \ge 0$.

The process $\mathbf{Z}^{(t_0)}$ is a Markov process with state-space $S_{\mathbf{v}} = \{\theta\mathbf{v} | \theta \ge 0\}$ and can be formally regarded as the limit of $\hat{\mathbf{Z}}^{(n)}$. Indeed, if the support of μ_0 is contained in $S_{\mathbf{v}}$, then, noting that $\mathbf{u}'\mathbf{v} = 1$, we see that the law of $\mathbf{v}\zeta_0$ equals μ_0 and that in fact $\hat{\mathbf{Z}}^{(n)}$

converges weakly to $\mathbf{v}\zeta$, where ζ is as in Theorem 2.10. We will be concerned with the Yaglom distribution of the $S_\mathbf{v}$-valued Markov process $\mathbf{Z}^{(t_0)}$ and its relation to the Yaglom distribution of $\hat{\mathbf{Z}}^{(n)}$. For that it will be convenient to regard a probability measure on $S_\mathbf{v}$ as one on $\mathbb{R}^k_+$. Denote by $\tilde{\nu}$ the exponential distribution with density $f(x) = -cQ^{-1}\exp(cQ^{-1}x)$, $x \geq 0$. Theorem 2.2 says that the Yaglom distribution of $\zeta^{(t_0)}$ is given by $\tilde{\nu}$. Since $\mathbf{Z}^{(t_0)} = \mathbf{v}\zeta^{(t_0)}$, the Yaglom distribution of $\mathbf{Z}^{(t_0)}$ exists as well and equals the distribution of $\mathbf{v}Y$, where Y has distribution $\tilde{\nu}$. Thus, we have the following:

Theorem 2.11. *The Yaglom distribution of $\zeta^{(t_0)}$ exists and equals $\tilde{\nu}$. Furthermore, the Yaglom distribution of $\mathbf{Z}^{(t_0)}$, denoted by $\bar{\nu}$, exists and equals the distribution of $\mathbf{v}Y$, where Y has distribution $\tilde{\nu}$.*

The following result relates the qsd's and Yaglom distributions of $\hat{\mathbf{Z}}^{(n)}$ to that of its "diffusion limit" $\mathbf{Z}^{(t_0)}$. Probability distributions similar to $\bar{\nu}$ have previously been noted in the study of qsd's of multitype BGW processes. In [1] (p. 191), a single critical BGW process $\mathbf{Z}$ (rather than a sequence of near critical BGW processes) is considered and it is shown that $\mathbf{Z}_n/n$ conditioned on non-extinction converges to a random variable that is concentrated on the ray $\{x\mathbf{v_Z} | x \geq 0\}$, where $\mathbf{v_Z}$ is the left eigenvector of the mean matrix of $\mathbf{Z}$ corresponding to the eigenvalue 1. In [17] (see Theorem 3 therein) the case where $\mathbf{Z}$ is near critical and a somewhat differently (component wise) scaled process $\mathbf{Z}^*$ is considered. The asymptotic behavior of $\mathbf{Z}^*_n$ conditioned on non-extinction, as $n \to \infty$ and the offspring distribution approaches criticality, is related to the limiting distributions considered here. In fact, we use an estimate from [17] to prove Theorem 2.12, below. We remark that none of these results concern the setting of diffusion approximation, where time and space are scaled and one starts with a large number of particles.

Theorem 2.12. *The Yaglom distribution $\nu^{(n)}$ of $\hat{\mathbf{Z}}^{(n)}$ converges weakly to the Yaglom distribution $\bar{\nu}$ of $\mathbf{Z}^{(t_0)}$.*

3 Catalyst-Reactant Branching Processes with Controlled Immigration

The particles in the Bienaymé–Galton–Watson processes considered in the previous section evolved independently of each other. In this section, we consider catalytic branching processes that model the dynamics of catalyst-reactant populations in which the activity level of the reactant depends on the amount of catalyst present. Branching processes in catalytic environment have been studied extensively and are motivated, for instance by biochemical reactions (see [4,7,9,15], and references therein). A typical setting consists of populations of multiple types such that the rate of growth (depletion) of one population type is directly affected by population sizes of other types. The simplest such model consists of a continuous-time countable-state branching process describing the evolution of the catalyst population and a second branching process for which the branching rate is proportional to the total mass of the catalyst population modeling the evolution of reactant particles. Such processes were introduced in [4] in the setting of super-Brownian motions [15]. For classical catalyst-reactant branching processes, the catalyst population dies out with positive probability and subsequent to the catalyst extinction, the reactant population stays unchanged and therefore the population dynamics are modeled until the time the catalyst becomes extinct. In this work, we consider a setting where the catalyst population is maintained above a positive threshold through a specific form of controlled immigration. Branching process models with immigration have also been well studied in literature (see [1, 15] and references therein). However, typical mechanisms that have been considered correspond to adding an independent Poisson component (see, e.g., [10]). Here, instead, we consider a model where immigration takes place only when the population drops below a certain threshold. Roughly speaking, we consider a sequence $\{X^{(n)}\}_{n\in\mathbb{N}}$ of continuous-time branching processes, where $X^{(n)}$ starts with n particles. When the population drops below n, it is instantaneously restored to the level n.

There are many settings where controlled immigration models of the above form arise naturally. One class of examples arise from predator-prey models in ecology, where one may be concerned with the restoration of populations that are close to extinction by reintroducing species when they fall below a certain threshold. In our work, the motivation for the study of such controlled immigration models comes from problems in chemical reaction networks where one wants to keep the levels of certain types of molecules above a threshold in order to maintain a desired level of production (or inhibition) of other chemical species in the network. Such questions are of interest in the study of control and regulation of chemical reaction networks. A control action where one minimally adjusts the levels of one chemical type to keep it above a fixed threshold is one of the simplest regulatory mechanism and the goal of this research is to study system behavior under such mechanisms with the long-term objective of designing optimal control policies. The specific goal of the current work is to derive simpler approximate and reduced models, through the theory of diffusion approximations and stochastic averaging techniques, that are more tractable for simulation and mathematical treatment than the original branching process models. In order to keep the presentation simple, we consider the setting of one catalyst and one reactant. However similar limit theorems can be obtained for a more general chemical reaction network in which the levels of some of the chemical species are regulated in a suitable manner. Settings where some of the chemical species act as inhibitors rather than catalysts are also of interest and can be studied using similar techniques. These extensions will be pursued elsewhere.

Our main goal is to establish diffusion approximations for such regulated catalyst-reactant systems under suitable scalings. We consider two different scaling regimes; in the first setting the catalyst and reactant evolve on "comparable timescales," while in the second setting the catalyst evolves "much faster" than the reactant. In the former setting, the limit model is described through a coupled system of reflected stochastic differential equations with reflection in the space

$[1,\infty) \times \mathbb{R}$. The precise result (Theorem 3.2) is stated in Sect. 3.1. Such limit theorems are of interest for various analytic and computational reasons. It is simpler to simulate (reflected) diffusions than branching processes, particularly for large network settings. Analytic properties such as hitting time probabilities and steady-state behavior are more easily analyzed for the diffusion models than for their branching process counterparts. In general, such diffusion limits give parsimonious model representations and provide useful qualitative insight to the underlying stochastic phenomena.

For the second scaling regime, where the catalyst evolution is much faster, we establish a stochastic averaging limit theorem. The limit evolution of the reactant population is given through an autonomous one-dimensional SDE with coefficients that depend on the stationary distribution of a reflected diffusion in $[1,\infty)$. Such model reductions are important in that they not only help in better understanding the dynamics of the system but also help in reducing computational costs in simulations. Indeed, since in the model considered here the invariant distribution is explicit, the coefficients in the one-dimensional averaged diffusion model are easily computed and consequently this model is significantly easier to analyze and simulate than the original two-dimensional model. We refer the reader to [9] and references therein for similar results in the setting of (nonregulated) chemical reaction networks. It will be of interest to see if similar model reductions can be obtained for general multidimensional regulated chemical reaction networks.

3.1 Diffusion Limit under Comparable Timescales

Consider a sequence of pairs of continuous-time, countable-state Markov branching processes $(X^{(n)},Y^{(n)})$, where $X^{(n)}$ and $Y^{(n)}$ represent the number of catalyst and reactant particles, respectively. The dynamics are described as follows. Each of the $X_t^{(n)}$ particles alive at time t has an exponentially distributed life time with parameter $\lambda_1^{(n)}$ (mean life time $1/\lambda_1^{(n)}$). When it dies, each such particle gives rise to a number of offspring, according to the offspring distribution $\mu_1^{(n)}(\cdot)$. Additionally, if the catalyst population drops below n, it is instantaneously replenished back to the level n (*controlled immigration*). The branching rate of the reactant process $Y^{(n)}$ is of the order of the current total mass of the catalyst population, i.e., $X^{(n)}/n$, and we denote the offspring distribution of $Y^{(n)}$ by $\mu_2^{(n)}(\cdot)$. A precise definition of the pair $(X^{(n)},Y^{(n)})$ will be given below. We are interested in the study of asymptotic behavior of $(X^{(n)},Y^{(n)})$, under suitable scaling, as $n \to \infty$.

We now give a precise description of the various processes and the scaling that is considered. Roughly speaking, time is accelerated by a factor of n, and mass is scaled down by a factor of n. Define RCLL processes

$$\hat{\mathbf{W}}_t^{(n)} := \left(\hat{X}_t^{(n)}, \hat{Y}_t^{(n)}\right) := \left(\frac{X_{nt}^{(n)}}{n}, \frac{Y_{nt}^{(n)}}{n}\right), \quad t \in \mathbb{R}_+. \tag{9}$$

Let $\mathbb{S}_X^{(n)} := \{\frac{l}{n} | l \in \mathbb{N}_0\} \cap [1, \infty)$, $\mathbb{S}_Y^{(n)} := \{\frac{l}{n} | l \in \mathbb{N}_0\}$, $\mathbb{W}^{(n)} := \mathbb{S}_X^{(n)} \times \mathbb{S}_Y^{(n)}$, $\mathbb{W} = [1, \infty) \times \mathbb{R}_+$, and

$$D(\mathbb{R}_+ : S) := \{f : \mathbb{R}_+ \to S | f \text{ is right continuous and has left limits}\}$$

endowed with the usual Skorohod topology. Assume that $(\hat{X}_0^{(n)}, \hat{Y}_0^{(n)}) = (x_0^{(n)}, y_0^{(n)}) \in \mathbb{W}^{(n)}$. Then $\{\hat{\mathbf{W}}_t^{(n)}\}_{t \in \mathbb{R}_+}$ is characterized as the $\mathbb{W}^{(n)}$-valued Markov process with sample paths in $D(\mathbb{R}_+ : \mathbb{W}^{(n)})$, starting at $\hat{\mathbf{W}}_0^{(n)} = (x_0^{(n)}, y_0^{(n)})$, and having infinitesimal generator $\hat{\mathcal{A}}^{(n)}$ given as

$$\begin{aligned}\hat{\mathcal{A}}^{(n)} \phi(x,y) &= \lambda_1^{(n)} n^2 x \sum_{k=0}^{\infty} \left[\phi\left(\left(1 \vee x + \frac{k-1}{n}, y\right) - \phi(x,y)\right] \mu_1^{(n)}(k) \\ &\quad + \lambda_2^{(n)} n^2 xy \sum_{k=0}^{\infty} \left[\phi\left(x, y + \frac{k-1}{n}\right) - \phi(x,y)\right] \mu_2^{(n)}(k),\end{aligned} \tag{10}$$

where $(x,y) \in \mathbb{W}^{(n)}$ and $\phi \in \mathrm{BM}(\mathbb{W})$ with $\mathrm{BM}(\mathbb{W})$ being the space of bounded, measurable, real-valued functions $\mathbb{W}$. From the definition of the generator, we see that, for each $k \geq 0$, given $\hat{\mathbf{W}}_t^{(n)} = (x,y) \in \mathbb{W}^{(n)}$, the process jumps to $(x, y + \frac{k-1}{n})$ with rate $\lambda_2^{(n)} n^2 xy \mu_2^{(n)}(k)$ and to $(x + \frac{k-1}{n}, y)$ with rate $\lambda_1^{(n)} n^2 x \mu_1^{(n)}(k)$, except when $k = 0$ and $x = 1$, in which case the latter jump is to (x,y) with rate $\lambda_1^{(n)} n^2 \mu_1^{(n)}(0)$. This property of the generator at $x = 1$ accounts for the instantaneous replenishment of the (unscaled) catalyst population to level n, whenever the catalyst drops below n.

For $i = 1, 2$, let

$$m_i^{(n)} := \sum_{k=0}^{\infty} k \mu_i^{(n)}(k) \quad \text{and} \quad \alpha_i^{(n)} = \sum_{k=0}^{\infty} (k-1)^2 \mu_i^{(n)}(k).$$

We make the following basic assumption on the parameters of the branching rates and offspring distributions as well as on the initial configurations of the catalyst and reactant populations.

Condition 3.1. (i) For $i = 1, 2$ and for $n \in \mathbb{N}$, $\alpha_i^{(n)}, \lambda_i^{(n)} \in (0, \infty)$ and $m_i^{(n)} = 1 + \frac{c_i^{(n)}}{n}$, $c_i^{(n)} \in (-n, \infty)$.

(ii) For $i = 1, 2$, as $n \to \infty$, $c_i^{(n)} \to c_i \in \mathbb{R}$, $\alpha_i^{(n)} \to \alpha_i \in (0, \infty)$, and $\lambda_i^{(n)} \to \lambda_i \in (0, \infty)$.

(iii) For $i = 1, 2$ and for every $\varepsilon \in (0, \infty)$,

$$\lim_{n\to\infty} \sum_{l:l>\varepsilon\sqrt{n}} (l - m_i^{(n)})^2 \mu_i^{(n)}(l) = 0.$$

(iv) As $n \to \infty$, $(x_0^{(n)}, y_0^{(n)}) \to (x_0, y_0) \in [1,\infty) \times \mathbb{R}_+$.

Condition 3.1 and the form of the generator in (10) ensure that the scaled catalyst and reactant processes transition on comparable timescales, namely, $\mathcal{O}(n^2)$.

In order to state the limit theorem for $(\hat{X}^{(n)}, \hat{Y}^{(n)})$, we need some notation and definitions associated with the one-dimensional Skorohod map with reflection at 1. Let $D_1(\mathbb{R}_+ : \mathbb{R}) := \{f \in D(\mathbb{R}_+ : \mathbb{R}) | f(0) \geq 1\}$, and let $\Gamma : D_1(\mathbb{R}_+ : \mathbb{R}) \to D(\mathbb{R}_+ : [1,\infty))$ be defined as

$$\Gamma(\psi)(t) := (\psi(t) + 1) - \inf_{0\leq s\leq t} \{\psi(s) \wedge 1\}, \quad \text{for } \psi \in D(\mathbb{R}_+ : \mathbb{R}). \tag{11}$$

The function Γ, known as Skorohod map, can be characterized as follows: If $\psi, \phi, \eta^* \in D(\mathbb{R}_+ : \mathbb{R})$ are such that (i) $\psi(0) \geq 1$, (ii) $\phi = \psi + \eta^*$, (iii) $\phi \geq 1$, (iv) η^* is nondecreasing, $\int_{[0,\infty)} 1_{\{\phi(s)\neq 1\}} d\eta^*(s) = 0$, and $\eta^*(0) = 0$, then $\phi = \Gamma(\psi)$ and $\eta^* = \phi - \psi$. The process η^* can be regarded as the reflection term that is applied to the original trajectory ψ to produce a trajectory ϕ that is constrained to $[1,\infty)$. From the definition of the Skorohod map and using the triangle inequality, we get the following Lipschitz property: For $\psi, \tilde{\psi} \in D_1(\mathbb{R}_+ : \mathbb{R})$,

$$\sup_{s\leq t} |\Gamma(\psi)(s) - \Gamma(\tilde{\psi})(s)| \leq 2 \sup_{s\leq t} |\psi(s) - \tilde{\psi}(s)|. \tag{12}$$

Let

$$\hat{\eta}_t^{(n)} := \lambda_1^{(n)} n \mu_1^{(n)}(0) \int_0^t 1_{\{\hat{X}_s^{(n)}=1\}} ds. \tag{13}$$

This process will play the role of the reflection term, in the dynamics of the catalyst, arising from the controlled immigration. The diffusion limit of $(\hat{X}^{(n)}, \hat{Y}^{(n)})$ will be the process (X,Y) which is characterized in the following proposition through a system of stochastic integral equations.

The diffusion limit of $(\hat{X}^{(n)}, \hat{Y}^{(n)})$ will be the process (X,Y), starting at (x_0, y_0), which is given through a system of stochastic integral equations as in the following proposition.

Proposition 3.1. *Let $(\bar{\Omega}, \bar{\mathcal{F}}, \bar{P}, \{\bar{\mathcal{F}}_t\})$ be a filtered probability space on which are given independent standard $\{\bar{\mathcal{F}}_t\}$ Brownian motions B^X and B^Y. Let X_0, Y_0 be square integrable $\bar{\mathcal{F}}_0$ measurable random variables with values in $[1,\infty)$ and $\mathbb{R}_+$, respectively. Then the following system of stochastic integral equations has a unique strong solution:*

$$X_t = \Gamma\left(X_0 + \int_0^{\cdot} c_1 \lambda_1 X_s ds + \int_0^{\cdot} \sqrt{\alpha_1 \lambda_1 X_s} dB_s^X\right)(t), \tag{14}$$

$$Y_t = Y_0 + \int_0^t c_2\lambda_2 X_s Y_s ds + \int_0^t \sqrt{\alpha_2\lambda_2 X_s Y_s} dB_s^Y, \tag{15}$$

$$\eta_t = X_t - X_0 - \int_0^t c_1\lambda_1 X_s ds - \int_0^t \sqrt{\alpha_1\lambda_1 X_s} dB_s^X, \tag{16}$$

where Γ is the Skorohod map defined in (11).

In the above proposition, by a strong solution of (14)–(16) we mean an $\bar{\mathcal{F}}$-adapted continuous process (X,Y,η) with values in $[1,\infty)\times\mathbb{R}_+\times\mathbb{R}_+$ that satisfies (14)–(16). The following is the main result of this subsection.

Theorem 3.2. *Suppose Condition 3.1 holds. The process $(\hat{X}^{(n)},\hat{Y}^{(n)})$ converges weakly in $D(\mathbb{R}_+ : \mathbb{W})$ to the process (X,Y) given in Proposition 3.1 with $(X_0,Y_0) = (x_0,y_0)$.*

3.2 Asymptotic Behavior of the Catalyst Population

Stochastic averaging results in this work rely on understanding the time asymptotic behavior of the catalyst process. Such behavior, of course, is also of independent interest. We begin with the following result on the stationary distribution of X, where X is the reflected diffusion from Proposition 3.1, approximating the catalyst dynamics (Theorem 3.2). The proof uses an extension of the Echeverria criterion for stationary distributions of diffusions to the setting of constrained diffusions (see [2, 18] and references therein). We will make the following additional assumption. Recall the constants $c_1^{(n)} \in (-n,\infty)$ and $c_1 \in \mathbb{R}$ introduced in Condition 3.1.

Condition 3.2. For all $n \in \mathbb{N}$, $c_1^{(n)} < 0$ and $c_1 < 0$.

Proposition 3.2. *Suppose Condition 3.2 holds. The process X defined through* (14) *has a unique stationary distribution, ν_1, which has density*

$$p(x) := \begin{cases} \frac{\theta}{x}\exp(2\frac{c_1}{\alpha_1}x), & \text{if } x \geq 1 \\ 0, & \text{if } x < 1, \end{cases} \tag{17}$$

where $\theta := \left(\int_1^\infty (\frac{1}{x}\exp(2\frac{c_1}{\alpha_1}x))dx\right)^{-1}$.

The following result shows that the time asymptotic behavior of the catalyst population is well approximated by that of its diffusion approximation given through (14). We make the following additional assumption on the moment generating function of the offspring distribution, which will allow us to construct certain "uniform Lyapunov functions" that play a key role in the analysis.

Condition 3.3. For some $\bar{\delta} > 0$,

$$\sup_{n\in\mathbb{N}} \sum_{k=0}^{\infty} e^{\bar{\delta}k} \mu_1^{(n)}(k) < \infty. \tag{18}$$

Theorem 3.2. *Suppose Conditions 3.1, 3.2, and 3.3 hold. Then, for each $n \in \mathbb{N}$, the process $\hat{X}^{(n)}$ has a unique stationary distribution $\nu_1^{(n)}$, and the family $\{\nu_1^{(n)}\}_{n\in\mathbb{N}}$ is tight. As $n \to \infty$, $\nu_1^{(n)}$ converges weakly to ν_1.*

3.3 *Diffusion Limit of the Reactant under Fast Catalyst Dynamics*

As noted in Sect. 3.1, the catalyst and reactant populations whose scaled evolution is described through (10) transition on comparable timescales. In situations in which the catalyst evolves "much faster" than the reactant, one can hope to find a simplified model that captures the dynamics of the reactant population in a more economical fashion. One would expect that the reactant population can be approximated by a diffusion whose coefficients depend on the catalyst only through the catalyst's stationary distribution. Indeed, we will show that the (scaled) reactant population can be approximated by the solution of

$$\check{Y}_t = \check{Y}_0 + \int_0^t c_2\lambda_2 m_X \check{Y}_s ds + \int_0^t \sqrt{\alpha_2\lambda_2 m_X \check{Y}_s} dB_s, \quad \check{Y}_0 = y_0, \tag{19}$$

where $m_X = \int_1^\infty x\nu_1(dx) = -\frac{\alpha_1\theta}{2c_1}\exp(2c_1/\alpha_1)$.

Such model reductions (see [9] and references therein for the setting of chemical reaction networks) not only help in better understanding the dynamics of the system but also help in reducing computational costs in simulations. In this section, we will consider such stochastic averaging results in two model settings. First, in Sect. 3.3.1, we consider the simpler setting where the population mass evolutions are described through (reflected) stochastic integral equations and a scaling parameter in the coefficients of the model distinguishes the timescales of the two processes. In Sect. 3.3.2, we will consider a setting which captures the underlying physical dynamics more accurately in the sense that the mass processes are described in terms of continuous-time branching processes, rather than diffusions.

3.3.1 Stochastic Averaging in a Diffusion Setting

In this section, we consider the setting where the catalyst and reactant populations evolve according to (reflected) diffusions similar to X and Y from Proposition 3.1, but where the evolution of the catalyst is accelerated by a factor of a_n such that $a_n \uparrow \infty$ as $n \uparrow \infty$ (i.e. drift and diffusion coefficients are scaled by a_n). More precisely, we consider a system of catalyst and reactant populations that are given as solutions of the following system of stochastic integral equations: For $t \geq 0$,

$$\check{X}_t^{(n)} = \Gamma\left(\check{X}_0^{(n)} + \int_0^{\cdot} a_n c_1 \lambda_1 \check{X}_s^{(n)} ds + \int_0^{\cdot} \sqrt{a_n \alpha_1 \lambda_1 \check{X}_s^{(n)}} dB_s^X\right)(t)$$

$$\check{Y}_t^{(n)} = \check{Y}_0^{(n)} + \int_0^t c_2 \lambda_2 \check{X}_s^{(n)} \check{Y}_s^{(n)} ds + \int_0^t \sqrt{\alpha_2 \lambda_2 \check{X}_s^{(n)} \check{Y}_s^{(n)}} dB_s^Y,$$

where $(\check{X}_0^{(n)}, \check{Y}_0^{(n)}) = (x_0, y_0)$, $c_1, c_2 \in \mathbb{R}$, $\alpha_i, \lambda_i \in (0, \infty)$, B^X and B^Y are independent standard Brownian motions and Γ is the Skorohod map described above Proposition 3.1.

The following result says that if $c_1 < 0$, then the reactant population process $\check{Y}^{(n)}$, which is given through a coupled two-dimensional system, can be well approximated by the one-dimensional diffusion $\check{Y}$ in (19), whose coefficients are given in terms of the stationary distribution of the catalyst process.

Theorem 3.3. *Suppose Condition 3.2 holds. The process $\check{Y}^{(n)}$ converges weakly in $C(\mathbb{R}_+ : \mathbb{R}_+)$ to the process $\check{Y}$.*

3.3.2 Stochastic Averaging for Scaled Branching Processes

We now consider stochastic averaging for the setting where the catalyst and reactant populations are described through branching processes. Consider catalyst and reactant populations evolving according to the branching processes introduced in Sect. 3.1, but where the catalyst evolution is sped up by a factor of a_n such that $a_n \uparrow \infty$ monotonically as $n \uparrow \infty$. That is, we consider a sequence of catalyst populations $\tilde{X}_t^{(n)} := X_{a_n t}^{(n)}$, $t \geq 0$, where $X^{(n)}$ are the branching processes introduced in Sect. 3.1. The reactant population evolves according to a branching process, $\tilde{Y}^{(n)}$, whose branching rate, as before, is of the order of the current total mass of the catalyst population, $\tilde{X}^{(n)}/n$. The infinitesimal generator $\check{\mathcal{G}}^{(n)}$ of the scaled process

$$\left(\check{X}_t^{(n)}, \check{Y}_t^{(n)}\right) := \left(\frac{1}{n}\tilde{X}_{nt}^{(n)}, \frac{1}{n}\tilde{Y}_{nt}^{(n)}\right), \quad t \geq 0,$$

is given as

$$\check{\mathcal{G}}^{(n)}\phi(x,y) = \lambda_1^{(n)} n^2 a_n x \sum_{k=0}^{\infty} \left[\phi\left(1 \vee \left(x + \frac{k-1}{n}\right), y\right) - \phi(x,y)\right] \mu_1^{(n)}(k)$$
$$+ \lambda_2^{(n)} n^2 xy \sum_{k=0}^{\infty} \left[\phi\left(x, y + \frac{k-1}{n}\right) - \phi(x,y)\right] \mu_2^{(n)}(k), \qquad (20)$$

where $(x,y) \in \mathbb{S}_X^{(n)} \times \mathbb{S}_Y^{(n)}$ and $\phi \in \text{BM}([1,\infty) \times \mathbb{R}_+)$.

We note that a key difference between the generators $\check{\mathcal{G}}^{(n)}$ above and $\hat{\mathcal{A}}^{(n)}$ in (10) is the extra factor of a_n in the first term of (20), which says that, for large n, the catalyst dynamics are much faster than that of the reactant.

We will show in Theorem 3.4 that the reactant population process $\check{Y}^{(n)}$ can be well approximated by the one-dimensional diffusion $\check{Y}$ in (19). Once again, the result provides a model reduction that is potentially useful for simulations and also for a general qualitative understanding of reactant dynamics near criticality.

Theorem 3.4. *Suppose Conditions 3.1, 3.2, and 3.3 hold. Then, as $n \to \infty$, $\check{Y}^{(n)}$ converges weakly in $D(\mathbb{R}_+ : \mathbb{R}_+)$ to the process $\check{Y}$.*

Acknowledgements We gratefully acknowledge the valuable feedback from the referee which, in particular, led to a simplification of Condition 2.1.

References

1. K. B. Athreya and P. Ney. *Branching Processes*. Springer-Verlag, 1972.
2. A. Budhiraja and D. Reinhold. Near critical catalyst reactant branching processes with controlled immigration. *Annals of Applied Probability (to appear),* arXiv:1203.6879 [math.PR].
3. A. Budhiraja and D. Reinhold. Some asymptotic results for near critical branching processes. *Communications on Stochastic Analysis*, Vol. 4, No. 1:91–113, 2010.
4. D. A. Dawson and K. Fleischmann. A continuous super-brownian motion in a super-brownian medium. *Journal of Theoretical Probability*, 10, No. 1:213–276, 1997.
5. S. N. Ethier and T. G. Kurtz. *Markov Processes: Characterization and Convergence*. Wiley, 1986.
6. W. Feller. Diffusion processes in genetics. *Proc. Second Berkeley Symp. on Math. Statist. and Prob.*, pages 227–246, 1951.
7. A. Greven, L. Popovic, and A. Winter. Genealogy of catalytic branching models. *The Annals of Applied Probability*, 19:1232–1272, 2009.
8. A. Joffe and M. Metivier. Weak convergence of sequences of semimartingales with applications to multitype branching processes. *Advances in Applied Probability*, Vol. 18:pp. 20–65, 1986.
9. H.-W. Kang and T. G. Kurtz. Separation of time-scales and model reduction for stochastic reaction networks. *Annals of Applied Probability (to appear),* arXiv:1011.1672 [math.PR].
10. S. Karlin. *A First Course in Stochastic Processes*. Academic Press, 1966.
11. A. Lambert. Quasi-stationary distributions and the continuous branching process conditioned to be never extinct. *Electronic Journal of Probability*, 12:420–446, 2007.
12. A. Lambert. Population dynamics and random genealogies. *Stochastic Models*, 24:45–163, 2008.
13. J. Lamperti and P. Ney. Conditioned branching processes and their limiting diffusions. *Theory of Probbility and its Applications*, 13:128–139, 1968.
14. Z. Li. A limit theorem for discrete Galton-Watson branching processes with immigration. *Journal of Applied Probability*, 43:289–295, 2006.
15. Z. Li and C. Ma. Catalytic discrete state branching models and related limit theorems. *Journal of Theoretical Probability*, 21:936–965, 2008.
16. Z.-H. Li. Asymptotic behaviour of continuous time and state branching processes. *J. Austral. Math. Soc.*, (Series A) 68:68–84, 2000.
17. M. P. Quine. The multitype Galton-Watson process with ρ near 1. *Advances in Applied Probability*, 4:pp.429–452, 1972.
18. D. Reinhold. *Asymptotic Behavior of Near Critical Branching Processes and Modeling of Cell Growth Data*. PhD thesis, University of North Carolina at Chapel Hill, NC, USA, 2011.

Some Path Large-Deviation Results for a Branching Diffusion

Robert Hardy and Simon C. Harris

Abstract We give an intuitive proof of a path large-deviations result for a typed branching diffusion as found in Git, J.Harris and S.C.Harris (Ann. App. Probab. 17(2):609-653, 2007). Our approach involves an application of a change of measure technique involving a distinguished infinite line of descent, or *spine*, and we follow the spine set-up of Hardy and Harris (Séminaire de Probabilités XLII:281–330, 2009). Our proof combines simple martingale ideas with applications of Varadhan's lemma and is successful mainly because a "spine decomposition" effectively reduces otherwise difficult calculations on the whole collection of branching diffusion particles down to just a single particle (the spine) whose large-deviations behaviour is well known. A similar approach was used for branching Brownian motion in Hardy and Harris (Stoch. Process. Appl. 116(12):1992–2013, 2006). Importantly, our techniques should be applicable in a much wider class of branching diffusion large-deviations problems.

Keywords Branching diffusions • Spatial branching process • Path large deviations • Spine decomposition • Spine change of measure • Additive martingales

AMS subject classification: 60J80.

R. Hardy
Department of Mathematical Sciences, University of Bath, Bath, BA2 7AY, UK
Currently at VTB Capital, London
e-mail: Robert.Hardy@vtbcapital.com

S.C. Harris (✉)
Department of Mathematical Sciences, University of Bath, Bath, BA2 7AY, UK
e-mail: S.C.Harris@bath.ac.uk

J. Englander and B. Rider (eds.), *Advances in Superprocesses and Nonlinear PDEs*, Springer Proceedings in Mathematics & Statistics 38,
DOI 10.1007/978-1-4614-6240-8_5,

1 Overview

Harris and Williams [7] introduced a model of a branching diffusion in which the diffusion and breeding rate of particles is controlled by their type process which moves as an Ornstein–Uhlenbeck process on $\mathbb{R}$, independently of the particle's position, associated with the generator

$$Q_\theta := \frac{\theta}{2}\Big(\frac{\partial^2}{\partial y^2} - y\frac{\partial}{\partial y}\Big), \qquad \text{with } \theta > 0 \text{ considered as the } \textit{temperature}. \qquad (1)$$

Throughout this chapter we shall refer to an OU process with generator $\frac{\theta}{2}\frac{\partial^2}{\partial y^2} - \mu y \frac{\partial}{\partial y}$ as an $\mathrm{OU}(\theta,\mu)$.

More precisely, the spatial movement of a particle of type y is a driftless Brownian motion with instantaneous variance

$$A(y) := ay^2, \qquad \text{for some fixed } a \geq 0.$$

The breeding of a particle of type y occurs at a rate

$$R(y) := ry^2 + \rho, \qquad \text{where } r, \rho > 0,$$

and binary splitting occurs at the fission times. The model has very different behaviour for low-temperature values (i.e., low θ), but throughout we consider that $\theta > 8r$—the high-temperature regime.

We can suppose that the probabilities of this are $\{P^{x,y} : x,y \in \mathbb{R}\}$ so that $P^{x,y}$ is a measure defined on the natural filtration $(\mathcal{F}_t)_{t\geq 0}$ such that it is the law of this branching diffusion process initiated from a single particle positioned at the space-type location (x,y). The configuration of this branching diffusion at time t is to be given by the $\mathbb{R}^2$-valued point process $\mathbb{X}_t := \{(X_u(t), Y_u(t)) : u \in N_t\}$ where N_t is the set of individuals alive at time t, and without loss of generality, we can assume that the initial ancestor starts out at the space-type origin—henceforth, we use P to mean $P^{0,0}$.

The main aim of this chapter is to prove upper and lower bounds for the probability of finding at least one of the branching particles very far from the space-type origin at large times, in a suitable large-deviations sense. The question of the lower bound was originally motivated by Git et al. [4] in order to determine the exponential growth rates and asymptotic shape of the branching diffusion. We briefly discuss this result in Sect. 2. In fact, our approach naturally gives rise to a stronger result where particles not only arrive at a very large space-type location $(\beta t, \kappa\sqrt{t})$ at a fixed time τ but are also known to have stayed "near" a specific space-type trajectory throughout the whole time interval $[0,\tau]$. The spine techniques used in this chapter involve a change of measure that makes a single "spine" particle "follow" a given trajectory. Our "spine" methods (for both bounds) should also prove useful to obtain large-deviations results in more general branching diffusions.

Although Git et al. [4] also used a spine change of measure, their original approach was far more model specific and quite different in flavour.

With $\theta > 8r$, let

$$\overline{\lambda} := \sqrt{\frac{\beta^2\theta(\theta-8r)}{a^2\kappa^4 + 4a\theta\beta^2}}, \qquad \overline{\mu} := \frac{\kappa^2\sqrt{\theta(\theta-8r)}}{2\sqrt{\kappa^4+4\theta\beta^2/a}} \tag{2}$$

and define a space-type trajectory $(x_s, y_s)_{s\in[0,\tau]}$ by

$$\overline{y}_s := \kappa\frac{\sinh\overline{\mu}s}{\sinh\overline{\mu}\tau}, \qquad \overline{x}_s := a\overline{\lambda}\int_0^s y_w^2\,\mathrm{d}w, \qquad s\in[0,\tau]. \tag{3}$$

Note that the path endpoints are $y_\tau = \kappa$ and $x_\tau = \beta$. Also define,

$$\Theta(\beta,\kappa) := \frac{\kappa^2}{4} + \frac{\sqrt{\theta(\theta-8r)(a^2\kappa^4+4\theta\beta^2)}}{4a\theta}. \tag{4}$$

Theorem 1.1 (The Short-Climb Probability). *Let $\beta < 0$, $\kappa\in\mathbb{R}$ and $\varepsilon > 0$.*
(a) If $\tau > 0$ is sufficiently large, then for all $\delta, \delta' > 0$

$$\liminf_{t\to\infty} t^{-1}\log P\Big(\exists u\in N_\tau : \forall s\in[0,\tau], \big|t^{-1}X_u(s)-\overline{x}_s\big| < \delta, \big|t^{-\frac{1}{2}}Y_u(s)-\overline{y}_s\big| < \delta'\Big)$$
$$\geq -\Theta(\beta,\kappa)-\varepsilon.$$

(b) If $\delta, \delta' > 0$ are sufficiently small, then for all $\tau > 0$

$$\limsup_{t\to\infty} t^{-1}\log P\Big(\exists u\in N_\tau : \forall s\in[0,\tau], \big|t^{-1}X_u(s)-\overline{x}_s\big| < \delta, \big|t^{-\frac{1}{2}}Y_u(s)-\overline{y}_s\big| < \delta'\Big)$$
$$\leq \limsup_{t\to\infty} t^{-1}\log P\Big(\exists u\in N_\tau : \big|t^{-1}X_u(\tau)-\beta\big| < \delta, \big|t^{-\frac{1}{2}}Y_u(\tau)-\kappa\big| < \delta'\Big)$$
$$\leq -\Theta(\beta,\kappa)+\varepsilon.$$

Note that the above new upper bound result shows that the trajectory followed really is "optimal" in order to achieve the required large position at time τ. We will actually prove a more general result for *any* (as opposed to sufficiently large) fixed time τ with a rate of decay $J(\tau)$, where $J(\tau)\downarrow\theta(\beta,\kappa)$. We state this stronger result as Theorem 3.2. We also note that the above lower bound on the probability of following the optimal paths to reach $(\beta t, \kappa\sqrt{t})$ is for a large *fixed* time τ. In contrast, the method used for the analogous result in [4, Theorem 7] dictated that $\tau = \tau(t) \propto \log t$. Our stronger result would enable a corresponding simplification in the proof of the application in [4].

The principle behind the proof of the lower bound is to design new measures $\mathbb{Q}_t$ for the branching diffusion such that one of the particles (the spine) will closely follow a specific space-type path. Our spine approach, which we briefly lay out in Sect. 4 (and which is fully presented in Hardy and Harris [5]), will allow us to explicitly find the Radon–Nikodym derivatives (martingales) of these new measures with respect to the original measure P. Then, using the spine decomposition together with Doob's submartingale inequality, we shall show that the growth rate of these martingales under $\mathbb{Q}_t$ is exactly the correct rate for the large-deviations lower bound.

In such branching diffusion settings, we comment that a large-deviations upper bound is usually easier to obtain than the lower bound. Generally, we can overestimate the probability that any particle succeeds in performing a certain "rare" event by the expected number of particles performing that event, which then reduces to a single-particle (large deviation) calculation. In the present context, the upper bound of Theorem 1.1 can also be proved directly using some fundamental "additive" martingales.

The layout of this article is as follows: in the next section we discuss the results of Git et al. [4] in order to give a context to our work. Section 3 contains a heuristic discussion of the large deviations for the model which motivates the choice of the subsequent martingales. A statement of a stronger path large-deviations result (Theorem 3.2) is also found in this section. In Sect. 4, we briefly present the foundations of our spine approach, giving definitions of the underlying space and its filtrations, measures and the fundamental martingales of interest. These strictly positive martingales, Z_t, are defined in terms of specific paths as suggested by our heuristic arguments. As Radon–Nikodym derivatives, these martingales can define the new measures $\mathbb{Q}_t$ (under which they become *sub*martingales), and we state a key result (Theorem 4.3) on their growth under the measures $\mathbb{Q}_t$; this growth result leads directly to the proof of the large-deviations lower bounds which we present in Sect. 5. Section 6 contains proofs for the upper bounds of the two main large-deviations results. Section 7 is devoted to proving the martingale growth Theorem 4.3; the proof is not particularly short, but neither is it difficult given the spine technology. It should be noted that this result is the main application of spines in this chapter. We use the so-called *spine decomposition* to simplify a computation involving the branching particle martingale Z_t into a computation involving a single "spine" particle that permits a standard application of Varadhan's lemma.

2 The Git et al.'s Almost-Sure Result

We first give some key parameter definitions for this model. Let

$$\lambda_{\min} := -\sqrt{\frac{\theta - 8r}{4a}}. \tag{5}$$

Let $\lambda \in \mathbb{R}$, with the following convention which we *always* use for λ:

$$\lambda_{\min} < \lambda < 0. \tag{6}$$

Also, define

$$\mu_\lambda := \frac{1}{2}\sqrt{\theta(\theta - 8r - 4a\lambda^2)} \qquad \psi_\lambda^\pm := \frac{1}{4} \pm \frac{\mu_\lambda}{2\theta} \qquad E_\lambda^\pm := \rho + \theta\psi_\lambda^\pm, \quad c_\lambda^\pm := -E_\lambda^\pm/\lambda \tag{7}$$

Note, $\lambda_{\min}$ is the point beyond which μ_λ is no longer a real number. The parameters $E_\lambda^\pm \in \mathbb{R}$ are in fact certain key eigenvalues, as will be described in the next section.

Before we move on to prove the above theorem, we summarise the main results from Git et al. [4] and the earlier Harris and Williams [7] so that the reader might understand how Theorem 1.1 fits into the picture.

Work on large-deviations results for this type of branching diffusion began in the paper by Harris and Williams [7], where they considered the behaviour in expectation of the counting function

$$N_t(\gamma) = \sum_{u \in N_t} \mathbf{1}_{(X_u(t) \le -\gamma t)}$$

(not to be confused with N_t, the set of individuals alive at time t) for each $\gamma \in \mathbb{R}$. In Git *et al.* [4] it was shown that

$$\lim_{t\to\infty} t^{-1} \log N_t(\gamma) = \Delta(\gamma)$$

exists almost surely and is *finite* for all $0 \le \gamma < \tilde{c}(\theta)$, for some constant $\tilde{c}(\theta)$; in the case that $\gamma \ge \tilde{c}(\theta)$ the limit is $-\infty$ since *no* particles will be as far out as the ray $-\gamma t$ at large times. In other words, this result says that we almost surely have exponential growth in numbers of particles following close to rays that are *not too steep*.

For later reference, the almost-sure growth rate is given explicitly by

$$\Delta(\gamma) = \inf_{\lambda \in (\lambda_{\min}, 0)} \{E_\lambda^- + \lambda\gamma\} = \rho + \frac{\theta}{4} - \frac{1}{4}\sqrt{\theta(\theta - 8r)(1 + 4\gamma^2/(\theta a))},$$

and it is also found that

$$\tilde{c}(\theta) := \sup\{\gamma : \Delta(\gamma) > 0\} = \sqrt{2a\left(r + \rho + \frac{2(2r+\rho)^2}{\theta - 8r}\right)}.$$

In fact, the work of Git et al. [4] improves this to obtain the almost-sure rate of growth in numbers of particles at certain *spatial and type* positions at large times. They study the following function that counts how many particles occupy a particular region in the type-space domain:

$$N_t(\gamma, \kappa) := \sum_{u \in N(t)} \mathbf{1}\{X_u(t) \le -\gamma t, Y_u(t)^2 \ge \kappa^2 t\}.$$

Theorem 2.1. *Under each $P^{x,y}$ law, the limit*

$$D(\gamma, \kappa) := \lim_{t\to\infty} t^{-1} \log N_t(\gamma, \kappa)$$

exists almost surely and is given by

$$D(\gamma,\kappa) = \begin{cases} \Delta(\gamma,\kappa) & \text{if } \Delta(\gamma,\kappa) > 0, \\ -\infty & \text{otherwise.} \end{cases}$$

Here,

$$\Delta(\gamma,\kappa) = \inf_{\lambda \in (\lambda_{min},0)} \{E_\lambda^- + \lambda\gamma - \kappa^2 \psi_\lambda^+\},$$
$$= \rho + \frac{\theta - \kappa^2}{4} - \frac{1}{4a\theta}\sqrt{\theta(\theta - 8r)(4a\theta\gamma^2 + a^2(\theta + \kappa^2)^2)}. \tag{8}$$

2.1 The Almost-Sure Upper Bound

Harris and Williams [7] and Harris [8] showed that there are *two* strictly positive martingales Z_λ^- and Z_λ^+ defined as

$$Z_\lambda^\pm(t) := \sum_{k=1}^{N(t)} v_\lambda^\pm(Y_k(t)) e^{\lambda X_k(t) - E_\lambda^\pm t}, \tag{9}$$

where v_λ^- and v_λ^+ are strictly positive eigenfunctions of the self-adjoint operator $\frac{1}{2}\lambda^2 A + R + Q_\theta$, with corresponding eigenvalues $E_\lambda^- < E_\lambda^+$. The explicit form for these eigenfunctions is

$$v_\lambda^\pm(y) = e^{\psi_\lambda^\pm y^2}$$

where $\psi_\lambda^\pm$ and μ_λ are given at (7) and $\psi_\lambda^\pm$ are both positive for all $\lambda \in (\lambda_{\min}, 0)$.

A useful trick to obtain upper bounds is to overestimate indicator functions by an exponential and optimise over parameters. It is often the case that this will bring in one of the martingales of the model: for $\lambda \in (\lambda_{\min}, 0)$,

$$\sum_{k=1}^{N(t)} \mathbf{1}\{X_k(t) \le -\gamma t, Y_k(t)^2 \ge \kappa^2 t\} \le \sum_{k=1}^{N(t)} \exp\{\psi_\lambda^+ (Y_k(t)^2 - \kappa^2 t)\} \exp\{\lambda(X_k(t) + \gamma t)\}$$
$$= e^{-\lambda(c_\lambda^+ - c_\lambda^-)t} Z_\lambda^+(t) e^{(E_\lambda^- + \lambda\gamma - \kappa^2\psi_\lambda^+)t}. \tag{10}$$

(Importantly for this, the parameter ψ_λ^+ is positive and λ is negative; the functions c_λ^- and c_λ^+ are defined as $c_\lambda^\pm := E_\lambda^\pm/(-\lambda)$.)

The expression for $\Delta(\gamma,\kappa)$ as a Legendre conjugate—see (8)—explains why $\Delta(\gamma,\kappa)$ relates to (10) above: by choosing λ at the infimum we get

$$N_t(\gamma,\kappa) \le e^{-\lambda(c_\lambda^+ - c_\lambda^-)t} Z_\lambda^+(t) e^{\Delta(\gamma,\kappa)t}. \tag{11}$$

We remember that $N_t(\gamma,\kappa)$ takes only integer values, and a separate theorem by Harris and Git states that

$$\limsup_{t\to\infty} e^{-\lambda(c_\lambda^+ - c_\lambda^-)t} Z_\lambda^+(t) \leq 0, \qquad \text{for each } \lambda \in (\lambda_{\min}, 0). \tag{12}$$

Thus if $\Delta(\gamma,\kappa) < 0$ we deduce that almost surely

$$N_t(\gamma,\kappa) = 0, \quad \text{eventually,}$$

whence $\lim_{t\to\infty} t^{-1} \log N_t(\gamma,\kappa) = -\infty$, as required.

On the other hand, if $\Delta(\gamma,\kappa) \geq 0$, (11) and (12) immediately imply that

$$\lim_{t\to\infty} t^{-1} \log N_t(\gamma,\kappa) \leq \Delta(\gamma,\kappa).$$

2.2 *A Two-Phase Mechanism for the Lower Bound*

For their proof of the almost-sure lower bound of Theorem 2.1, Git et al. [4] propose an explicit mechanism by which a sufficient number of particles will obtain a position near $(\gamma T, \kappa\sqrt{T})$ in the type-space domain at large times T. It is made up of two phases:

The long tread: Over a long period $[0,t]$, taking up nearly all of the time, a large number of particles will drift spatially with speed $\gamma\theta/(\theta+\kappa^2)$—as if their type has had a modified occupation measure, as described by Harris and Williams [7];
The short climb: Following this, over a short period of time $[t, t+\tau]$ with τ a fixed time ($\tau \ll t$), each of the particles from this group will have a small probability of further rushing to the large type position $\kappa\sqrt{t}$ whilst additionally gaining $\{\gamma\kappa^2/(\theta+\kappa^2)\}t$ in spatial position.The combination of these two phases will present us with *sufficiently many* particles at the space-type position $\left(\gamma T, \kappa\sqrt{T}\right)$ at the large time $T = t+\tau$, as Git et al. [4] show in their proof of the lower bound of Theorem 2.1 – we refer the reader to their work for further details. This lower bound requires a substantial amount of technical work, mainly focussed on the short climb in which, in particular, they required $\tau = O(\log t)$. Our Theorem 1.1 includes an alternative *short-climb* lower bound for τ fixed, and using this would slightly simplify the combination of phases. Our current proof will also provide a cleaner, more intuitive and more generic approach to such path large-deviations results in branching diffusions.

3 Large Deviations Heuristics

We now present some heuristic arguments concerning the large-deviations behaviour of the branching diffusion which will serve as the *intuition* behind our later *rigourous* proofs.

Under a measure $\tilde{P}$ let (ξ_s, η_s) satisfy

$$\mathrm{d}\eta_s = \sqrt{\theta}\mathrm{d}B_s - \frac{\theta}{2}\eta_s\mathrm{d}s, \quad \text{and} \quad \mathrm{d}\xi_s = \sqrt{a}\eta_s\,\mathrm{d}W_s,$$

for two independent $\tilde{P}$-Brownian motions B_s and W_s. Under $\tilde{P}$, (ξ_s, η_s) moves like a single particle within the branching diffusion. For a large-deviations analysis we observe that for any $t > 0$,

$$\mathrm{d}\left(\frac{\eta_s}{\sqrt{t}}\right) = \sqrt{\theta}\left(\frac{\mathrm{d}B_s}{\sqrt{t}}\right) - \frac{\theta}{2}\left(\frac{\eta_s}{\sqrt{t}}\right)\mathrm{d}s, \quad \text{and} \quad \mathrm{d}\left(\frac{\xi_s}{t}\right) = \sqrt{a}\left(\frac{\eta_s}{\sqrt{t}}\right)\left(\frac{\mathrm{d}W_s}{\sqrt{t}}\right),$$

and it will be natural to work with the *rescaled* processes $(\xi_s/t, \eta_s/\sqrt{t})$ since in this way we obtain a variance coefficient of $1/\sqrt{t}$ on the driving Brownian motions.

Definition 3.1. For each $t > 0$ we define the *rescaled* single-particle motion (ξ_s^t, η_s^t) by

$$\xi_s^t := \xi_s/t, \quad \text{and} \quad \eta_s^t := \eta_s/\sqrt{t},$$

We note that under $\tilde{P}$ we have for $s \in [0, \tau]$:

$$\mathrm{d}\eta_s^t = \frac{\sqrt{\theta}}{\sqrt{t}}\mathrm{d}B_s - \frac{\theta}{2}\eta_s^t\mathrm{d}s, \quad \text{and} \quad \mathrm{d}\xi_s^t = \frac{\sqrt{a}\eta_s^t}{\sqrt{t}}\mathrm{d}W_s,$$

for two independent $\tilde{P}$-Brownian motions B_s and W_s.

Throughout the remainder of this chapter and different from the earlier parts, the variable t will not be a time parameter but will bring about this large-deviations scaling; typically we shall use either w or s to denote the time parameter from the time interval $[0, \tau]$ where $\tau > 0$ is considered as fixed.

Suppose that we are given two paths: a type path $y : [0, \tau] \to \mathbb{R}$ and a spatial path $x : [0, \tau] \to \mathbb{R}$. On a *heuristic* level we can say that the probability of the type diffusion η_s^t closely following y and the space diffusion ξ_s^t closely following x is roughly

$$\exp\left(-\frac{t}{2\theta}\int_0^\tau \left(\dot{y}_s + \frac{\theta}{2}y_s\right)^2 \mathrm{d}s - \frac{t}{2}\int_0^\tau \frac{\dot{x}_s^2}{ay_s^2}\,\mathrm{d}s\right), \tag{13}$$

for large enough t. See [2], for example, for the large-deviations theory of Wentzell–Freidlin.

The reader who is familiar with the large-deviations principle for branching Brownian motion (see Hardy and Harris [6] for a spine proof or Lee [14] for a classical proof) might guess that the *probability at least one* of the *rescaled* branching particles $(X_u(s)/t, Y_u(s)/\sqrt{t})$ follows the *difficult* space-type path (x_s, y_s) 'closely' over the time interval $[0, \tau]$ is roughly

$$\exp\left\{-\sup_{w\in[0,\tau]}\left[\left(\int_0^w \frac{1}{2\theta}\left(\dot{y}_s + \frac{\theta}{2}y_s\right)^2 + \frac{1}{2}\frac{\dot{x}_s^2}{ay_s^2} - ry_s^2\,\mathrm{d}s\right)t - \rho w\right]\right\},$$

when t is large. This *guess* can be obtained by upper estimating the *probability* by the *expected number of particles* following the path, then using the "one-particle picture" of Sect. 4 and large-deviations theory for one particle before optimising by choosing the most difficult time along the path.

By standard optimisation arguments (Git et al. [4] give some details of how this can be carried out) this implies that the probability of at least one of the rescaled branching particles being near the space-type position (β, κ) at a fixed time τ (which is also the event that the non-rescaled particles arrive near $(\beta t, \kappa\sqrt{t})$ of course) should be roughly

$$\exp\left\{-\inf_{x,y}\sup_{w\in[0,\tau]}\left[\left(\int_0^w \frac{1}{2\theta}\left(\dot{y}_s+\frac{\theta}{2}y_s\right)^2+\frac{1}{2}\frac{\dot{x}_s^2}{ay_s^2}-ry_s^2\,\mathrm{d}s\right)t-\rho w\right]\right\}, \tag{14}$$

when t is large and where the infimum is taken over all paths $x, y \in C[0,\tau]$ satisfying

$$y(0)=0, y(\tau)=\kappa, x(0)=0, x(\tau)=\beta. \tag{15}$$

This is typical in a large-deviations setting: although there are many possible trajectories that the (rescaled) particles could travel along to get to a position (β, κ), the *dominant number* will have followed *optimal* paths.

Git et al. [4] state that for any given type path y, the optimal space path x for (14) under the constraint $x(\tau) = \beta$ will always be given by

$$x_s = \lambda \int_0^s ay_w^2\,\mathrm{d}w, \qquad \text{for } s \in [0,\tau], \tag{16}$$

for some value $\lambda \in \mathbb{R}$. Briefly, their arguments rely on the fact that in the definition of our model the spatial diffusion $X_u(s)$ of the branching particles can be seen as a time-changed Brownian motion where the time scaling is determined by its type process $Y_u(s)$:

$$X_u(s) = \hat{B}\Big(\int_0^s aY_u(w)^2\,\mathrm{d}w\Big)$$

for a Brownian motion $\hat{B}(\cdot)$ on $[0,\tau]$. A measure change that introduces a linear drift of λ to this Brownian motion will give

$$X_u(s) = \tilde{B}\Big(\int_0^s aY_u(w)^2\,\mathrm{d}w\Big) + \lambda\int_0^s aY_u(w)^2\,\mathrm{d}w,$$

where $\tilde{B}(\cdot)$ is a Brownian motion under the new measure—this clearly relates to (16). Linear drifts are the optimal path (in a large-deviations sense) for a Brownian motion to be at a given point at a given time, and the constraint $x(\tau) = \beta$ for our problem will determine the value of λ in terms of the type path y:

$$\lambda = \frac{\beta}{a\int_0^\tau y_s^2\,\mathrm{d}s}. \tag{17}$$

Thus for the event being considered in Theorem 1.1, the optimal spatial path x is determined *uniquely* by (16) together with (17). Therefore, an equivalent but easier statement of our large-deviations result is that the probability of at least one of the rescaled branching particles being near the space-type position (β, κ) at a fixed time τ is roughly

$$\exp\left\{ -\inf_{y} \sup_{w\in[0,\tau]} \left[\left(\int_0^w \frac{1}{2\theta}\left(\dot{y}_s + \frac{\theta}{2}y_s\right)^2 + \frac{a\lambda^2}{2}y_s^2 - ry_s^2\,\mathrm{d}s \right) t - \rho w \right] \right\}, \tag{18}$$

when t is large and where the infimum is taken over all paths $y \in C[0,\tau]$ and all $\lambda \in (\lambda_{\min}, 0)$ satisfying

$$y(0) = 0, y(\tau) = \kappa, \quad \lambda = \frac{\beta}{a\int_0^\tau y_s^2\,\mathrm{d}s}. \tag{19}$$

Git et al. [4] presented alternative heuristic arguments based on birth-death processes to arrive at the expression (18). Using Euler–Lagrange techniques, they showed that the specific path

$$y_s = \kappa \frac{\sinh \mu_\lambda s}{\sinh \mu_\lambda \tau}, \quad s \in [0,\tau] \tag{20}$$

is optimal for this expression, where

$$\mu_\lambda = \frac{\sqrt{\theta(\theta - 8r - 4a\lambda^2)}}{2}, \tag{21}$$

and $\lambda \in (\lambda_{\min}, 0)$ is dependent on the choice of τ (which we are anyway considering as fixed throughout) and is chosen to satisfy

$$\frac{\beta}{a\lambda} = \kappa^2 \left(\frac{\coth \mu_\lambda \tau}{\mu_\lambda} - \frac{\tau}{2\sinh^2 \mu_\lambda \tau} \right). \tag{22}$$

We refer the reader to Git et al. [4] for details of these relationships between the parameters, but note that particles staying close to this path will arrive near $y(\tau) = \kappa$ at time τ in agreement with the heuristics.

As we mentioned just before the statement of Theorem 1.1, our spine techniques will naturally use the path y_s defined at (20) together with x_s defined at (16), since they are the optimal paths (in a large-deviations sense) for accumulating particles near the point $(\beta t, \kappa\sqrt{t})$ at time τ. In fact, our spine proof of Theorem 1.1 will result in a proof of the following stronger result, from which Theorem 1.1(a) would follow as a corollary.

Theorem 3.2. *Let $\tau > 0$ be fixed and suppose $\beta < 0$ and $\kappa \in \mathbb{R}$. Define two paths on $[0,\tau]$ by*

$$y_s := \kappa \frac{\sinh \mu_\lambda s}{\sinh \mu_\lambda \tau}, \quad x_s := a\lambda \int_0^s y_w^2 \, \mathrm{d}w, \quad s \in [0,\tau], \tag{23}$$

where $\lambda \in (\lambda_{\min}, 0)$ is chosen so that

$$\beta = a\lambda \int_0^\tau y_w^2 \, \mathrm{d}w = a\lambda\kappa^2 \left(\frac{\coth \mu_\lambda \tau}{2\mu_\lambda} - \frac{\tau}{2\sinh^2 \mu_\lambda \tau} \right). \tag{24}$$

Note that the path endpoints are $y_\tau = \kappa$ and $x_\tau = \beta$. Define

$$J(\tau) := \int_0^\tau \left[\frac{1}{2\theta}(\dot{y}_s + \frac{\theta}{2} y_s)^2 + \frac{a\lambda^2}{2} y_s^2 - r y_s^2 \right] \mathrm{d}s = \lambda\beta + \kappa^2 \Big(\frac{1}{4} + \frac{\mu_\lambda}{2\theta} \coth \mu_\lambda \tau \Big). \tag{25}$$

(a) For all $\delta, \delta' > 0$,

$$\liminf_{t\to\infty} t^{-1} \log P\Big(\exists u \in N_\tau : \forall s \in [0,\tau], |X_u(s) - t x_s| < \delta t, |Y_u(s) - \sqrt{t} y_s| < \delta' \sqrt{t} \Big) \geq -J(\tau),$$

(b) Let $\varepsilon > 0$. For all sufficiently small $\delta, \delta' > 0$,

$$\limsup_{t\to\infty} t^{-1} \log P\left(\exists u \in N_\tau : \forall s \in [0,\tau], |X_u(s) - t x_s| < \delta t, |Y_u(s) - \sqrt{t} y_s| < \delta' \sqrt{t} \right)$$
$$\leq -J(\tau) + \varepsilon.$$

It can be easily checked that $J(\tau) \downarrow \Theta(\beta,\kappa)$, $\lambda \to \overline{\lambda}$ and $(x_s, y_s) \to (\overline{x}_s, \overline{y}_s)$ as $\tau \to \infty$. Theorem 1.1 can now easily be deduced from Theorem 3.2 by choosing τ sufficiently large.

Although some additional work would be required to prove as much, the paths (x_s, y_s) above are chosen as they are the "best" ones for particles to follow in order to reach position $(\beta t, \kappa\sqrt{t})$ at (fixed) time τ, as found in the large-deviations heuristics discussed above and in [4].

Importantly, we emphasise that it should be possible to develop the ideas and techniques used in this chapter to obtain proofs of large-deviations principles for many other branching diffusion models, essentially because we can reduce the branching particles down to the spine and in general this gives a technique for deriving large-deviations principles for the branching diffusion from those of the single diffusing particle (the spine) which are already well studied.

4 The Spine Approach, Martingales and Measures

In this section we will introduce some of the key concepts used in the proofs of the main results. We shall construct the branching diffusion with a distinguished infinite line of decent, the *spine*, and then perform a change of measure that will make the spine "closely" follow a given path. Estimates on the martingale associated with this change of measure can then give a lower bound for large-deviations events in the branching diffusion. The heuristics of the previous section will serve as an important guide. However, although they have already indicated a specific path at (20), it should be noted that in our proofs we use properties of this path only at a few points—elsewhere the techniques can be applied in general to any path. Therefore the reader may suppose that $y:[0,\tau]\to\mathbb{R}$ is any given and fixed path, and we shall be very careful to highlight those points when we use specific properties of the path defined at (20). Also, to keep notational complexity to a reasonable minimum, we tend not to make the dependencies of the martingales and action functionals on the underlying chosen paths explicit in the notation.

The Spine Set-Up. Recall that the original branching diffusion $\mathbb{X}_t := \{(X_u(t),Y_u(t)) : u\in N_t\}$ where N_t is the set of individuals alive at time t has associated probability measures $P^{x,y}$ with natural filtration $\{\mathcal{F}_t\}_{t\geq 0}$. We label all particles according to the Ulam–Harris convention. For example, "$\emptyset 213$" represents "the 3rd child of the 1st child of the 2nd child of the initial ancestor." For two labels $v,u\in\Omega$ the notation $v<u$ means that v is an *ancestor* of u, $|u|$ is the generation of particle u, and so forth.

A *spine* ξ is a *distinguished* infinite line of descent starting with the initial ancestor, where $\xi=\{\xi_0,\xi_1,\xi_2,\ldots\}$ with $\xi_0=\emptyset$, ξ_n the label of the spine at the n-th generation and $u\in\xi$ means that $u=\xi_i$ for some $i\geq 0$. Let $\{(\xi(t),\eta(t))\}_{t\geq 0}$ represent the space-type path of spine, that is, the time position of the spine at time t is $(\xi(t),\eta(t)) := (X_u(t),Y_u(t))$ for $u\in N(t)\cap\xi$. Define $n=\{n_t : t\geq 0\}$ to be the counting process for the number of fissions that have occurred along the path of the spine by time t, with the actual fission times along the spine denoted by $\{S_i\}_{\geq 1}$. Note that ξ_{n_t} is the label of the spine at time t.

We will make important use of a variety of filtrations for the process with a distinguished spine. Let the *enriched* filtration for the branching diffusion with distinguished spine be $\tilde{\mathcal{F}}_t=\sigma(\mathcal{F}_t,\{\xi_{n_s}\}_{s\leq t})$. Then $\tilde{\mathcal{F}}_t$ knows everything up to time t, all particle paths, genealogy and identification of spine, whereas $\mathcal{F}_t$ knows about the paths and genealogy of all particles up to the time t, but does not know the identity of the spine. In addition, let $\tilde{\mathcal{G}}_t := \sigma(\{(\xi(s),\eta(s),n_s,\xi_{n_s})\}_{s\leq t})$ and $\mathcal{G}_t := \sigma(\{(\xi(s),\eta(s))\}_{s\leq t})$. Then $\tilde{\mathcal{G}}_t$ knows everything *along* the spine up to time t - the *spine's* motion, the *spine's* genealogy and the *spine's* fission times. It does not know about any information "off" the spine. On the other hand, $\tilde{\mathcal{G}}_t$ only knows about the *spine's* motion but not about the births along the spine, the spine's genealogical information, nor any information "off" the spine.

The Spine Construction. Under a measure $\tilde{P}^{x,y}$, the branching diffusion $(\mathbb{X}_s)_{s\geq 0}$ with distinguished spine ξ is constructed as follows:

- The spine process (ξ_s,η_s) starts at (x,y) and diffuses as a solution to

$$d\eta_s=\sqrt{\theta}dB_s-\frac{\theta}{2}\eta_s ds,\quad\text{and}\quad d\xi_s=\sqrt{a}\eta_s dW_s,\tag{26}$$

where B_s and W_s are standard Brownian motions.

- At rate $R(\eta_s)$ the spine undergoes fission producing two particles.
- With equal probability, one of these two particles is selected to continue the spine ξ.
- The other particle initiates, from its birth space-type position, an independent copy of the original P branching diffusion with branching rate $R(\cdot)$.

We note that $\tilde{P}^{x,y}$ is an extension of the original measure $P^{x,y}$, with $P = \tilde{P}|_{\mathcal{F}_\infty}$. In fact, it is easy to see that an alternative way to construct the process under $\tilde{P}$ is to first construct the entire tree of the branching diffusion according to P and, secondly, choose the spine by starting from the initial ancestor then following the spine path forward in time with independent uniform choices made from the particles produced at each fission. In particular, for $u \in N(t)$, we note that $\tilde{P}(u \in \xi | \mathcal{F}_t) = \prod_{v<u} 2^{-1} = 2^{-|u|}$.

We also have the very useful and intuitive *'one-particle picture' (OPP)*. For example, for any measurable function f of single-particle paths on $[0,t]$,

$$\tilde{P}^{x,y}\left(\sum_{u\in N(t)} f(X_u(s), Y_u(s); s \le t)\right) = \tilde{P}^{x,y}\left(e^{\int_0^t \beta(\eta(s))\,ds} f(\xi(s), \eta(s); s \le t)\right)$$

For further details of this spine set-up and various key results, see Hardy and Harris [5]. Also see Lyons et al. [11, 15, 16] and other recent work based on these (examples are Kyprianou [12], Kyprianou and Sani [13], Athreya [1], Olofsson [18], amongst others) for similar spine-based approaches in branching processes.

Changes of Measure. We can now able to perform our "spine change of measure." For any $t > 0$ and any given $y : [0,\tau] \to \mathbb{R}$ that is square integrable along with its derivative

$$\exp\left(\frac{\sqrt{t}}{\sqrt{\theta}} \int_0^w \left(\dot{y}_s + \frac{\theta}{2} y_s\right) \mathrm{d}B_s - \frac{t}{2\theta} \int_0^w \left(\dot{y}_s + \frac{\theta}{2} y_s\right)^2 \mathrm{d}s\right) \times \exp\left(\sqrt{at}\lambda \int_0^w y_s \,\mathrm{d}W_s - \frac{a\lambda^2 t}{2} \int_0^w y_s^2 \,\mathrm{d}s\right),$$

is a strictly-positive $\tilde{P}$-martingale over the time period $w \in [0,\tau]$ (see, e.g., Øksendal [17]). As one part of the change of measure defined below, this martingale will introduce drift terms into the diffusions η_s and ξ_s such that $\eta_s^t \sim y_s$ and $\xi_s^t \sim a\lambda y_s^2$ when t is large, and we note a comparison between this martingale and the expression (18) above.

The process n_w which counts the number of fission times on the spine up to time w is a Cox process of rate $R(\eta_s)$ and therefore for $w \in [0,\tau]$,

$$w \mapsto e^{-\int_0^w R(\eta_s)\,\mathrm{d}s} 2^{n_w}$$

is also a $\tilde{P}$-martingale. We can use the product of these two martingales to define a new measure:

Theorem 4.1 (Spine Change of Measure). *Let $\tau > 0$ be fixed. For $t > 0$, we define a measure $\tilde{\mathbb{Q}}_t$ on $\tilde{\mathcal{F}}_\tau$ where*

$$\left.\frac{\mathrm{d}\tilde{\mathbb{Q}}_t}{\mathrm{d}\tilde{P}}\right|_{\tilde{\mathcal{F}}_w} := \tilde{\zeta}_t(w) := \exp\left(\frac{\sqrt{t}}{\sqrt{\theta}}\int_0^w\left(\dot{y}_s+\frac{\theta}{2}y_s\right)\mathrm{d}B_s - \frac{t}{2\theta}\int_0^w\left(\dot{y}_s+\frac{\theta}{2}y_s\right)^2\mathrm{d}s\right)$$
$$\times\exp\left(\sqrt{at}\lambda\int_0^w y_s\,\mathrm{d}W_s - \frac{a\lambda^2 t}{2}\int_0^w y_s^2\,\mathrm{d}s\right)\times e^{-\int_0^w R(\eta_s)\,\mathrm{d}s}2^{n_w}, \qquad (27)$$

for $w\in[0,\tau]$. Under the measure $\tilde{\mathbb{Q}}_t^{x,y}$ we can give a pathwise construction of the branching diffusion $(\mathbb{X}_s)_{s\in[0,\tau]}$ with distinguished spine ξ:

- *The spine process (ξ_s,η_s) starts at (x,y) and diffuses as a solution to*

$$\mathrm{d}(\eta_s-\sqrt{t}y_s)=\sqrt{\theta}\mathrm{d}\tilde{B}_s-\frac{\theta}{2}(\eta_s-\sqrt{t}y_s)\mathrm{d}s \qquad (28)$$

and

$$\mathrm{d}\xi_s=\sqrt{a}\eta_s\,\mathrm{d}\tilde{W}_s+a\lambda\sqrt{t}y_s\eta_s\,\mathrm{d}s, \qquad (29)$$

where $\tilde{B}$ and $\tilde{W}$ are standard Brownian motions under $\tilde{\mathbb{Q}}_t$, with

$$\mathrm{d}\tilde{B}_s=\mathrm{d}B_s-\frac{\sqrt{t}}{\sqrt{\theta}}\left(\dot{y}_s+\frac{\theta}{2}y_s\right)\mathrm{d}s, \qquad \mathrm{d}\tilde{W}_s=\mathrm{d}W_s-\sqrt{at}\lambda y_s\,\mathrm{d}s.$$

- *At the accelerated rate $2R(\eta_s)$, the spine undergoes fission producing two particles.*
- *With equal probability, one of these two particles is selected to continue the spine.*
- *The other particle initiates, from its birth space-type position, an independent copy of the original P branching diffusion with normal branching rate $R(\cdot)$.*

Due to our formulation of the underlying spine foundations in terms of filtrations and sub-filtrations, we can project this new measure $\tilde{\mathbb{Q}}_t$ down onto the branching diffusion particles and define a measure $\mathbb{Q}_t$ on $\mathcal{F}_\tau$ by $\mathbb{Q}_t := \tilde{\mathbb{Q}}_t|_{\mathcal{F}_\tau}$.

Theorem 4.2. *Let $\tau>0$ be fixed. For each fixed $t>0$, define the $((\mathcal{F}_w)_{0\le w\le\tau},P)$-martingale $Z_t(w)$ for $w\in[0,\tau]$ by*

$$Z_t(w) := \left.\frac{\mathrm{d}\mathbb{Q}_t}{\mathrm{d}P}\right|_{\mathcal{F}_w} = \tilde{P}\big(\tilde{\zeta}_t(w)|\mathcal{F}_w\big).$$

Then

$$Z_t(w)=\sum_{u\in N(w)} f_{t,w}(u)$$

where

$$f_{t,w}(u) := e^{-\int_0^w R(Y_u(s))\,\mathrm{d}s}\exp\left(\frac{\sqrt{t}}{\sqrt{\theta}}\int_0^w\left(\dot{y}_s+\frac{\theta}{2}y_s\right)\mathrm{d}B_u(s)-\frac{t}{2\theta}\int_0^w\left(\dot{y}_s+\frac{\theta}{2}y_s\right)^2\mathrm{d}s\right)$$

$$\times \exp\left(\sqrt{at}\lambda\int_0^w y_s \,\mathrm{d}W_u(s) - \frac{a\lambda^2 t}{2}\int_0^w y_s^2 \,\mathrm{d}s\right) \tag{30}$$

and B_u and W_u to denote the P-Brownian motions driving the type and spatial processes of u.

See Hardy and Harris [5], or Engländer and Kyprianou [3], for more details and proofs.

The Spine Decomposition. Consider those particles alive at time τ and group them together according to the time that they first branched off the spine's path. Since, under $\tilde{\mathbb{Q}}_t$ particles "off" the spine behave as if under P and since Z_t is a P-martingale, it is easy to see the following "spine decomposition":

$$\tilde{\mathbb{Q}}_t(Z_t(\tau)|\tilde{\mathcal{G}}_\infty) = f_{t,\tau}(\xi) + \sum_{i=1}^{n_t} f_{t,S_i}(\xi)$$

where

$$f_{t,w}(\xi) := e^{-\int_0^w R(\eta(s))\,\mathrm{d}s} \exp\left(\frac{\sqrt{t}}{\sqrt{\theta}}\int_0^w \left(\dot{y}_s + \frac{\theta}{2}y_s\right)\mathrm{d}B(s) - \frac{t}{2\theta}\int_0^w \left(\dot{y}_s + \frac{\theta}{2}y_s\right)^2 \mathrm{d}s\right)$$

$$\times \exp\left(\sqrt{at}\lambda\int_0^w y_s \,\mathrm{d}W(s) - \frac{a\lambda^2 t}{2}\int_0^w y_s^2 \,\mathrm{d}s\right) \tag{31}$$

The spine decomposition will prove essential for our key martingale growth estimates.

The Growth of Martingale Z_t. For our proof of Theorem 1.1 (and its stronger version of Theorem 3.2) it is important to know how quickly $Z_t(\tau)$ grows under the measure $\mathbb{Q}_t$. The following key result is the main application of spines in this article:

Theorem 4.3. *For the specific path y defined at* (20) *and for any $\alpha \in [0,1]$ we have*

$$\limsup_{t\to\infty} t^{-1} \log \tilde{\mathbb{Q}}_t\left(Z_t(\tau)^\alpha\right) \le \alpha J(\tau) + \alpha^2 M(\tau),$$

where we define

$$J(w) := \int_0^w \left[\frac{1}{2\theta}\left(\dot{y}_s + \frac{\theta}{2}y_s\right)^2 + \frac{a\lambda^2}{2}y_s^2 - r y_s^2\right] \mathrm{d}s, \tag{32}$$

and

$$M(w) := \int_0^w \left[\frac{1}{2\theta}\left(\dot{y}_s + \frac{\theta}{2}y_s\right)^2 + \frac{a\lambda^2}{2}y_s^2\right] \mathrm{d}s. \tag{33}$$

We emphasise that without the technology of spines, the proof of this result would be far more difficult. The spine decomposition gives us a methodology for reducing the additive structure of these martingales essentially to a single-particle problem, and since it does this through a conditional-expectation operation rather than with an inequality, it is *exact* and therefore can lead to tight estimates that are useful. Due to its length, we dedicate the whole of Sect. 7 to the spine proof of this above theorem and now proceed to show how this result can be used to obtain the upper bound on $Z_t(\tau)$ that we require for Theorem 1.1.

It is not difficult to verify that for any $\alpha \in [0,1]$, $Z_t(w)^\alpha$ is a submartingale with respect to the measure $\mathbb{Q}_t$. Given Theorem 4.3, we can therefore use Doob's submartingale inequality to prove the following:

Theorem 4.4. *Let $\tau > 0$ be fixed. Then for all $\varepsilon > 0$,*

$$\lim_{t\to\infty} \mathbb{Q}_t\left(\sup_{s\in[0,\tau]} Z_t(s) \le e^{(J(\tau)+\varepsilon)t}\right) \to 1.$$

Proof. For a given $\varepsilon > 0$ and for any $\alpha \in [0,1]$, Doob's inequality gives

$$\mathbb{Q}_t\Big(\sup_{s\in[0,\tau]} Z_t(s) > e^{(J(\tau)+\varepsilon)t}\Big) = \mathbb{Q}_t\Big(\sup_{s\in[0,\tau]} Z_t(s)^\alpha > e^{\alpha(J(\tau)+\varepsilon)t}\Big) \le \frac{\mathbb{Q}_t\left(Z_t(\tau)^\alpha\right)}{e^{\alpha(J(\tau)+\varepsilon)t}}.$$

From Theorem 4.3, we know that for each $\alpha \in [0,1]$ and for all large t, we have

$$\mathbb{Q}_t\Big(\sup_{s\in[0,\tau]} Z_t(s) > e^{(J(\tau)+\varepsilon)t}\Big) \le e^{(\alpha M(\tau)-\varepsilon)\alpha t}.$$

If we also have $\alpha \in (0, \varepsilon/M(\tau))$ then clearly this above is a decaying exponential and so it follows that

$$\lim_{t\to\infty} \mathbb{Q}_t\Big(\sup_{s\in[0,\tau]} Z_t(s) > e^{(J(\tau)+\varepsilon)t}\Big) \to 0.$$

□

For the specific y defined at (20), it can be shown that

$$J(\tau) = \int_0^\tau \left[\frac{1}{2\theta}\left(\dot{y}_s + \frac{\theta}{2}y_s\right)^2 + \frac{a\lambda^2}{2}y_s^2 - ry_s^2\right] \mathrm{d}s = \lambda\beta + \kappa^2\left(\frac{1}{4} + \frac{\mu_\lambda}{2\theta}\coth\mu_\lambda\tau\right),$$

where we recall that this $\lambda \in (\lambda_{\min}, 0)$ was specifically determined by (22). In fact, Git et al. [4] explain that this choice of λ was optimal in that

$$\lambda\beta + \kappa^2\left(\frac{1}{4} + \frac{\mu_\lambda}{2\theta}\coth\mu_\lambda\tau\right) = \sup_\gamma\left\{\gamma\beta + \kappa^2\left(\frac{1}{4} + \frac{\mu_\gamma}{2\theta}\coth\mu_\gamma\tau\right)\right\}. \tag{34}$$

On the other hand we can find a similar representation for the parameter $\Theta(\beta,\kappa)$: if we define

$$\bar{\lambda} := \sqrt{\frac{\beta^2\theta(\theta-8r)}{a^2\kappa^4 + 4a\theta\beta^2}}, \quad \text{so that } \mu_{\bar{\lambda}} = \frac{\kappa^2\sqrt{\theta(\theta-8r)}}{2\sqrt{\kappa^4 + 4\theta\beta^2/a}},$$

then

$$\Theta(\beta,\kappa)=\bar{\lambda}\beta+\kappa^2\psi^+_{\bar{\lambda}}=\sup_{\gamma}\left\{\gamma\beta+\kappa^2\Big(\frac{1}{4}+\frac{\mu_\gamma}{2\theta}\Big)\right\}=\lim_{\tau\to\infty}\sup_{\gamma}\left\{\gamma\beta+\kappa^2\Big(\frac{1}{4}+\frac{\mu_\gamma}{2\theta}\coth\mu_\gamma\tau\Big)\right\},$$

where we recall that

$$\psi^+_{\bar{\lambda}}:=\frac{1}{4}+\frac{\mu_{\bar{\lambda}}}{2\theta}.$$

In this way it can be deduced from (34) that $J(\tau)>\Theta(\beta,\kappa)$ with

$$J(\tau)\downarrow\Theta(\beta,\kappa),\quad \text{as } \tau\to\infty.$$

It is now easy to deduce the following corollary to Theorem 4.4:

Corollary 4.5. *Given $\varepsilon>0$, for $\tau>0$ chosen sufficiently large,*

$$\lim_{t\to\infty}\mathbb{Q}_t\Big(\sup_{s\in[0,\tau]}Z_t(s)\le e^{(\Theta(\beta,\kappa)+\varepsilon)t}\Big)\to 1.$$

5 Proving the Large-Deviations Lower Bound

Barring the proof of Theorem 4.3 which we cover fully in Sect. 7, we now have all the ingredients required to prove the large-deviations lower bound for the short-climb event of Theorem 3.2.

Throughout this proof we are focussing on the specific path

$$y_s:=\kappa\frac{\sinh\mu_\lambda s}{\sinh\mu_\lambda\tau},\qquad s\in[0,\tau]$$

where $\lambda\in(\lambda_{\min},0)$ satisfies

$$\frac{\beta}{a\lambda}=\kappa^2\Big(\frac{\coth\mu_\lambda\tau}{\mu_\lambda}-\frac{\tau}{2\sinh^2\mu_\lambda\tau}\Big).$$

as discussed at (22). We define the event that the space-type location $(X_u(s),Y_u(s))$ of a particular particle $u\in N_\tau$ remains near $(a\lambda t\int_0^s y_w^2\,\mathrm{d}w,\sqrt{t}y_s)$ throughout the interval $s\in[0,\tau]$:

$$A_t(u):=\left\{\forall s\in[0,\tau],\left|X_u(s)-a\lambda t\int_0^s y_w^2\,\mathrm{d}w\right|<\delta t,\left|Y_u(s)-\sqrt{t}y_s\right|<\delta'\sqrt{t}\right\},$$

where $\delta,\delta'>0$ are given and fixed. In addition, we define the event that any of the particles performs this event (whilst emphasising the parameter dependence) by

$$A_{t,\tau}^{\delta,\delta'} := \bigcup_{u \in N(\tau)} A_t(u)$$

Noting that this event is $\mathcal{F}_\tau$-measurable since it depends only on the branching particles and does not refer to the spine, it follows that on this event the change of measure is carried out by Z_t, as noted in Theorem 4.2. The upper bound that we have derived for Z_t at Corollary 4.5 will serve as a lower bound for $1/Z_t(\tau)$ in this change of measure and will combine with the fact that under the measure $\tilde{\mathbb{Q}}_t$ (for large t) we know that the spine will carry out the large-deviations behaviour that we want.

Then for any $\varepsilon > 0$,

$$\begin{aligned} P\left(A_{t,\tau}^{\delta,\delta'}\right) &= \mathbb{Q}_t\left(\frac{1}{Z_t(\tau)}; \exists u \in N_\tau, A_t(u)\right) \\ &\geq \mathbb{Q}_t\left(\frac{1}{Z_t(\tau)}; \exists u \in N_\tau, A_t(u);\ \sup_{s\in[0,\tau]} Z_t(s) \leq e^{(J(\tau)+\varepsilon)t}\right) \\ &\geq e^{-(J(\tau)+\varepsilon)t}\, \mathbb{Q}_t\left(\exists u \in N_\tau, A_t(u);\ \sup_{s\in[0,\tau]} Z_t(s) \leq e^{J(\tau)+\varepsilon)t}\right) \\ &\geq e^{-(J(\tau)+\varepsilon)t}\, \tilde{\mathbb{Q}}_t\left(A_t(\xi);\ \sup_{s\in[0,\tau]} Z_t(s) \leq e^{(J(\tau)+\varepsilon)t}\right). \end{aligned} \tag{35}$$

Given (28) and (29), standard theory says that under the measure $\tilde{\mathbb{Q}}_t$ (with t large) the rescaled spine (ξ_s^t, η_s^t) will tend to stay close to the space-type paths $(a\lambda \int_0^s y_w^2\,\mathrm{d}w, y_s)$ over the whole time interval $[0,\tau]$:

$$\xi_s^t \sim a\lambda \int_0^s y_w^2\,\mathrm{d}w, \quad \text{and} \quad \eta_s^t \sim y_s,$$

by which we mean that for a fixed $\tau > 0$ and any $\delta, \delta' > 0$,

$$\lim_{t\to\infty} \tilde{\mathbb{Q}}_t\left(\left|\xi_s^t - a\lambda \int_0^s y_w^2\,\mathrm{d}w\right| < \delta, \left|\eta_s^t - y_s\right| < \delta', \text{for all } s \in [0,\tau]\right) \to 1,$$

which can equally be written as

$$\lim_{t\to\infty} \tilde{\mathbb{Q}}_t\left(\left|\xi_s - a\lambda t \int_0^s y_w^2\,\mathrm{d}w\right| < \delta t, \left|\eta_s - y_s\sqrt{t}\right| < \delta'\sqrt{t}, \text{for all } s \in [0,\tau]\right) \to 1,$$

equivalently,

$$\lim_{t\to\infty} \tilde{\mathbb{Q}}_t(A_t(\xi)) = 1.$$

At the same time, Theorem 4.4 says,

$$\lim_{t\to\infty} \tilde{\mathbb{Q}}_t\left(\sup_{s\in[0,\tau]} Z_t(s) \leq e^{(J(\tau)+\varepsilon)t}\right) = 1.$$

Recalling that $P(A \cap B) = 1 - P(A^c \cup B^c) \geq 1 - P(A^c) - P(B^c)$, we find

$$\lim_{t\to\infty} \tilde{\mathbb{Q}}_t \left(A_t(\xi);\ \sup_{s\in[0,\tau]} Z_t(s) \leq e^{(J(\tau)+\varepsilon)t} \right) = 1.$$

Since $\varepsilon > 0$ was arbitrary, it follows from (35) that for all fixed $\tau > 0$ and $\delta, \delta' > 0$,

$$\liminf_{t\to\infty} t^{-1} \log P\left(A_{t,\tau}^{\delta,\delta'} \right) \geq -J(\tau),$$

which gives the proof of the lower bound of Theorem 3.2. □

6 Proving the Large-Deviations Upper Bounds

We first give a quick and direct martingale proof of the upper bound of Theorem 1.1 that identifies the optimal path to reach $(\beta t, \kappa\sqrt{t})$ at large time τ. This follows along similar lines to the almost-sure upper bound of Sect. 2.1.

Let $\gamma, \kappa > 0$. Then for $\theta \in (0, -\lambda_{min})$ and $\phi \in (0, \psi_\theta^+]$ to ensure that expectations remain finite for all time and recalling the martingale Z_θ^+ at Eq. (9),

$$\begin{aligned} &P\left(\exists u \in N_\tau : X_u(\tau) \leq -\gamma t, Y_u(\tau)^2 \geq \kappa^2 t\right) \\ &\leq P\Big(\sum_{u\in N_\tau} \mathbf{1}\{X_u(\tau) + \gamma t \leq 0, Y_u(\tau)^2 - \kappa^2 t \geq 0\}\Big) \\ &\leq P\Big(\sum_{u\in N_\tau} e^{-\theta(X_u(\tau)+\gamma t)+\phi(Y_u(\tau)^2-\kappa^2 t)}\Big) \\ &= e^{-\theta\gamma t-\kappa^2\phi t+E_\theta^+\tau} P\Big(\sum_{u\in N_\tau} e^{-\theta X_u(\tau)+\phi Y_u(\tau)^2-E_\theta^+\tau}\Big) \\ &\leq e^{-\theta\gamma t-\kappa^2\phi t+E_\theta^+\tau} P\left(Z_\theta^+(\tau)\right) = e^{-(\theta\gamma+\kappa^2\phi)t+E_\theta^+\tau}. \end{aligned}$$

Hence, taking t limits and then optimising over θ we find

$$\limsup_{t\to\infty} t^{-1} \log P\left(\exists u \in N_\tau : X_u(\tau) \leq -\gamma t, Y_u(\tau)^2 \geq \kappa^2 t\right) \leq -\sup_{\theta>0}\{\theta\gamma + \kappa^2\phi\} = \Theta(\gamma,\kappa) \tag{36}$$

which is then sufficient to give part (b) of Theorem 1.1.

Whilst this upper bound is good enough for sufficiently large τ, for fixed τ the expectation upper bound can be sharpened with a bit more work to yield part (b) of Theorem 3.2. Firstly, recalling the one-particle picture and that recalling (30), $Z_t(w) = \sum_{u\in N(w)} f_{t,w}(u)$ we have,

$$
\begin{aligned}
P(\exists u \in N_\tau \text{ such that } A_t(u)) &\leq P\left(\sum_{u\in N(\tau)} \mathbf{1}\{A_t(u)\}\right) \\
&= \tilde{P}\left(e^{\int_0^\tau R(\eta_s)\,ds}; A_t(\xi)\right) = \tilde{\mathbb{Q}}_t\left(\frac{1}{f_{t,\tau}(\xi)}; A_t(\xi)\right) \qquad (37)
\end{aligned}
$$

Then if a good lower bound for $f_{t,\tau}(\xi)$ can be found on the event $A_t(\xi)$, a suitable upper bound for the required probability will follow.

Consider y and τ fixed and let us restrict attention to the event $A_t(\xi)$. It is now relatively straightforward making use of Itõ calculus and the constraints from the event, to consider terms in the expression for the spine term $f_{t,\tau}(\xi)$ to derive the following bounds (we omit the proof):

Proposition 6.1. *(a) There exists $K > 0$ depending only on y and τ such that, for all $w \in [0, \tau]$,*

$$
\left|\frac{\sqrt{t}}{\sqrt{\theta}}\int_0^w (\dot{y}_s + \frac{\theta}{2}y_s)\,\mathrm{d}B_u(s) - \frac{t}{\theta}\int_0^w (\dot{y}_s + \frac{\theta}{2}y_s)^2\,\mathrm{d}s\right| < \frac{t}{\theta}K\delta'
$$

for almost every path in the event $A_t(\xi)$.

(b) There exists a $K' > 0$ depending only on y and τ, such that for all $w \in [0, \tau]$,

$$
\left|\int_0^w \eta_s^2\,ds - t\int_0^w y_s^2 ds\right| \leq t\delta'(2K' + \tau)
$$

for almost every path in the event $A_t(\xi)$.

(c)

$$
\tilde{\mathbb{Q}}_t\left(e^{-\sqrt{at}\lambda\int_0^\tau y_s\,dW_s}; A_t(\xi)\right) \leq e^{-a\lambda^2 t\int_0^\tau y_s^2\,ds + t(\frac{1}{2}a\lambda^2\delta'^2\tau + \delta\lambda)}
$$

Then,

$$
\begin{aligned}
\tilde{\mathbb{Q}}_t\left(\frac{1}{f_{t,\tau}(\xi)}; A_t(\xi)\right) &= \tilde{\mathbb{Q}}_t\left(e^{\int_0^\tau R(\xi_s)\,\mathrm{d}s} \times e^{-\frac{\sqrt{t}}{\sqrt{\theta}}\int_0^\tau (\dot{y}_s + \frac{\theta}{2}y_s)\,\mathrm{d}B_s} \times e^{\sqrt{at}\lambda\int_0^\tau y_s\,\mathrm{d}W_s}; A_t(\xi)\right) \\
&\quad \times e^{\frac{t}{2\theta}\int_0^\tau (\dot{y}_s + \frac{\theta}{2}y_s)^2\,\mathrm{d}s} \times e^{t\frac{a\lambda^2}{2}\int_0^\tau y_s^2\,\mathrm{d}s} \\
&\leq e^{t\int_0^\tau r y_s^2\,\mathrm{d}s + \rho\tau} \times e^{t\delta'(2K'+\tau)} \times e^{-\frac{t}{\theta}\int_0^\tau (\dot{y}_s + \frac{\theta}{2}y_s)^2\,\mathrm{d}s} \times e^{t\frac{K}{\theta}\delta'} \\
&\quad \times e^{-ta\lambda^2\int_0^\tau y_s^2\,\mathrm{d}s} \times e^{t(\frac{1}{2}a\lambda^2\delta'\tau + \delta\lambda)} \times e^{-\frac{t}{2\theta}\int_0^\tau (\dot{y}_s + \frac{\theta}{2}y_s)^2\,\mathrm{d}s + t\frac{a\lambda^2}{2}\int_0^\tau y_s^2\,\mathrm{d}s}
\end{aligned}
$$

Then, given $\varepsilon > 0$, we can choose $\delta, \delta' > 0$ sufficiently small such that

$$
\tilde{\mathbb{Q}}_t\left(\frac{1}{f_{t,\tau}(\xi)}; A_t(\xi)\right) \leq e^{t\left\{\int_0^\tau r y_s^2\,\mathrm{d}s + \rho\tau - \frac{1}{2\theta}\int_0^\tau (\dot{y}_s + \frac{\theta}{2}y_s)^2\,\mathrm{d}s - \frac{a\lambda^2}{2}\int_0^\tau y_s^2\mathrm{d}s + \varepsilon\right\}} = e^{-tJ(\tau) + \rho\tau + \varepsilon\tau}
$$

□

7 A Spine Proof of the Martingale Upper-Bound

In this section we use the spine decomposition of the martingale Z_t to prove Theorem 4.3.

It is Jensen's inequality that immediately allows us to concentrate on the spine decomposition since

$$\mathbb{Q}_t\left(Z_t(\tau)^\alpha\right) \le \tilde{\mathbb{Q}}_t\left(\tilde{\mathbb{Q}}_t\left(Z_t(\tau)|\tilde{\mathcal{G}}_\infty\right)^\alpha\right), \qquad \text{for } \alpha \in [0,1].$$

Similar tricks to compute the moments of martingales with Jensen's inequality and the spine decomposition can be found in Iksanov [9] and in Hardy and Harris [5].

The spine decomposition of $Z_t(\tau)$ is

$$\tilde{\mathbb{Q}}_t\left(Z_t(\tau)|\tilde{\mathcal{G}}_\infty\right) = e^{-r\int_0^\tau \eta_s^2\,\mathrm{d}s - \rho\tau} e^{\left[\sqrt{a}\lambda\int_0^\tau y_s\,\mathrm{d}W_s - \frac{a\lambda^2}{2}\int_0^\tau y_s^2\,\mathrm{d}s\right] + \left[\frac{1}{\sqrt{\theta}}\int_0^\tau(\dot{y}_s+\frac{\theta}{2}y_s)\,\mathrm{d}B_s - \frac{1}{2\theta}\int_0^\tau(\dot{y}_s+\frac{\theta}{2}y_s)^2\,\mathrm{d}s\right]}$$
$$+ \sum_{k=1}^{n_\tau} e^{-r\int_0^{S_k}\eta_s^2\,\mathrm{d}s - \rho S_k} e^{\left[\sqrt{a}\lambda\int_0^{S_k} y_s\,\mathrm{d}W_s - \frac{a\lambda^2}{2}\int_0^{S_k} y_s^2\,\mathrm{d}s\right] + \left[\frac{1}{\sqrt{\theta}}\int_0^{S_k}(\dot{y}_s+\frac{\theta}{2}y_s)\,\mathrm{d}B_s - \frac{1}{2\theta}\int_0^{S_k}(\dot{y}_s+\frac{\theta}{2}y_s)^2\,\mathrm{d}s\right]}.$$

We consider the two parts of this spine decomposition separately—the *spine term* and then the *sum term*—and aim to show that they both have exponential growth of the same order.

Definition 7.1. We define

$$\textbf{spine term} := e^{-r\int_0^\tau \eta_s^2\,\mathrm{d}s - \rho\tau} e^{[\sqrt{a}\lambda\int_0^\tau y_s\,\mathrm{d}W_s - \frac{a\lambda^2}{2}\int_0^\tau y_s^2\,\mathrm{d}s] + [\frac{1}{\sqrt{\theta}}\int_0^\tau(\dot{y}_s+\frac{\theta}{2}y_s)\,\mathrm{d}B_s - \frac{1}{2\theta}\int_0^\tau(\dot{y}_s+\frac{\theta}{2}y_s)^2\,\mathrm{d}s]},$$

and

$$\textbf{sum term} :=$$
$$\sum_{k=1}^{n_\tau} e^{-r\int_0^{S_k}\eta_s^2\,\mathrm{d}s - \rho S_k} e^{[\sqrt{a}\lambda\int_0^{S_k} y_s\,\mathrm{d}W_s - \frac{a\lambda^2}{2}\int_0^{S_k} y_s^2\,\mathrm{d}s] + [\frac{1}{\sqrt{\theta}}\int_0^{S_k}(\dot{y}_s+\frac{\theta}{2}y_s)\,\mathrm{d}B_s - \frac{1}{2\theta}\int_0^{S_k}(\dot{y}_s+\frac{\theta}{2}y_s)^2\,\mathrm{d}s]}.$$

In each case we first use some martingale techniques to factor out exponential terms that give us the correct growth rate (and here we are guided by the heuristics) and then use Varadhan's lemma to show that the remaining terms do not contribute any further exponential growth. The spine term is simpler to deal with and is considered first.

7.1 Factoring Out the Spine Term

Girsanov's theorem (see Øksendal [17]) states that under the new measure $\tilde{\mathbb{Q}}_t$ we have

$$\mathrm{d}B_s = \mathrm{d}\tilde{B}_s + \frac{\sqrt{t}}{\sqrt{\theta}}\Big(\dot{y}_s + \frac{\theta}{2}y_s\Big)\mathrm{d}s, \quad \text{and} \quad \mathrm{d}W_s = \mathrm{d}\tilde{W}_s + \sqrt{at}\lambda y_s\,\mathrm{d}s, \tag{38}$$

where $\tilde{B}$ and $\tilde{W}$ are BMs under $\tilde{\mathbb{Q}}_t$, and these representations can be substituted into the spine term to give

$$\begin{aligned}\textbf{Spine term} &= e^{t\int_0^\tau \frac{1}{2\theta}(\dot{y}_s+\frac{\theta}{2}y_s)^2+\frac{a\lambda^2}{2}y_s^2\,\mathrm{d}s-\rho\tau} \times e^{-r\int_0^\tau \eta_s^2\,\mathrm{d}s}\, e^{[\sqrt{at}\lambda\int_0^\tau y_s\,\mathrm{d}\tilde{W}_s]+[\frac{\sqrt{t}}{\sqrt{\theta}}\int_0^\tau(\dot{y}_s+\frac{\theta}{2}y_s)\,\mathrm{d}\tilde{B}_s]} \\ &= e^{tJ(\tau)-\rho\tau} \times e^{rt\int_0^\tau[(\eta_s^t)^2-y_s^2]\,\mathrm{d}s}\, e^{[\sqrt{at}\lambda\int_0^\tau y_s\,\mathrm{d}\tilde{W}_s]+[\frac{\sqrt{t}}{\sqrt{\theta}}\int_0^\tau(\dot{y}_s+\frac{\theta}{2}y_s)\,\mathrm{d}\tilde{B}_s]}.\end{aligned} \tag{39}$$

Using the standard martingale

$$e^{\alpha\sqrt{at}\lambda\int_0^\tau y_s\,\mathrm{d}\tilde{W}_s-\alpha^2\frac{a\lambda^2 t}{2}\int_0^\tau y_s^2\,\mathrm{d}s},$$

we can factor out one of the terms of the expectation:

$$\begin{aligned}\tilde{\mathbb{Q}}_t(\textbf{spine term}^\alpha) &= e^{\alpha tJ(\tau)-\alpha\rho\tau}\,\tilde{\mathbb{Q}}_t\Big(e^{\alpha rt\int_0^\tau[y_s^2-(\eta_s^t)^2]\,\mathrm{d}s}\, e^{\alpha[\sqrt{at}\lambda\int_0^\tau y_s\,\mathrm{d}\tilde{W}_s]+\alpha[\frac{\sqrt{t}}{\sqrt{\theta}}\int_0^\tau(\dot{y}_s+\frac{\theta}{2}y_s)\,\mathrm{d}\tilde{B}_s]}\Big) \\ &= e^{\alpha tJ(\tau)-\alpha\rho\tau}\, e^{\alpha^2\frac{a\lambda^2 t}{2}\int_0^\tau y_s^2\,\mathrm{d}s}\,\tilde{\mathbb{Q}}_t\Big(e^{\alpha r\int_0^\tau[y_s^2-(\eta_s^t)^2]\,\mathrm{d}s}\, e^{\alpha[\frac{\sqrt{t}}{\sqrt{\theta}}\int_0^\tau(\dot{y}_s+\frac{\theta}{2}y_s)\,\mathrm{d}\tilde{B}_s]}\Big).\end{aligned}$$

This final expectation can be dealt with by another change of measure:

$$\begin{aligned}&\tilde{\mathbb{Q}}_t\Big(e^{\alpha rt\int_0^\tau[y_s^2-(\eta_s^t)^2]\,\mathrm{d}s}\, e^{\alpha[\frac{\sqrt{t}}{\sqrt{\theta}}\int_0^\tau(\dot{y}_s+\frac{\theta}{2}y_s)\,\mathrm{d}\tilde{B}_s]}\Big) \\ &\quad= e^{\frac{\alpha^2 t}{2\theta}\int_0^\tau(\dot{y}_s+\frac{\theta}{2}y_s)^2\,\mathrm{d}s} \times \tilde{\mathbb{Q}}_t\Big(e^{\alpha rt\int_0^\tau[y_s^2-(\eta_s^t)^2]\,\mathrm{d}s}\, e^{\frac{\alpha\sqrt{t}}{\sqrt{\theta}}\int_0^\tau(\dot{y}_s+\frac{\theta}{2}y_s)\,\mathrm{d}\tilde{B}_s-\frac{\alpha^2 t}{2\theta}\int_0^\tau(\dot{y}_s+\frac{\theta}{2}y_s)^2\,\mathrm{d}s}\Big), \\ &\quad= e^{\frac{\alpha^2 t}{2\theta}\int_0^\tau(\dot{y}_s+\frac{\theta}{2}y_s)^2\,\mathrm{d}s} \times \tilde{\mathbb{Q}}_t^\alpha\Big(e^{\alpha rt\int_0^\tau[y_s^2-(\eta_s^t)^2]\,\mathrm{d}s}\Big),\end{aligned}$$

where we have used the martingale

$$e^{\frac{\alpha\sqrt{t}}{\sqrt{\theta}}\int_0^\tau \dot{y}_s+\frac{\theta}{2}y_s\,\mathrm{d}\tilde{B}_s-\frac{\alpha^2 t}{2\theta}\int_0^\tau(\dot{y}_s+\frac{\theta}{2}y_s)^2\,\mathrm{d}s}$$

to change the measure from $\tilde{\mathbb{Q}}_t$ to $\tilde{\mathbb{Q}}_t^\alpha$. Another application of the Girsanov theorem implies that under the measure $\tilde{\mathbb{Q}}_t^\alpha$, the rescaled process η_s^t satisfies

$$\mathrm{d}\big(\eta_s^t - (1+\alpha)y\big) = \frac{\sqrt{\theta}}{\sqrt{t}}\mathrm{d}\bar{B}_s - \frac{\theta}{2}\big(\eta_s^t - (1+\alpha)y_s\big)\mathrm{d}s \tag{40}$$

where $\bar{B}_s$ is a Brownian motion, which is to say that η^t is an $\mathrm{OU}(\frac{\theta}{t}, \frac{\theta}{2})$ along the *perturbed* path $(1+\alpha)y$.

Putting this all together we are left with a neat factorisation expressed in terms of the rescaled-type process η_s^t:

$$\begin{aligned}\tilde{\mathbb{Q}}_t(\textbf{spine term}^\alpha) &= e^{\alpha t J(\tau) - \alpha\rho\tau} e^{\alpha^2 t M(\tau)} \times \tilde{\mathbb{Q}}_t^\alpha\left(e^{\alpha r t \int_0^\tau [y_s^2 - (\eta_s^t)^2]\,\mathrm{d}s}\right),\\ &\le e^{\alpha t J(\tau)} e^{\alpha^2 t M(\tau)} \times \tilde{\mathbb{Q}}_t^\alpha\left(e^{\alpha r t \int_0^\tau [y_s^2 - (\eta_s^t)^2]\,\mathrm{d}s}\right), \end{aligned} \tag{41}$$

where we remember that $M(\tau) := \int_0^\tau \left[\frac{1}{2\theta}(\dot{y}_s + \frac{\theta}{2}y_s)^2 + \frac{a\lambda^2}{2}y_s^2\right]\mathrm{d}s$. The term $\alpha\rho\tau$ becomes insignificant in the large-deviations limit (for which $t \to \infty$), and therefore it is convenient to have removed it here.

The martingale techniques have now played their part, and we move on to use Varadhan's lemma to show that the term $\tilde{\mathbb{Q}}_t^\alpha\left(e^{\alpha r t \int_0^\tau [y_s^2 - (\eta_s^t)^2]\,\mathrm{d}s}\right)$ decays exponentially as $t \to \infty$.

7.2 A First Application of Varadhan's Lemma

Under the measure $\tilde{\mathbb{Q}}_t^\alpha$ the process η^t is an $\mathrm{OU}(\frac{\theta}{t}, \frac{\theta}{2})$ along the perturbed path $(1+\alpha)y$ (or equivalently we can say that $[\eta_s^t - (1+\alpha)y_s]$ is an $\mathrm{OU}(\frac{\theta}{t}, \frac{\theta}{2})$), and therefore it satisfies a large-deviations principle:

Theorem 7.2. *If we use the notation η^t to refer to the element (path) in $C[0,\tau]$ defined by*

$$\eta^t(s) := \eta_s^t, \quad \textit{for } s \in [0,\tau]$$

then there is a large-deviations principle for η^t with respect to the measure $\tilde{\mathbb{Q}}_t^\alpha$:

- *Upper bound: If C is a closed subset of $C[0,\tau]$ then*

$$\limsup_{t\to\infty} t^{-1}\log\tilde{\mathbb{Q}}_t^\alpha(\eta_s^t \in C) \le -\inf_{g\in C} I(g,\tau),$$

- *Lower bound: If V is an open subset of $C[0,\tau]$ then*

$$\liminf_{t\to\infty} t^{-1}\log\tilde{\mathbb{Q}}_t^\alpha(\eta_s^t \in V) \ge -\inf_{g\in V} I(g,\tau),$$

where

$$I(g,w) := \int_0^w \frac{1}{2\theta}\left[\dot{g}_s + \frac{\theta}{2}g_s - (1+\alpha)\left(\dot{y}_s + \frac{\theta}{2}y_s\right)\right]^2 \mathrm{d}s.$$

if $g \in C[0,\tau]$ *with* $g(0) = 0$ *is square integrable along with its derivative; otherwise, we define* $I(g) = \infty$.

Given the upper bound (41), we now want to understand the behaviour of the expectation term $\tilde{\mathbb{Q}}_t^\alpha\left(e^{\alpha r t \int_0^\tau [y_s^2 - (\eta_s^t)^2]\,\mathrm{d}s}\right)$ for large t. Varadhan's lemma is a common way to deal with expectations of this form, and we quote the following from Dembo and Zeitouni [2]:

Theorem 7.3 (Varadhan). *Let* $(X^t)_{t\geq 0}$ *be a family of random variables taking values in the space* $\mathcal{X}$, *and let* μ_t *denote the probability measures associated with* $(X^t)_{t\geq 0}$.

Suppose that the measures μ_t *satisfy the LDP with a good rate function* $I : \mathcal{X} \to [0,\infty]$, *and let* $\phi : \mathcal{X} \to \mathbb{R}$ *be any continuous function. Assume further that the following moment condition holds for some* $\gamma > 1$*:*

$$\limsup_{t\to\infty} t^{-1} \log \mathbb{E}\left[e^{\gamma t \phi(X^t)}\right] < \infty. \tag{42}$$

Then

$$\lim_{t\to\infty} t^{-1} \log \mathbb{E}\left[e^{t\phi(X^t)}\right] = \sup_{x\in\mathcal{X}}\left[\phi(x) - I(x)\right].$$

This powerful result will confirm our hopes that the expectation decays as $t \to \infty$.

Theorem 7.4. *For each* $\alpha > 0$ *the expectation decays exponentially to 0:*

$$\lim_{t\to\infty} t^{-1} \log \tilde{\mathbb{Q}}_t^\alpha\left(e^{\alpha r t \int_0^\tau [y_s^2 - (\eta_s^t)^2]\,\mathrm{d}s}\right) < 0. \tag{43}$$

For small α *we can give more precise expression of the exponential decay:*

$$\lim_{t\to\infty} t^{-1} \log \tilde{\mathbb{Q}}_t^\alpha\left(e^{\alpha r t \int_0^\tau [y_s^2 - (\eta_s^t)^2]\,\mathrm{d}s}\right)$$

$$= -\alpha^2\left\{k_1\left[\int_0^\tau r y_s^2\,\mathrm{d}s\right] + k_2\left[\frac{1}{2\theta}\int_0^\tau\left(\dot{y}_s + \frac{\theta}{2}y_s\right)^2 \mathrm{d}s\right]\right\} + o(\alpha^2), \quad \text{as } \alpha \to 0,$$

where k_1 *and* k_2 *are strictly positive.*

Proof. Given the large-deviations principle stated in Theorem 7.2, we shall be equating $\mathcal{X} = C[0,\tau]$, $X^t = \eta^t$ and $\mu_t = \tilde{\mathbb{Q}}_t^\alpha$ and have $\phi(\eta^t) = \int_0^\tau [y_s^2 - (\eta_s^t)^2]\,\mathrm{d}s$; the moment condition (42) is satisfied because

$$\tilde{\mathbb{Q}}_t^\alpha\left(e^{2\alpha r t \int_0^\tau [y_s^2 - (\eta_s^t)^2]\,\mathrm{d}s}\right) < e^{2\alpha r t \int_0^\tau y_s^2\,\mathrm{d}s}.$$

Varadhan's lemma implies that

$$\lim_{t\to\infty} t^{-1}\log\tilde{\mathbb{Q}}_t^{\alpha}\left(e^{\alpha rt\int_0^\tau [y_s^2-(\eta_s^t)^2]\,\mathrm{d}s}\right) = \sup_{z\in C_0[0,\tau]}\left\{\left(\int_0^\tau \alpha r\left[y_s^2 - z_s^2\right]\mathrm{d}s\right) - I(z,\tau)\right\}. \tag{44}$$

Standard Euler–Lagrange techniques for maximising the right-hand integral lead to the following differential equation for z:

$$\ddot{z}_s - \left(\frac{\theta^2}{4} + 2\theta\alpha r\right)z_s = (1+\alpha)\ddot{y}_s - \frac{\theta^2}{4}(1+\alpha)y_s, \tag{45}$$

which in general will give the optimal path as a solution in terms of the given path y.

With the *specific* path (20) that resulted from the Harris and Git optimisations of the large-deviations heuristics, it is relatively simple to solve (45) and find that the optimal path z is just a constant multiple of the path y:

$$z_s = K_\alpha y_s, \qquad \text{where } K_\alpha := \frac{\mu_\lambda^2 - \theta^2/4}{\mu_\lambda^2 - \theta^2/4 - 2\theta\alpha r}(1+\alpha). \tag{46}$$

Substituting for z into (44) we find that

$$\lim_{t\to\infty} t^{-1}\log\tilde{\mathbb{Q}}_t^{\alpha}\left(e^{\alpha rt\int_0^\tau [y_s^2-(\eta_s^t)^2]\,\mathrm{d}s}\right)$$

$$= \alpha(1-K_\alpha^2)\left[\int_0^\tau r y_s^2\,\mathrm{d}s\right] - (K_\alpha-(1+\alpha))^2\left[\frac{1}{2\theta}\int_0^\tau \left(\dot{y}_s + \frac{\theta}{2}y_s\right)^2\mathrm{d}s\right], \tag{47}$$

and the following simple bound on K_α implies that this is a negative quantity

Lemma 7.5. *For all* $\alpha > 0$,

$$1 < K_\alpha < 1+\alpha. \tag{48}$$

This small lemma can be proved with simple algebra from the definition of μ_λ given at (21): we can use this to show that $\mu_\lambda^2 - \theta^2/4 = -2\theta r - a\theta\lambda^2 < 0$, from which it follows that

$$\frac{1}{1+\alpha} < \frac{\mu_\lambda^2 - \theta^2/4}{\mu_\lambda^2 - \theta^2/4 - 2\theta\alpha r} < 1.$$

If we make a Taylor expansion about $\alpha = 0$:

$$\frac{\mu_\lambda^2 - \theta^2/4}{\mu_\lambda^2 - \theta^2/4 - 2\theta\alpha r} = \frac{1}{1-k\alpha} = 1 + k\alpha + k^2\alpha^2 + o(\alpha^2) + \cdots$$

where $k := \frac{2\theta r}{\mu_\lambda^2 - \theta^2/4}$, it follows that for strictly positive constants k_1 and k_2,

$$\alpha(1-K_\alpha^2) = -k_1\alpha^2 + o(\alpha^2), \quad \text{and} \quad (K_\alpha - (1+\alpha))^2 = k_2\alpha^2 + o(\alpha^2) \quad \text{as } \alpha\to 0,$$

completing the proof □

7.3 Dealing with the Sum Term

Focusing on the sum term, we can again substitute for $\mathrm{d}W_s$ and $\mathrm{d}B_s$ with (38) and immediately factor out the term $J(S_k)$ by overestimating

$$\textbf{sum term} = \sum_{k=1}^{n_\tau} e^{tJ(S_k)-\rho S_k} e^{r\int_0^{S_k}[y_s^2-(\eta_s^t)^2]\,\mathrm{d}s} e^{[\sqrt{at}\lambda\int_0^{S_k} y_s\,\mathrm{d}\tilde{W}_s]+[\frac{\sqrt{t}}{\sqrt{\theta}}\int_0^{S_k}(\dot{y}_s+\frac{\theta}{2}y_s)\,\mathrm{d}\tilde{B}_s]}$$

$$\leq e^{t\left(\sup_{0\leq w\leq\tau} J(w)\right)} \sum_{k=1}^{n_\tau} e^{r\int_0^{S_k}\eta_s^2-y_s^2\,\mathrm{d}s} e^{[\sqrt{a}\lambda\int_0^{S_k} y_s\,\mathrm{d}\tilde{W}_s]+[\frac{1}{\sqrt{\theta}}\int_0^{S_k}(\dot{y}_s+\frac{\theta}{2}y_s)\,\mathrm{d}\tilde{B}_s]}.$$

For the particular path y that we chose at (20), it was shown by Git et al. [4] that

$$\sup_{0\leq w\leq\tau} J(w) = J(\tau)$$

and therefore we have

$$\textbf{sum term} \leq e^{tJ(\tau)} \sum_{k=1}^{n_\tau} e^{r\int_0^{S_k}\eta_s^2-y_s^2\,\mathrm{d}s} e^{[\sqrt{a}\lambda\int_0^{S_k} y_s\,\mathrm{d}\tilde{W}_s]+[\frac{1}{\sqrt{\theta}}\int_0^{S_k}(\dot{y}_s+\frac{\theta}{2}y_s)\,\mathrm{d}\tilde{B}_s]}.$$

The following small result is very useful for dealing with the sum term:

Proposition 7.6. *If* $\alpha\in(0,1]$ *and* $u,v>0$ *then* $(u+v)^\alpha\leq u^\alpha+v^\alpha$.

This proposition implies that for $0\leq\alpha\leq 1$,

$$\tilde{\mathbb{Q}}_t\left(\textbf{sum term}^\alpha\right) \leq e^{\alpha tJ(\tau)}\tilde{\mathbb{Q}}_t\left(\sum_{k=1}^{n_\tau} e^{\alpha rt\int_0^{S_k}[y_s^2-(\eta_s^t)^2]\,\mathrm{d}s} e^{\alpha[\sqrt{at}\lambda\int_0^{S_k} y_s\,\mathrm{d}\tilde{W}_s]+\alpha[\frac{\sqrt{t}}{\sqrt{\theta}}\int_0^{S_k}(\dot{y}_s+\frac{\theta}{2}y_s)\,\mathrm{d}\tilde{B}_s]}\right),$$

and we can transform the sum into an integral by standard techniques (see, e.g., Kallenberg [10]), since the fission times on the spine form a Cox process of rate $2(r\eta_w^2+\rho)$, as explained in Theorem 4.1:

$$= 2e^{\alpha tJ(\tau)}\tilde{\mathbb{Q}}_t\left(\int_0^\tau e^{\alpha rt\int_0^w[y_s^2-(\eta_s^t)^2]\,\mathrm{d}s} e^{\alpha[\sqrt{at}\lambda\int_0^w y_s\,\mathrm{d}\tilde{W}_s]+\alpha[\frac{\sqrt{t}}{\sqrt{\theta}}\int_0^w(\dot{y}_s+\frac{\theta}{2}y_s)\,\mathrm{d}\tilde{B}_s]}\left[r\eta_w^2+\rho\right]\mathrm{d}w\right);$$

Fubini's theorem can be applied to this, and the transformations that worked on the spine term to give (41) can here too be applied to arrive at

$$= 2e^{\alpha tJ(\tau)}\int_0^\tau e^{\alpha^2\int_0^w\frac{a\lambda^2}{2}y_s^2\,\mathrm{d}s} e^{\frac{\alpha^2}{2\theta}\int_0^w(\dot{y}_s+\frac{\theta}{2}y_s)^2\,\mathrm{d}s}\times\tilde{\mathbb{Q}}_t^\alpha\left(\left[rt(\eta_w^t)^2+\rho\right]e^{\alpha rt\int_0^w[y_s^2-(\eta_s^t)^2]\,\mathrm{d}s}\right)\mathrm{d}w,$$

$$\leq 2e^{\alpha tJ(\tau)}e^{\alpha^2 tM(\tau)}\times\int_0^\tau\tilde{\mathbb{Q}}_t^\alpha\left(\left[rt(\eta_w^t)^2+\rho\right]e^{\alpha rt\int_0^w[y_s^2-(\eta_s^t)^2]\,\mathrm{d}s}\right)\mathrm{d}w.$$

We want to take advantage of the fact that the terms in the integral look similar to those already dealt with for the spine term. A first step in this direction is to replace the random factor $rt(\eta_w^t)^2$ at the front of the expectation with the deterministic rty_w^2, and since the value of α will eventually be chosen and fixed the following estimate is sufficient for our purposes:

Lemma 7.7. *For all $\alpha > 0$ and for all large enough t,*

$$\int_0^\tau \tilde{\mathbb{Q}}_t^\alpha\left(\left[rt(\eta_w^t)^2+\rho\right]e^{\alpha rt\int_0^w [y_s^2-(\eta_s^t)^2]\,\mathrm{d}s}\right)\mathrm{d}w \leq \frac{1}{\alpha}+\int_0^\tau \left[rty_w^2+\rho\right]\tilde{\mathbb{Q}}_t^\alpha\left(e^{\alpha rt\int_0^w [y_s^2-(\eta_s^t)^2]\,\mathrm{d}s}\right)\mathrm{d}w.$$

Proof. Since

$$\frac{\partial}{\partial w}\tilde{\mathbb{Q}}_t^\alpha\left(e^{\alpha rt\int_0^w [y_s^2-(\eta_s^t)^2]\,\mathrm{d}s}\right) = \tilde{\mathbb{Q}}_t^\alpha\left(\alpha rt[y_w^2-(\eta_w^t)^2]e^{\alpha rt\int_0^w [y_s^2-(\eta_s^t)^2]\,\mathrm{d}s}\right), \tag{49}$$

it follows that

$$\tilde{\mathbb{Q}}_t^\alpha\left(\left[rt(\eta_w^t)^2+\rho\right]e^{-\alpha rt\int_0^w [y_s^2-(\eta_s^t)^2]\,\mathrm{d}s}\right) = \left[rty_w^2+\rho\right]\tilde{\mathbb{Q}}_t^\alpha\left(e^{\alpha rt\int_0^w [y_s^2-(\eta_s^t)^2]\,\mathrm{d}s}\right) - \frac{1}{\alpha}\frac{\partial}{\partial w}\tilde{\mathbb{Q}}_t^\alpha\left(e^{\alpha rt\int_0^w [y_s^2-(\eta_s^t)^2]\,\mathrm{d}s}\right).$$

Integration by parts now proves

$$\int_0^\tau \tilde{\mathbb{Q}}_t^\alpha\left(\left[rt(\eta_w^t)^2+\rho\right]e^{\alpha rt\int_0^w [y_s^2-(\eta_s^t)^2]\,\mathrm{d}s}\right)\mathrm{d}w = \int_0^\tau \left[rty_w^2+\rho\right]\tilde{\mathbb{Q}}_t^\alpha\left(e^{\alpha rt\int_0^w [y_s^2-(\eta_s^t)^2]\,\mathrm{d}s}\right)\mathrm{d}w + \frac{1}{\alpha}\left[1-\tilde{\mathbb{Q}}_t^\alpha\left(e^{\alpha rt\int_0^\tau [y_s^2-(\eta_s^t)^2]\,\mathrm{d}s}\right)\right].$$

The exponential decay proved in Theorem 7.4 implies $\lim_{t\to\infty}\tilde{\mathbb{Q}}_t^\alpha\left(e^{\alpha rt\int_0^\tau [y_s^2-(\eta_s^t)^2]\,\mathrm{d}s}\right) = 0$, and this completes the proof □

It follows therefore that for all large enough t,

$$\tilde{\mathbb{Q}}_t\left(\textbf{sum term}^\alpha\right) \leq 2e^{\alpha tJ(\tau)}e^{\alpha^2 tM(\tau)}\times\left(\frac{1}{\alpha}+\int_0^\tau \left[rty_w^2+\rho\right]\tilde{\mathbb{Q}}_t^\alpha\left(e^{\alpha rt\int_0^w [y_s^2-(\eta_s^t)^2]\,\mathrm{d}s}\right)\mathrm{d}w\right).$$

We now make some simple overestimates of the integral. Firstly, it is immediate that

$$\int_0^\tau \left[rty_w^2+\rho\right]\tilde{\mathbb{Q}}_t^\alpha\left(e^{\alpha rt\int_0^w [y_s^2-(\eta_s^t)^2]\,\mathrm{d}s}\right)\mathrm{d}w \leq \left[rt\kappa^2+\rho\right]\int_0^\tau \tilde{\mathbb{Q}}_t^\alpha\left(e^{\alpha rt\int_0^w [y_s^2-(\eta_s^t)^2]\,\mathrm{d}s}\right)\mathrm{d}w$$

since $\left(\sup_{0\leq w\leq\tau} y_w^2\right) = \kappa^2$. Then, for each $w \in [0,\tau]$, it is true by definition that

$$e^{\alpha rt\int_0^w [y_s^2-(\eta_s^t)^2]\,\mathrm{d}s} \leq e^{\alpha rt(\sup_{0\leq w\leq\tau}\int_0^w [y_s^2-(\eta_s^t)^2]\,\mathrm{d}s)},$$

and therefore

$$\tilde{\mathbb{Q}}_t^\alpha\left(e^{\alpha rt\int_0^w [y_s^2-(\eta_s^t)^2]\,\mathrm{d}s}\right) \leq \tilde{\mathbb{Q}}_t^\alpha\left(e^{\alpha rt\left(\sup_w\int_0^w [y_s^2-(\eta_s^t)^2]\,\mathrm{d}s\right)}\right).$$

Since this holds for all $w \in [0,\tau]$ we can deduce

$$\sup_{0\le w\le\tau} \tilde{\mathbb{Q}}_t^\alpha\left(e^{\alpha rt\int_0^w[y_s^2-(\eta_s^t)^2]\,\mathrm{d}s}\right) \le \tilde{\mathbb{Q}}_t^\alpha\left(e^{\alpha rt\left(\sup_w\int_0^w[y_s^2-(\eta_s^t)^2]\,\mathrm{d}s\right)}\right),$$

which we can use to get

$$\int_0^\tau \tilde{\mathbb{Q}}_t^\alpha\left(e^{\alpha rt\int_0^w[y_s^2-(\eta_s^t)^2]\,\mathrm{d}s}\right)\mathrm{d}w \le \tau\times\sup_{0\le w\le\tau} \tilde{\mathbb{Q}}_t^\alpha\left(e^{\alpha rt\int_0^w[y_s^2-(\eta_s^t)^2]\,\mathrm{d}s}\right),$$
$$\le \tau\times\tilde{\mathbb{Q}}_t^\alpha\left(e^{\alpha rt\left(\sup_w\int_0^w[y_s^2-(\eta_s^t)^2]\,\mathrm{d}s\right)}\right).$$

Thus we arrive at a simple upper bound for the sum term: for all $\alpha\in[0,1]$ and all large t,

$$\tilde{\mathbb{Q}}_t\left(\textbf{sum term}^\alpha\right)\le 2e^{\alpha tJ(\tau)}e^{\alpha^2tM(\tau)}\left\{\frac{1}{\alpha}+\left[rt\kappa^2+\rho\right]\tau\tilde{\mathbb{Q}}_t^\alpha\left(e^{\alpha rt\left(\sup_w\int_0^w[y_s^2-(\eta_s^t)^2]\,\mathrm{d}s\right)}\right)\right\}. \tag{50}$$

7.4 A Second Application of Varadhan's Lemma

We already applied Varadhan's lemma to the term $\tilde{\mathbb{Q}}_t^\alpha\left(e^{\alpha rt\int_0^\tau[y_s^2-(\eta_s^t)^2]\,\mathrm{d}s}\right)$, and now we show how it can in fact deal with the more complex term $\tilde{\mathbb{Q}}_t^\alpha\left(e^{\alpha rt\sup_w\int_0^w[y_s^2-(\eta_s^t)^2]\,\mathrm{d}s}\right)$ without much more effort.

Once again the observation

$$\tilde{\mathbb{Q}}_t^\alpha\left(e^{2\alpha rt\left(\sup_w\int_0^w[y_s^2-(\eta_s^t)^2]\,\mathrm{d}s\right)}\right) < \tilde{\mathbb{Q}}_t^\alpha\left(e^{2\alpha rt\tau(\sup_w y_w^2)}\right)$$

shows that the moment condition (42) is satisfied, and therefore from Varadhan's lemma, Theorem 7.3, it follows that

$$\lim_{t\to\infty} t^{-1}\log\tilde{\mathbb{Q}}_t^\alpha\left(e^{\alpha rt\left(\sup_w\int_0^w[y_s^2-(\eta_s^t)^2]\,\mathrm{d}s\right)}\right) = \sup_{z\in C_0[0,\tau]}\left\{\left(\sup_{0\le w\le\tau}\int_0^w \alpha r\left[y_s^2-z_s^2\right]\mathrm{d}s\right)-I(z,\tau)\right\}.$$

For any path z, the action functional $I(z,w)$ is nondecreasing in w and therefore

$$\left(\int_0^w \alpha r\left[y_s^2-z_s^2\right]\mathrm{d}s\right)-I(z,\tau)\le\left(\int_0^w \alpha r\left[y_s^2-z_s^2\right]\mathrm{d}s\right)-I(z,w),$$

and taking the supremum over $w\in[0,\tau]$ of both sides we deduce

$$\left(\sup_w\int_0^w \alpha r\left[y_s^2-z_s^2\right]\mathrm{d}s\right)-I(z,\tau)\le\sup_w\left\{\left(\int_0^w \alpha r\left[y_s^2 z_s^2\right]\mathrm{d}s\right)-I(z,w)\right\},$$

We now take the supremum of both sides over the set of paths $z \in C_0[0,\tau]$ and interchange the order to obtain

$$\sup_z\left\{\left(\sup_w \int_0^w \alpha r\left[y_s^2 - z_s^2\right] \mathrm{d}s\right) - I(z,\tau)\right\} \le \sup_z \sup_w \left\{\left(\int_0^w \alpha r\left[y_s^2 - z_s^2\right] \mathrm{d}s\right) - I(z,w)\right\}$$
$$= \sup_{0\le w\le\tau} \sup_z \left\{\left(\int_0^w \alpha r\left[y_s^2 - z_s^2\right] \mathrm{d}s\right) - I(z,w)\right\}. \tag{51}$$

If we compare the term

$$\sup_z\left\{\left(\int_0^w \alpha r\left[y_s^2 - z_s^2\right] \mathrm{d}s\right) - I(z,w)\right\}$$

with (44) from our first application of Varadhan's lemma, it is clear that Euler–Lagrange optimisation techniques will result in exactly the same optimal path for this integral, namely $z_s = K_\alpha y_s$ as at (46). Furthermore, evaluating the left-hand side of (51) shows that we actually have the equality

$$\sup_z\left\{\left(\int_0^\tau \alpha r\left[y_s^2 - z_s^2\right] \mathrm{d}s\right) - I(z,\tau)\right\} = \sup_z\left\{\left(\sup_w \int_0^w \alpha r\left[y_s^2 - z_s^2\right] \mathrm{d}s\right) - I(z,\tau)\right\},$$
$$= \alpha(1-K_\alpha^2)\left[\int_0^\tau r y_s^2 \,\mathrm{d}s\right] - (K_\alpha - (1+\alpha))^2\left[\frac{1}{2\theta}\int_0^\tau (\dot{y}_s + \frac{\theta}{2} y_s)^2 \mathrm{d}s\right],$$
$$< 0 \ \left(\text{and } = O(\alpha^2) \text{ as } \alpha \to 0\right).$$

Consequently we see that there is no difference in the growth rate between the remaining terms of the spine term and the sum term:

$$\lim_{t\to\infty} t^{-1} \log \tilde{\mathbb{Q}}_t^\alpha\left(e^{\alpha r t\left(\sup_w \int_0^w [y_s^2 - (\eta_s^t)^2]\,\mathrm{d}s\right)}\right) = \lim_{t\to\infty} t^{-1} \log \tilde{\mathbb{Q}}_t^\alpha\left(e^{\alpha r t \int_0^\tau [y_s^2 - (\eta_s^t)^2]\,\mathrm{d}s}\right) < 0. \tag{52}$$

7.5 *Concluding the Upper-Bound for* $Z_t(\tau)$

We have shown that

$$\tilde{\mathbb{Q}}_t(\textbf{spine term}^\alpha) \le e^{\alpha t J(\tau)} e^{\alpha^2 t M(\tau)} \times \tilde{\mathbb{Q}}_t^\alpha\left(e^{\alpha r t \int_0^\tau [y_s^2 - (\eta_s^t)^2]\,\mathrm{d}s}\right),$$

and since we clearly have $\tilde{\mathbb{Q}}_t^\alpha\left(e^{\alpha r t \int_0^\tau [y_s^2-(\eta_s^t)^2]\,\mathrm{d}s}\right) \le \tilde{\mathbb{Q}}_t^\alpha\left(e^{\alpha r t\left(\sup_w \int_0^w [y_s^2-(\eta_s^t)^2]\,\mathrm{d}s\right)}\right)$, it follows that

$$\tilde{\mathbb{Q}}_t\left(Z_t(\tau)^\alpha\right) \le \tilde{\mathbb{Q}}_t\left(\textbf{spine term}^\alpha\right) + \tilde{\mathbb{Q}}_t\left(\textbf{sum term}^\alpha\right)$$
$$\le e^{\alpha t J(\tau)} e^{\alpha^2 t M(\tau)} \left\{\left(1+2\left[rt\kappa^2+\rho\right]\tau\right)\tilde{\mathbb{Q}}_t^\alpha\left(e^{\alpha r t\left(\sup_w \int_0^w [y_s^2-(\eta_s^t)^2]\mathrm{d}s\right)}\right)+\frac{2}{\alpha}\right\}. \tag{53}$$

Thus

$$\lim_{t\to\infty} t^{-1}\log\tilde{\mathbb{Q}}_t\left(Z_t(\tau)^\alpha\right) \le \alpha J(\tau) + \alpha^2 M(\tau),$$

and the proof of Theorem 4.3 is completed. □

8 An Alternative Approach to the Lower Bound

Finally, we suggest an alternative approach to gaining the lower bound of Theorem 3.2. The key tool is still the spine decomposition and this approach should work more generally. In particular, it would not require calculation of the $\tilde{\mathbb{Q}}_t(Z_t(\tau)^\alpha)$ in order to control the size of the martingale $Z_t(s)$, which may be advantageous in some models. However, in this instance, there appears little to choose between the two approaches.

Making use of some simple estimation, the tower property, recalling that $\mathcal{G}_\infty$ contains information only about the spine's spatial trajectory and using the conditional form of Jensen's inequality, we have

$$\begin{aligned}
P\left(\exists u \in N_\tau \text{ such that } A_t(u)\right) &= \tilde{\mathbb{Q}}_t\left(\frac{1}{Z_t(\tau)}; \exists u \in N_\tau, A_t(u)\right)\\
&\ge \tilde{\mathbb{Q}}_t\left(\frac{1}{Z_t(\tau)}; A_t(\xi)\right) = \tilde{\mathbb{Q}}_t\left\{\tilde{\mathbb{Q}}_t\left(\frac{1}{Z_t(\tau)}\,\middle|\,\mathcal{G}_\infty\right); A_t(\xi)\right\}\\
&\ge \tilde{\mathbb{Q}}_t\left\{\frac{1}{\tilde{\mathbb{Q}}_t\left(Z_t(\tau)\,|\,\mathcal{G}_\infty\right)}; A_t(\xi)\right\}\\
&= \tilde{\mathbb{Q}}_t\left\{\frac{1}{f_{t,\tau}(\xi)+\int_0^\tau 2R(\xi_s) f_{t,s}(\xi)\,\mathrm{d}s}; A_t(\xi)\right\}.
\end{aligned}$$

If a good upper bound for $f_{t,\tau}(\xi)$ and $\int_0^\tau 2R(\xi_s) f_{s,\tau}(\xi)\,\mathrm{d}s$ can be found on the event $A_t(\xi)$, a suitable lower bound for the required probability will follow. This is a similar idea as used in the upper bound approach (see Eq. (37)), except some additional work would be required to control the integral over time (although in this situation, it will be dominated in exponential order by the final value of $f_{t,\tau}(\xi)$ to match the upper bound).

References

1. Athreya, K.B. *Change of measures for Markov chains and the* $L\log L$ *theorem for branching processes*, Bernoulli **6** (2000), no. 2, 323–338.
2. Dembo, A. and Zeitouni, O. *Large deviations techniques and applications*, Springer, 1998.
3. Engländer, J. and Kyprianou, A.E. *Local extinction versus local exponential growth for spatial branching processes*, Ann. Probab. **32** (2004), no. 1A, 78–99.
4. Git, Y., Harris, J.W. and Harris, S.C *Exponential growth rates in a typed branching diffusion*, Ann. App. Probab., **17** (2007), no. 2, 609-653.
5. Hardy, R., and Harris, S. C. *A spine approach to branching diffusions with applications to* L^p*-convergence of martingales*, Séminaire de Probabilités, **XLII**, (2009), 281–330.
6. Hardy, R., and Harris, S. C. *A conceptual approach to a path result for branching Brownian motion*, Stochastic Processes and their Applications, **116** (2006), no. 12, 1992–2013.
7. Harris, S. C. and Williams, D. *Large deviations and martingales for a typed branching diffusion. I*, Astérisque (1996), no. 236, 133–154, Hommage à P. A. Meyer et J. Neveu.
8. Harris, S.C. *Convergence of a 'Gibbs-Boltzmann' random measure for a typed branching diffusion*, Séminaire de Probabilités, **XXXIV**, Lecture Notes in mathematics, vol. 1729, Springer, (2000), 133–154.
9. Iksanov, A.M. *Elementary fixed points of the BRW smoothing transforms with infinite number of summands*, Stochastic Process. Appl. **114** (2004), no. 1, 27–50.
10. Kallenberg, O. *Foundations of modern probability*, Springer-Verlag, 2002.
11. Kurtz, T., Lyons, R., Pemantle, R. and Peres, Y. *A conceptual proof of the Kesten-Stigum theorem for multi-type branching processes*, Classical and modern branching processes (Minneapolis, MN, 1994), IMA Vol. Math. Appl., vol. 84, Springer, New York, 1997, pp. 181–185.
12. Kyprianou, A.E. *Travelling wave solutions to the K-P-P equation: alternatives to Simon Harris's probabilistic analysis*, **40** (2004), no. 1, 53–72.
13. Kyprianou, A.E. and Rahimzadeh Sani, A. *Martingale convergence and the functional equation in the multi-type branching random walk*, Bernoulli **7** (2001), no. 4, 593–604.
14. Lee, T. *Some large-deviation theorems for branching diffusions*, Ann. Probab. **20** (1992), no. 3, 1288–1309.
15. Lyons, R. *A simple path to Biggins' martingale convergence for branching random walk*, Classical and modern branching processes (Minneapolis, MN, 1994), IMA Vol. Math. Appl., vol. 84, Springer, New York, 1997, pp. 217–221.
16. Lyons, R., Pemantle, R. and Peres, Y. *Conceptual proofs of* $L\log L$ *criteria for mean behavior of branching processes*, Ann. Probab. **23** (1995), no. 3, 1125–1138.
17. Øksendal, B. *Stochastic differential equations*, fifth ed., Springer, 2000.
18. Olofsson, P. *The* $x\log x$ *condition for general branching processes*, J. Appl. Probab. **35** (1998), no. 3, 537–544.

Longtime Behavior for Mutually Catalytic Branching with Negative Correlations

Leif Döring and Leonid Mytnik

Abstract In several examples, dualities for interacting diffusion and particle systems permit the study of the longtime behavior of solutions. A particularly difficult model in which many techniques collapse is a two-type model with mutually catalytic interaction introduced by Dawson/Perkins for which they proved under some assumptions a dichotomy between extinction and coexistence directly from the defining equations.

In the present chapter we show how to prove a precise dichotomy for a related model with negatively correlated noises. The proof uses moment bounds on exit times of correlated Brownian motions from the first quadrant and explicit second moment calculations. Since the uniform integrability bound is independent of the branching rate our proof can be extended to infinite branching rate processes.

Keywords Longtime Behavior • Branching process • Planar Brownian Motion • Duality

Mathematics Subject Classification (2000): Primary 60J80; Secondary 60J85.

L. Döring (✉)
LPMA, Université Paris VI, Tours 16/26, 4 Place Jussieu, 75005 Paris, France

LM is partly supported by the Israel Science Foundation and B. and G. Greenberg Research Fund (Ottawa)
e-mail: leif.doering@googlemail.com

L. Mytnik
Faculty of Industrial Engineering and Management Technion Israel Institute of Technology, Haifa 32000, Israel

LD was supported by the Fondation Science Matématiques de Paris
e-mail: leonid@ie.technion.ac.il

J. Englander and B. Rider (eds.), *Advances in Superprocesses and Nonlinear PDEs*, Springer Proceedings in Mathematics & Statistics 38,
DOI 10.1007/978-1-4614-6240-8_6,

1 Introduction

A classical task in the theory of interacting particle systems and interacting diffusion models is a precise understanding of the longtime behavior of the system. For many models dichotomies between extinction and survival have been revealed. In most cases proofs are based on clever duality constructions or explicit representations of the particular process. A good example is the *voter model* for which a graphical construction can be applied that reduces extinction problems to hitting problems of random walks (see, e.g., Chap. V of [L05] for many beautiful results).

In this chapter we aim at giving simple proofs for the dichotomy in a related class of mutually interacting diffusion processes with negatively correlated driving noises. Interestingly, the particular structure of the processes permits a second moment calculation that leads to the precise characterization of survival/extinction via recurrence/transience of the underlying migration mechanism. The approach is more direct than the previously used extension of arguments due to [DP98] for which regularity assumption on the underlying migration mechanism was imposed.

1.1 Finite Rate Symbiotic Branching Processes

In 2004, Etheridge and Fleischmann [EF04] introduced a stochastic spatial model of two interacting populations known as the (finite rate) *symbiotic branching model*, parametrized by a parameter $\rho \in [-1,1]$ governing the correlations between the driving noises and a branching parameter $\gamma > 0$ amplifying the strength of the noises. For a discrete spatial version of their stochastic heat equations, we consider the system of interacting diffusions on a countable set S with values in $\mathbb{R}_{\geq 0}$, defined by the coupled stochastic differential equations

$$\mathrm{SBM}_\gamma = \mathrm{SBM}(\rho,\gamma)_{u_0,v_0}: \begin{cases} du_t(i) = \mathcal{A}u_t(i)\,dt + \sqrt{\gamma u_t(i)v_t(i)}\,dB_t^1(i), \\ dv_t(i) = \mathcal{A}v_t(i)\,dt + \sqrt{\gamma u_t(i)v_t(i)}\,dB_t^2(i), \\ u_0(i) \geq 0, \quad i \in S, \\ v_0(i) \geq 0, \quad i \in S, \end{cases}$$

where $\{B^1(i), B^2(i)\}_{i\in S}$ is a $(\mathbb{R}^2)^S$-valued centered Gaussian process with covariance structure

$$\mathbb{E}\left[B_t^n(i)B_t^m(j)\right] = \begin{cases} \rho t & : i = j \text{ and } n \neq m, \\ t & : i = j \text{ and } n = m, \\ 0 & : \text{otherwise}. \end{cases}$$

The migration operator $\mathcal{A}$ is defined as

$$\mathcal{A}w(i) = \sum_{j \in S} a(i,j)w(j),$$

where $\big(a(i,j)\big)_{i,j \in S}$ will always be assumed to be the Q-matrix of a symmetric continuous-time S-valued Markov process with bounded jump rate, i.e.,

$$\sup_{k \in S} |a(k,k)| < \infty.$$

Some care is needed to define properly the state-space of solutions. Here, we consider solutions in the space $L^\beta \subset (\mathbb{R}_+^2)^S$, usually referred to as Liggett–Spitzer space in the theory of interacting particle systems. Fixing a test-sequence $\beta \in (0,\infty)^S$ such that

$$\sum_{i \in S} \beta(i) < \infty \quad \text{and} \quad \sum_{i \in S} \beta(i)|a(i,k)| < M\beta(k)$$

for all $k \in S$, the state-space of solutions becomes

$$L^\beta := \big\{(u,v) : S \to \mathbb{R}^+ \times \mathbb{R}^+, \langle u, \beta \rangle, \langle v, \beta \rangle < \infty\big\},$$

where $\langle f, g \rangle = \sum_{k \in S} f(k)g(k)$. The existence of the test-sequence β is ensured by Lemma IX.1.6 of [L05]. For $u \in \mathbb{R}^S$, let

$$\|u\|_{\beta,1} = \sum_{k \in S} |u(k)|\beta(k).$$

Furthermore, for $w = (u,v) \in L^\beta$, let $\|w\|_\beta = \|u\|_{\beta,1} + \|v\|_{\beta,1}$. Note that $\|\cdot\|_\beta$ defines a topology on L^β. We will henceforth assume that L^β is equipped with this topology.

We shall say that a pair (u_t^γ, v_t^γ) of continuous L^β-valued adapted processes is a solution of SBM_γ on the stochastic basis $(\Omega, \mathcal{F}, \mathcal{F}_t, \mathbb{P})$ if there is a family $\big\{B^1(i), B^2(i)\big\}_{i \in S}$ of $\mathcal{F}_t$-Brownian motions with the aforementioned correlation structure and the stochastic equation is fulfilled. Hence, the solutions are defined in the weak sense. They are usually first built on finite subsets of S via standard SDE theory and then transferred to S via a weak-limiting procedure. To avoid long repetitions we refer the reader to Sect. 3 of [BDE11] for a summary of existence of weak solutions, uniqueness and duality results for symbiotic branching processes in the case of $\mathcal{A} = \Delta$ being the discrete Laplacian; note that these results can be easily translated for the general case of $\mathcal{A}$ treated here. We also refer the reader to [DM11] for a general review of the subject.

Let us be more specific about the migration operator $\mathcal{A}$. There are a number of typical choices for $\mathcal{A}$ which the reader might keep in mind. First of all it is the case of the discrete Laplacian

$$\mathcal{A}w(i) = \Delta w(i) = \sum_{|k-i|=1} \frac{1}{2d}(w(k) - w(i)),\ i \in \mathbb{Z}^d,$$

which describes nearest neighbor interaction on $S = \mathbb{Z}^d$. There are also the cases of complete graph interaction on a finite set S corresponding to $a(i,j) = |S|^{-1}$, $i \neq j$, and the trivial migration $\mathcal{A}w = 0$ on a single point set $S = \{s\}$ leading to the nonspatial symbiotic branching SDE

$$\begin{cases} du_t = \sqrt{\gamma u_t v_t}\, dB_t^1, \\ dv_t = \sqrt{\gamma u_t v_t}\, dB_t^2, \end{cases} \tag{1}$$

driven by correlated Brownian motions.

So far we did not motivate the reason to study this particular set of stochastic differential equations. Interestingly, symbiotic branching models relate spatial models from different branches of probability theory of the type

$$dw_t(i) = \mathcal{A}w_t(i)\,dt + \sqrt{\gamma f(w_t(i))}\,dB_t(i), \quad i \in S, \tag{2}$$

that are usually referred to as interacting diffusions models. Here are some particular examples:

Example 1.1. The *stepping stone model* from mathematical genetics: $f(x) = x(1-x)$ (see for instance [SS80]).

Example 1.2. The *parabolic Anderson model (with Brownian potential)* from mathematical physics: $f(x) = x^2$ (see for instance [S92]).

Example 1.3. The *super-random walk* from the theory of branching processes: $f(x) = x$ (see for instance Sect. 2.2.4 of [E00] for the continuum analogue).

For the super-random walk, γ is the branching rate which in this case is time-space independent. In [DP98], a two-type model based on two super-random walks with time-space dependent branching was introduced. The branching rate for one species at a given site is proportional to the size of the other species at the same site. More precisely, the authors considered for $i \in S$

$$\begin{cases} du_t(i) = \mathcal{A}u_t(i)\,dt + \sqrt{\gamma u_t(i) v_t(i)}\,dB_t^1(i), \\ dv_t(i) = \mathcal{A}v_t(i)\,dt + \sqrt{\gamma u_t(i) v_t(i)}\,dB_t^2(i), \end{cases} \tag{3}$$

where now $\{B^1(i), B^2(i)\}_{i\in S}$ is a family of independent standard Brownian motions. Solutions are called *mutually catalytic branching* processes. The interest in mutually catalytic branching processes originates from the fact that it (resp. its continuum analogue) constitutes a version of two interacting super-processes with random branching environment. Many of the classical tools from the study of superprocesses fail since the branching property breaks down. Nonetheless, a detailed study is possible due to the symmetric nature of the equations. During the last decade, properties of this model were well studied (see for instance [CK00] and [CDG04]).

Let us now see how the above examples relate to the symbiotic branching model SBM_γ. For correlation $\rho = 0$, solutions of the symbiotic branching model are obviously solutions of the mutually catalytic branching model. The case $\rho = -1$ with the additional assumption $u_0 + v_0 \equiv 1$ corresponds to the stepping stone model. To see this, observe that in this perfectly negatively correlated case of $B^1(i) = -B^2(i)$, the sum $u+v$ solves a discrete heat equation, and with the further assumption $u_0 + v_0 \equiv 1$, $u+v$ stays constant for all time. Hence, for all $t \geq 0$, $u_t \equiv 1 - v_t$ which shows that u_t is a solution of the stepping stone model with initial condition u_0 and v_t is a solution with initial condition v_0. Finally, suppose w is a solution of the parabolic Anderson model, then, for $\rho = 1$, the pair $(u,v) := (w,w)$ is a solution of the symbiotic branching model with initial conditions $u_0 = v_0 = w_0$.

This unifying property motivated the study in [BDE11] of the influence of varying ρ on the longtime behavior for γ being a fixed constant. Since the stepping stone model and the parabolic Anderson model have very different longtime properties it could be expected to recover some of those properties in disjoint regions of the parameter range. Restricting to $\mathcal{A} = \Delta$ on $\mathbb{Z}^d$ and $d = 1,2$, the longtime behavior of laws and moments has been analyzed. An important observation of [BDE11], which holds equally for quite general $\mathcal{A}$ generating a recurrent Markov process, is the following moment transition for solutions SBM_γ started in homogeneous initial conditions $u_0 = v_0 \equiv 1$:

$$\rho < 0 \quad \Longleftrightarrow \quad \text{There is } \varepsilon > 0 \text{ such that } \mathbb{E}\big[u_t^\gamma(k)^{2+\varepsilon}\big] \text{ is bounded in } t \text{ for all } k \in S. \tag{4}$$

A property of this kind could of course be expected: for $\rho = -1$ solutions correspond to the stepping stone model which is bounded by 1 and hence has bounded moments of all order. The other extremal case $\rho = 1$ corresponds to the parabolic Anderson model that has exponentially increasing second moments (see for instance Theorem 1.6 of [GdH07]).

1.2 Infinite Rate Symbiotic Branching Processes

Looking more closely at the proofs of [BDE11], one observes that (4) can be strengthened as follows:

$$\rho < 0 \quad \Longrightarrow \quad \text{There is } \varepsilon > 0 \text{ such that } \mathbb{E}\big[u_t^\gamma(k)^{2+\varepsilon}\big] \text{ is bounded in } t \textbf{ and } \gamma \text{ for all } k \in S, \tag{5}$$

and this holds for any $\mathcal{A}$ satisfying the assumptions mentioned in the introduction (recall that (u^γ, v^γ) is a solution to SBM_γ). This observation shall be combined in the following with a recent development for mutually catalytic branching processes which we now briefly outline.

In [KM10b], for $\rho = 0$, existence of the *infinite rate mutually catalytic branching processes*, appearing as limits in

$$(u^\gamma, v^\gamma) \overset{\gamma\to\infty}{\Longrightarrow} (u^\infty, v^\infty) \tag{6}$$

has been shown. This process and its properties have been further studied in [KM10a] and [KO10]. In [DM11] the *infinite rate symbiotic branching processes* have been constructed for the whole range of parameters $\rho \in (-1,1)$. Below we state the result from [DM11] that introduces the infinite rate symbiotic branching process via the limiting procedure (6). Before doing this, let us shortly discuss why this procedure can lead to an exciting process.

To understand the effect of sending γ to infinity, one might take a closer look at the nonspatial symbiotic branching SDE (1). Due to the symmetric structure and the Dubins–Schwartz theorem, solutions (u^γ, v^γ) can be regarded as time-changed correlated Brownian motions with time-change $\gamma \int_0^t u_s^\gamma v_s^\gamma \, ds$. The boundary of the first quadrant

$$E = [0,\infty) \times [0,\infty) \backslash (0,\infty) \times (0,\infty)$$

is absorbing as the noise is multiplicative with diffusion coefficients proportional to the product of both coordinates. Hence, the pair of time-changed Brownian motions stops once hitting E. Increasing γ only has the effect that the process follows the Brownian paths with higher speed so that $\gamma = \infty$ corresponds to immediately picking a point in E according to the exit-point measure of the two-dimensional Brownian motion at the boundary of the first quadrant and the process stays at this point forever.

Incorporating space, a second effect appears: both types are distributed in space according to the heat flow corresponding to $\mathcal{A}$. This smoothing effect immediately tries to lift a zero coordinate if it was pushed by the exit measure to zero. Interestingly, none of these two effects dominates and a nontrivial limiting process is obtained via the limiting procedure (6).

Before stating the existence theorem, we need to refine the state-space:

$$L^{\beta,E} := L^\beta \cap E^S.$$

The space $L^{\beta,E}$ consists of sequences of pairs of points in E (i.e. one coordinate is zero, one coordinate is nonnegative) with restricted growth condition. The space $L^{\beta,E}$ is equipped with the topology induced from the topology on L^β introduced in Sect. 1.1. From the motivation given above for the infinite branching rate limiting behavior of the one-dimensional version (1), the occurrence of the state-space $L^{\beta,E}$ is not surprising: the infinite rate symbiotic branching process, abbreviated as SBM$_\infty$, takes values in E at each fixed site $k \in S$. Since in this chapter we will be dealing with finite total mass processes, let us also define

$$L^1 := \{(u,v) : S \to \mathbb{R}^+ \times \mathbb{R}^+, \langle u,1\rangle, \langle v,1\rangle < \infty\},$$
$$L^{1,E} := L^1 \cap E^S.$$

Additionally, we will denote by $D_{L^{\beta,E}}$ the space of RCLL functions on $L^{\beta,E}$. Before we state the result from [DM11] on existence and uniqueness of the infinite rate symbiotic branching processes, we need to introduce some additional notation. For $\rho \in (-1,1)$, any $(x_1,x_2) \in L^\beta$ and any compactly supported $(y_1,y_2) \in L^{1,E}$, set

$$\langle\langle x_1,x_2,y_1,y_2\rangle\rangle_\rho = \sum_{k\in S}\Big[-\sqrt{1-\rho}\big(x_1(k)+x_2(k)\big)\big(y_1(k)+y_2(k)\big)$$
$$+ i\sqrt{1+\rho}\big(x_1(k)-x_2(k)\big)\big(y_1(k)-y_2(k)\big)\Big],$$

and define $F(x_1,x_2,y_1,y_2) \equiv \exp(\langle\langle x_1,x_2,y_1,y_2\rangle\rangle_\rho)$. Then we have the following result.

Theorem 1.1. *Suppose $\rho \in (-1,1)$ and $\{(u^\gamma,v^\gamma)\}_{\gamma>0}$ is a family of solutions to* SBM$_\gamma$ *with initial conditions $(u_0^\gamma,v_0^\gamma) = (u_0,v_0) \in L^{\beta,E}$ that do not depend on γ. For any sequence γ_n tending to infinity, we have the convergence in law*

$$(u^{\gamma_n},v^{\gamma_n}) \Longrightarrow (u^\infty,v^\infty), \quad n\to\infty,$$

in $D([0,\infty),L^\beta)$ equipped with the Meyer–Zheng "pseudo-path" topology. The limiting process (u^∞,v^∞) is almost surely in $D_{L^{\beta,E}}$ and it is the unique solution in $D_{L^{\beta,E}}$ to the following martingale problem:

$$\begin{cases} \text{For any compactly supported } (y_1,y_2) \in L^{1,E}, \\ F(u_t^\infty,v_t^\infty,y_1,y_2) - F(u_0,v_0,y_1,y_2) - \int_0^t \langle\langle Au_s^\infty, Av_s^\infty, y_1,y_2\rangle\rangle_\rho F(u_s^\infty,v_s^\infty,y_1,y_2)\,ds \\ \textit{is a martingale null at zero.} \end{cases}$$

Since the theorem is a direct combination of Theorems 3.4 and 3.6 from [DM11] its proof is omitted and we will make just a few comments. The convergence of (u^γ,v^γ) to (u^∞,v^∞) is not stated in the Skorohod topology on D_{L^β} but instead in the Meyer–Zheng pseudo-path topology. In fact, the weak convergence is impossible in the Skorohod topology since (u^γ,v^γ) are continuous processes whereas the limiting process (u^∞,v^∞) only has RCLL paths (non-continuity does not directly become apparent but follows from a jump-type diffusion characterization), and in the Skorohod topology, the subspace of continuous functions is closed. For the Meyer–Zheng pseudo-path topology this, in fact, is possible (for more details, see [MZ84] and for other interesting developments, see also [J97]).

For the next theorem, we assume that $\rho < 0$ and, in this case, some useful properties of (u^∞, v^∞) with initial conditions in $L^{1,E}$ are established.

Theorem 1.2. *Suppose $\rho < 0$ and let (u^∞, v^∞) be the infinite rate process from Theorem 1.1 with initial conditions $(u_0^\infty, v_0^\infty) \in L^{1,E}$.*

(a) *The total mass processes $\langle u_t^\infty, 1\rangle$ and $\langle v_t^\infty, 1\rangle$ are well-defined square-integrable martingales for $t \geq 0$.*
(b) *There is an $\varepsilon > 0$ such that*

$$\sup_{t>0} \mathbb{E}\left[\langle u_t^\infty, 1\rangle^{2+\varepsilon}\right] < \infty \quad \textit{and} \quad \sup_{t>0} \mathbb{E}\left[\langle v_t^\infty, 1\rangle^{2+\varepsilon}\right] < \infty.$$

(c) *If ξ_t^1, ξ_t^2 are independent continuous-time Markov processes with generator $\mathcal{A}$, then*

$$\mathbb{E}\left[u_t^\infty(a) v_t^\infty(b)\right] = \mathbb{E}^{a,b}\left[u_0(\xi_t^1) v_0(\xi_t^2) \mathbf{1}_{\{\xi_s^1 \neq \xi_s^2, \forall s \leq t\}}\right]$$

for any $t \geq 0$ and $a, b \in S$.

The second moment expression in part c) of the theorem emphasizes the fact that $(u_t^\infty(a), v_t^\infty(a)) \in E$ for any $a \in S$, $t \geq 0$. This becomes clear since for $a = b$, $\xi_0^1 = \xi_0^2$ which gives $\mathbb{E}[u_t^\infty(a) v_t^\infty(a)] = 0$ and this, in turn, implies that $u_t^\infty(a) v_t^\infty(a) = 0$ almost surely, for any $t \geq 0$.

2 The Longtime Behavior for Negative Correlations

We now turn to the results of this chapter that address the longtime behavior of finite and infinite rate symbiotic branching processes. The question we address is classical for many particle systems on $[0,\infty)^S$: suppose a system has initial condition w_0 that is either infinite (e.g. $w_0 \equiv 1$) or summable (e.g. $w_0 = \mathbf{1}_{\{k\}}$ for some $k \in S$), what can be said about limits of w_t as t tends to infinity? In many situations it turns out that equivalence holds between almost sure extinction for the (finite) total mass process $\langle w_t, 1\rangle$ when $w_0 = \mathbf{1}_k$ and weak convergence of w_t from constant initial states to the absorbing states of the system. Using different duality techniques, this can be shown, for instance, for SBM_γ, the stepping stone model, the parabolic Anderson model, and for the voter process. It is not known yet whether this property holds for SBM_∞.

Extinction/survival dichotomies are typically of the following type depending only on the recurrence/transience of the migration mechanism:

$$\begin{aligned} &\lim_{t\to\infty} \langle 1, w_t\rangle \overset{a.s.}{=} 0 \text{ for all summable initial conditions} \\ &\iff \quad \mathbb{P}^i(\xi_t = j \text{ for some } t \geq 0) = 1 \quad \forall i, j \in S, \end{aligned} \tag{7}$$

where ξ_t is a continuous-time Markov process with generator $\mathcal{A}$.

2.1 Some Known Results

Before stating our main theorem on the recurrence/transience dichotomy we recall some known results.

For $\gamma < \infty$, by writing the stochastic equations for SBM_γ in mild form (see Lemma 3.1 below), it can be shown that if $(u_0^\gamma, v_0^\gamma) \in L^1$, then $\langle u_t^\gamma, 1\rangle$ and $\langle v_t^\gamma, 1\rangle$ are well-defined martingales which by nonnegativity of solutions are nonnegative. Also for $\gamma = \infty$, if $\rho < 0$ and additionally $(u_0^\infty, v_0^\infty) \in L^{1,E}$, the total mass processes $\langle u_t^\infty, 1\rangle$ and $\langle v_t^\infty, 1\rangle$ are nonnegative martingales by Theorem 1.2a). Hence, for $\gamma \in (0,\infty]$, by the martingale convergence theorem, $\langle u_t^\gamma, 1\rangle$ and $\langle v_t^\gamma, 1\rangle$ converge almost surely, as $t \to \infty$, and we denote by $\tilde{u}_\infty^\gamma, \tilde{v}_\infty^\gamma \in [0,\infty)$ their almost sure limits. Therefore,

$$\lim_{t\to\infty} \langle u_t^\gamma, 1\rangle \langle v_t^\gamma, 1\rangle = \tilde{u}_\infty^\gamma \tilde{v}_\infty^\gamma \in [0,\infty)$$

irrespectively of γ being finite or infinite. In our two-type model this convergence is used to adapt the notion of existence/survival from one-type models.

Definition 2.1. Let $\gamma \in (0,\infty]$, then we say coexistence of types is possible for SBM_γ if there are summable initial conditions (u_0^γ, v_0^γ) such that $\tilde{u}_\infty^\gamma \tilde{v}_\infty^\gamma > 0$ with positive probability. Otherwise, we say that coexistence is impossible.

We define the Green function of $\mathcal{A}$ by

$$g_t(j,k) = \int_0^t p_s(j,k)\,ds,$$

where $p_s(j,k) = \mathbb{P}(\xi_s = k \,|\, \xi_0 = j)$ and ξ is a continuous-time Markov process with generator $\mathcal{A}$. The next theorem was the starting point for the longtime analysis of two-type models that we consider in this chapter.

Theorem 2.1 (Theorem 1.2 of [DP98]). *Suppose $\gamma < \infty$ and $\rho = 0$, then the following dichotomy holds:*

(a) *Transient case: If $\sup_{k\in S} g_\infty(k,k) < \infty$, then coexistence of types is possible.*
(b) *Recurrent case: Assuming additionally the uniformity condition*

$$g_T(j,j) \geq C \sup_{k\in S} g_T(k,k), \quad \forall j \in S, T \geq T_0(j), \tag{8}$$

on $\mathcal{A}$, a criterion for impossibility of coexistence of types is the following:

$$\mathbb{P}^j(\xi_t = i \text{ for some } t \geq 0) = 1 \quad \forall j,i \in S.$$

The additional assumption (8) is fulfilled, for instance, for $\mathcal{A} = \Delta$ on $S = \mathbb{Z}^d$ so that coexistence in this case occurs if and only if $d \geq 3$.

Remark 2.1. The result of Theorem 2.1 has already been partially generalized to $\rho \neq 0$. For $\mathcal{A} = \Delta$ on $S = \mathbb{Z}^d$, the recurrent case (i.e., $d = 1,2$) was dealt with in the proof of Proposition 2.1 of [BDE11]. The proof tacitly uses the condition (8).

In the sequel, we will present a different approach that only works for $\rho < 0$ and proves the full coexistence/non-coexistence dichotomy for general $\mathcal{A}$. The additional assumption (8) on $\mathcal{A}$ is not necessary in this case.

Now let us switch to infinite rate processes. For $\rho = 0$ the longtime behavior was analyzed in [KM10a] based on the approach of [DP98] for $\gamma < \infty$. The proofs required more caution than the proofs for $\gamma < \infty$ as all second moment arguments had to be avoided since the infinite rate mutually catalytic branching process does not possess finite second moments. Such spatial systems are rare but interesting as their scaling properties change. The classical recurrence/transience dichotomy for finite variance models can break down in such a way that the criticality between survival and extinction is shifted to higher orders of the Green function (see, for instance, [DF85]). To extend Theorem 1.2 of [DP98], the log-Green function is needed:

$$g_{\infty,\log}(j,k) = \int_0^\infty p_s(j,k)(1 + |\log(p_s(j,k))|)\,ds.$$

Note that the log-Green function is infinite if the Green-function is infinite so that in the recurrent regime both are infinite.

Theorem 2.2 (Theorem 1, Theorem 2 of [KM10a]). *Suppose that* $\rho = 0$, $(u_0^\infty, v_0^\infty) \in L^{1,E}$ *and let* $\mathcal{A}$ *be such that*

$$\sup_{k\in S} |a(k,k)| < 1 \quad \textit{and} \quad \inf_{k\in S} |a(k,k)| > 0.$$

(a) If $g_{\infty,\log}(k,l)$ *is "small enough," then* $\langle u_\infty^\infty, 1\rangle\langle v_\infty^\infty, 1\rangle > 0$ *with positive probability for localized initial conditions, i.e., there are* $k,l \in S$ *such that*

$$(u_0^\infty, v_0^\infty)(i) = \begin{cases} (1,0) & : i = k, \\ (0,1) & : i = l, \\ (0,0) & : \textit{otherwise}. \end{cases}$$

In particular, for log-Green function "small enough," coexistence of types occurs.

(b) The "recurrent" regime holds as in part (b) of Theorem 2.1.

Unfortunately, the description of the recurrence/transience dichotomy is even less precise than in Theorem 2.1. It remains open what happens in the case when $g_\infty(k,l) < \infty$ and $g_{\infty,\log}(k,l) = \infty$. Again, the simple random walk serves as an example for which the results can be clarified. Suppose $\mathcal{A} = \Delta$, then coexistence is impossible if $d = 1,2$. For $d \geq 3$ one can use the local central limit theorem to show that $g_{\infty,\log}(k,j) \approx \|k-j\|^{2-d}$ as $\|k-j\| \to \infty$. Hence, the assumption of (a) is fulfilled if initially the two populations are sufficiently far apart.

2.2 Main Result

The main result of this chapter is a precise recurrence/transience dichotomy for $\rho < 0$ in both cases $\gamma < \infty$ and $\gamma = \infty$. The possibility of a dichotomy in terms of the Green function for $\gamma = \infty$ stems from the fact that for $\rho < 0$ the process has finite variance in contrast to infinite variance for $\rho = 0$.

Theorem 2.3. *Suppose $\rho < 0$ and γ is either finite or infinite, then*

$$\textit{coexistence of types is impossible} \quad \Longleftrightarrow \quad \mathbb{P}^j(\xi_t{=}i \textit{ for some } t{\geq}0) = 1 \quad \forall j,i \in S.$$

Interestingly, the negative correlations that seem to worsen the problem as symmetry gets lost simplify the problem a lot so that transparent proofs for this precise dichotomy are possible.

2.3 An Open Problem: Longtime Behavior for Positive Correlations

The reason we skip the longtime behavior for positive correlations is simple: the longtime behavior even for the finite branching rate processes SBM_γ is unknown for $\rho > 0$. Here we briefly describe what might be expected for $\mathcal{A} = \Delta$ and $S = \mathbb{Z}^d$ but we do not have proofs.

The particular case $\rho = 1$ corresponds to the classical parabolic Anderson problem (see Example 2). Having an explicit representation of the solution as Feynman–Kac functional this problem can be analyzed more easily. It is known that, started at localized initial conditions $w_0 = \mathbf{1}_{\{k\}}$, almost sure extinction $\lim_{t\to\infty}\langle w_t, 1\rangle = 0$ occurs if $d = 1,2$ (the recurrent case). In [S92], see also [GdH07], it was proved that for $d \geq 3$ (the transient case) there are critical values $\gamma^1(d) > \gamma^2(d) > 0$ such that for $\gamma > \gamma^1(d)$ extinction occurs almost surely and for $\gamma < \gamma^2(d)$, survival occurs with positive probability. This is one example where a "more noise kills" effect can be proved rigorously.

For SBM_γ and $d \geq 3$, we conjecture the following: there should be a strictly decreasing critical curve $\gamma(\cdot,d) : (0,1] \to \mathbb{R}^+$ coming down from infinity at zero and converging towards the critical threshold for the parabolic Anderson model at 1 such that coexistence of types for SBM_γ is impossible if $\gamma > \gamma(\rho,d)$ and coexistence is possible if $\gamma < \gamma(\rho,d)$. If the conjecture for $\gamma < \infty$ holds, it is furthermore natural to conjecture that for $\gamma = \infty$ coexistence of types is always impossible if $\rho > 0$.

3 Proofs

3.1 Some Properties of SBM_γ

To prepare for the proofs of the longtime behavior, we start with some lemmas for the finite branching rate processes SBM_γ. In order to avoid confusions with the notion $\langle f,g\rangle$, we denote the quadratic variation of square-integrable martingales by $[\cdot,\cdot]$.

Lemma 3.1. *Suppose that* $\rho\in(-1,1)$, $\gamma\in(0,\infty)$ *and* $(u_0^\gamma,v_0^\gamma)\in L^1$, *then* $\langle u_t^\gamma,1\rangle$ *and* $\langle v_t^\gamma,1\rangle$ *are nonnegative martingales with cross variations*

$$[\langle u_\cdot^\gamma,1\rangle,\langle u_\cdot^\gamma,1\rangle]_t=[\langle v_\cdot^\gamma,1\rangle,\langle v_\cdot^\gamma,1\rangle]_t=\gamma\int_0^t\langle u_s^\gamma,v_s^\gamma\rangle\,ds$$

and

$$[\langle u_\cdot^\gamma,1\rangle,\langle v_\cdot^\gamma,1\rangle]_t=\rho\gamma\int_0^t\langle u_s^\gamma,v_s^\gamma\rangle\,ds.$$

Proof. We only sketch a proof as it is rather straightforward. The proof is based on the stochastic variation of constant representation for stochastic heat equations. For bounded test functions ϕ, one can derive as in the proof of Theorem 2.2 of [DP98] that

$$\langle u_t^\gamma,\phi\rangle=\langle u_0^\gamma,P_t\phi\rangle+\sum_{k\in S}\int_0^t P_{t-s}\phi(k)\sqrt{\gamma u_s^\gamma(k)v_s^\gamma(k)}\,dB_s^1(k),$$

$$\langle v_t^\gamma,\phi\rangle=\langle v_0^\gamma,P_t\phi\rangle+\sum_{k\in S}\int_0^t P_{t-s}\phi(k)\sqrt{\gamma u_s^\gamma(k)v_s^\gamma(k)}\,dB_s^2(k),$$

where $P_tf(k)=\sum_{j\in S}p_t(k,j)f(j)$ and the Brownian motions are as in the definition of SBM_γ. The infinite sums can be shown to converge in $L^2(\mathbb{P})$. Setting $\phi\equiv1$ shows that $\langle u_t^\gamma,1\rangle$ and $\langle v_t^\gamma,1\rangle$ are infinite sums of martingales and for any finite subset Γ of S the correlation structure of the stochastic integrals can be easily calculated. As the martingale property is conserved under $L^2(\mathbb{P})$-convergence and the cross variations converge, we obtain that the total mass processes are square-integrable martingales with cross variation

$$\begin{aligned}
&\big[\langle u_\cdot^\gamma,1\rangle,\langle v_\cdot^\gamma,1\rangle\big]_t\\
&=\lim_{|\Gamma|\to\infty}\left[\sum_{k\in\Gamma}\int_0^\cdot\sqrt{\gamma u_s^\gamma(k)v_s^\gamma(k)}\,dB_s^1(k),\sum_{j\in\Gamma}\int_0^\cdot\sqrt{\gamma u_s^\gamma(j)v_s^\gamma(j)}\,dB_s^2(j)\right]_t\\
&=\lim_{|\Gamma|\to\infty}\sum_{j,k\in\Gamma}\int_0^t\sqrt{\gamma u_s^\gamma(k)v_s^\gamma(k)}\sqrt{\gamma u_s^\gamma(j)v_s^\gamma(j)}\,d\left[B_\cdot^1(k),B_\cdot^2(j)\right]_s
\end{aligned}$$

which by the presumed correlation structure equals

$$\rho \lim_{|\Gamma|\to\infty} \sum_{k\in\Gamma} \int_0^t \gamma u_s^\gamma(k) v_s^\gamma(k)\, ds = \rho\gamma \int_0^t \langle u_s^\gamma, v_s^\gamma\rangle\, ds.$$

The derivation of the quadratic variations of $\langle u_t^\gamma, 1\rangle$ and $\langle v_t^\gamma, 1\rangle$ is similar, by dealing with the same sets of driving Brownian motions instead. □

The next lemma gives a refinement of Theorem 2.5 of [BDE11] uniformly in γ.

Lemma 3.2. *For any $\rho < 0$, there is an $\varepsilon > 0$ such that*

$$\sup_{\gamma<\infty, T>0} \mathbb{E}\big[\sup_{t\le T} \langle u_t^\gamma, 1\rangle^{2+\varepsilon}\big] < \infty,$$

$$\sup_{\gamma<\infty, T>0} \mathbb{E}\big[\sup_{t\le T} \langle v_t^\gamma, 1\rangle^{2+\varepsilon}\big] < \infty.$$

Proof. The cross-variation structure found for the martingales $\langle u_t^\gamma, 1\rangle, \langle v_t^\gamma, 1\rangle$ in the previous lemma allows us to obtain an upper bound for arbitrary (not only integer) moments by representing $(\langle u_t^\gamma, 1\rangle, \langle v_t^\gamma, 1\rangle)$ as a pair of time-changed correlated Brownian motions. A version of the Dubins–Schwartz theorem shows that

$$(W_t^1, W_t^2) := (\langle u_{A_t}^\gamma, 1\rangle, \langle v_{A_t}^\gamma, 1\rangle) \tag{9}$$

is a pair of correlated Brownian motions started in $(W_0^1, W_0^2) = (\langle u_0^\gamma, 1\rangle, \langle v_0^\gamma, 1\rangle)$ with covariance $\mathbb{E}[W_t^1 W_t^2] = \rho t$. Here, A_t is the generalized inverse of $[\langle v_\cdot^\gamma, 1\rangle, \langle v_\cdot^\gamma, 1\rangle]_t = [\langle u_\cdot^\gamma, 1\rangle, \langle u_\cdot^\gamma, 1\rangle]_t$. Furthermore, as the total masses are nonnegative, the time-change A_t is bounded by τ, the first hitting time of (W_t^1, W_t^2) at the boundary of the first quadrant (otherwise one of the total masses would become negative). Applying the Burkholder–Davis–Gundy inequality, this shows that

$$\sup_{\gamma<\infty, T>0} \mathbb{E}\big[\sup_{t\le T} \langle u_t^\gamma, 1\rangle^p\big] \le cE\Big[\tau^{p/2}\,\Big|\, W_0^1 = \langle u_0^\gamma, 1\rangle, W_0^2 = \langle v_0^\gamma, 1\rangle\Big],$$

which is finite if $p = 2+\varepsilon$ and ε is sufficiently small. For independent Brownian motions the number of finite moments of the first exit time τ has been determined by Spitzer [S58] (see also Burkholder [B77]); for the simple transformation to correlated Brownian motions, we refer to Theorem 2.5 of [BDE11]. □

The final tool is a second moment formula for SBM_γ for which we introduce the abbreviation $L_t = \int_0^t \mathbf{1}_{\{\xi_s^1 = \xi_s^2\}}\, ds$ for the collision time of two independent continuous-time Markov processes ξ^1, ξ^2 with generator $\mathcal{A}$.

Lemma 3.3. *If L_t denotes the collision time, then the second moment formula*

$$\mathbb{E}[u_t^\gamma(a) v_t^\gamma(b)] = \mathbb{E}^{a,b}\big[u_0^\gamma(\xi_t^1) v_0^\gamma(\xi_t^2) e^{\rho\gamma L_t}\big], \quad a, b \in S, \tag{10}$$

holds.

Proof. There are several ways to see this expression. For instance, this follows as a particularly simple application of the moment duality for SBM_γ (derived in Proposition 9 of [EF04]; see also the explanation in Lemma 3.3 of [BDE11]). □

Formula (10) shows the significant difference of negative, null or positive correlations: only in the case of negative correlations one can hope to have finite second moment for the $\gamma = \infty$ limiting process.

With these preparations, we can proceed with the proof of Theorem 1.2.

3.2 Proof of Theorem 1.2

Let $\gamma_n \to \infty$, then we know from Theorem 1.1 that

$$(u^{\gamma_n}, v^{\gamma_n}) \Rightarrow (u^\infty, v^\infty)$$

in D_{L^β} in the Meyer–Zheng pseudo-path topology.

Proof of (a), (b): First, we would like to show that the limiting total mass processes $\langle u^\infty_\cdot, 1\rangle$ and $\langle v^\infty_\cdot, 1\rangle$ are martingales. One should be a little bit careful, since convergence of $(u^{\gamma_n}, v^{\gamma_n})$ is in D_{L^β} and not in D_{L^1}, and hence not necessarily $(\langle u^{\gamma_n}_\cdot, 1\rangle, \langle v^{\gamma_n}_\cdot, 1\rangle)$ converges to $(\langle u^\infty_\cdot, 1\rangle, \langle v^\infty_\cdot, 1\rangle)$. However, without loss of generality we may and will assume that for our chosen subsequence $\{(u^{\gamma_n}, v^{\gamma_n})\}_{n\geq 1}$, at least,

$$\left(u^{\gamma_n}_\cdot, v^{\gamma_n}_\cdot, \langle u^{\gamma_n}_\cdot, 1\rangle, \langle v^{\gamma_n}_\cdot, 1\rangle\right) \Rightarrow \left(u^\infty_\cdot, v^\infty_\cdot, \bar{u}^\infty_\cdot, \bar{v}^\infty_\cdot\right) \tag{11}$$

in $D_{L^\beta} \times D_\mathbb{R} \times D_\mathbb{R}$ in the Meyer–Zheng pseudo-path topology. Let us show that indeed

$$\bar{u}^\infty_t = \langle u^\infty_t, 1\rangle, \quad \bar{v}^\infty_t = \langle v^\infty_t, 1\rangle, \quad \forall t \geq 0. \tag{12}$$

By Theorem 5 of [MZ84] the convergence in the Meyer–Zheng topology implies convergence (along a further subsequence which, again, we denote by γ_n for convenience) of one-dimensional distributions on a set of full Lebesgue measure, say $\mathbb{T}$. Fix any $t \in \mathbb{T}$. Since u_0 and v_0 are summable, for any $\varepsilon > 0$, one can fix a sufficiently large compact $\Gamma \subset S$ such that

$$\mathbb{E}\left[\langle u^{\gamma_n}_t, \mathbf{1}_{\Gamma^c}\rangle + \langle v^{\gamma_n}_t, \mathbf{1}_{\Gamma^c}\rangle\right] = \mathbb{E}\left[\langle u^{\gamma_n}_0, P_t\mathbf{1}_{\Gamma^c}\rangle + \langle v^{\gamma_n}_0, P_t\mathbf{1}_{\Gamma^c}\rangle\right] \leq \varepsilon, \quad \forall \gamma_n > 0.$$

For the equality we have used the fact $\mathbb{E}[\langle u^\gamma_t, \phi\rangle] = \langle u^\gamma_0, P_t\phi\rangle$ which follows from the stochastic variation of constant representation utilized in the proof of Lemma 3.1. This implies that, in fact, $(u^{\gamma_n}_t, u^{\gamma_n}_t) \Rightarrow (u^\infty_t, u^\infty_t)$ in $M_F(S) \times M_F(S)$—the product space of finite measures on S equipped with the weak topology. This immediately implies

$$\left(\langle u^{\gamma_n}_t, 1\rangle, \langle v^{\gamma_n}_t, 1\rangle\right) \Rightarrow \left(\langle u^\infty_t, 1\rangle, \langle v^\infty_t, 1\rangle\right), \quad \forall t \in \mathbb{T}, \tag{13}$$

and combined with (11) this gives

$$(\bar{u}_t^\infty, \bar{v}_t^\infty) = (\langle u_t^\infty, 1\rangle, \langle v_t^\infty, 1\rangle), \ \forall t \in \mathbb{T}.$$

Since both processes, on the right- and on the left-hand sides, are right-continuous, we get that in fact the equality holds for all t, and hence (12) follows.

By the above Theorem 11 of [MZ84] and Lemma 3.2 we immediately get that the limiting total mass processes $\langle u_t^\infty, 1\rangle, \langle v_t^\infty, 1\rangle$, $t \geq 0$, are martingales. We will get the square integrability of these martingales (and thus finish the proof of a)) when we prove b). For b) we use Fatou's lemma combined with (13) and Lemma 3.2 to obtain that

$$\sup_{t\in\mathbb{T}} \mathbb{E}\left[\langle u_t^\infty, 1\rangle^{2+\varepsilon} + \langle v_t^\infty, 1\rangle^{2+\varepsilon}\right] < \infty.$$

Recalling that $t \mapsto (\langle u_t^\infty, 1\rangle, \langle v_t^\infty, 1\rangle)$ is right-continuous, one can take the supremum over $t \in \mathbb{R}^+$, and b) follows.

Proof of (c): The representation is first deduced for $t \in \mathbb{T}$. By the choice of $\mathbb{T}$ and (11), we get

$$u_t^{\gamma_n}(a)v_t^{\gamma_n}(b) \Rightarrow u_t^\infty(a)v_t^\infty(b). \tag{14}$$

To get convergence of the first moments of $\{u_t^{\gamma_n}(a)v_t^{\gamma_n}(b)\}_{n\geq 1}$, it is enough to check the uniform integrability of $\{u_t^{\gamma_n}(a)v_t^{\gamma_n}(b)\}_{n\geq 1}$. However this follows from Lemma 3.2 and Hölder's inequality:

$$\sup_{n\geq 1} \mathbb{E}\left[\left(u_t^{\gamma_n}(a)v_t^{\gamma_n}(b)\right)^{\frac{2+\varepsilon}{2}}\right] \leq \sup_{n\geq 1} \sqrt{\mathbb{E}\left[\langle u_t^{\gamma_n}, 1\rangle^{2+\varepsilon}\right]\mathbb{E}\left[\langle v_t^{\gamma_n}, 1\rangle^{2+\varepsilon}\right]} < \infty,$$

if ε is chosen sufficiently small. The uniform integrability, Lemma 3.3 and the dominated convergence then imply

$$\begin{aligned}
\mathbb{E}[u_t^\infty(a)v_t^\infty(b)] &= \lim_{n\to\infty} \mathbb{E}[u_t^{\gamma_n}(a)v_t^{\gamma_n}(b)] \\
&= \lim_{n\to\infty} \mathbb{E}^{a,b}\left[u_0^{\gamma_n}(\xi_t^1)v_0^{\gamma_n}(\xi_t^2)e^{\rho\gamma_n L_t}\right] \\
&= \mathbb{E}^{a,b}\left[u_0^\infty(\xi_t^1)v_0^\infty(\xi_t^2)\mathbf{1}_{\{L_t=0\}}\right] + \mathbb{E}^{a,b}\left[u_0^\infty(\xi_t^1)v_0^\infty(\xi_t^2)e^{\lim_{n\to\infty}\gamma_n\rho L_t}\mathbf{1}_{\{L_t>0\}}\right] \\
&= \mathbb{E}^{a,b}\left[u_0^\infty(\xi_t^1)v_0^\infty(\xi_t^2)\mathbf{1}_{\{L_t=0\}}\right],
\end{aligned}$$

where we also used that by definition $u_0^{\gamma_n} = u_0^\infty$ and $v_0^{\gamma_n} = v_0^\infty$.

3.3 *Proof of Theorem 2.3*

With the preparations of Sect. 3.1 we can now prove the extinction/coextinction dichotomy.

3.3.1 Proof of Theorem 2.3, $\gamma < \infty$

Recall that due to the martingale convergence theorem the product $M_t^\gamma := \langle u_t^\gamma, 1\rangle\langle v_t^\gamma, 1\rangle$ of the two nonnegative martingales converges almost surely to $\tilde{u}_\infty^\gamma \tilde{v}_\infty^\gamma$, and our task is to determine when the limit equals zero almost surely. As the limit is nonnegative, the most straightforward approach is to deduce a formula for $\mathbb{E}[M_\infty^\gamma]$ and to determine when this is strictly positive or null. For this to work, the assumption $\rho < 0$ is crucial.

Luckily, using the results from Sect. 3.1, the convergence of $\mathbb{E}[M_t^\gamma]$ to $\mathbb{E}[M_\infty^\gamma]$ comes almost for granted. The convergence of M_t^γ holds almost surely, so it suffices to show uniform integrability in t. Choosing ε small enough, we obtain the uniform integrability from Hölder's inequality and Lemma 3.2:

$$\mathbb{E}\left[(M_t^\gamma)^{\frac{2+\varepsilon}{2}}\right] \leq \sqrt{\mathbb{E}[\langle u_t^\gamma, 1\rangle^{2+\varepsilon}]\mathbb{E}[\langle v_t^\gamma, 1\rangle^{2+\varepsilon}]} \leq C.$$

Combined with nonnegativity of $\tilde{u}_\infty^\gamma \tilde{v}_\infty^\gamma$ this implies that

$$\tilde{u}_\infty^\gamma \tilde{v}_\infty^\gamma = 0 \quad a.s. \quad \Longleftrightarrow \quad \lim_{t\to\infty} \mathbb{E}[M_t^\gamma] = 0.$$

The first moment of M_t^γ can be calculated from the moment duality for SBM_γ in such a way that the criterion of the dichotomy directly drops out; to finish the proof, we aim at using Lemma 3.3 to show that

$$\lim_{t\to\infty} \mathbb{E}[M_t^\gamma] = 0 \text{ for all } (u_0^\gamma, v_0^\gamma) \in L^1 \quad \Longleftrightarrow \quad \mathbb{P}^{i,j}(\xi_t^1 = \xi_t^2 \text{ for some } t \geq 0) = 1 \quad \forall i,j \in S.$$

Taking into account Lemma 3.3, we get, for any $(u_0^\gamma, v_0^\gamma) \in L^1$ with $\langle u_0^\gamma, 1\rangle\langle v_0^\gamma, 1\rangle > 0$,

$$\begin{aligned}
\mathbb{E}[M_t^\gamma] &= \sum_{a,b\in S} \mathbb{E}[u_t^\gamma(a) v_t^\gamma(b)] \\
&= \sum_{a,b\in S} \mathbb{E}^{a,b}\left[u_0^\gamma(\xi_t^1) v_0^\gamma(\xi_t^2) e^{\gamma\rho L_t}\right] \\
&= \sum_{a,b\in S} \sum_{i,j\in S} u_0^\gamma(i) v_0^\gamma(j) \mathbb{E}^{a,b}\left[\mathbf{1}_i(\xi_t^1)\mathbf{1}_j(\xi_t^2) e^{\gamma\rho L_t}\right]
\end{aligned}$$

$$= \sum_{i,j\in S}\sum_{a,b\in S} u_0^\gamma(i)v_0^\gamma(j)\mathbb{E}^{i,j}\left[\mathbf{1}_a(\xi_t^1)\mathbf{1}_b(\xi_t^2)e^{\gamma\rho L_t}\right]$$

$$= \sum_{i,j\in S} u_0^\gamma(i)v_0^\gamma(j)\mathbb{E}^{i,j}\left[e^{\gamma\rho L_t}\right].$$

For the fourth equation we used reversibility of the paths of ξ^1,ξ^2 since we assumed $\mathcal{A}$ to be symmetric with bounded jump rate. As ρ is negative, by the dominated convergence, and taking into account $\sum_{i,j} u_0^\gamma(i)v_0^\gamma(j)\in(0,\infty)$, we see that the right-hand side converges to zero if and only if $\lim_{t\to\infty}\mathbb{E}^{i,j}\left[e^{\gamma\rho L_t}\right]=0$ for all $i,j\in S$. The proof can now be finished via the dominated convergence again and the Markov property:

$$\begin{aligned}\lim_{t\to\infty}\mathbb{E}^{i,j}\left[e^{\gamma\rho L_t}\right]=0\quad\forall i,j\in S\quad &\Longleftrightarrow\quad \mathbb{P}^{i,j}(L_t\to\infty)=1\quad\forall i,j\in S\\ &\Longleftrightarrow\quad \mathbb{P}^{i,j}(\xi_t^1=\xi_t^2\text{ for some }t\geq 0)=1\\ &\qquad\quad \forall i,j\in S.\end{aligned}$$

3.3.2 Proof of Theorem 2.3, $\gamma=\infty$

The proof of the dichotomy on the level of infinite branching rate can now be deduced similarly to the $\gamma<\infty$ case for which we define

$$M_t^\infty := \langle u_t^\infty,1\rangle\langle v_t^\infty,1\rangle$$

for $t\geq 0$.

Lemma 3.4. *Suppose $\rho<0$ and for the initial conditions $(u_0^\infty,v_0^\infty)\in L^{1,E}$, then*

$$\mathbb{E}[M_t^\infty] = \sum_{i,j\in S, i\neq j} u_0^\infty(i)v_0^\infty(j)\mathbb{P}^{i,j}(\xi_s^1\neq\xi_s^2,\forall s\leq t) \tag{15}$$

for $t\geq 0$.

Proof. We use Theorem 1.2(c), Fubini's theorem and the time-reversion trick used in the proof of Theorem 2.3 for $\gamma<\infty$:

$$\mathbb{E}[M_t^\infty] = \sum_{a,b\in S}\mathbb{E}[u_t^\infty(a)v_t^\infty(b)]$$

$$= \sum_{a,b\in S}\sum_{i,j\in S}\mathbb{E}^{a,b}\left[u_0^\infty(i)\mathbf{1}_i(\xi_t^1)v_0^\infty(j)\mathbf{1}_j(\xi_t^2)\mathbf{1}_{L_t=0}\right]$$

$$= \sum_{i,j \in S} \sum_{a,b \in S} u_0^\infty(i) v_0^\infty(j) \mathbb{E}^{i,j}\left[\mathbf{1}_a(\xi_t^1)\mathbf{1}_b(\xi_t^2)\mathbf{1}_{L_t=0}\right]$$

$$= \sum_{i,j \in S} u_0^\infty(i) v_0^\infty(j) \mathbb{P}^{i,j}\left[\xi_s^1 \neq \xi_s^2, \forall s \leq t\right].$$

By assumption $(u_0^\infty, v_0^\infty) \in E^S$ so that the terms with $i = j$ vanish as then $u_0^\infty(i) v_0^\infty(i) = 0$. □

Now we have to verify necessary and sufficient conditions for the limit

$$\lim_{t \to \infty} \langle u_t^\infty, 1 \rangle \langle v_t^\infty, 1 \rangle = \tilde{u}_\infty^\infty \tilde{v}_\infty^\infty$$

being equal to zero almost surely. This can be done similarly as in the case $\gamma < \infty$.

Lemma 3.5. *Suppose $\rho < 0$ and for the initial conditions $(u_0^\infty, v_0^\infty) \in L^{1,E}$, then*

$$\tilde{u}_\infty^\infty \tilde{v}_\infty^\infty = 0 \quad a.s. \qquad \Longleftrightarrow \qquad \lim_{t \to \infty} \mathbb{E}[M_t^\infty] = 0.$$

Proof. Almost sure convergence of M_t^∞ to $\tilde{u}_\infty^\infty \tilde{v}_\infty^\infty$ is due to the martingale convergence theorem, so that it suffices to show that M_t^∞ is uniformly integrable in t. By Hölder's inequality, we reduce the question to the total mass processes:

$$\mathbb{E}\left[(M_t^\infty)^{\frac{2+\varepsilon}{2}}\right] \leq \sqrt{\mathbb{E}\left[\langle u_t^\infty, 1 \rangle^{2+\varepsilon}\right] \mathbb{E}\left[\langle v_t^\infty, 1 \rangle^{2+\varepsilon}\right]}, \tag{16}$$

and the result follows from Theorem 1.2(c) if ε is chosen small enough. □

Now we can proceed with the proof of Theorem 2.3 for $\gamma = \infty$ exactly as for $\gamma < \infty$.

Proof of Theorem 2.3, $\gamma = \infty$. The theorem follows directly from Lemma 3.4 and Lemma 3.5. Using the dominated convergence, justified by $\sum_{i,j \in S} u_0^\infty(i) v_0^\infty(j) < \infty$, and monotonicity of measures we obtain

$$\lim_{t \to \infty} \sum_{i,j \in S, i \neq j} u_0^\infty(i) v_0^\infty(j) \mathbb{P}^{i,j}(\xi_s^i \neq \xi_s^2, \forall s \leq t) = 0, \quad \forall (u_0^\infty, v_0^\infty) \in L^{1,E}$$

$$\Longleftrightarrow \quad \mathbb{P}^{i,j}(\xi_t^1 = \xi_t^2 \text{ for some } t \geq 0) = 1 \quad \forall j, i \in S, i \neq j.$$

This finishes the proof of Theorem 2.3 also for infinite branching rate. □

Acknowledgements The authors thank an anonymous referee for a very careful reading of the manuscript.

References

BDE11. J. Blath, L. Döring, A. Etheridge "On the Moments and the Wavespeed of the Symbiotic Branching Model." The Annals of Probability, 2011, Vol. 39, No. 1, pp. 252–290.

B77. D. L. Burkholder. Exit times of Brownian motion, harmonic majorization, and Hardy spaces. *Advances in Math.*, 26(2):182–205, 1977.

CDG04. J.T. Cox, D. Dawson, A. Greven "Mutually Catalytic Super branching Random Walks: Large Finite Systems and Renormalization Analysis." AMS Memoirs, 2004, Vol. 171.

CK00. J.T. Cox, A. Klenke "Recurrence and Ergodicity of Interacting Particle Systems." Probab. Theory Relat. Fields, 2000, Vol. 116, pp. 239–255.

DF85. D. Dawson, K. Fleischmann "Critical dimension for a model of branching in a random medium." Z. Wahrsch. Verw. Gebiete, 70(3), pp. 315–334, 1985.

DM11. L. Döring, L. Mytnik "Mutually Catalytic Branching Processes on the lattice and Voter Processes with Strength of Opinion." arXiv:1109.6106v1.

DP98. D. Dawson, E. Perkins "Long-Time Behavior and Coexistence in a Mutually Catalytic branching model." The Annals of Probability, 1998, Vol. 26, No.3. pp. 1088–1138.

E00. A. Etheridge "An Introduction to Superprocesses." University Lecture Series, Amer. Math. Soc., 2000.

EF04. A. Etheridge, K. Fleischmann "Compact interface property for symbiotic branching." Stochastic Processes and their Applications, 2004, Vol. 114, pp. 127–160.

GdH07. A. Greven, F. den Hollander "Phase transitions for the longtime behavior of interacting diffusions." The Annals of Probability, 2007, Vol. 35, No.4., pp. 1250–1306.

J97. A. Jakubowski "A non-Skorohod topology on the Skorohod space." Electronic Journal of Probability, 1997, Vol. 2, No.4., pp. 1–21.

KM10a. A. Klenke, L. Mytnik "Infinite Rate Mutually Catalytic Branching in Infinitely Many Colonies: The Longtime Behaviour." to appear in Annals of Probability.

KM10b. A. Klenke, L. Mytnik "Infinite Rate Mutually Catalytic Branching in Infinitely Many Colonies: Construction, Characterization and Convergence." to appear in Probab. Theory Relat. Fields.

KO10. A. Klenke, M. Oeler "A Trotter Type Approach to Infinite Rate Mutually Catalytic Branching." The Annals of Probability, 2010, Vol. 38, 479–497.

L05. T. Liggett "Interacting Particle Systems." Reprint of the 1985 original. Classics in Mathematics. Springer-Verlag, Berlin, 2005.

MZ84. P.A. Meyer, W.A. Zheng "Tightness criteria for laws of semimartingales." Ann. Inst. H. Poincaré Probab. Statist. 20 (1984), no. 4, 353–372.

S92. T. Shiga "Ergodic Theorems and Exponential Decay of Sample Paths for Certain Interacting Diffusion Systems." Osaka J. Math., 1992, Vol. 29, p. 789–807.

SS80. T. Shiga, A. Shimizu "Infinite dimensional stochastic differential equations and their applications." J. Math. Kyoto Univ., 1980, Vol. 20, No.3., pp. 395–416.

S58. F. Spitzer "Some theorems concerning 2-dimensional Brownian motion." Trans. Amer. Math. Soc., 1958, No. 87, pp. 187–197.

Super-Brownian Motion: L^p-Convergence of Martingales Through the Pathwise Spine Decomposition

A.E. Kyprianou and A. Murillo-Salas

Abstract Evans [7] described the semigroup of a superprocess with quadratic branching mechanism under a martingale change of measure in terms of the semigroup of an immortal particle and the semigroup of the superprocess prior to the change of measure. This result, commonly referred to as the spine decomposition, alludes to a pathwise decomposition in which independent copies of the original process "immigrate" along the path of the immortal particle. For branching particle diffusions, the analogue of this decomposition has already been demonstrated in the pathwise sense; see, for example, [10, 11]. The purpose of this short note is to exemplify a new *pathwise* spine decomposition for supercritical super-Brownian motion with general branching mechanism (cf. [13]) by studying L^p-convergence of naturally underlying additive martingales in the spirit of analogous arguments for branching particle diffusions due to Harris and Hardy [10]. Amongst other ingredients, the Dynkin–Kuznetsov $\mathbb{N}$-measure plays a pivotal role in the analysis.

Keywords Super-Brownian motion • Additive martingales • $\mathbb{N}$-measure • Spine decomposition • L^p-convergence

MSC subject classifications (2010): 60J68, 60F25.

On the occasion of the 60th birthday of Sergei Kuznetsov

A.E. Kyprianou (✉)
Department of Mathematical Sciences, University of Bath, Claverton Down,
Bath BA2 7AY, United Kingdom
e-mail: a.kyprianou@bath.ac.uk

A. Murillo-Salas
Departamento de Matemáticas, Universidad de Guanajuato, Jalisco S/N Mineral de Valenciana,
Guanajuato, Gto. C.P. 36240, México
e-mail: amurillos@ugto.mx

J. Englander and B. Rider (eds.), *Advances in Superprocesses and Nonlinear PDEs*,
Springer Proceedings in Mathematics & Statistics 38,
DOI 10.1007/978-1-4614-6240-8_7, © Springer Science+Business Media, LLC 2013

1 Introduction

Suppose that $X = \{X_t : t \geq 0\}$ is a (one-dimensional) ψ-super-Brownian motion with general branching mechanism ψ taking the form

$$\psi(\lambda) = -\alpha\lambda + \beta\lambda^2 + \int_{(0,\infty)} (\mathrm{e}^{-\lambda x} - 1 + \lambda x)\nu(\mathrm{d}x), \tag{1}$$

for $\lambda \geq 0$ where $\alpha = -\psi'(0+) \in (0,\infty)$, $\beta \geq 0$ and ν is a measure concentrated on $(0,\infty)$ which satisfies $\int_{(0,\infty)}(x \wedge x^2)\nu(\mathrm{d}x) < \infty$. Let $\mathcal{M}_F(\mathbb{R})$ be the space of finite measures on $\mathbb{R}$ and note that X is a $\mathcal{M}_F(\mathbb{R})$-valued Markov process under $\mathbb{P}_\mu$ for each $\mu \in \mathcal{M}_F(\mathbb{R})$, where $\mathbb{P}_\mu$ is the law of X with initial configuration μ. We shall use standard inner product notation, for $f \in C_b^+(\mathbb{R})$, the space of positive, uniformly bounded, continuous functions on $\mathbb{R}$, and $\mu \in \mathcal{M}_F(\mathbb{R})$,

$$\langle f, \mu \rangle = \int_{\mathbb{R}} f(x)\mu(\mathrm{d}x).$$

Accordingly we shall write $||\mu|| = \langle 1, \mu \rangle$. Recall that the total mass of the process X, $\{||X_t|| : t \geq 0\}$ is a continuous-state branching process with branching mechanism ψ. Such processes may exhibit explosive behaviour; however, under the conditions assumed above, $||X||$ remains finite at all times. We insist moreover that $\psi(\infty) = \infty$ which means that with positive probability the event $\lim_{t\uparrow\infty} ||X_t|| = 0$ will occur. Equivalently this means that the total mass process does not have monotone increasing paths; see, for example, the summary in Chap. 10 of Kyprianou [12]. The existence of these superprocesses is guaranteed by [1,3,4].

The following standard result from the theory of superprocesses describes the evolution of X as a Markov process. For all $f \in C_b^+(\mathbb{R})$ and $\mu \in \mathcal{M}_F(\mathbb{R})$,

$$-\log \mathbb{E}_\mu(\mathrm{e}^{-\langle f, X_t \rangle}) = \int_{\mathbb{R}} u_f(x,t)\mu(\mathrm{d}x),\ \mu \in \mathcal{M}_F(\mathbb{R}),\ t \geq 0, \tag{2}$$

where $u_f(x,t)$ is the unique positive solution to the evolution equation for $x \in \mathbb{R}$ and $t > 0$

$$\frac{\partial}{\partial t} u_f(x,t) = \frac{1}{2}\frac{\partial^2}{\partial x^2} u_f(x,t) - \psi(u_f(x,t)), \tag{3}$$

with initial condition $u_f(x,0) = f(x)$. The reader is referred to Theorem 1.1 of Dynkin [2], Proposition 2.3 of Fitzsimmons [8] and Proposition 2.2 of Watanabe [15] for further details; see also Dynkin [3,4] and Engländer and Pinsky [6] for a general overview.

Associated to the process X is the following martingale $Z(\lambda) = \{Z_t(\lambda), t \geq 0\}$, where

$$Z_t(\lambda) := \mathrm{e}^{\lambda c_\lambda t}\langle \mathrm{e}^{\lambda \cdot}, X_t \rangle,\ t \geq 0, \tag{4}$$

where $c_\lambda = \psi'(0+)/\lambda - \lambda/2$ and $\lambda \in \mathbb{R}$ (cf. [13] Lemma 2.2). To see why this is a martingale, note the following steps. Define for each $x \in \mathbb{R}$, $g \in C_b^+(\mathbb{R})$ and $\theta, t \geq 0$, $u_g^\theta(x,t) = -\log \mathbb{E}_{\delta_x}(\mathrm{e}^{-\theta\langle g, X_t\rangle})$. With limits understood as $\theta \downarrow 0$, we have $u_g(x,t)|_{\theta=0} = 0$; moreover $v_g(x,t) := \mathbb{E}_{\delta_x}(\langle g, X_t\rangle) = \partial u_g^\theta(x,t)/\partial\theta|_{\theta=0}$. Differentiating in θ in (3) shows that v_g solves the equation

$$\frac{\partial}{\partial t} v_g(x,t) = \frac{1}{2}\frac{\partial^2}{\partial x^2} v_g(x,t) - \psi'(0+) v_g(x,t), \tag{5}$$

with $v_g(x,0) = g(x)$. Classical Feynman–Kac theory tells us that (5) has a unique solution which is necessarily equal to $\Pi_x(\mathrm{e}^{-\psi'(0+)t} g(\xi_t))$ where $\{\xi_t : t \geq 0\}$ is a Brownian motion issued from $x \in \mathbb{R}$ under the measure Π_x. The above procedure also works for $g(x) = \mathrm{e}^{\lambda x}$ in which case we easily conclude that for all $x \in \mathbb{R}$ and $t \geq 0$, $\mathrm{e}^{\lambda c_\lambda t}\mathbb{E}_{\delta_x}(\langle \mathrm{e}^{\lambda\cdot}, X_t\rangle) = \mathrm{e}^{\lambda x}$. Finally, the martingale property follows using the previous equality together with the Markov branching property associated with X. Note that as a positive martingale, it is automatic that the limit

$$Z_\infty(\lambda) := \lim_{t\uparrow\infty} Z_t(\lambda)$$

exists $\mathbb{P}_\mu$ almost surely for all $\mu \in \mathcal{M}_F(\mathbb{R})$ such that $\langle \mathrm{e}^{\lambda\cdot}, \mu\rangle < \infty$.

The purpose of this note is to demonstrate the robustness of a new path decomposition of our ψ-super-Brownian motion by studying the L^p-convergence of the martingales $Z(\lambda)$. Specifically we shall prove the following theorem.

Theorem 1.1. *Assume that $p \in (1,2]$, $\int_{(0,\infty)} r^p \nu(\mathrm{d}r) < \infty$ and $p\lambda^2 < -2\psi'(0+)$. Then $Z_t(\lambda)$ converges to $Z_\infty(\lambda)$ in $L^p(\mathbb{P}_\mu)$, for all $\mu \in \mathcal{M}_F(\mathbb{R})$ such that $\langle \mathrm{e}^{\lambda p\cdot}, \mu\rangle$ are finite.*

The method of proof we use is quite similar to the one used in Harris and Hardy [10] for branching Brownian motion, where a pathwise spine decomposition functions as the key instrument of the proof. Roughly speaking, in that setting, the spine decomposition says that under a change of measure, the law of the branching Brownian motion has the same law as an immortal Brownian diffusion (with drift) along the path of which independent copies of the original branching Brownian motion immigrate at times which form a Poisson process. Until recently, such a spine decomposition for superdiffusions was only available in the literature in a weak form, meaning that it takes the form of a semigroup decomposition. See the original paper of Evans [7] as well as, for example, amongst others, Engländer and Kyprianou [9]. Recently, however, Kyprianou et al. [13] give a pathwise spine decomposition which provides a natural analogue to the pathwise spine decomposition for branching Brownian motion. Amongst other ingredients, the Dynkin–Kuznetsov $\mathbb{N}$-measure plays a pivotal role in describing the immigration off an immortal particle. We give a description of this new spine decomposition in the next section and thereafter we proceed to the proof of Theorem 1.1 in Sect. 3.

2 Spine Decomposition

For each $\lambda \in \mathbb{R}$ and $\mu \in \mathcal{M}_F(\mathbb{R})$ satisfying $\langle e^{\lambda\cdot}, \mu\rangle < \infty$, we introduce the following martingale change of measure

$$\left.\frac{\mathrm{d}\mathbb{P}_\mu^\lambda}{\mathrm{d}\mathbb{P}_\mu}\right|_{\mathcal{F}_t} = \frac{Z_t(\lambda)}{\langle e^{\lambda\cdot}, \mu\rangle}, t \geq 0, \tag{6}$$

where $\mathcal{F}_t := \sigma(X_s, s \leq t)$. The preceding change of measure induces the *spine decomposition* of X alluded to above. To describe it in detail, we need some more ingredients.

According to Dynkin and Kuznetsov [5], there exists a collection of measures $\{\mathbb{N}_x, x \in \mathbb{R}\}$, defined on the same probability space as X, such that

$$\mathbb{N}_x\left(1 - \mathrm{e}^{-\langle f, X_t\rangle}\right) = u_f(x,t),\ x \in \mathbb{R}, t \geq 0. \tag{7}$$

Roughly speaking, the branching property tells us that for each $n \in \mathbb{N}$, the measures $\mathbb{P}_{\delta_x}$ can be written as the n-fold convolution of $\mathbb{P}_{\frac{1}{n}\delta_x}$ which indicates that, on the trajectory space of the superprocess, $\mathbb{P}_x$ is infinitely divisible. Hence, the role of $\mathbb{N}_x$ in (7) is analogous to that of the Lévy measure for positive real-valued random variables.

From identity (7) and Eq. (2), it is straightforward to deduce that

$$\mathbb{N}_x(\langle f, X_t\rangle) = \mathbb{E}_{\delta_x}(\langle f, X_t\rangle), \tag{8}$$

whenever $f \in C_b^+(\mathbb{R})$.

For each $x \in \mathbb{R}$, let Π_x be the law of a Brownian motion $\xi := \{\xi_t : t \geq 0\}$ issued from x. If Π_x^λ is the law under which ξ is a Brownian motion with drift $\lambda \in \mathbb{R}$ and issued from $x \in \mathbb{R}$, then for each $t \geq 0$,

$$\left.\frac{\mathrm{d}\Pi_x^\lambda}{\mathrm{d}\Pi_x}\right|_{\mathcal{G}_t} = \mathrm{e}^{\lambda(\xi_t - x) - \frac{1}{2}\lambda^2 t}, t \geq 0, \tag{9}$$

where $\mathcal{G}_t := \sigma(\xi_s, s \leq t)$. For convenience, we shall also introduce the measure

$$\Pi_\mu^\lambda(\cdot) := \frac{1}{\langle \mathrm{e}^{\lambda\cdot}, \mu\rangle}\int \mathrm{e}^{\lambda x}\mu(\mathrm{d}x)\Pi_x^\lambda(\cdot), \tag{10}$$

for all $\lambda \in \mathbb{R}$. In other words, Π_μ^λ has the law of a Brownian motion with drift at rate λ with an initial position which has been independently randomised in a way that is determined by μ.

Now fix $\mu \in \mathcal{M}_F(\mathbb{R})$ and $x \in \mathbb{R}$ and let us define a measure-valued process $\Lambda := \{\Lambda_t, t \geq 0\}$ as follows:

(i) Take a copy of the process $\xi = \{\xi_t, t \geq 0\}$ under Π_x^λ; we shall refer to this process as the *spine*.
(ii) Suppose that $\mathbf{n}$ is a Poisson point process such that, for $t \geq 0$, given the spine ξ, $\mathbf{n}$ issues superprocess $X^{\mathbf{n},t}$ at space-time position (ξ_t, t) with rate $\mathrm{d}t \times 2\beta \mathrm{d}\mathbb{N}_{\xi_t}$.
(iii) Suppose that $\mathbf{m}$ is a Poisson point process such that, independently of $\mathbf{n}$, given the spine ξ, $\mathbf{m}$ issues a superprocess $X^{\mathbf{m},t}$ at space-time point (ξ_t, t) with initial mass r at rate $\mathrm{d}t \times r\nu(\mathrm{d}r) \times \mathrm{d}\mathbb{P}_{r\delta_{\xi_t}}$.

Note in particular that, when $\beta > 0$, the rate of immigration under the process $\mathbf{n}$ is infinite and moreover, each process that immigrates is issued with zero initial mass. One may therefore think of $\mathbf{n}$ as a process of continuous immigration. In contrast, when ν is a non-zero measure, processes that immigrate under $\mathbf{m}$ have strictly positive initial mass and therefore contribute to path discontinuities of $||X||$.

Now, for each $t \geq 0$, we define

$$\Lambda_t = X'_t + X_t^{(\mathbf{n})} + X_t^{(\mathbf{m})}, \tag{11}$$

where $\{X'_t : t \geq 0\}$ is an independent copy of $(X, \mathbb{P}_\mu)$,

$$X_t^{(\mathbf{n})} = \sum_{s \leq t:\mathbf{n}} X_{t-s}^{\mathbf{n},s}, t \geq 0 \quad \text{and} \quad X_t^{(\mathbf{m})} = \sum_{s \leq t:\mathbf{m}} X_{t-s}^{\mathbf{m},s}, t \geq 0.$$

In the last two equalities, we understand the first sum to be over times for which $\mathbf{n}$ experiences points and the second sum is understood similarly. Note that since the processes $X^{(\mathbf{n})}$ and $X^{(\mathbf{m})}$ are initially zero valued, it is clear that since $X'_0 = \mu$, then $\Lambda_0 = \mu$. In that case, we use the notation $\tilde{\mathbb{P}}_{\mu,x}^\lambda$ to denote the law of the pair (Λ, ξ). Note also that the pair (Λ, ξ) is a time-homogenous Markov process. We are interested in the case that the initial position of the spine ξ is randomised using the measure μ via (10). In that case, we shall write

$$\tilde{\mathbb{P}}_\mu^\lambda(\cdot) = \frac{1}{\langle \mathrm{e}^{\lambda \cdot}, \mu \rangle} \int_{\mathbb{R}} \mathrm{e}^{\lambda x} \mu(\mathrm{d}x) \tilde{\mathbb{P}}_{\mu,x}^\lambda(\cdot)$$

for short. The next theorem identifies the process Λ as the *pathwise* spine decomposition of $(X, \mathbb{P}_\mu^\lambda)$, and in particular, it shows that as a process on its own, Λ is Markovian.

Theorem 2.1 (Theorem 5.1, [13]). *For all $\mu \in \mathcal{M}_F(\mathbb{R})$ such that $\langle \mathrm{e}^{\lambda \cdot}, \mu \rangle < \infty$, $(X, \mathbb{P}_\mu^\lambda)$ and $(\Lambda, \tilde{\mathbb{P}}_\mu^\lambda)$ are equal in law.*

3 Proof of Theorem 1.1

First, note that when $\mu \in \mathcal{M}_F(\mathbb{R})$, the assumption $\langle e^{\lambda p\cdot}, \mu\rangle < \infty$ implies $\langle e^{\lambda\cdot}, \mu\rangle < \infty$. From the last section, we have the following spine decomposition of the martingale (4),

$$Z_t^\lambda(\lambda) = Z_t'(\lambda) + \sum_{s \le t:\mathbf{n}} e^{\lambda c_\lambda s} Z_{t-s}^{\mathbf{n},s}(\lambda) + \sum_{s \le t:\mathbf{m}} e^{\lambda c_\lambda s} Z_{t-s}^{\mathbf{m},s}(\lambda), \tag{12}$$

where $Z'(\lambda)$ is an independent copy of $Z(\lambda)$ under $\mathbb{P}_\mu$,

$$Z_{t-s}^{\mathbf{n},s} := e^{\lambda c_\lambda (t-s)} \langle e^{\lambda\cdot}, X_{t-s}^{\mathbf{n},s}\rangle,$$

and

$$Z_{t-s}^{\mathbf{m},s} := e^{\lambda c_\lambda (t-s)} \langle e^{\lambda\cdot}, X_{t-s}^{\mathbf{m},s}\rangle.$$

Since $\{Z_t(\lambda), t \ge 0\}$ is a martingale and we assume that $p \in (1,2]$, then Doob's submartingale inequality tells us that $Z(\lambda)$ converges in $L^p(\mathbb{P}_\mu)$ as soon as we can show that $\sup_{t\ge 0} \mathbb{E}_\mu(Z_t(\lambda)^p) < \infty$. To this end, and with the above pathwise spine decomposition in hand, we may now proceed to address the analogue of the proof for branching Brownian motion given in [10].

First note that, for all $p \in (1,2]$,

$$\mathbb{E}_\mu(Z_t(\lambda)^p) = \langle e^{\lambda\cdot}, \mu\rangle \mathbb{E}_\mu^\lambda(Z_t(\lambda)^q) = \langle e^{\lambda\cdot}, \mu\rangle \tilde{\mathbb{E}}_\mu^\lambda(Z_t^\lambda(\lambda)^q), \text{ for all } t \ge 0, \tag{13}$$

where $q = p - 1$. Now let $m = \{m_t : t \ge 0\}$ where for $t \ge 0$, $m_t = ||X_0^{\mathbf{m},t}||$. In particular, note that the process $\{m_t : t \ge 0\}$ is a Poisson point process on $(0,\infty)^2$, independent of ξ, with intensity $dt \times r\nu(dr)$. By Jensen's inequality, we have that, for all $q \in (0,1]$

$$\begin{aligned}
&\tilde{\mathbb{E}}_\mu^\lambda\left(Z_t^\lambda(\lambda)^q \mid \xi, m\right) \\
&\le \left[\tilde{\mathbb{E}}_\mu^\lambda\left(Z_t^\lambda(\lambda) \mid \xi, m\right)\right]^q \\
&\le \langle e^{\lambda\cdot}, \mu\rangle^q + \left[\tilde{\mathbb{E}}_\mu^\lambda\left(\sum_{s\le t:\mathbf{n}} e^{\lambda c_\lambda s} Z_{t-s}^{\mathbf{n},s}(\lambda) \middle| \xi\right)\right]^q + \left[\tilde{\mathbb{E}}_\mu^\lambda\left(\sum_{s\le t:\mathbf{m}} e^{\lambda c_\lambda s} Z_{t-s}^{\mathbf{m},s}(\lambda) \middle| \xi, m\right)\right]^q,
\end{aligned} \tag{14}$$

to get the last inequality we have used the fact that $(\sum_i u_i)^q \le \sum_i u_i^q$ with $u_i \ge 0$. On the one hand, recalling from (8) that $\mathbb{N}_{\xi_s}[Z_{t-s}(\lambda)] = \mathbb{E}_{\xi_s}[Z_{t-s}(\lambda)]$, we obtain

$$\tilde{\mathbb{E}}^{\lambda}_{\mu}\left(\sum_{s\le t:\mathbf{n}} e^{\lambda c_\lambda s} Z^{\mathbf{n},s}_{t-s}(\lambda)\middle|\xi\right) = \int_0^t e^{\lambda c_\lambda s}\mathbb{N}_{\xi_s}[Z_{t-s}(\lambda)]\mathrm{d}s$$
$$= \int_0^t e^{\lambda(\xi_s + c_\lambda s)}\mathrm{d}s. \tag{15}$$

On the other hand, we have that

$$\tilde{\mathbb{E}}^{\lambda}_{\mu}\left(\sum_{s\le t:\mathbf{m}} e^{\lambda c_\lambda s} Z^{\mathbf{m},s}_{t-s}(\lambda)\middle|\xi,m\right) = \sum_{s\le t:\mathbf{m}} e^{\lambda c_\lambda s}\mathbb{E}_{m_s\delta_{\xi_s}}[Z^{\mathbf{m},s}_{t-s}(\lambda)]$$
$$= \sum_{s\le t:\mathbf{m}} m_s e^{\lambda(\xi_s + c_\lambda s)} \tag{16}$$

Then, putting (15) and (16) into (14), making use again of the inequality $(\sum_i u_i)^q \le \sum_i u_i^q$ where $u_i \ge 0$ for all i, we obtain

$$\tilde{\mathbb{E}}^{\lambda}_{\mu}\left(Z^{\lambda}_t(\lambda)^q \mid \xi, m\right) \le \langle e^{\lambda\cdot},\mu\rangle^q + \left(\int_0^t e^{\lambda(\xi_s+c_\lambda s)}\mathrm{d}s\right)^q + \left(\sum_{s\le t:\mathbf{m}} m_s e^{\lambda(\xi_s+c_\lambda s)}\right)^q$$
$$\le \langle e^{\lambda\cdot},\mu\rangle^q + \left(\int_0^\infty e^{\lambda(\xi_s+c_\lambda s)}\mathrm{d}s\right)^q + \sum_{s\ge 0:\mathbf{m}} m_s^q e^{q\lambda(\xi_s+c_\lambda s)}. \tag{17}$$

Taking expectations again in (17) gives us that, for all $t \ge 0$,

$$\tilde{\mathbb{E}}^{\lambda}_{\mu}(Z^{\lambda}_t(\lambda)^q) \le \langle e^{\lambda\cdot},\mu\rangle^q + \Pi^{\lambda}_{\mu}\left[\left(\int_0^\infty e^{\lambda(\xi_s+c_\lambda s)}\mathrm{d}s\right)^q\right] + \tilde{\mathbb{E}}^{\lambda}_{\mu}\left(\sum_{s\ge 0:\mathbf{m}} m_s^q e^{q\lambda(\xi_s+c_\lambda s)}\right). \tag{18}$$

We know that, under Π^{λ}_{μ}, the process ξ is a Brownian motion with drift λ. Thus, with respect to the same measure, $\xi_s + c_\lambda s$ is a Brownian motion with drift $\lambda + c_\lambda$ which is strictly negative for $\lambda \in (0, \sqrt{-2\psi'(0+)})$. Note that this latter condition holds in particular under assumption that $p\lambda^2 < -2\psi'(0+)$ and $p > 1$. From Sect. 2 of Maulik and Zwart [14] we can conclude that

$$\Pi^{\lambda}_0\left(\int_0^\infty e^{\lambda(\xi_s+c_\lambda s)}\mathrm{d}s\right) < \infty,$$

which in turn implies that, for all $q \in (0,1]$,

$$\Pi^{\lambda}_0\left[\left(\int_0^\infty e^{\lambda(\xi_s+c_\lambda s)}\mathrm{d}s\right)^q\right] < \infty,$$

and hence

$$\Pi_\mu^\lambda\left[\left(\int_0^\infty e^{\lambda(\xi_s+c_\lambda s)}ds\right)^q\right] = \frac{1}{\langle e^{\lambda\cdot},\mu\rangle}\int e^{\lambda x}\mu(dx)\Pi_0^\lambda\left[\left(\int_0^\infty e^{\lambda(x+\xi_s+c_\lambda s)}ds\right)^q\right]$$
$$= \frac{\langle e^{\lambda p\cdot},\mu\rangle}{\langle e^{\lambda\cdot},\mu\rangle}\Pi_0^\lambda\left[\left(\int_0^\infty e^{\lambda(\xi_s+c_\lambda s)}ds\right)^q\right] < \infty. \qquad (19)$$

It remains to prove that the last term in (18) is finite. This can be done by computing the expectation directly. We obtain

$$\tilde{\mathbb{E}}_\mu^\lambda\left(\sum_{s\geq 0:\mathbf{m}} m_s^q e^{q\lambda(\xi_s+c_\lambda s)}\right) = \int_0^\infty ds\int_0^\infty r\nu(dr)r^q\Pi_\mu^\lambda\left(e^{q\lambda(\xi_s+c_\lambda s)}\right)$$
$$= \int_0^\infty ds\int_0^\infty r^p\nu(dr)\frac{1}{\langle e^{\lambda\cdot},\mu\rangle}\int e^{\lambda x}\mu(dx)\Pi_0^\lambda\left(e^{q\lambda(x+\xi_s+c_\lambda s)}\right)$$
$$= \frac{\langle e^{\lambda p\cdot},\mu\rangle}{\langle e^{\lambda\cdot},\mu\rangle}\int_0^\infty r^p\nu(dr)\int_0^\infty \Pi_0^\lambda\left(e^{q\lambda(\xi_s+c_\lambda s)}\right)ds.$$

Note that

$$\Pi_0^\lambda\left(e^{q\lambda(\xi_s+c_\lambda s)}\right) = \exp\{qs\lambda^2+s(q\lambda)^2/2+qs\psi'(0+)-qs\lambda^2/2\}$$
$$= \exp\{qs(p\lambda^2/2+\psi'(0+))\}$$

for all $s\geq 0$. Moreover, this expectation has a negative exponent as soon as $p\lambda^2 < -2\psi'(0+)$. Together with the assumption $\int_0^\infty r^p\nu(dr) < \infty$ we conclude that

$$\tilde{\mathbb{E}}_\mu^\lambda\left(\sum_{s\geq 0:\mathbf{m}} m_s^q e^{q\lambda(\xi_s+c_\lambda s)}\right) < \infty. \qquad (20)$$

Finally, from (18) to (20) we get that

$$\sup_{t\geq 0}\tilde{\mathbb{E}}_\mu^\lambda\left(Z_t^\lambda(\lambda)^q\right) < \infty,$$

which, in combination with (13), completes the proof. □

Remark 3.1. Following the reasoning in the above proof, one can also adapt the argument to show a slightly different version of Theorem 1.1. Specifically, assume that $p\in(1,2]$, $\int_{[1,\infty)} r^p\nu(dr) < \infty$ and $\lambda^2 < -\psi'(0+)$. Then $Z_t(\lambda)$ converges to $Z_\infty(\lambda)$ in $L^p(\mathbb{P}_\mu)$, for all $\mu\in\mathcal{M}_F(\mathbb{R})$ such that $\langle e^{2\lambda\cdot},\mu\rangle$ are finite.

The way to do this is to split the second Poisson sum on the right-hand side of the first inequality in (17) according to whether the value of m_s exceeds unity or not. One then uses a simple estimate $a^q\leq 1+a$, for $a\geq 0$, to deal with the part of the

Poisson sum that contains the small values of m_s. The part of the Poisson sum that contains the large values of m_s is dealt with the same way as before. The details are left to the reader.

Acknowledgements The second author would like to thank the University of Bath, where most of this research was done. He also acknowledges the financial support of CONACYT-Mexico grant number 129076. Both authors would like to thank an anonymous referee for their comments on an earlier draft of this chapter.

References

1. E.B. Dynkin (1991): Branching particle systems and superprocesses, *Ann. Probab.* **19(3)**, 1157–1194.
2. E.B. Dynkin (1991): A probabilistic approach to one class of non-linear differential equations. *Probab. Th. Rel. Fields*. **89**, 89–115.
3. E.B. Dynkin (1993): Superprocesses and Partial Differential Equations. *Ann. Probab.* **21**, 1185–1262.
4. E.B. Dynkin (2002): *Diffusions, Superdiffusions and Partial Differential Equations.* AMS, Providencem R.I.
5. E.B. Dynkin and S.E. Kuznetsov (2004): ℕ-measures for branching Markov exit systems and their applications to differential equations. *Probab. Th. Rel. Fields*. **130**, 135–150.
6. Engländer, J. and Pinsky, R. (1999): On the construction and support properties of measure-valued diffusions on $D \subseteq R^d$ with spatially dependent branching. *Ann. Probab.* **27**, 684–730.
7. S.N. Evans (1993): Two representations of a conditioned superprocess. *Proc. Royal. Soc. Edin.* **123A**, 959–971.
8. P.J. Fitzsimmons (1988): Construction and regularity of measure-valued Markov branching processes. *Israeli J. Math.* **64**, 337–361.
9. J. Engländer and A.E. Kyprianou (2004): Local extinction versus local exponential growth for spatial branching processes. *Ann. Probab.* **32**, 78–99.
10. R. Hardy and S.C. Harris (2009): A spine approach to branching diffusions with applications to $\mathcal{L}^p$-convergence of martingales. *Séminaire de Probabilités*, **XLII**, 281–330.
11. Kyprianou, A.E. (2004): Travelling wave solutions to the K-P-P equation: alternitives to Simon Harris' probabilistic analysis. *Ann. Inst. H. Poincaré*. **40**, 53–72.
12. Kyprianou, A.E. (2006): *Introductory lectures on fluctuations of Lévy processes with applications*. Springer.
13. A.E. Kyprianou, R.-L. Liu, A. Murillo-Salas and Y.-X. Ren. (2011): Supercritical super-Brownian motion with a general branching mechanism and travelling waves. *To appear in Ann. Inst. H. Poincaré.*
14. K. Maulik and B. Zwart (2006): Tail asymptotics for exponential functionals of Lévy processes. *Stoch. Proc. Appl.* **116**, 156–177.
15. S. Watanabe (1968): A limit theorem of branching processes and continuous-state branching processes. *J. Math. Kyoto Univ.* **8**, 141–167.

Index

J. Englander and B. Rider (eds.), *Advances in Superprocesses and Nonlinear PDEs*, Springer Proceedings in Mathematics & Statistics 38,
DOI 10.1007/978-1-4614-6240-8, © Springer Science+Business Media, LLC 2013